Plant Secondary Metabolites: Natural Defenses in Disease Management

Plant Secondary Metabolites: Natural Defenses in Disease Management

Pratima Raypa
Department of Biochemistry, H.N.B. Garhwal University, Srinagar, Uttarakhand, India

Susheel Kumar
Plant Molecular Virology Laboratory, CPMB Division, CSIR-National Botanical Research Institute, Rana Pratap Marg, Lucknow, Uttar Pradesh, India

Reshma Jadhav
Department of Biochemistry, Mahatma Phule Krishi Vidyapeeth, Rahuri, Ahmednagar, Maharashtra, India

Zuby Gohar Ansari
Department of Plant Physiology,
B.M. College of Agriculture, Jaswadi Road Khandwa, Madhya Pradesh, India

Praneet Chauhan
Department of Plant Pathology, Dr. Khem Singh Gill Akal College of Agriculture, Eternal University, Baru Sahib, Sirmour, Himachal Pradesh, India

Vinod Chouhan
Division of Plant Pathology, Indian Council of Agricultural Research-Indian Agricultural Research Institute (ICAR-IARI), New Delhi, India

Sheetanshu Gupta
Department of Biochemistry, G.B. Pant University of Agriculture & Technology, Pantnagar, Uttarakhand, India

-EPH-
Elite Publishing House

CRC Press
Taylor & Francis Group
Boca Raton London New York

CRC Press is an imprint of the
Taylor & Francis Group, an **informa** business

First published 2026
by CRC Press
4 Park Square, Milton Park, Abingdon, Oxon, OX14 4RN

and by CRC Press
2385 NW Executive Center Drive, Suite 320, Boca Raton FL 33431

CRC Press is an imprint of Informa UK Limited

British Library Cataloguing-in-Publication Data
A catalogue record for this book is available from the British Library

Print edition not for sale in India.

ISBN: 9781041205661 (hbk)
ISBN: 9781041206651 (pbk)
ISBN: 9781003717454 (ebk)

DOI: 10.1201/9781003717454

Typeset in A Caslon Pro
by Elite Publishing House, Delhi

-EPH-

Contents

Preface

Immobile organisms and plants have evolved an incredible variety of strategies for survival in variable and often hostile environments. During this remarkable evolutionary process, certain strategies, such as the production of secondary metabolites, are compelling examples of nature's intricate biochemical ingenuity. While not directly involved in the core physiological processes of growth and reproduction, these compounds act as the plant's natural defense against attack from pathogens and herbivores, as well as against environmental stressors.

The observation that secondary metabolites are of paramount importance in the respective ecological niches in which the plant kingdom flourished, but also in numerous human activities (agriculture, medicine, biotechnology) compelled us to at least attempt to compile a comprehensive text that may serve as a reference guide to this unique class of plant products. Secondary metabolites are ubiquitous compounds that represent a nexus between the fields of plant ecology and applied sciences. But as interest in sustainable methods increases, so does interest in better understanding these natural compounds that have the potential to supplant synthetic chemicals in pest management, disease control, and crop improvement.

This book, Plant Secondary Metabolites: Natural Defenses in Disease Management, is intended to impart thorough knowledge about these important compounds. It provides an overall overview of their importance in plant defense and human innovation by discussing their biosynthetic pathways, ecological roles, and applications. The chapters are aimed at taking the readers from the basic principles of secondary metabolites to more recent studies and applications.

This book is not only a union of knowledge but a sign to explore the world of plant chemistry. We hope this book facilitates their work to develop better and innovative solutions on how we can still utilize these secondary metabolites for sustainable disease control management and more.

Editors

About the Editors

Dr. Pratima Raypa was born in Tehsil, Dharchula, Pithoragarh Uttarakhand. Dr Raypa accomplished her higher secondary from Gov. Girls Inter College, Pithoragarh, Uttarakhand in 2006. In 2010, She was awarded with B.Sc. from L.S.M.Gov. P.G.College, Pithoragarh. Dr. Raypa was awarded M.Sc. (Biochemistry) and Ph.D. (Biochemistry) from the prestigious G. B. Pant University of Agriculture and Technology, Pantnagar, Uttarakhand. Dr. Raypa is presently holding the post of guest faculty in the Department of Biochemistry at H.N.B. Garhwal University, Srinagar, Uttarakhand.

Dr. Raypa has received a fellowship and honor in recognition to her excellent research contributions. She was awarded with UGC-NFST fellowship during her Ph.D. programme. She was awarded with Young Scientist award. Additionally, she qualified for national-level competitive exams such as CSIR NET, GATE, and U-SET. Dr. Raypa has published numerous research papers and articles and contributed to several book chapters in peer-reviewed international journals. She presented her research papers at several national and international symposia/ seminars/ conferences. She has two years and four months of professional experience. She specializes in metabolite analysis, *in vitro* micropropagation of medicinal plants, and protein characterization of poultry sleeves. Previously, she had worked at ICAR-IVRI in Izatnagar, Bareilly.

Dr. Susheel Kumar holds a doctorate in Plant Virology and currently serves as a Senior Scientist at the CSIR-National Botanical Research Institute, Lucknow. His research focuses on host-virus interactions, viral disease development, and virus management. His laboratory specializes in the molecular identification of DNA and RNA viruses and their management through transgenic techniques and genome editing. Additionally, he investigates the physicochemical disruptions caused by viral infections, including changes in primary and secondary metabolites in economically important crops.

Dr. Kumar has received numerous fellowships and awards and is a life member of the Indian Virological Society and the Indian Phytopathological Society. He is currently mentoring three Ph.D. students at AcSIR and has authored over 83 research articles in peer-reviewed national and international journals, along with several book chapters and a book. He also serves as an editor for several prestigious journals, including Frontiers in Plant Science, Frontiers in Molecular Biosciences, Molecular Biology Reports, Microbiology Research, BMC Plant Biology, 3 Biotech, Virus Disease, and BMC Genomics.

Dr. Reshma Jadhav is a Senior Research Assistant in the Department of Biochemistry, Mahatma Phule Krishi Vidyapeeth, Rahuri, Ahmednagar, Maharashtra, where they specialize in Plant Biochemistry. She obtain a Bachelor's in Agriculture science and a Master's in Biochemistry. She received ICAR-SRF for Ph. D program in Biochemistry from GB. Pant University of Agriculture and Technology Pantnagar. She is a fellow of the Indian Society of Agriculture Biochemists. She brings over 18 years of experience in research, and teaching in the biochemistry area. Their work has been published in leading peer-reviewed journals and has made significant contributions to biotic and abiotic stress tolerance and nutritional biochemistry.

In this book, Dr. Jadhav explores Plant Secondary Metabolites: Natural Defenses in Disease Management, aiming to bridge the gap between complex scientific concepts and public understanding. Their approach combines rigorous research with engaging narratives, making the subject accessible to both scholars and researchers. By using real-world examples and clear explanations, they encourage curiosity and critical thinking.

Dr. Zuby Gohar Ansari holds a Doctorate in Plant Physiology and currently serves as an Assistant Professor (Contractual Teacher) at B.M. College of Agriculture, Khandwa under RVSKVV University, Gwalior. Her research focuses on abiotic stress in legume crops during research work. Dr. Zuby has received numerous fellowships and awards and is a life member of the Andhra Pradesh Journal of Agricultural Sciences (Society of Agricultural Scientists, SVAC, Tirupati), Astha Foundation, Meerut (U.P.), Environment and Social Development Association (ESDA), New Delhi, All India Agricultural Students Association. She is currently mentoring one M.Sc. student at the B.M. College of Agriculture under the Plant Pathology Department and has authored over 14 research articles in peer-reviewed national and international journals, along with several book chapters and a book.

Dr. Praneet Chauhan is working as an Assistant Professor in Plant Pathology and has made significant contributions to various areas, including spawn production, composting, mushroom cultivation, detection and diagnosis of plant diseases, molecular biology, and plant disease management. He has taken on the role of Incharge of the Mushroom Production Unit at Dr. Khem Singh Gill Akal College of Agriculture, Eternal University, Baru Sahib, Sirmour, Himachal Pradesh, India. With a robust publication record, Dr. Chauhan has authored 20 research papers in esteemed national and international journals, contributed to 7 book chapters, and written 4 review papers along with 7 popular articles. He has also guided 6 M.Sc. students and participated in the advisory committees for several other students in the Plant Pathology department. Notably, Dr. Chauhan was honored with the 'Best Ph.D. Thesis Award' during the 1st International Conference (GIRISDA-2022), organized by Just Agriculture at Guru Kashi University, Bathinda, an event recognized by ICAR and AEEFWS, highlighting his exceptional contributions to his field.

Mr. Vinod Chouhan is a committed scholar in the field of agricultural sciences with a strong academic background and notable achievements. He completed his graduation from Agriculture University, Jodhpur, where he developed a solid foundation in agricultural principles and practices. To further specialize, he pursued his post-graduation in Plant Pathology from CCS Haryana Agricultural University (CCSHAU), Hisar. His postgraduate studies allowed him to delve deeper into the study of plant diseases, pathogen biology, and disease management, equipping him with advanced knowledge and skills in this critical field. Currently, Mr. Chouhan is pursuing his Ph.D. in the Division of Plant Pathology at the Indian Council of Agricultural Research-Indian Agricultural Research Institute (ICAR-IARI), New Delhi, one of India's premier institutes for agricultural research. His dedication and academic excellence are evident through his success in competitive examinations. He qualified for the prestigious ICAR-JRF/SRF in 2020, which is a testament to his exceptional understanding of the subject. Additionally, he cleared the National Eligibility Test (NET) in 2023, further showcasing his expertise and commitment to the discipline. With a strong academic trajectory and a clear focus on research, Mr. Chouhan is well on his way to contributing meaningfully to the advancement of plant pathology and agricultural sciences.

Dr. Sheetanshu Gupta working in Biochemistry, Molecular Biology, and Biotechnology for the last 12 years, an alumnus of G.B. Pant University of Agriculture & Technology, Pantnagar, Uttarakhand, India, ICAR ARS NET (Plant Biochemistry and Plant Physiology) and CSIR NET. Dr. Gupta gained valuable experience in the areas of environmental problems, aeroponics and hydroponics, and protein biochemistry. His work showcased his expertise in applying innovative farming techniques and exploring novel approaches to plant protection. Dr. Gupta researched tissue culture hydroponics and mushroom farming, contributing to the advancement of these fields. Dr. Gupta's academic qualifications, research experience, and expertise in biochemistry and biotechnology make him a valuable asset in the field. With his passion for teaching and research, dedication to innovative farming techniques, and student advocacy, Dr. Gupta continues to contribute significantly to the advancement of agricultural and biochemical sciences with more than 20 published books, 250 Book Chapters, and nearly 20 Research/ Review papers.

Chapter - 1

Plant Secondary Metabolites and their Role in Defense

Susheel Kumar[1], Prajna Samal[2], Shivangi S. Kansara[3] and Pratima Raypa[4]

[1]Plant Molecular Virology Laboratory, CPMB Division, CSIR-National Botanical Research Institute, Rana Pratap Marg, Lucknow, Uttar Pradesh, India
[2]Department of Plant Pathology, Uttar Banga Krishi Vishwavidyalaya, West Bengal, India
[3]Department for Plant Pathology, N. M. College of Agriculture, Navsari Agricultural University, Navsari, Gujarat, India
[4]Department of Biochemistry, H.N.B. Garhwal University, Srinagar, Uttarakhand, India

Abstract

Plant secondary metabolites are pivotal in plants' ecological survival and defense mechanisms. These bioactive compounds, distinct from primary metabolites essential for basic cellular functions, serve as a crucial interface between plants and their environment. Secondary metabolites, including phenolics, terpenoids, and alkaloids, are produced as a response to various biotic and abiotic stresses. They have evolved to protect plants against herbivores, pathogens, and competing plants through chemical deterrence and toxicity. These compounds also contribute to a plant's resilience in extreme environmental conditions, such as ultraviolet radiation, drought, and oxidative stress. The classification of secondary metabolites is based on their diverse chemical structures and roles in plant defense, ranging from mechanical barriers to producing toxic substances that inhibit feeding or disrupt the growth of pathogens. For instance, phenolics like flavonoids and tannins have antioxidant properties that protect plant tissues, while alkaloids such as nicotine act as potent deterrents against herbivores due to their toxicity. Terpenoids, another significant class, provide both direct defense by deterring herbivores and indirect defense by attracting predators of herbivores, thus creating a multi-layered defense system. Moreover, secondary metabolites play a critical role in plant-microbe interactions, either by inhibiting pathogen growth or by promoting beneficial microbial relationships that enhance plant health and productivity. This dual role demonstrates the adaptability of these compounds in maintaining plant health and survival. As plants face increasing environmental stress due to climate

DOI: 10.1201/9781003717454-1

change and anthropogenic activities, the study of secondary metabolites becomes even more relevant.

Keywords: Plant secondary metabolites, plant defense, phenolics, terpenoids, alkaloids, ecological roles

1.1 Primary vs. Secondary Metabolites in Plants

Introduction to Plant Metabolites

Plants are remarkable chemical factories, capable of synthesizing an immense diversity of organic compounds, essential for their survival, reproduction, and interaction with the environment. These chemical compounds are classified into two major categories: primary metabolites and secondary metabolites. Primary metabolites are indispensable for plant growth, cellular maintenance, and reproduction, while secondary metabolites, although not directly essential for survival, play significant roles in defense, ecological interactions, and adaptation to environmental stresses (Bourgaud et al., 2016). Primary metabolites are ubiquitous and found in all plants, contributing to the most basic functions such as respiration, growth, and development. In contrast, secondary metabolites tend to be more species-specific and often accumulate in response to environmental factors such as herbivore pressure, pathogen attacks, and nutrient availability (Mithöfer & Boland, 2012). These compounds have significant ecological roles and also hold great potential for human applications, particularly in the fields of medicine, agriculture, and biotechnology (Wink, 2010). To fully appreciate the role of secondary metabolites in disease management and ecological balance, it is crucial to distinguish them from primary metabolites.

Defining Primary Metabolites

Primary metabolites are, therefore, universally present across all plant species and are integral to the plant's survival. Primary metabolites are compounds directly involved in essential metabolic processes that are critical for plant growth and development. These include carbohydrates, proteins, lipids, and nucleic acids, all of which contribute to cellular processes such as photosynthesis, respiration, and cellular division (Saito & Matsuda, 2010). The production of primary metabolites is a fundamental aspect of plant metabolism, ensuring that the plant has the necessary resources for survival and reproduction.

1. **Amino Acids**: Amino acids are the building blocks of proteins, which are critical for numerous cellular functions, including enzymatic reactions, cellular signaling, and structural integrity. Amino acids are synthesized through pathways like the Calvin cycle and the tricarboxylic acid (TCA) cycle, both of which are essential for cellular metabolism (Fiehn, 2002).
2. **Carbohydrates**: Carbohydrates are a primary energy source for plants. Simple sugars like glucose are produced through photosynthesis and are subsequently used in respiration to generate ATP, the energy currency of the cell. Carbohydrates also play a structural role, particularly in the formation of cellulose, a key component of the plant cell wall (Taiz et al., 2015).
3. **Lipids**: Lipids, which include fats, oils, and waxes, play both structural and functional roles in plant cells. They form the primary components of cell membranes and are crucial for maintaining cellular integrity and signaling. Lipids also function as energy storage molecules, which are particularly important during seed germination (Chapman et al., 2012).

4. **Nucleotides**: Nucleotides are the building blocks of nucleic acids (DNA and RNA), which are essential for genetic information storage and transmission. They also play a role in energy transfer, particularly through ATP, which powers many cellular processes (Nelson & Cox, 2017).

Defining Secondary Metabolites

The production of secondary metabolites is often triggered by biotic or abiotic stresses, making these compounds vital for plant survival in a competitive and hostile environment. Their roles extend beyond defense; they are also involved in allelopathy, symbiotic relationships, and communication with other organisms (Hartmann, 2007). Secondary metabolites, on the other hand, are not directly involved in the primary metabolic processes that sustain life. Instead, they serve more specialized functions, often related to the plant's interactions with its environment. These compounds are typically produced as a response to ecological pressures, such as herbivory, pathogen attacks, or competition with other plants. Secondary metabolites are diverse and include alkaloids, terpenoids, phenolics, and glycosides, each with its own set of ecological roles (Dewick, 2009).

1. **Alkaloids**: Alkaloids are nitrogen-containing compounds that are often involved in plant defense. Many alkaloids are toxic to herbivores and pathogens, making them effective deterrents. For example, nicotine from tobacco (*Nicotiana tabacum*) acts as a potent insecticide, protecting the plant from pest attacks (Wink, 2010). Alkaloids also have significant pharmacological properties and are used in medicine for their analgesic, stimulant, and antimalarial effects (Kutchan, 1995).
2. **Terpenoids**: Terpenoids are one of the largest groups of secondary metabolites, with over 40,000 different compounds identified. These metabolites play roles in plant defense, pollinator attraction, and protection from environmental stresses. A well-known terpenoid is menthol, derived from mint species (*Mentha* spp.), which has insect-repellent properties and is widely used in pharmaceuticals and flavoring agents (Gershenzon & Dudareva, 2007).
3. **Phenolics**: Phenolic compounds are a large and diverse group of secondary metabolites that contribute to plant structural integrity and defense. They are involved in protecting plants against UV radiation, herbivory, and pathogen infections. Flavonoids, a subgroup of phenolics, are known for their antioxidant properties and contribute to the coloration of flowers and fruits, aiding in pollination and seed dispersal (Quideau et al., 2011).
4. **Glycosides**: Glycosides are molecules that consist of a sugar bound to a non-sugar substance, often a secondary metabolite. These compounds can serve as deterrents to herbivores and pathogens due to their toxicity when broken down by enzymes during digestion (Robinson, 1991). For example, cyanogenic glycosides release hydrogen cyanide upon enzymatic degradation, providing a potent defense mechanism against herbivores.

Non-Essential but Crucial Ecological Functions of Secondary Metabolites in Plant Survival

Plants face constant challenges from a variety of environmental stressors, including herbivores, pathogens, and competitors. While secondary metabolites are not directly required for the plant's primary metabolic functions, such as growth, reproduction, and energy production, they play indispensable roles in ensuring a plant's survival in its environment. These compounds serve as chemical defenses, signaling agents, and ecological regulators that allow plants to adapt to their surroundings and thrive despite the many biotic and abiotic stresses they face. In this section, we will explore the critical ecological functions of secondary metabolites in plant survival, emphasizing their roles in defense, pollination, symbiotic relationships, and allelopathy.

1. Defense Against Herbivores and Pathogens

One of the most well-recognized roles of secondary metabolites is in plant defense. Unlike primary metabolites, which are essential for the plant's core life processes, secondary metabolites are often synthesized in response to external threats such as herbivory or pathogen attack. These compounds can deter herbivores, inhibit pathogen growth, or even kill the organisms that pose a threat to the plant.

1. **Chemical Defense Against Herbivores**: Secondary metabolites such as alkaloids, terpenoids, and phenolics act as chemical deterrents that protect plants from herbivores. For instance, alkaloids like nicotine, found in tobacco plants (*N. tabacum*), are neurotoxins that interfere with the nervous systems of herbivores, preventing them from feeding on the plant (Steppuhn et al., 2004). Similarly, phenolic compounds such as tannins, found in oak trees (*Quercus* spp.), reduce the digestibility of plant tissues, making them less attractive to herbivores (Barbehenn & Constabel, 2011). These chemical defenses can be constitutive, meaning they are always present in the plant, or inducible, produced in response to an attack.
2. **Antimicrobial Defense**: Many secondary metabolites also have antimicrobial properties that protect plants from pathogens such as bacteria, fungi, and viruses. For example, phytoalexins are antimicrobial compounds that are synthesized by plants in response to pathogen invasion. Resveratrol, a phytoalexin produced by grapevines (*Vitis vinifera*), inhibits the growth of fungal pathogens like *Botrytis cinerea*, preventing them from causing disease (Jeandet et al., 2014). In this way, secondary metabolites contribute to a plant's immune system, enhancing its ability to survive in environments where pathogens are prevalent.

2. Attraction of Pollinators and Seed Dispersers

In addition to their defensive roles, secondary metabolites are crucial for facilitating reproduction by attracting pollinators and seed dispersers. Many plants rely on animals, such as bees, birds, and bats, to transfer pollen from one flower to another or to disperse seeds across a wide area. Secondary metabolites play a key role in these interactions by producing the scents, colors, and flavors that attract these organisms.

1. **Floral Scents**: Terpenoids and phenolic compounds are responsible for the production of floral scents that attract pollinators. For example, the monoterpenoid linalool, found in many flowering plants, produces a pleasant floral scent that attracts pollinators such as bees and butterflies (Raguso, 2008). The specific combination of volatile secondary metabolites emitted by a flower can determine which pollinators are attracted to it, thereby enhancing the plant's reproductive success.
2. **Fruit Flavors**: Secondary metabolites also influence the flavor and color of fruits, which attract animals that act as seed dispersers. Anthocyanins, a type of phenolic compound, give fruits their vibrant red, blue, or purple colors, making them more visible and appealing

to animals (Delgado-Vargas & Paredes-López, 2002). In turn, animals such as birds and mammals consume the fruit, and the seeds are dispersed through their feces, helping the plant to spread its offspring over a larger area. Without these secondary metabolites, many plants would struggle to reproduce successfully.

3. Allelopathy: Inhibiting Competitor Growth

Plants not only have to contend with herbivores and pathogens but also with competition from other plants for resources such as light, water, and nutrients. Allelopathy is a process by which plants release secondary metabolites into the environment to inhibit the growth of nearby plants, giving them a competitive advantage.

1. **Phenolic Acids and Terpenoids**: Many allelopathic plants produce phenolic acids and terpenoids that are released into the soil through root exudation or decaying leaf litter. These compounds inhibit seed germination and root growth in neighboring plants, reducing competition for resources. For example, juglone, a phenolic compound produced by black walnut trees (*Juglans nigra*), inhibits the growth of many herbaceous plants and trees that might otherwise compete with the walnut for sunlight and soil nutrients (Hejl & Koster, 2004). Similarly, cineole, a terpenoid produced by eucalyptus trees (*Eucalyptus* spp.), is released into the environment where it acts as an allelopathic agent, inhibiting the growth of nearby plants (May & Ash, 1990).
2. **Cyanogenic Glycosides: Cyanogenic glycosides** are another group of secondary metabolites that serve an allelopathic function. When these compounds are broken down, they release toxic hydrogen cyanide, which can inhibit the growth of neighboring plants. For example, sorghum (*Sorghum bicolor*) produces cyanogenic glycosides that are released into the soil, suppressing the germination and growth of weed species (Tibugari & Chiduza, 2024). This allows sorghum to outcompete other plants and secure more resources for its own growth.

4. Symbiotic Relationships with Microorganisms

Secondary metabolites also play a crucial role in the establishment and maintenance of symbiotic relationships between plants and microorganisms. These relationships can provide plants with additional nutrients, enhance their resistance to environmental stresses, or improve their overall fitness.

1. **Flavonoids in Legume-Rhizobia Symbiosis**: Flavonoids, a group of phenolic compounds, are essential for the establishment of symbiotic relationships between leguminous plants (such as soybeans and peas) and rhizobia bacteria. Flavonoids released by the plant roots serve as chemical signals that attract rhizobia to the roots, where they form nitrogen-fixing nodules (Mathesius et al., 1998). In exchange for the fixed nitrogen, which the plant can use for growth, the rhizobia receive carbohydrates and a protected environment within the root nodules.
2. **Saponins and Mycorrhizal Fungi**: Saponins, another class of secondary metabolites, play a role in the interaction between plants and mycorrhizal fungi. These fungi form symbiotic associations with plant roots, enhancing the plant's ability to absorb water and nutrients, particularly phosphorus. In turn, the fungi receive carbohydrates from the plant. Avenacin, a

saponin produced by oats (*Avena sativa*), helps protect the plant from pathogenic fungi while simultaneously supporting the growth of beneficial mycorrhizal fungi (Osbourn, 1996). This symbiotic relationship enhances the plant's overall fitness by improving nutrient acquisition and resistance to pathogens.

5. Defense Against Abiotic Stresses

Secondary metabolites help plants combat not only biotic factors but also abiotic stresses such as drought, salinity, extreme temperatures, and radiation. These compounds act as protectants, stabilizing cellular structures and reducing oxidative stress.

1. **Phenolic Compounds as UV Protectants**: Exposure to ultraviolet (UV) radiation can lead to significant damage in plant tissues, affecting DNA, proteins, and lipids. Flavonoids, a major class of phenolic compounds, absorb harmful UV radiation and thus function as natural sunscreens for plants (Rozema et al., 1997). Plants growing in environments with high UV radiation, such as alpine or desert regions, typically produce higher levels of these compounds to prevent oxidative damage. For example, plants such as *Artemisia tridentata* (sagebrush), which grows in high UV environments, exhibit elevated flavonoid levels in their tissues, offering protection from radiation (Cockell & Knowland, 1999).
2. **Terpenes and Heat Stress Tolerance**: Certain terpenes, such as isoprene, are emitted by plants as a protective mechanism against heat stress. These compounds stabilize cellular membranes and protect plant cells from thermal damage. Isoprene emission, for example, is a well-documented response in plants like *Populus* (poplar trees) to elevated temperatures (Sharkey et al., 2008). This emission helps mitigate the damage to the photosynthetic machinery, ensuring that the plant can maintain normal functioning under extreme conditions.
3. **Osmoprotectants in Drought Stress**: During drought, plants produce secondary metabolites that function as osmoprotectants, helping to retain water and stabilize cell membranes. Proline and trehalose are classic examples of osmoprotectants. Proline, an amino acid derivative, accumulates in plant tissues during periods of water stress and plays a critical role in maintaining turgor pressure and protecting proteins from denaturation (Ashraf & Foolad, 2007). Similarly, trehalose, a disaccharide produced in some plant species, protects cells against desiccation by stabilizing membranes and proteins (Iturriaga et al., 2009).
4. **Carotenoids in Photoprotection**: Carotenoids are pigments that serve dual functions in plants, contributing to photosynthesis and acting as antioxidants. Carotenoids quench reactive oxygen species (ROS) generated during high light intensity or oxidative stress. For example, plants exposed to intense sunlight, such as those in desert environments, rely on carotenoids like β-carotene to protect the photosynthetic apparatus from oxidative damage (Demmig-Adams & Adams, 2006).
5. **Betaines in Salinity Stress**: Plants growing in saline environments, such as mangroves and halophytes, synthesize compounds known as betaines, which help to mitigate the toxic effects of salt by maintaining osmotic balance. Glycine betaine, for example, accumulates in plants like *Atriplex* (saltbush) and enhances tolerance to high salinity by stabilizing the structure of enzymes and proteins in the presence of salt (Chen & Murata, 2008).

Historical Perspective on Research Regarding Secondary Metabolites

The study of secondary metabolites has a rich history that dates back thousands of years, intertwined with the development of medicinal, agricultural, and scientific practices. Although early civilizations lacked a modern understanding of plant biochemistry, they recognized the practical utility of certain plants for their therapeutic, toxicological, and preservative properties. Over time, these observations laid the groundwork for the formal study of secondary metabolites, which gained momentum in the late 19th and 20th centuries with advances in chemistry, biology, and pharmacology. Today, secondary metabolites remain a key focus of research due to their ecological significance and their potential applications in various fields, including pharmaceuticals, agriculture, and biotechnology.

Early Observations and Traditional Uses

The recognition of plants for their healing and poisonous properties has been an integral part of human history, predating recorded knowledge. Ancient civilizations, such as those in Egypt, India, and China, made extensive use of plant-derived compounds for treating various ailments, and these practices were passed down through generations (Jiang et al., 2016). For example, the use of opium (derived from *Papaver somniferum*) as an analgesic and sedative can be traced back to the Sumerians around 3400 BCE (Booth, 1996). Similarly, the Ebers Papyrus, an Egyptian medical text from around 1550 BCE, documents the use of over 700 plant species for medicinal purposes, including plants known to contain alkaloids, terpenoids, and phenolics (Nunn, 2002).

In India, the Ayurvedic system of medicine, dating back over 3,000 years, made extensive use of plant-based treatments. The classic Ayurvedic text, Charaka Samhita, contains detailed descriptions of various herbs and their therapeutic effects. For instance, *Rauvolfia serpentina*, a plant used in Ayurvedic medicine, was traditionally employed to treat hypertension and mental disorders. It wasn't until the 20th century that the active compound, reserpine, was isolated from the plant and used in modern medicine to treat high blood pressure (Warrier et al., 1996).

The Chinese traditional medicine system also relied heavily on plant secondary metabolites. One of the earliest Chinese texts, the Shennong Bencao Jing (The Divine Farmer's Materia Medica), written around 200 CE, classifies 365 medicinal plants, many of which are known today for their bioactive compounds. *Artemisia annua*, from which the anti-malarial compound artemisinin was later derived, has been used in China for over 2,000 years to treat fevers and other ailments (Tu, 2016). These early civilizations, though unaware of the chemical basis for the efficacy of these plants, laid the foundation for future scientific inquiry into the nature of secondary metabolites.

The Birth of Modern Phytochemistry (19th Century)

The formal scientific study of secondary metabolites began in earnest in the 19th century with the development of organic chemistry and the ability to isolate, identify, and synthesize specific compounds from plants. The isolation of morphine from opium by the German pharmacist Friedrich Sertürner in 1806 is widely regarded as the birth of alkaloid chemistry (Sneader, 2005). Sertürner's work demonstrated that plants contained active compounds that could be isolated and used therapeutically. This discovery opened the door for the isolation of other alkaloids, such as quinine from *Cinchona* bark (used to treat malaria) and strychnine from *Strychnos nux-vomica* (used as a stimulant and poison).

By the mid-19th century, research into plant-derived chemicals had gained momentum, with notable advancements in the isolation of terpenoids and phenolics. The essential oils from plants, primarily composed of terpenoids, were identified and characterized for their aromatic and medicinal properties. Eucalyptus oil and peppermint oil, both rich in monoterpenes, were commonly used for their antiseptic and digestive benefits (Guenther, 1972). Additionally, the isolation of salicylic acid from willow bark by Raffaele Piria in 1838 eventually led to the synthesis of aspirin, marking a major breakthrough in both medicine and plant biochemistry (Sneader, 2005). The work of 19th-century scientists, particularly in Europe, laid the groundwork for phytochemistry, the branch of chemistry concerned with plant-derived compounds. During this period, the concept of secondary metabolites as "waste products" of primary metabolism was still prevalent, though their ecological and medicinal importance was becoming increasingly clear.

Advances in the 20th Century: The Ecological Role of Secondary Metabolites

The early 20th century marked a shift in the understanding of secondary metabolites, particularly with regard to their ecological roles. Julius von Sachs, a prominent plant physiologist in the late 19th century, proposed that these compounds were mere by-products of plant metabolism, with no significant role in the plant's life (Sachs, 1887). However, research in the 20th century began to refute this idea, demonstrating that secondary metabolites play active roles in plant defense, pollination, and allelopathy. In the 1950s and 1960s, the chemical ecology of plants gained recognition, with researchers like Gottfried Fraenkel proposing that secondary metabolites evolved primarily as defensive compounds to protect plants from herbivores and pathogens (Fraenkel, 1959). Fraenkel's work was pivotal in shifting the view of secondary metabolites from metabolic waste to essential defensive chemicals. His research showed that alkaloids, terpenoids, and phenolics serve as chemical deterrents to herbivores by being toxic, bitter, or indigestible. Another significant advancement in this period was the recognition of phytoalexins, secondary metabolites produced by plants in response to pathogen attacks. The term phytoalexin was coined by Müller and Börger in 1939 after they observed the accumulation of antimicrobial compounds in plants infected with fungi (Müller & Börger, 1939). This discovery led to further research into the role of secondary metabolites in plant immune responses. By the 1970s, the field of chemical ecology was flourishing, with researchers like Paul Ehrlich and Peter Raven exploring the co-evolution of plants and herbivores, emphasizing the arms race between plant defenses (secondary metabolites) and herbivore adaptations (Ehrlich & Raven, 1964). This era firmly established the ecological significance of secondary metabolites beyond their traditional medicinal uses.

Recent Developments (21st Century): Biotechnology and Beyond

The turn of the 21st century has brought about remarkable advancements in the study of secondary metabolites, particularly with the advent of molecular biology, genomics, and biotechnology. Researchers now have a greater understanding of the biosynthetic pathways responsible for the production of these compounds, as well as the genes that regulate their synthesis. One major area of focus has been the use of metabolic engineering to enhance the production of valuable secondary metabolites in plants or microbial systems. For example, the anti-cancer compounds vinblastine and vincristine, originally isolated from *Catharanthus roseus*, have been successfully produced in higher yields through genetic modification and fermentation processes (El-Sayed & Verpoorte, 2007). This allows for a more sustainable production of these critical pharmaceuticals. Similarly, the biotechnological production of artemisinin, a secondary metabolite derived from *Artemisia annua* and used in treating malaria, has been significantly improved through synthetic biology. The research led by Jay Keasling at the University of California, Berkeley, used genetically engineered yeast to produce artemisinin precursors, reducing reliance on traditional agricultural sources (Paddon et al., 2013). In recent years, the application of omics technologies, such as metabolomics and transcriptomics, has revolutionized the study of secondary metabolites. These techniques allow for the comprehensive profiling of plant metabolites and the identification of biosynthetic gene clusters, leading to a deeper understanding of how plants produce and regulate these compounds under various environmental conditions (Weckwerth, 2003). Moreover, the role of secondary metabolites in promoting human health through nutraceuticals and functional foods is another growing area of research. Compounds like resveratrol, curcumin, and flavonoids have gained attention for their potential in preventing chronic diseases such as cancer, cardiovascular diseases, and neurodegenerative disorders (Pan et al., 2010).

1.2 Ecological Roles of Secondary Metabolites

Plants are not passive organisms; they interact constantly with their environment through a wide array of biochemical signals. One of the primary mechanisms through which plants manage these interactions is the production of secondary metabolites. Unlike primary metabolites, which are essential for basic metabolic functions, secondary metabolites serve crucial ecological functions that help plants respond to biotic and abiotic factors. These compounds play essential roles in defense against herbivores and pathogens, competition with other plants (allelopathy), attraction of pollinators, and symbiotic relationships with microorganisms.

Defense Against Herbivores and Pathogens

One of the most vital ecological roles of secondary metabolites is in the defense against herbivores and pathogens. Plants cannot physically escape from predators or infectious agents, so they rely on chemical defenses to deter or inhibit these threats. These chemical compounds, often toxic, unpalatable, or inhibitory, can either deter herbivores and pathogens or reduce their ability to attack the plant.

1. **Alkaloids as Defense Compounds**: Many plants produce alkaloids, nitrogen-containing secondary metabolites, as a defense mechanism against herbivores and pathogens. Alkaloids are often toxic, and they can affect the nervous systems of herbivores, making them highly effective deterrents. For example, nicotine from the tobacco plant (*N. tabacum*) is an alkaloid that acts as a neurotoxin, affecting the nervous system of insects and mammals alike (Steppuhn et al., 2004). The rapid onset of neurotoxic effects deters further feeding, protecting the plant from damage.
2. **Phenolic Compounds in Disease Resistance**: Secondary metabolites also play an essential role in resisting pathogens. Phenolic compounds, such as flavonoids and tannins, have antimicrobial properties and help protect plants from bacterial and fungal infections (Barbehenn & Constabel, 2011). For instance, phytoalexins, which are phenolic secondary metabolites produced by plants in response to pathogen attack, can inhibit the growth and spread of fungal and bacterial pathogens (Jeandet et al., 2014). The rapid production of these compounds upon infection allows plants to mount an effective chemical defense, reducing disease incidence.

3. **Terpenoids as Herbivore Deterrents**: Another class of secondary metabolites, terpenoids, plays a significant role in herbivore deterrence. Terpenoids, particularly monoterpenes and sesquiterpenes, are responsible for the strong odors emitted by many plants. These volatile compounds serve as repellents to herbivores. For example, the essential oil of rosemary (*Rosmarinus officinalis*), which contains the terpenoid cineole, deters insects and herbivores from feeding on the plant (Usseglio et al., 2023). The strong scent and taste associated with these terpenoids discourage herbivores from consuming the plant, thereby minimizing damage.

Competition with Other Plants: Allelopathy

Plants are not only under pressure from herbivores and pathogens but also from other plants competing for resources such as sunlight, water, and nutrients. Allelopathy refers to the process by which a plant releases chemical compounds into the environment that inhibit the growth and development of neighboring plants, thus reducing competition. Many secondary metabolites play a central role in allelopathic interactions.

1. **Juglone in Black Walnut Trees**: A well-known example of allelopathy involves the black walnut tree (*Juglans nigra*), which produces juglone, a phenolic compound released through its roots, leaves, and bark. Juglone inhibits the growth of surrounding plants by interfering with their photosynthesis and respiration processes, giving the black walnut tree a competitive advantage (Hejl & Koster, 2004). This chemical exclusion reduces competition for resources and ensures the tree's dominance in its habitat.
2. **Sorgoleone in Sorghum**: Another allelopathic example is sorgoleone, a secondary metabolite produced by the roots of sorghum (*S. bicolor*). Sorgoleone inhibits the germination and growth of weed species in the vicinity of sorghum plants, making it an effective natural herbicide (Czarnota et al., 2001). This process not only reduces competition for water and nutrients but also helps maintain crop productivity in agricultural systems by limiting weed growth.
3. **Cineole in Eucalyptus Trees**: Eucalyptus trees (*Eucalyptus spp.*) are known to release cineole, a monoterpenoid found in their essential oils, which inhibits the germination and growth of competing plants in the surrounding soil (May & Ash, 1990). This allelopathic action creates a chemical barrier that limits the growth of other species, allowing eucalyptus trees to dominate their environment and access more resources for their growth and survival.

Pollination and Seed Dispersal

Secondary metabolites also play an essential role in facilitating pollination and seed dispersal, which are critical for plant reproduction. Many plants rely on animals to carry their pollen or seeds to new locations, and secondary metabolites can attract these pollinators and seed dispersers by influencing the scent, color, or taste of flowers and fruits.

1. **Floral Scents and Pollinators**: Terpenoids and phenolic compounds are responsible for the floral scents that attract pollinators such as bees, butterflies, and birds. For example, the monoterpenoid linalool, which produces a pleasant floral scent, is commonly found in the flowers of plants like lavender (*Lavandula spp.*) and is known to attract pollinators (Raguso, 2008). By emitting these aromatic compounds, plants increase their chances of successful pollination, ensuring the transfer of pollen between flowers.

2. **Anthocyanins and Fruit Coloration**: Secondary metabolites also influence the color of fruits, making them more attractive to animals that act as seed dispersers. Anthocyanins, a class of phenolic compounds, give fruits their red, blue, and purple colors. For example, the bright red coloration of cherries (*Prunus avium*) is due to anthocyanins, which signal ripeness to frugivorous birds and mammals (Delgado-Vargas & Paredes-López, 2002). These animals consume the fruit and disperse the seeds through their feces, helping the plant spread its progeny to new locations.
3. **Fruit Flavors and Seed Dispersal**: The flavor of fruits, influenced by secondary metabolites such as alkaloids, terpenoids, and glycosides, can encourage animals to consume the fruits and disperse seeds. For example, the strong flavor of citrus fruits, derived from limonene and other terpenoids, attracts a variety of animals that play a role in seed dispersal (Defilippi et al., 2009). By making their fruits palatable and nutritious, plants incentivize animals to assist in spreading their seeds, ensuring reproduction and survival across generations.

Symbiotic Relationships with Microorganisms

Another critical ecological role of secondary metabolites is their involvement in symbiotic relationships between plants and microorganisms. These relationships are often mutualistic, with both the plant and microorganism benefiting from the exchange of nutrients or protection.

1. **Flavonoids in Legume-Rhizobia Symbiosis**: In leguminous plants, flavonoids act as signaling molecules that facilitate the establishment of symbiotic relationships with rhizobia, nitrogen-fixing bacteria. The flavonoids released by legume roots attract rhizobia to the root zone, where they form nodules and begin fixing atmospheric nitrogen into a form the plant can use for growth (Oldroyd & Downie, 2008). In return, the plant provides the bacteria with carbohydrates, creating a mutualistic relationship that enhances plant nutrition and growth.
2. **Mycorrhizal Fungi and Secondary Metabolites**: Secondary metabolites also mediate interactions between plants and mycorrhizal fungi. These fungi form symbiotic relationships with plant roots, helping the plant absorb nutrients such as phosphorus and water from the soil. In return, the plant supplies the fungi with carbohydrates. Secondary metabolites like terpenoids and phenolic compounds help regulate this symbiotic relationship by influencing fungal growth and maintaining the health of the association (Balestrini & Bonfante, 2005). This partnership enhances plant resilience and nutrient acquisition, especially in nutrient-poor environments.

Role of Primary and Secondary Metabolites in Plant Defense Against Herbivores, Pathogens, and Competitors

Plants are constantly exposed to biotic stresses such as herbivory, pathogens, and competition from other plants. To survive and thrive in such hostile environments, they have evolved a multitude of defense mechanisms. Among the most critical tools in their defense arsenal are secondary metabolites, which serve as chemical deterrents against herbivores, pathogens, and competing plants. These compounds, though not essential for the primary metabolic functions of plants, play a pivotal role in enhancing a plant's fitness and survival. On the other hand, primary metabolites, while mainly responsible for essential life processes, also contribute indirectly to defense by supporting the plant's growth and resilience under stress conditions (Mithöfer & Boland, 2012). The production of secondary metabolites is often triggered by environmental stresses or attacks by herbivores and pathogens. These metabolites not only deter herbivores and inhibit pathogen growth but also act as allelopathic agents, giving plants a competitive advantage by inhibiting the growth of neighboring plants.

Defense Against Herbivores

Herbivory poses a significant threat to plants, as herbivores, ranging from insects to large mammals, feed on plant tissues, causing damage and reducing fitness. To mitigate this threat, plants have evolved a range of chemical defenses, many of which involve the production of secondary metabolites. These metabolites can act as toxins, feeding deterrents, or digestibility reducers, thereby protecting plants from herbivore attacks (Agrawal, 2011).

1. **Alkaloids**: One of the most well-known classes of defensive compounds, alkaloids are nitrogen-containing secondary metabolites that can be highly toxic to herbivores. For example, nicotine, produced by the tobacco plant (*N. tabacum*), acts as a neurotoxin in many insects, disrupting their nervous systems and effectively deterring them from feeding on the plant (Baldwin, 2010). Similarly, quinine, an alkaloid from the cinchona tree (*Cinchona* spp.), has been used as a natural insecticide due to its toxicity to herbivores (Wink, 2010).
2. **Terpenoids**: Terpenoids, particularly monoterpenes, and sesquiterpenes, are volatile compounds that can repel herbivores by producing strong odors or tastes that are unpleasant to them. For instance, limonene, found in citrus plants, has been shown to deter herbivorous insects such as aphids (Kessler & Baldwin, 2007). In addition to their repellent properties, terpenoids can also act as direct toxins to herbivores by interfering with their metabolic processes (Gershenzon & Dudareva, 2007).
3. **Phenolic Compounds**: Phenolics, such as tannins and flavonoids, reduce the digestibility of plant material by binding to proteins in the herbivore's gut, making it more difficult to digest. This not only decreases the nutritional value of the plant tissue but also discourages further feeding (Barbehenn & Constabel, 2011). For example, the high tannin content in oak leaves (*Quercus* spp.) is a key factor in reducing herbivore damage, as it makes the leaves less palatable to many insect species (Salminen & Karonen, 2011).

Defense Against Pathogens

Pathogens such as bacteria, fungi, and viruses pose another significant threat to plant health. To defend against these invaders, plants produce a variety of secondary metabolites with antimicrobial and antifungal properties. These compounds can inhibit pathogen growth, disrupt pathogen cell membranes, or interfere with the pathogen's ability to infect plant cells.

1. **Phytoalexins**: Phytoalexins are antimicrobial compounds that are synthesized by plants in response to pathogen attacks. For example, resveratrol, a stilbene-type phytoalexin produced by grapevines (*V. vinifera*), has been shown to inhibit the growth of fungal pathogens such as *B. cinerea* (Jeandet et al., 2014). Similarly, camalexin, produced by *Arabidopsis thaliana*, is known to protect the plant against bacterial and fungal pathogens (Bednarek et al., 2010).
2. **Saponins**: Saponins are glycoside compounds with surfactant properties that disrupt the cell

membranes of pathogens, leading to cell lysis and death. These compounds are particularly effective against fungal pathogens, which rely on intact cell membranes for survival (Osbourn, 1996). Avenacin, a saponin found in oat (*Avena* spp.), provides resistance against fungal diseases by disrupting the pathogen's ability to invade the plant (Papadopoulou et al., 1999).

3. **Glucosinolates**: Glucosinolates are sulfur-containing compounds that, upon hydrolysis, produce isothiocyanates, which have strong antimicrobial properties. These compounds are especially effective against bacterial pathogens and have been studied for their role in protecting cruciferous plants (*Brassicaceae*) from bacterial and fungal infections (Connolly et al., 2021).

Defense Against Competitors (Allelopathy)

In natural ecosystems, plants often compete with each other for resources such as light, water, and nutrients. To gain an advantage over their competitors, some plants produce allelopathic chemicals that inhibit the growth of neighboring plants. These compounds, often secondary metabolites, are released into the environment through root exudation, leaf litter, or volatilization, where they affect the germination and growth of competing species (Bertin et al., 2003).

1. **Phenolic Acids**: Phenolic acids, such as ferulic acid and p-coumaric acid, are commonly found in the leaf litter of plants like rice (*Oryza sativa*) and wheat (*Triticum aestivum*). These compounds are known to inhibit the germination of weed seeds, giving the crop an advantage in resource competition (Gao et al., 2017).
2. **Cyanogenic Glycosides**: These compounds are released when plant tissues are damaged, and they produce toxic hydrogen cyanide. In addition to their role in deterring herbivores, cyanogenic glycosides also serve an allelopathic function by inhibiting the growth of nearby plants. Sorghum (*S. bicolor*) releases cyanogenic glycosides through its roots, which suppress the germination of weed seeds (Tibugari & Chiduza, 2024).
3. **Terpenoids and Essential Oils**: Terpenoids, particularly those found in essential oils, can also act as allelopathic agents. For example, cineole, a monoterpene found in eucalyptus (*Eucalyptus* spp.), is released into the soil through leaf litter, where it inhibits the germination and growth of nearby plants (May & Ash, 1990). This allelopathic effect helps eucalyptus trees dominate their environment by reducing competition from other plant species.

Examples of Plant Species and Their Specific Secondary Metabolite Defenses

Plants have evolved a diverse array of chemical defenses that enable them to survive in environments where herbivores, pathogens, and competitors are present. These defenses are often in the form of secondary metabolites, which are unique to each plant species and vary widely in their mode of action. The following examples highlight plant species that use specific secondary metabolites to protect themselves against biotic stressors.

1. Tobacco (*Nicotiana tabacum*): Nicotine as a Neurotoxin

The tobacco plant is one of the most well-known examples of a plant that uses secondary metabolites for defense. Nicotine, an alkaloid, serves as a potent neurotoxin that protects the plant from herbivores. Nicotine interferes with the nervous system of insects and other herbivores, leading to paralysis or death. Insects feeding on tobacco leaves are exposed to high levels of nicotine, which acts quickly to immobilize them (Steppuhn et al., 2004). This defensive mechanism is crucial for tobacco plants growing in environments where insect herbivores are prevalent. Moreover, studies have shown that the production of nicotine can be induced by herbivore damage, which suggests that tobacco plants can increase their chemical defenses in response to attack (Ohnmeiss & Baldwin, 2000). This inducible defense mechanism allows tobacco plants to conserve energy by producing large amounts of nicotine only when necessary.

2. Periwinkle (*Catharanthus roseus*): Vinblastine and Vincristine as Anticancer Agents

Periwinkle is a plant renowned for producing secondary metabolites with significant medicinal properties. The plant synthesizes vinblastine and vincristine, two alkaloids that have potent anticancer properties and are used in chemotherapy treatments (Goswami et al., 2024). While these compounds are valuable in human medicine, they serve a defensive function for the plant by deterring herbivores and pathogens. Vinblastine and vincristine disrupt microtubule formation in the cells of herbivores and pathogens, preventing cell division and ultimately leading to cell death. This makes them highly effective defenses against a wide range of organisms that might otherwise exploit the plant (Liu et al., 2021).

3. Willow (Salix spp.): Salicylic Acid as a Signal Molecule

The willow tree (*Salix* spp.) produces salicylic acid, a phenolic compound that plays a dual role in plant defense. In addition to its well-known medicinal properties in humans (as the precursor to aspirin), salicylic acid functions as a signal molecule in plants, activating systemic acquired resistance (SAR) against pathogens (Durner et al., 1997). SAR is a defense response in which plants increase their resistance to future attacks after an initial infection. Salicylic acid triggers the production of pathogenesis-related proteins, which inhibit the growth of pathogens such as fungi and bacteria. This defense mechanism is particularly important for perennial plants like willows that are regularly exposed to pathogens over their long lifespans (Vlot et al., 2009).

4. Neem (Azadirachta indica): Azadirachtin as an Insecticide

Neem (*Azadirachta indica*) is a tree native to the Indian subcontinent that produces azadirachtin, a triterpenoid compound with strong insecticidal properties. Azadirachtin disrupts the hormonal systems of insects, inhibiting their ability to feed, grow, and reproduce (Schmutterer, 1990). It is particularly effective against insect pests such as aphids, caterpillars, and whiteflies, making neem oil a popular organic pesticide in agriculture. The ecological role of azadirachtin extends beyond pest deterrence. It also has antimicrobial properties, helping the plant resist infections from bacteria and fungi. The wide range of defensive actions provided by azadirachtin ensures that neem trees can thrive in environments where both herbivores and pathogens pose a threat (Mordue & Nisbet, 2000).

5. Garlic (Allium sativum): Allicin as an Antimicrobial Agent

Garlic (*Allium sativum*) is a bulbous plant that produces allicin, a sulfur-containing secondary metabolite with potent antimicrobial properties. When garlic is damaged, such as when a clove is crushed or chewed, the enzyme alliinase converts the compound alliin into allicin (Borlinghaus et al., 2014). Allicin acts as a defense mechanism by inhibiting the growth of a wide range of pathogens, including bacteria, fungi, and viruses. Allicin's antimicrobial activity is due to its ability to interfere with cellular processes in microorganisms, including the inhibition of enzyme activity and the disruption of cell membranes (Ankri & Mirelman, 1999). This defense not only protects garlic from soilborne pathogens but also helps it compete with other plants by reducing microbial activity in the surrounding environment.

6. Eucalyptus (Eucalyptus spp.): Cineole as an Allelopathic Agent

Eucalyptus trees produce cineole, a monoterpene that serves both as a chemical defense against herbivores and as an allelopathic agent. Cineole has insecticidal properties, making it effective at deterring herbivores such as beetles and leafhoppers (Zhan et al., 2010). It works by disrupting the nervous systems of insects, leading to paralysis or death. In addition to its role in herbivore deterrence, cineole is released into the soil through leaf litter, where it acts as an allelopathic agent, inhibiting the germination and growth of nearby plants (Bais et al., 2003). This helps eucalyptus trees maintain dominance in their habitats by reducing competition for resources like water and nutrients.

1.3 Chemical Classes of Secondary Metabolites

Plants produce a wide range of secondary metabolites, which can be broadly classified into three main chemical groups: phenolics, terpenoids, and alkaloids. These compounds, though not essential for the basic metabolic processes that sustain plant life, play critical roles in plant survival, particularly in defense against herbivores, pathogens, and competitors. Each of these classes of secondary metabolites has a unique biosynthesis pathway, diverse chemical structures, and specific ecological functions.

1. Phenolics: Structure, Biosynthesis, and Functions

Phenolic compounds are a broad class of secondary metabolites characterized by the presence of one or more hydroxyl groups attached to an aromatic ring. They are synthesized primarily through the shikimate pathway, which converts simple carbohydrates derived from photosynthesis into aromatic amino acids, such as phenylalanine. These amino acids are then used to produce phenolic compounds, including flavonoids, tannins, lignins, and coumarins (Cheynier, 2012).

1. **Biosynthesis of Phenolics**: The shikimate pathway leads to the production of phenylalanine, which is converted into a wide range of phenolic compounds through the action of phenylalanine ammonia-lyase (PAL), a key enzyme in phenolic biosynthesis (Boerjan et al., 2003). The diversity of phenolic compounds arises from various modifications to the basic phenylpropanoid structure, such as methylation, glycosylation, and hydroxylation.
2. **Diverse Chemical Structures**: Phenolics can range from simple molecules like salicylic acid to complex polymers such as lignin. Flavonoids, a major subclass of phenolics, consist of two aromatic rings connected by a three-carbon bridge, forming a variety of structures like anthocyanins, flavonols, and isoflavones (Cheynier, 2012). Tannins are larger phenolic compounds that bind to proteins, reducing the digestibility of plant material for herbivores. Lignin, a complex polymer, provides structural support to plant cell walls, making them more resistant to pathogens and herbivores.
3. **Role in Plant Defense**: Phenolic compounds play several roles in plant defense. Flavonoids act as antioxidants, protecting plants from damage caused by UV radiation and oxidative stress (Rozema et al., 1997). Tannins deter herbivores by reducing the palatability and digestibility of plant tissues. Phytoalexins, phenolic compounds produced in response to

pathogen attack, inhibit the growth of bacteria and fungi (Jeandet et al., 2014). Lignin deposition in cell walls strengthens plant tissues and makes them more resistant to invasion by pathogens (Boerjan et al., 2003).

2. Terpenoids: Structure, Biosynthesis, and Functions

Terpenoids (also known as isoprenoids) are one of the largest and most diverse classes of secondary metabolites, composed of units of isoprene (C5H8). They are derived from two main biosynthetic pathways: the mevalonate pathway in the cytoplasm and the methylerythritol phosphate (MEP) pathway in plastids (Vranová et al., 2013). Terpenoids serve a variety of ecological roles, from deterring herbivores and pathogens to attracting pollinators.

1. **Biosynthesis of Terpenoids**: The biosynthesis of terpenoids begins with the formation of isopentenyl diphosphate (IPP) and dimethylallyl diphosphate (DMAPP), the building blocks of terpenoids. These precursors undergo polymerization and cyclization reactions to form a wide range of terpenoids, classified into monoterpenes (C10), sesquiterpenes (C15), diterpenes (C20), triterpenes (C30), and tetraterpenes (C40) (Tholl, 2015).
2. **Diverse Chemical Structures**: Terpenoids exhibit immense structural diversity. Monoterpenes (C10), such as limonene, are small, volatile compounds that often serve as insect repellents. Sesquiterpenes (C15), such as farnesene, are larger and often act as signaling molecules or antimicrobial agents. Diterpenes, such as gibberellins, play a role in plant growth regulation. Triterpenes, such as saponins, are often involved in plant defense against herbivores, while carotenoids (tetraterpenes) contribute to photosynthesis and antioxidant defense (Vranová et al., 2013).
3. **Role in Plant Defense**: Terpenoids have diverse defensive roles in plants. Monoterpenes and sesquiterpenes are often volatile and serve as insect repellents or pheromones, deterring herbivores from feeding on the plant (Dudareva et al., 2004). Some terpenoids, like taxol, a diterpene produced by the yew tree (*Taxus spp.*), have potent anti-cancer and anti-fungal properties (Croteau et al., 2000). Saponins, a class of triterpenoids, disrupt the cell membranes of herbivores and pathogens, providing a chemical barrier to invasion (Osbourn, 1996).

3. Alkaloids: Structure, Biosynthesis, and Functions

Alkaloids are nitrogen-containing compounds that are primarily synthesized from amino acids such as tryptophan, tyrosine, and lysine. They are among the most potent plant secondary metabolites, often exhibiting strong biological activity. Alkaloids are known for their role in plant defense, particularly in deterring herbivores due to their toxicity.

1. **Biosynthesis of Alkaloids**: Alkaloid biosynthesis begins with the decarboxylation of amino acids to form the nitrogen-containing precursors, which are then modified through methylation, oxidation, and cyclization reactions to form the final alkaloid structures. Tryptophan, for example, serves as the precursor for indole alkaloids like vinblastine, while tyrosine is the precursor for isoquinoline alkaloids like morphine (Kutchan, 1995).
2. **Diverse Chemical Structures**: Alkaloids exhibit a wide range of chemical structures, from simple molecules like nicotine to complex compounds like quinine. They are typically

classified based on their nitrogen-containing ring structures, with major classes including indole alkaloids, pyridine alkaloids, tropane alkaloids, and quinoline alkaloids. For example, morphine and codeine are opium alkaloids that act on the nervous system, while atropine, a tropane alkaloid, is a powerful anticholinergic compound (Wink, 2010).

3. **Role in Plant Defense**: Alkaloids are among the most effective chemical deterrents in plants. Nicotine, for instance, is a powerful neurotoxin that disrupts the nervous systems of insects, deterring them from feeding on tobacco plants (*N. tabacum*) (Steppuhn et al., 2004). Morphine and codeine, both opium alkaloids, deter herbivores due to their strong effects on the central nervous system. Quinine, produced by the cinchona tree (*Cinchona officinalis*), has been historically important for its anti-malarial properties, but it also serves as a chemical defense against herbivores. The presence of alkaloids in plant tissues often makes them highly toxic to both insects and larger herbivores.

Significance in Plant Defense Strategies

The diverse chemical structures of phenolics, terpenoids, and alkaloids allow plants to deploy a range of defense strategies. These compounds protect plants in multiple ways:

1. **Deterrence of Herbivores**: Many secondary metabolites serve as chemical deterrents that reduce the palatability of plant tissues or cause toxic effects in herbivores. Alkaloids, in particular, are potent toxins that interfere with the nervous systems of herbivores, while terpenoids such as limonene and cineole function as insect repellents.
2. **Antimicrobial Activity**: Secondary metabolites like phenolic phytoalexins and terpenoid saponins provide plants with strong defenses against microbial pathogens. These compounds can inhibit bacterial and fungal growth, reducing the risk of infection and disease.
3. **Structural Defense**: Phenolics such as lignin reinforce plant cell walls, making it more difficult for pathogens to penetrate plant tissues. This structural defense is particularly important in woody plants, where lignin provides both physical strength and resistance to microbial attack.
4. **Allelopathy**: Secondary metabolites like juglone and sorgoleone enable plants to outcompete neighboring species by inhibiting their growth. These allelopathic interactions reduce competition for resources such as water and nutrients, ensuring the plant's survival.

1.4 Defensive Strategies: Chemical Warfare in Plants

Plants have evolved a sophisticated array of chemical defense strategies to protect themselves from herbivores, pests, and pathogens. Unlike animals, plants are sessile and cannot escape predators or environmental threats. Instead, they rely on the production of secondary metabolites, which function as chemical weapons to deter, repel, or even kill herbivores and pests. These metabolites vary in structure and mode of action, providing plants with multiple layers of defense. Secondary metabolites such as alkaloids, tannins, and terpenoids are particularly effective in this "chemical warfare," enabling plants to survive and reproduce in hostile environments.

Chemical Mechanisms in Plant Defense

The production of secondary metabolites as chemical defenses is a complex process that is often triggered by external stimuli, such as herbivore attack or pathogen invasion. Plants utilize various biosynthetic pathways to produce these compounds, which can have direct toxic effects, inhibit digestion, or disrupt the physiological processes of herbivores and pests.

1. **Direct Toxicity**: Many secondary metabolites exert direct toxic effects on herbivores and pathogens. These compounds can interfere with the nervous system, disrupt cellular membranes, or inhibit key metabolic enzymes. For instance, alkaloids often function as neurotoxins that impair the nervous systems of herbivores, leading to paralysis or death (Wink, 2010). Some terpenoids also exhibit strong insecticidal properties, killing or deterring pests through similar mechanisms (Tholl, 2015).
2. **Digestibility Reduction**: Another defense strategy involves making plant tissues less digestible for herbivores. Tannins, for example, bind to proteins in the digestive systems of herbivores, making it difficult for them to break down the plant material. This reduces the nutritional value of the plant tissue and discourages further feeding (Barbehenn & Constabel, 2011).
3. **Repellent Properties**: Some secondary metabolites act as repellents, deterring herbivores and pests from feeding on the plant. Volatile compounds such as terpenoids are often released in response to herbivory, producing strong odors that repel insects. The release of these compounds can also signal nearby plants to increase their own chemical defenses, creating a community-wide defense response (Dudareva et al., 2004).

Specific Compounds and Their Mode of Action

Plants produce a wide variety of secondary metabolites, each with unique chemical structures and modes of action. The following are key classes of secondary metabolites involved in plant defense:

1. **Alkaloids**: Potent Neurotoxins. Alkaloids are nitrogen-containing compounds that serve as highly effective chemical defenses. Many alkaloids target the nervous systems of herbivores, disrupting neurotransmission and leading to paralysis or death. For instance, nicotine from tobacco plants (*N. tabacum*) binds to nicotinic acetylcholine receptors in the nervous systems of insects, causing overstimulation and eventual paralysis (Steppuhn et al., 2004). Similarly, morphine and codeine, alkaloids derived from the opium poppy (*Papaver somniferum*), affect the central nervous system by interacting with opioid receptors. Quinine, an alkaloid found in the bark of the cinchona tree (*C. officinalis*), disrupts the digestive processes of herbivores and is also a potent anti-malarial compound. Its bitter taste discourages herbivores from consuming the bark, providing a chemical shield against damage (Wink, 2010).
2. **Tannins**: Digestibility Reducers. Tannins are polyphenolic compounds that defend plants by reducing the digestibility of their tissues. These compounds bind to proteins in the saliva and digestive enzymes of herbivores, forming insoluble complexes that are difficult to digest (Barbehenn & Constabel, 2011). This reduces the nutritional quality of the plant material and discourages herbivores from feeding. Plants like oak trees (*Quercus spp.*) are rich in tannins, which make their leaves unpalatable to many herbivores, particularly insects. The high tannin content of oak leaves has been shown to reduce feeding by caterpillars and other herbivorous insects (Salminen & Karonen, 2011).
3. **Terpenoids**: Repellents and Insecticides. Terpenoids (or isoprenoids) are the largest class

of secondary metabolites, with over 40,000 known compounds. Many terpenoids serve as insecticides or repellents, protecting plants from herbivore attack. For example, limonene, a monoterpenoid found in citrus fruits, repels a wide range of insect pests, including aphids and mites (Usseglio et al., 2023). The strong odor of limonene deters herbivores, preventing them from feeding on the plant. Another example is pyrethrins, terpenoid-derived insecticides produced by chrysanthemum flowers (*Chrysanthemum cinerariifolium*). Pyrethrins target the nervous systems of insects, leading to paralysis and death, making them highly effective as natural pesticides (Casida, 1980).

Examples of Plants Known for Their Defensive Chemical Profiles

1. **Tobacco (*Nicotiana tabacum*)**: Alkaloid Defense. Tobacco plants are famous for their production of nicotine, a powerful alkaloid that serves as a chemical defense against herbivores. Nicotine is synthesized in the roots and transported to the leaves, where it is stored in vacuoles. When herbivores, particularly insects, feed on tobacco leaves, they ingest nicotine, which disrupts their nervous systems and deters further feeding (Steppuhn et al., 2004). The production of nicotine is often upregulated in response to herbivore attack, making it a key inducible defense mechanism.
2. **Oak (*Quercus* spp.)**: Tannin Defense. Oak trees are well-known for their high tannin content, particularly in their leaves and bark. Tannins act as both a structural and chemical defense, making oak tissues unpalatable to many herbivores. In addition to binding to proteins and reducing digestibility, tannins also act as antioxidants, protecting plant tissues from oxidative stress caused by herbivory (Salminen & Karonen, 2011). The effectiveness of tannins as a defense is evident in the reduced herbivore feeding on oak leaves compared to other plants with lower tannin content.
3. **Citrus (Citrus spp.)**: Terpenoid Defense. Citrus fruits are rich in terpenoids, particularly limonene, which serves as both a repellent and insecticide. Limonene is found in the essential oils of citrus peels and is responsible for the strong citrus scent. This compound deters a variety of insect pests, including aphids, whiteflies, and mites (Usseglio et al., 2023). The presence of limonene in citrus plants helps protect the fruits from herbivore damage, ensuring successful reproduction and seed dispersal.
4. **Chrysanthemum (*C. cinerariifolium*)**: Pyrethrin Production. The chrysanthemum flower produces pyrethrins, a group of terpenoid-based compounds that are highly effective insecticides. Pyrethrins act by disrupting the nervous systems of insects, causing paralysis and death. This defense mechanism makes chrysanthemums resistant to insect herbivores and has also been harnessed by humans as a natural pesticide in agriculture (Casida, 1980). Pyrethrins are considered environmentally friendly due to their rapid degradation in sunlight, minimizing their impact on non-target species.

1.5 Secondary Metabolites and Plant-Microbe Interactions

Plants are constantly exposed to microbial pathogens, including bacteria, fungi, and viruses, which can cause diseases and significantly impact plant health. To protect themselves, plants have evolved a range of defense mechanisms, including the production of secondary metabolites that act as antimicrobial agents. These compounds play a critical role in limiting pathogen growth, preventing infection, and facilitating recovery from microbial attacks. Among the most important antimicrobial secondary metabolites are phytoalexins, which are produced in response to pathogen invasion, as well as constitutively expressed compounds that contribute to the plant's innate immunity.

Role of Secondary Metabolites in Defending Plants Against Microbial Pathogens

Secondary metabolites serve as the first line of chemical defense against microbial pathogens. These compounds can prevent pathogen entry, inhibit pathogen growth, and trigger plant immune responses. The biosynthesis of antimicrobial secondary metabolites is often initiated upon recognition of pathogen-associated molecular patterns (PAMPs) by plant cells, leading to a cascade of defense responses.

1. **Inhibition of Pathogen Growth**: Many secondary metabolites have direct antimicrobial properties, targeting the cellular structures or metabolic processes of pathogens. For instance, phenolic compounds can disrupt the cell membranes of bacteria and fungi, leading to cell lysis and death (Lattanzio et al., 2006). Terpenoids, particularly sesquiterpenes and diterpenes, are known to inhibit fungal spore germination and mycelial growth by interfering with fungal cell walls and metabolic pathways (Tholl, 2015). Alkaloids, such as berberine and quinine, also exhibit strong antimicrobial activity by disrupting DNA replication and protein synthesis in bacterial and fungal pathogens (Wink, 2010).
2. **Barrier Formation**: Secondary metabolites such as lignin and suberin contribute to the formation of physical barriers that limit pathogen entry. These phenolic-based polymers are deposited in the cell walls and outer layers of plant tissues, creating a structural defense that pathogens find difficult to penetrate (Boerjan et al., 2003). The reinforcement of cell walls through lignin deposition is a critical component of plant defense against both bacterial and fungal infections.
3. **Induced Defense Responses**: Secondary metabolites also play a key role in activating induced defense responses. When plants detect pathogen invasion, they rapidly produce antimicrobial compounds, known as phytoalexins, which help limit pathogen spread. This induced response is part of the plant's systemic acquired resistance (SAR) mechanism, which strengthens the plant's defenses against future infections (Jeandet et al., 2014).

Phytoalexins and Their Role in Plant-Pathogen Interactions

Phytoalexins are antimicrobial secondary metabolites that are synthesized de novo by plants in response to microbial infection. These compounds are not present in healthy plant tissues but are rapidly produced at the site of infection following pathogen recognition. Phytoalexins have diverse chemical structures, including flavonoids, terpenoids, and alkaloids, and their production is specific to plant species and the type of pathogen involved.

1. **Biosynthesis of Phytoalexins**: The production of phytoalexins is triggered by the activation of plant immune receptors that recognize specific molecules from pathogens. This recognition activates signaling pathways, such as the salicylic acid and jasmonic acid pathways, which lead to the expression of defense-related genes and the biosynthesis of phytoalexins (Glazebrook,

2005). For example, the production of camalexin, a key phytoalexin in *A. thaliana*, is upregulated in response to fungal and bacterial infections (Bednarek et al., 2010).

2. **Antimicrobial Activity of Phytoalexins**: Phytoalexins exhibit broad-spectrum antimicrobial activity, inhibiting the growth of bacteria, fungi, and even viruses. These compounds interfere with various cellular processes in pathogens, including membrane integrity, enzyme activity, and DNA replication. For example, resveratrol, a stilbene-type phytoalexin produced by grapevines (*V. vinifera*), has been shown to inhibit the growth of *B. cinerea*, a fungal pathogen responsible for gray mold disease in grapevines (Jeandet et al., 2014).
3. **Specific Phytoalexins in Plant Defense**: Different plant species produce unique phytoalexins with specialized functions. For instance, pisatin, a phytoalexin produced by pea plants (*Pisum sativum*), is highly effective against fungal pathogens, particularly *Fusarium* species (Cruickshank & Perrin, 1960). In soybeans (*Glycine max*), the phytoalexin glyceollin is synthesized in response to infection by the fungus *P. sojae*, a major cause of root rot disease in soybeans (Ebel & Mithöfer, 1998).

Case Studies of Plants with Notable Microbe-Resistant Traits

Several plants have gained attention for their ability to resist microbial pathogens due to their production of secondary metabolites. The following case studies highlight plants that use specific secondary metabolites to fend off bacterial and fungal infections.

1. **Rice (*O. sativa*)**: Phytoalexin-Based Resistance to Rice Blast. Rice is a staple crop worldwide, but it is highly susceptible to rice blast disease, caused by the fungal pathogen *M. oryzae*. To defend against this pathogen, rice plants produce a class of phytoalexins known as momilactones, which are diterpenoids with potent antifungal properties (Hasegawa et al., 2010). Momilactones inhibit the growth of fungal spores and prevent the spread of infection, providing rice plants with a chemical defense against one of the most devastating rice diseases.
2. **Grapevine (*V. vinifera*)**: Stilbene Phytoalexins and Resistance to Gray Mold. Grapevines are commonly infected by the fungal pathogen *Botrytis cinerea*, which causes gray mold, a disease that can significantly reduce grape yields. In response to infection, grapevines produce resveratrol, a stilbene-type phytoalexin that inhibits fungal growth and prevents the development of gray mold (Jeandet et al., 2014). Resveratrol's antifungal activity is attributed to its ability to disrupt fungal cell membranes and inhibit spore germination.
3. **Tomato (*Solanum lycopersicum*)**: Terpenoid-Based Defense Against Fusarium Wilt. Tomato plants produce several secondary metabolites that provide defense against soilborne pathogens. One of the most important defense mechanisms against fusarium wilt, caused by the fungal pathogen *F. oxysporum*, is the production of saponins and sesquiterpenoids, which exhibit strong antifungal activity (Bennett & Wallsgrove, 1994). These compounds disrupt the fungal cell wall, leading to the collapse of the pathogen's cellular structure.
4. **Soybean (*Glycine max*)**: Isoflavonoid Phytoalexins in Resistance to Phytophthora. Soybean plants produce glyceollins, a class of isoflavonoid phytoalexins, in response to infection by *P. sojae*, a pathogen that causes root rot and stem rot in soybeans. Glyceollins accumulate at the infection site, where they inhibit the growth and spread of the pathogen by disrupting its

cell membranes and metabolic functions (Ebel & Mithöfer, 1998). This defense mechanism is crucial for the survival of soybean crops in regions where Phytophthora is prevalent.

References

Agrawal, A. A. (2011). Current trends in the evolutionary ecology of plant defence. *Functional Ecology*, 25(2), 420-432.

Ankri, S., & Mirelman, D. (1999). Antimicrobial properties of allicin from garlic. *Microbes and Infection*, 1(2), 125-129.

Ashraf, M., & Foolad, M. R. (2007). Roles of glycine betaine and proline in improving plant abiotic stress resistance. *Environmental and Experimental Botany*, 59(2), 206-216.

Bais, H. P., Vepachedu, R., Gilroy, S., Callaway, R. M., & Vivanco, J. M. (2003). Allelopathy and exotic plant invasion: From molecules and genes to species interactions. *Science*, 301(5638), 1377-1380.

Baldwin, I. T. (2010). Plant volatiles. *Current Biology*, 20(9), R392-R397.

Balestrini, R., & Bonfante, P. (2005). The interface compartment in arbuscular mycorrhizae: A special type of plant cell wall? *Plant Biosystems*, 139(1), 8-15.

Barbehenn, R. V., & Constabel, C. P. (2011). Tannins in plant-herbivore interactions. *Phytochemistry*, 72(13), 1551-1565.

Bednarek, P., Piślewska-Bednarek, M., Svatoš, A., Schneider, B., Doubsky, J., Mansurova, M., & Schulze-Lefert, P. (2010). A glucosinolate metabolism pathway in living plant cells mediates broad-spectrum antifungal defense. *Science*, 323(5910), 101-106.

Bennett, R. N., & Wallsgrove, R. M. (1994). Secondary metabolites in plant defense mechanisms. *New Phytologist*, 127(4), 617-633.

Bertin, C., Yang, X., & Weston, L. A. (2003). The role of root exudates and allelochemicals in the rhizosphere. *Plant and Soil*, 256(1), 67-83.

Boerjan, W., Ralph, J., & Baucher, M. (2003). Lignin biosynthesis. *Annual Review of Plant Biology*, 54, 519-546.

Booth, M. (1996). *Opium: A history*. Macmillan.

Borlinghaus, J., Albrecht, F., Gruhlke, M. C. H., Nwachukwu, I. D., & Slusarenko, A. J. (2014). Allicin: Chemistry and biological properties. *Molecules*, 19(8), 12591-12618.

Bourgaud, F., Gravot, A., Milesi, S., & Gontier, E. (2016). Production of plant secondary metabolites: A historical perspective. *Plant Science*, 161(5), 839-851.

Casida, J. E. (1980). Pyrethrum flowers and pyrethroid insecticides. *Environmental Health Perspectives*, 34, 189-102.

Chapman, K. D., Dyer, J. M., & Mullen, R. T. (2012). Biogenesis and functions of lipid droplets in plants: Thematic review series: Lipid droplet synthesis and metabolism: From yeast to man. *Journal of Lipid Research*, 53(2), 215-226.

Chen, T. H., & Murata, N. (2008). Glycine betaine: An effective protectant against abiotic stress in plants. *Trends in Plant Science*, 13(9), 499-505.

Cheynier, V. (2012). Phenolic compounds: From plants to foods. *Phytochemistry Reviews*, 11(2-3), 153-177.

Cockell, C. S., & Knowland, J. (1999). Ultraviolet radiation screening compounds. *Biological Reviews*, 74(3), 311-345.

Connolly, E. L., Sim, M., Travica, N., Marx, W., Beasy, G., Lynch, G. S., Bondonno, C. P., Lewis, J. R., Hodgson, J. M., & Blekkenhorst, L. C. (2021). Glucosinolates from cruciferous vegetables and their potential role in chronic disease: Investigating the preclinical and clinical evidence. *Frontiers in Pharmacology, 12*:767975.

Croteau, R., Kutchan, T. M., & Lewis, N. G. (2000). Natural products (secondary metabolites). In *Biochemistry and Molecular Biology of Plants* (24, 1250-1319). American Society of Plant Biologists.

Cruickshank, I. A. M., & Perrin, D. R. (1960). Isolation of pisatin, an antifungal substance from Pisum sativum L. *Nature*, 187(4742), 799-800.

Czarnota, M. A., Paul, R. N., Weston, L. A., & Duke, S. O. (2003). Anatomy of sorgoleone-secreting root hairs of Sorghum species. *International Journal of Plant Sciences*, 164(6), 861-866.

Defilippi, B. G., Manríquez, D., Luengwilai, K., González-Agüero, M. (2009). Aroma volatiles: Biosynthesis and mechanisms of modulation during fruit ripening. *Advances in Botanical Research* (Academic Press), 50(1), 1-37.

Delgado-Vargas, F., & Paredes-López, O. (2002). Natural colorants for food and nutraceutical uses. CRC Press.

Demmig-Adams, B., & Adams, W. W. (2006). Photoprotection in an ecological context: The remarkable complexity of thermal energy dissipation. *New Phytologist*, 172(1), 11-21.

Dewick, P. M. (2009). *Medicinal natural products: A biosynthetic approach* (3rd ed.). Wiley.

Dudareva, N., Pichersky, E., & Gershenzon, J. (2004). Biochemistry of plant volatiles. *Plant Physiology*, 135(4), 1893-1902.

Durner, J., Shah, J., & Klessig, D. F. (1997). Salicylic acid and disease resistance in plants. *Trends in Plant Science*, 2(7), 266-274.

Ebel, J., & Mithöfer, A. (1998). Early events in the elicitation of plant defense. *Planta*, 206(3), 335-348.

Ehrlich, P. R., & Raven, P. H. (1964). Butterflies and plants: A study in coevolution. *Evolution*, 18(4), 586-608.

El-Sayed, M., & Verpoorte, R. (2007). Catharanthus terpenoid indole alkaloids: Biosynthesis and regulation. *Phytochemistry Reviews*, 6(2), 277-305.

Fiehn, O. (2002). Metabolomics-the link between genotypes and phenotypes. *Plant Molecular Biology*, 48(1), 155-171.

Fraenkel, G. S. (1959). The raison d'être of secondary plant substances. *Science*, 129(3361), 1466-1470.

Gao, Y., Dong, J., Fu, Q., Wang, Y., Chen, C., Li, J., Li, R., Chuanjiang, Z. (2017). Allelopathic effects of submerged macrophytes on phytoplankton. *Allelopathy Journal*, *40*, 1-22.

Gershenzon, J., & Dudareva, N. (2007). The function of terpene natural products in the natural world. *Nature Chemical Biology*, 3(7), 408-414.

Glazebrook, J. (2005). Contrasting mechanisms of defense against biotrophic and necrotrophic pathogens. *Annual Review of Phytopathology*, 43, 205-227.

Goswami, S., Ali, A., Prasad, M. E., & Singh, P. (2024). Pharmacological significance of *Catharanthus roseus* in cancer management: A review. *Pharmacological Research - Modern Chinese Medicine*, *11*, 100444,

Guenther, E. (1972). *The essential oils*. Krieger Publishing. Company, New York, 453-454.

Hartmann, T. (2007). From waste products to ecochemicals: Fifty years research of plant secondary metabolism. *Phytochemistry*, 68(22-24), 2831-2846.

Hasegawa, M., Mitsuhara, I., Seo, S., Imai, T., Koga, J., Okada, K., Yamane, H., & Ohashi, Y. (2010). Phytoalexin accumulation in the interaction between rice and the blast fungus. *Molecular Plant-Microbe Interactions*, 23(8), 1000-1011.

Hejl, A. M., & Koster, K. L. (2004). Juglone disrupts root plasma membrane H+-ATPase activity and impairs water uptake, root respiration, and growth in soybean (*Glycine max*) and corn (*Zea mays*). *Journal of Chemical Ecology*, 30(3), 453-471.

Iturriaga, G., Suárez, R., & Nova-Franco, B. (2009). Trehalose metabolism: From osmoprotection to signaling. *International Journal of Molecular Sciences*, 10(9), 3793-3810.

Jeandet, P., Hébrard, C., Deville, M. A., Cordelier, S., Dorey, S., Aziz, A., & Crouzet, J. (2014). Deciphering the role of phytoalexins in plant-microorganism interactions and human health. *Molecules*, 19(11), 18033-18056.

Jiang, Z., Kempinski, C., & Chappell, J. (2016). Extraction and analysis of terpenes/terpenoids. *Current Protocols in Plant Biology*, 1:345-358.

Kessler, A., & Baldwin, I. T. (2007). Making sense of nectar scents: the effect of floral nectar on floral scent. *The Plant Journal*, *49*(4), 840-854.

Kutchan, T. M. (1995). Plant alkaloid biosynthesis: The basis for metabolic engineering of medicinal plants. *The Plant Cell*, *7*(7), 1059-1070.

Lattanzio, V., Lattanzio, V. M. T., & Cardinali, A. (2006). Role of phenolics in the resistance mechanisms of plants against fungal pathogens and insects. *Phytochemistry: Advances in Research*, 23-67.

Liu, Y., Patra, B., Singh, S. K., Paul, P., Zhou, Y., Li, Y., Wang, Y., Pattanaik, S., & Yuan, L. (2021). Terpenoid indole alkaloid biosynthesis in *Catharanthus roseus*: effects and prospects of environmental factors in metabolic engineering. *Biotechnology Letters*, *43*(11), 2085-2103.

Mathesius, U., Schlaman, H. R., Spaink, H. P., Of Sautter, C., Rolfe, B. G., & Djordjevic, M. A. (1998). Auxin transport inhibition precedes root nodule formation in white clover roots and is regulated by flavonoids and derivatives of chitin oligosaccharides. *Plant Journal*, *14*(1), 23-34.

May, F. E., & Ash, J. E. (1990). An assessment of the allelopathic potential of eucalyptus. *Australian Journal of Botany*, 38(3), 245-254.

Mithöfer, A., & Boland, W. (2012). Plant defense against herbivores: Chemical aspects. *Annual Review of Plant Biology*, 63, 431-450.

Mordue, A. J., & Nisbet, A. J. (2000). Azadirachtin from the neem tree *Azadirachta indica*: Its action against insects. *Anais da Sociedade Entomológica do Brasil*, 29(4), 615-632.

Müller, K. O., & Börger, H. (1939). Experimentelle Untersuchungen über die Phytophthora-Resistenz der Kartoffel. *Arbeiten der Biologischen Reichsanstalt für Land- und Forstwirtschaft*, 1-43.

Nelson, D. L., & Cox, M. M. (2017). *Lehninger principles of biochemistry* (7th ed.). W.H. Freeman and Sapling Learning.

Nunn, J. F. (2002). *Ancient Egyptian medicine*. University of Oklahoma Press.

Ohnmeiss, T. E., & Baldwin, I. T. (2000). Optimal defense theory predicts the ontogeny of an induced nicotine defense. *Ecology, 81*(7), 1765-1783.

Oldroyd, G. E. D., & Downie, J. A. (2008). Coordinating nodule morphogenesis with rhizobial infection in legumes. *Annual Review of Plant Biology, 59*(1), 519-546.

Osbourn, A. E. (1996). Saponins and plant defense: A soap story. *Trends in Plant Science*, 1(1), 4-9.

Paddon, C. J., Westfall, P. J., Pitera, D. J., Benjamin, K., Fisher, K., McPhee, D., Leavell, M. D., Tai, A., Main, A., Eng, D., Polichuk, D. R., Teoh, K. H., Reed, D. W., Treynor, T., Lenihan, J., Fleck, M., Bajad, S., Dang, G., Dengrove, D., Diola, D., Dorin, G., Ellens, K. W., Fickes, S., Galazzo, J., Gaucher, S. P., Geistlinger, T., Henry, R., Hepp, M., Horning, T., Iqbal, T., Jiang, H., Kizer, L., Lieu, B., Melis, D., Moss, N., Regentin, R., Secrest, S., Tsuruta, H., Vazquez, R., Westblade, L. F., Xu, L., Yu, M., Zhang, Y., Zhao, L., Lievense, J., Covello, P. S., Keasling, J. D., Reiling, K. K., Renninger, N. S., Newman, J. D. (2013). High-level semi-synthetic production of the potent antimalarial artemisinin. *Nature, 496*(7446), 528-532.

Pan, M. H., Lai, C. S., & Ho, C. T. (2010). Anti-inflammatory activity of natural dietary flavonoids. *Food & Function*, 1(1), 15-31.

Papadopoulou, K., Melton, R. E., Leggett, M., Daniels, M. J., & Osbourn, A. E. (1999). Compromised disease resistance in saponin-deficient plants. *Proceedings of the National Academy of Sciences*, 96(22), 12923-12928.

Quideau, S., Deffieux, D., Douat-Casassus, C., & Pouységu, L. (2011). Plant polyphenols: Chemical properties, biological activities, and synthesis. *Angewandte Chemie International Edition*, 50(3), 586-621.

Raguso, R. A. (2008). Wake up and smell the roses: The ecology and evolution of floral scent. *Annual Review of Ecology, Evolution, and Systematics*, 39(1), 549-569.

Robinson, T. (1991). The organic constituents of higher plants: Their chemistry and interrelationships. (6th ed.). Cordus Press.

Rozema, J., van de Staaij, J., Björn, L. O., & Caldwell, M. (1997). UV-B as an environmental factor in plant life: Stress and regulation. *Trends in Ecology & Evolution*, 12(1), 22-28.

Sachs, J. (1887). *Lectures on the physiology of plants*. Clarendon Press.

Saito, K., & Matsuda, F. (2010). Metabolomics for functional genomics, systems biology, and biotechnology. *Annual Review of Plant Biology*, *61*(1), 463-489.

Salminen, J. P., & Karonen, M. (2011). Chemical ecology of tannins and other phenolics: We need a change in approach. *Functional Ecology*, 25(2), 325-338.

Schmutterer, H. (1990). Properties and potential of natural pesticides from the neem tree, *Azadirachta indica*. *Annual Review of Entomology*, 35(1), 271-297.

Sharkey, T. D., Wiberley, A. E., Donohue, A. R. (2008). Isoprene emission from plants: Why and how. *Annals of Botany*, 101(1), 5-18.

Sneader, W. (2005). *Drug discovery: A history*. Wiley.

Steppuhn, A., Gase, K., Krock, B., Halitschke, R., & Baldwin, I. T. (2004). Nicotine's defensive function in nature. *PLoS Biology*, 2(8), e217.

Taiz, L., Zeiger, E., Møller, I. M., & Murphy, A. (2015). *Plant physiology and development* (6th ed.). Sinauer Associates, Sunderland, CT.

Tholl, D. (2015). Biosynthesis and biological functions of terpenoids in plants. *Advances in Biochemical Engineering/Biotechnology*, 148, 63-106.

Tibugari, H., & Chiduza, C. (2024). Sorghum allelopathy under field conditions may be caused by a combination of allelochemicals. Cogent Food & Agriculture, 10(1). 2324528

Tu, Y. (2016). Artemisinin-A gift from traditional Chinese medicine to the world (Nobel Lecture). *Angewandte Chemie International Edition*, 55(35), 10210-10226.

Usseglio, V. L., Dambolena, J. S., Zunino, M. P. (2023). Can essential oils be a natural alternative for the control of *Spodoptera frugiperda*? A review of toxicity methods and their modes of action. Plants, *12*(1), 3.

Vlot, A. C., Dempsey, D. A., & Klessig, D. F. (2009). Salicylic acid, a multifaceted hormone to combat disease. *Annual Review of Phytopathology*, 47(1), 177-206.

Vranová, E., Coman, D., & Gruissem, W. (2013). Network analysis of the MVA and MEP pathways for isoprenoid synthesis. *Annual Review of Plant Biology*, 64, 665-700.

Warrier, P. K., Nambiar, V. P. K., & Ramankutty, C. (1996). *Indian medicinal plants: A compendium of 500 species*. Universities Press.

Weckwerth, W. (2003). Metabolomics in systems biology. *Annual Review of Plant Biology*, 54(1), 669-689.

Wink, M. (2010). Functions and biotechnology of plant secondary metabolites. *Annual Plant Reviews*, 39(2), 195-258.

Zhang, J., An, M., Wu, H., Stanton, R., & Lemerle, D. (2010). Chemistry and bioactivity of Eucalyptus essential oils. *Allelopathy Journal*, *25*, 313-330.

Chapter - 2

Biosynthesis of Secondary Metabolites: Pathways and Mechanisms

Pratima Raypa[1], Reshma Jadhav[2], Vinay Pratap Singh[3] and Vinod Chouhan[4]

[1]Department of Biochemistry, H.N.B. Garhwal University, Srinagar, Uttarakhand, India

[2]Department of Biochemistry, Mahatma Phule Krishi Vidyapeeth, Rahuri, Ahmednagar, Maharashtra, India

[3]Division of Plant Pathology, Indian Council of Agricultural Research-Indian Agricultural Research Institute (ICAR-IARI), New Delhi, India

[4]Plant Physiology, College of Agriculture, Khurai, JNKVV, Jabalpur, M.P. India

Abstract

The biosynthesis of plant secondary metabolites involves complex pathways that are crucial for plant survival, adaptation, and defense. Unlike primary metabolites, which are essential for basic cellular functions, secondary metabolites are produced through specialized metabolic processes triggered by various biotic and abiotic stresses. These metabolites include phenolics, terpenoids, and alkaloids, each synthesized through distinct biochemical pathways. The Shikimate pathway, for instance, plays a fundamental role in phenolic biosynthesis, leading to the production of compounds like flavonoids and tannins that enhance plant resilience against herbivores and pathogens. Terpenoids, synthesized via the Mevalonate (MVA) and Methylerythritol Phosphate (MEP) pathways, contribute to plant defense by deterring herbivores and attracting predatory insects that serve as indirect protectors. Alkaloid biosynthesis, involving complex enzymatic reactions from amino acid precursors, results in a diverse group of compounds known for their toxicity and deterrent effects on herbivores. The regulation of secondary metabolite biosynthesis is highly dynamic, influenced by genetic and environmental factors, as well as plant developmental stages. Transcription factors and signaling molecules play critical roles in activating or suppressing these biosynthetic pathways in response to stress conditions, allowing plants to optimize their chemical defense strategies.

Keywords: Secondary metabolite biosynthesis, Shikimate pathway, terpenoid biosynthesis, alkaloid biosynthesis, plant defense mechanisms, metabolic regulation

DOI: 10.1201/9781003717454-2

2.1 Introduction to Biosynthetic Pathways

Overview of Primary and Secondary Metabolic Pathways in Plants

Plants, being highly complex organisms, rely on a series of intricate metabolic pathways to sustain their growth, reproduction, and adaptation to their environment. These metabolic processes are broadly categorized into two groups: primary metabolic pathways and secondary metabolic pathways. While primary metabolic processes are crucial for essential life functions such as growth, energy production, and cellular maintenance, secondary metabolic processes are involved in the production of specialized compounds that are not directly related to growth but play critical roles in plant survival, defense, and adaptation. Primary metabolic pathways generate fundamental compounds like carbohydrates, amino acids, lipids, and nucleic acids, which are indispensable for the plant's structure and physiological functions. On the other hand, secondary metabolic pathways lead to the formation of highly specialized compounds, known as secondary metabolites, which are vital for plant interaction with their environment, including defense against herbivores, microbial pathogens, competition with other plants, and even communication with symbiotic organisms. These compounds are not directly essential for the plant's basic life functions, but they are crucial for the plant's ability to adapt, survive, and thrive in its ecological niche. These metabolic pathways allowed researchers to explore the relationship between a plant's internal biochemistry and its external ecological interactions. This distinction between primary and secondary metabolism is essential for understanding how plants allocate resources to both basic life-sustaining processes and specialized functions necessary for environmental adaptation.

Primary Metabolic Pathways

Primary metabolic processes are ubiquitous and conserved across plant species, ensuring that plants can grow, reproduce, and sustain their cellular activities under normal environmental conditions. However, the primary metabolites also serve as precursors for the biosynthesis of secondary metabolites, linking these two metabolic categories. Primary metabolism encompasses the biochemical pathways that are directly involved in producing the essential building blocks necessary for plant survival and reproduction. These processes are fundamental to all living organisms and include pathways such as photosynthesis, respiration, and the synthesis of proteins, lipids, and nucleic acids.

1. **Photosynthesis**: At the heart of primary metabolism is photosynthesis, a process that occurs in the chloroplasts of plant cells. This process converts light energy into chemical energy, producing glucose from carbon dioxide and water. The glucose produced during photosynthesis serves as the primary energy source for the plant and is also a precursor for the synthesis of various primary and secondary metabolites. Photosynthesis not only sustains the plant's growth and energy needs but also forms the foundation for more complex metabolic processes (Buchanan et al., 2015).

2. **Respiration**: Once glucose is produced through photosynthesis, it undergoes cellular respiration, a metabolic pathway that converts glucose into ATP, the plant's energy currency. ATP is essential for powering various cellular processes, from growth to reproduction. The respiration process occurs in the mitochondria and involves several stages, including glycolysis, the Krebs cycle, and oxidative phosphorylation. This pathway ensures that the plant has the energy needed to carry out vital physiological processes and maintain cellular function.

3. **Protein Synthesis**: Proteins are the key structural and functional components of plant cells. The synthesis of proteins involves the translation of amino acids, which are produced through primary metabolic pathways such as the shikimate pathway. This pathway converts simple carbohydrates into aromatic amino acids like phenylalanine, tyrosine, and tryptophan, which are the building blocks for protein synthesis as well as secondary metabolites such as flavonoids and alkaloids (Herrmann & Weaver, 1999).
4. **Lipid Biosynthesis**: Lipids, including fats, oils, and phospholipids, are essential components of plant membranes and serve as energy storage molecules. The biosynthesis of lipids occurs through pathways like fatty acid synthesis, where acetyl-CoA is converted into long-chain fatty acids. These fatty acids are crucial for maintaining cell membrane integrity, fluidity, and function, particularly in plant cells that experience environmental stress such as drought or extreme temperatures.

Secondary Metabolic Pathways

Unlike primary metabolism, secondary metabolism is not essential for the plant's basic growth and development, but it plays an indispensable role in the plant's interactions with its environment. Secondary metabolites serve as chemical signals, defense compounds, attractants, and deterrents, enabling the plant to adapt to its surroundings, compete with other organisms, and defend itself from biotic and abiotic stresses. Secondary metabolic pathways are often induced in response to environmental stress or herbivore attack, making them highly dynamic and adaptable. Unlike primary metabolism, which is constitutive and essential for all plants, secondary metabolism is more specialized and variable among different plant species. The diversity of secondary metabolites is vast, but they are generally categorized into three major classes: phenolics, terpenoids, and alkaloids. Each of these classes has its own biosynthetic pathway and unique ecological function.

1. **Phenolics**: This group of compounds is derived from the shikimate pathway, which also plays a role in primary metabolism. Phenolics include compounds like flavonoids, tannins, lignins, and coumarins. These compounds are involved in protecting the plant from UV radiation, defending against pathogens and herbivores, and providing structural support to plant tissues. For example, lignin, a complex phenolic polymer, strengthens the cell walls of plants, making them more resistant to pathogens and physical damage (Vogt, 2010).
2. **Terpenoids**: These are the largest class of secondary metabolites, derived from the mevalonate pathway and the methylerythritol phosphate (MEP) pathway. Terpenoids, also known as isoprenoids, include compounds like carotenoids, saponins, and essential oils. These compounds serve various ecological functions, from acting as insect repellents to attracting pollinators. Terpenoids are also involved in plant defense, as many of them have antimicrobial properties (Vranová et al., 2013).
3. **Alkaloids**: Alkaloids are nitrogen-containing compounds synthesized from amino acids through complex biosynthetic pathways. They are known for their potent biological activity, often serving as neurotoxins or antiherbivore agents. Alkaloids like nicotine, morphine, and quinine are well-known for their defensive roles in plants, deterring herbivores and inhibiting pathogen growth (Wink, 2010). For instance, nicotine, produced by the tobacco plant, acts

as a neurotoxin that targets the nervous systems of herbivores, effectively protecting the plant from being eaten.

Differences Between Primary and Secondary Metabolic Processes

Though both primary and secondary metabolic processes contribute to plant survival, they differ in several key aspects:

Trait	Primary metabolism	Secondary metabolism
Function	Primary metabolism focuses on generating the essential biomolecules required for the plant's basic life processes, including growth, reproduction, and energy production. These processes are conserved across all plant species.	Secondary metabolism serves to enhance the plant's ability to interact with its environment, including defense against herbivores and pathogens, competition with other plants, and attracting pollinators. These processes are often more species-specific and ecologically driven
Products	Primary metabolites include essential compounds such as sugars, amino acids, lipids, and nucleic acids, which are necessary for all living cells.	Secondary metabolites include specialized compounds like alkaloids, phenolics, and terpenoids, which are not required for basic cellular functions but are crucial for ecological interactions, such as defense and communication.
Biosynthetic Pathways	Primary metabolic pathways are universal and include processes such as photosynthesis, glycolysis, and the citric acid cycle. These pathways produce the energy and precursors needed for cellular function.	Secondary metabolic pathways are more specialized and involve unique enzymes and reactions. For example, the mevalonate pathway produces terpenoids, while the shikimate pathway leads to the production of phenolics.
Tissue Distribution	Primary metabolites are produced and utilized in all parts of the plant as they are necessary for maintaining cellular function.	Secondary metabolites are often restricted to specific tissues or organs, such as leaves, roots, or flowers, and their production may be triggered by environmental factors like herbivory or pathogen attack.
Regulation	Primary metabolic processes are constitutively active, ensuring that the plant has a continuous supply of energy and essential biomolecules.	Secondary metabolic processes are more dynamic and are often inducible, meaning they are activated in response to external stimuli, such as herbivores or pathogens. This allows plants to conserve energy by producing secondary metabolites only when necessary.

Role of Biosynthesis in Plant Adaptation and Defense Mechanisms

Plants are constantly challenged by a variety of biotic and abiotic stressors, ranging from herbivore attacks and pathogen invasions to extreme environmental conditions like drought, UV radiation, and high salinity. To survive and thrive in these hostile environments, plants rely heavily on their ability to produce a vast array of specialized compounds through biosynthesis. The biosynthesis of secondary metabolites is a crucial component of plant adaptation and defense, enabling plants to cope with these challenges by synthesizing chemical compounds that serve protective and regulatory functions. These metabolites not only act as physical and chemical barriers against pests and diseases but also help plants adjust to changing environmental conditions.

Biosynthesis as a Key Component of Plant Defense Mechanisms

Plants have evolved an intricate system of defense mechanisms to protect themselves from herbivores, pathogens, and environmental stress. These defense mechanisms can be classified into constitutive defenses (those always present) and induced defenses (those activated upon stress). Both types of defenses rely on the biosynthesis of a wide range of secondary metabolites, each tailored to specific threats.

1. **Constitutive Defenses**: Constitutive defenses are always present in the plant, even in the absence of an immediate threat. These defenses include structural components such as lignin, cuticle layers, and waxes, which provide a physical barrier to herbivores and pathogens. In addition to physical barriers, plants produce secondary metabolites, such as phenolics and tannins, that deter herbivores and inhibit pathogen growth. For example, lignin biosynthesis in cell walls helps strengthen the plant tissue, making it more resistant to mechanical damage and microbial infection (Boerjan et al., 2003).
2. **Induced Defenses**: Induced defenses are activated in response to specific threats, such as herbivore attack or pathogen invasion. Upon detection of an external threat, plants rapidly activate the biosynthesis of secondary metabolites to protect themselves. For example, phytoalexins, a group of antimicrobial compounds, are synthesized de novo in response to pathogen infection (Jeandet et al., 2014). These compounds inhibit the growth of bacteria and fungi, preventing the spread of infection. Similarly, the production of alkaloids like nicotine increases in tobacco plants when they are attacked by herbivores, serving as an effective deterrent by disrupting the nervous system of the invaders (Steppuhn et al., 2004).
3. **Systemic Acquired Resistance (SAR)**: In addition to local responses, plants can activate systemic defense mechanisms, such as systemic acquired resistance (SAR), where a localized infection triggers the production of defensive compounds throughout the entire plant. SAR is mediated by signaling molecules like salicylic acid, which activate the biosynthesis of phytoalexins and other secondary metabolites in distant tissues (Dempsey et al., 2011). This systemic response helps protect the plant from future attacks and reinforces its overall immune system.

Biosynthesis and Plant Adaptation to Environmental Stress

In addition to biotic stressors like herbivores and pathogens, plants must also adapt to abiotic stresses such as drought, salinity, extreme temperatures, and UV radiation. Biosynthesis of secondary metabolites plays a key role in helping plants withstand these environmental challenges.

1. **Drought and Salinity Stress**: Under drought or high salinity conditions, plants activate the

biosynthesis of osmoprotectants such as proline and glycine betaine, which help maintain cellular turgor and protect proteins from denaturation (Ashraf & Foolad, 2007). These compounds are synthesized through specific metabolic pathways and accumulate in plant cells, allowing them to retain water and stabilize cellular structures during periods of water stress. Additionally, secondary metabolites like terpenoids and phenolics act as antioxidants, reducing oxidative stress caused by water deficiency and high salt concentrations.

2. **Heat and Cold Stress**: Plants exposed to extreme temperatures, whether high or low, rely on the biosynthesis of secondary metabolites to protect cellular components. Isoprene, a terpenoid synthesized through the MEP pathway, is known to protect plants from heat stress by stabilizing cell membranes and reducing thermal damage (Sharkey et al., 2008). On the other hand, the biosynthesis of flavonoids and carotenoids helps protect plants from cold-induced oxidative damage by neutralizing reactive oxygen species (ROS) generated during cold stress (Demmig-Adams & Adams, 2006).
3. **UV Radiation**: Flavonoids and other phenolic compounds play a crucial role in protecting plants from the harmful effects of UV radiation. These compounds absorb UV light and prevent it from damaging DNA, proteins, and other cellular structures. The biosynthesis of flavonoids is often upregulated in plants growing in high UV environments, such as alpine and desert ecosystems. For example, plants like *Artemisia tridentata* (sagebrush) increase the production of flavonoids in response to high UV exposure, acting as natural sunscreens (Rozema et al., 1997).

Case Studies of Biosynthesis in Plant Defense and Adaptation

The role of biosynthesis in plant defense and adaptation is illustrated by several key case studies that demonstrate how specific secondary metabolites contribute to plant survival.

1. **Rice (*Oryza sativa*) and Phytoalexin Biosynthesis**: Rice plants produce momilactones, a group of diterpenoid phytoalexins, in response to infection by the rice blast fungus *Magnaporthe oryzae*. The biosynthesis of momilactones is triggered upon pathogen detection, and these compounds inhibit fungal spore germination and growth (Hasegawa et al., 2010). This defense mechanism helps rice plants resist one of the most destructive pathogens in rice agriculture.
2. **Tomato (*Solanum lycopersicum*) and Terpenoid Biosynthesis**: In response to herbivore attacks, tomato plants synthesize saponins and terpenoid-based compounds, which act as both insecticides and repellents. For instance, the biosynthesis of α-tomatine, a glycoalkaloid, is upregulated when tomato plants are attacked by caterpillars. α-Tomatine disrupts the cell membranes of herbivores, leading to their death and providing an effective chemical defense (Bennett & Wallsgrove, 1994).
3. **Grapevine (*Vitis vinifera*) and Stilbene Biosynthesis**: Grapevine produces resveratrol, a stilbene-type secondary metabolite, in response to infection by fungal pathogens like Botrytis cinerea. Resveratrol is synthesized de novo through the phenylpropanoid pathway and has potent antifungal activity, preventing the development of gray mold disease in grapevines (Jeandet et al., 2014). This biosynthetic response is a key factor in the grapevine's ability to resist fungal infections and maintain crop productivity.

2.2 The Shikimate Pathway and Phenolic Biosynthesis

The Shikimate pathway is a critical metabolic route in plants responsible for the biosynthesis of aromatic amino acids and phenolic compounds. It serves as a major conduit through which plants produce a wide range of secondary metabolites, especially phenolics, which play a crucial role in plant defense, structural integrity, and interaction with the environment. The pathway is named after shikimic acid, an intermediate in the process, and it is primarily responsible for synthesizing essential aromatic amino acids such as phenylalanine, tyrosine, and tryptophan. These amino acids are not only key building blocks of proteins but also precursors for the vast array of secondary metabolites, including flavonoids, tannins, lignins, coumarins, and other phenolic compounds that are integral to plant physiology. The shikimate pathway is unique to plants, fungi, and bacteria, and is absent in animals, which makes it a target for herbicides and antibiotics, as it provides essential components that humans and other animals must obtain through their diet.

Detailed Mechanisms of the Shikimate Pathway

The shikimate pathway begins with the condensation of simple carbohydrate precursors derived from glycolysis and the pentose phosphate pathway. This series of seven metabolic steps leads to the production of chorismate, the precursor for aromatic amino acids and phenolic compounds. The process is outlined in several stages:

1. **Phosphoenolpyruvate (PEP) and Erythrose-4-Phosphate**: The pathway starts with the combination of phosphoenolpyruvate (PEP) from glycolysis and erythrose-4-phosphate from the pentose phosphate pathway. These two precursors undergo an enzymatic reaction catalyzed by DAHP synthase (3-deoxy-D-arabino-heptulosonate-7-phosphate synthase) to form DAHP, marking the first committed step of the shikimate pathway (Herrmann & Weaver, 1999).
2. **Formation of Shikimate**: Through a series of reactions involving dehydroquinase and shikimate dehydrogenase, DAHP is converted into shikimic acid, the compound that gives the pathway its name. This intermediate is critical for further transformations that lead to the production of essential metabolites (Boerjan et al., 2003).
3. **Chorismate Synthesis**: Shikimic acid undergoes further phosphorylation and reduction, eventually leading to the formation of chorismate, the central precursor for the biosynthesis of aromatic amino acids, phenylalanine, tyrosine, and tryptophan, via specific branch points in the pathway (Tohge et al., 2013). These aromatic amino acids serve as the starting materials for the biosynthesis of secondary metabolites, including phenolic compounds.
4. **Phenylalanine and Phenylpropanoid Pathway**: The branch leading to the production of phenylalanine is particularly important for phenolic biosynthesis. Once phenylalanine is synthesized, it is converted to cinnamic acid by the enzyme phenylalanine ammonia-lyase (PAL). This reaction marks the entry point into the phenylpropanoid pathway, which produces a wide range of phenolic compounds (Vogt, 2010). The phenylpropanoid pathway is the major route for synthesizing key secondary metabolites such as flavonoids, lignins, and tannins.

Phenolic Compound Biosynthesis via the Shikimate Pathway

Phenolic compounds derived from the **shikimate pathway** are some of the most important secondary metabolites in plants. These compounds play diverse roles, ranging from structural support to defense against herbivores and pathogens. The following are the major types of phenolic compounds synthesized through this pathway:

1. **Flavonoids**: Flavonoids are a large family of phenolic compounds synthesized through the phenylpropanoid pathway. They are involved in multiple plant functions, including UV protection, pigmentation, and defense against pathogens. Flavonoids are synthesized from p-coumaroyl-CoA, an intermediate in the phenylpropanoid pathway, and malonyl-CoA, which is derived from acetyl-CoA (Cheynier, 2012). The flavonoid biosynthesis pathway involves the activity of several enzymes, including chalcone synthase (CHS) and chalcone isomerase (CHI), which produce key flavonoid structures such as anthocyanins, flavonols, and isoflavones. These compounds provide essential UV protection by absorbing UV-B radiation, which would otherwise cause damage to cellular DNA and proteins. They are also involved in signaling during plant-pathogen interactions, acting as chemical deterrents to herbivores and inhibiting the growth of microbes.
2. **Lignin**: Lignin is a complex polymer that provides structural support to plant cell walls, particularly in vascular tissues such as xylem. The biosynthesis of lignin begins with the conversion of phenylalanine into monolignols via the phenylpropanoid pathway (Boerjan et al., 2003). Monolignols, such as p-coumaryl alcohol, coniferyl alcohol, and sinapyl alcohol, are then polymerized to form the lignin structure, which helps to reinforce the plant's structural integrity, making it resistant to physical stress and pathogen invasion. Lignin deposition is particularly important for the plant's defense, as it creates a barrier that prevents pathogen penetration and herbivore feeding.
3. **Tannins**: Tannins are another class of phenolic compounds, known for their role in plant defense. These compounds bind to proteins, reducing their digestibility and making plant tissues less palatable to herbivores. Tannins also possess antimicrobial properties, inhibiting the growth of bacterial and fungal pathogens (Barbehenn & Constabel, 2011). Tannin biosynthesis is also derived from the phenylpropanoid pathway, utilizing intermediates such as p-coumaric acid and ferulic acid. The presence of tannins in plant tissues, particularly in leaves, bark, and seeds, serves as a deterrent to herbivores, protecting the plant from being consumed.
4. **Coumarins**: Coumarins are a group of phenolic compounds known for their allelopathic and antimicrobial properties. They are synthesized from p-coumaric acid, which is a key intermediate in the phenylpropanoid pathway. Coumarins act as chemical inhibitors, preventing the germination and growth of competing plant species and reducing the incidence of pathogen infection (Wu et al., 2023).

Significance of the Shikimate Pathway in Plant Defense and Adaptation

The shikimate pathway is fundamental to the production of phenolic compounds, which are critical for plant defense and adaptation to environmental stresses. Phenolic compounds serve as the primary chemical weapons that protect plants from herbivores, pathogens, and UV radiation.

1. **Defense Against Herbivores**: Phenolic compounds such as tannins and flavonoids act as deterrents to herbivores by reducing the palatability of plant tissues. The bitter taste and reduced digestibility caused by tannins discourage herbivores from consuming plant tissues, while flavonoids protect plants from oxidative stress induced by herbivore attack.
2. **Resistance to Pathogens**: The biosynthesis of phytoalexins, a class of antimicrobial phenolic compounds, is activated in response to pathogen invasion. These compounds inhibit the growth of bacteria and fungi, preventing the spread of infection in plant tissues. For instance, resveratrol, a stilbene-type phytoalexin produced in grapevines, inhibits the growth of fungal pathogens like Botrytis cinerea, protecting the plant from gray mold disease (Jeandet et al., 2014).
3. **UV Protection**: Phenolic compounds such as flavonoids absorb harmful UV-B radiation, protecting plants from DNA damage and oxidative stress. This is particularly important for plants growing in high-UV environments, such as alpine or desert ecosystems, where UV exposure is intense. Plants increase the production of flavonoids in response to UV stress, effectively acting as natural sunscreens (Li et al., 2002).
4. **Structural Integrity**: The biosynthesis of lignin is crucial for maintaining the structural integrity of plants, particularly in woody tissues. Lignin strengthens the cell walls, making plants more resistant to physical damage and pathogen invasion. Additionally, lignin plays a role in water transport through the xylem, ensuring that plants can efficiently move water and nutrients from roots to leaves.

Mechanisms of Converting Primary Metabolites into Phenolic Secondary Metabolites

The conversion of primary metabolites into phenolic secondary metabolites is a crucial process in plants that enables the production of a wide array of phenolic compounds such as flavonoids, tannins, and lignin. These phenolic compounds are derived primarily from the shikimate pathway, which serves as the initial metabolic route, converting simple carbohydrate precursors into aromatic amino acids like phenylalanine, tyrosine, and tryptophan. These amino acids are then transformed into secondary metabolites through a series of enzymatic reactions, most notably within the phenylpropanoid pathway, which is a key branch of the shikimate pathway. The transition from primary metabolism to the production of secondary metabolites involves several key steps:

1. From Primary to Secondary Metabolites: The Phenylpropanoid Pathway

The phenylpropanoid pathway is the central biosynthetic route through which primary metabolites like phenylalanine are converted into phenolic secondary metabolites. The initial step of this pathway involves the enzyme phenylalanine ammonia-lyase (PAL), which deaminates phenylalanine, producing cinnamic acid. This reaction marks the transition from primary metabolism (protein synthesis) to secondary metabolism, as cinnamic acid is the precursor for a wide range of phenolic compounds. Once cinnamic acid is formed, it undergoes a series of hydroxylation, methylation, and glycosylation reactions that produce key intermediates such as p-coumaric acid, ferulic acid, and sinapic acid. These intermediates are further transformed into various phenolic secondary metabolites like flavonoids, tannins, and lignin. The specific enzymes involved in these transformations, such as cinnamate-4-

hydroxylase (C4H) and 4-coumarate ligase (4CL), dictates the specific phenolic products formed and their roles in the plant.

2. Shikimate Pathway: Feeding into the Phenylpropanoid Pathway

The shikimate pathway is responsible for providing the precursor compounds necessary for the phenylpropanoid pathway. It begins with the condensation of phosphoenolpyruvate (PEP) and erythrose-4-phosphate, leading to the formation of chorismate, the precursor to aromatic amino acids. These amino acids, particularly phenylalanine, are the key starting materials for phenolic biosynthesis. Once phenylalanine is formed, it feeds into the phenylpropanoid pathway, where PAL converts it into cinnamic acid, setting the stage for the synthesis of phenolic compounds. This interplay between the shikimate pathway and the phenylpropanoid pathway is essential for the plant's ability to produce a wide array of phenolic metabolites, each serving different ecological and physiological roles.

3. Biosynthesis of Flavonoids, Tannins, and Lignin

The phenylpropanoid pathway not only leads to the production of simple phenolic compounds but also serves as the starting point for more complex secondary metabolites, including flavonoids, tannins, and lignin.

Role of the Shikimate Pathway in Producing Flavonoids, Tannins, and Lignin for Plant Defense

The phenolic compounds derived from the shikimate pathway serve crucial roles in plant defense. These compounds not only act as physical and chemical barriers to herbivores and pathogens but also contribute to the plant's overall resilience against environmental stressors. Flavonoids, tannins, and lignin are the three major classes of phenolic compounds that play significant roles in plant defense mechanisms.

1. Flavonoids:

Flavonoids are a diverse group of phenolic compounds that are produced through the phenylpropanoid pathway and serve several functions related to plant defense. Flavonoids are derived from p-coumaroyl-CoA, an intermediate in the phenylpropanoid pathway, and undergo enzymatic reactions catalyzed by chalcone synthase (CHS) and chalcone isomerase (CHI) to form a wide range of flavonoid compounds. Flavonoids act as natural sunscreens, UV Protection, protecting plants from harmful UV-B radiation. They absorb UV light, preventing DNA damage and the formation of reactive oxygen species (ROS) in plant cells (Agati et al., 2012). This is particularly important for plants in high-altitude or desert environments, where UV exposure is intense. Flavonoids also play a significant role in defending plants against microbial pathogens. Many flavonoid compounds have been shown to inhibit the growth of bacteria, fungi, and viruses by disrupting microbial cell walls, inhibiting enzymes, or interfering with the replication of microbial DNA (Treutter, 2006). For example, quercetin, a common flavonoid, has strong antimicrobial properties and is known to be effective in inhibiting fungal pathogens in plants. The production of quercetin and other flavonoids is often upregulated in response to pathogen attack, providing a rapid and localized defense.

2. Tannins:

Tannins are another class of phenolic compounds derived from the phenylpropanoid pathway. These compounds are synthesized from intermediates such as p-coumaric acid and ferulic acid, and they play a crucial role in plant defense against herbivores as anti-herbivore defense. Tannins bind to proteins in the digestive systems of herbivores, reducing their ability to break down plant material. This makes the plant tissue less palatable and less nutritious for herbivores, discouraging them from feeding (Barbehenn & Constabel, 2011). For example, many tree species such as oaks and eucalypts produce high levels of tannins in their leaves and bark, making them highly resistant to herbivore damage. Tannins also have antimicrobial properties, as they can inhibit the growth of bacterial and fungal pathogens by binding to microbial proteins and enzymes, disrupting their function. The ability of tannins to both deter herbivores and inhibit pathogen growth makes them a crucial component of a plant's defense strategy.

3. Lignin:

Lignin is a complex phenolic polymer that is synthesized through the phenylpropanoid pathway. It is primarily found in the cell walls of vascular tissues, providing structural support and rigidity to the plant. The biosynthesis of lignin begins with the conversion of phenylalanine into monolignols (p-coumaryl alcohol, coniferyl alcohol, and sinapyl alcohol), which are then polymerized to form lignin. Lignin reinforces plant cell walls, making them more resistant to mechanical stress, pathogen invasion, and environmental factors such as wind and drought (Boerjan et al., 2003). The lignification of vascular tissues also enhances the plant's ability to transport water and nutrients efficiently. Lignin plays a critical role in forming a physical barrier that prevents pathogen penetration into plant tissues. When plants are attacked by pathogens, lignin deposition is often enhanced at the site of infection, limiting the spread of the pathogen (Miedes et al., 2014). For example, plants that produce higher amounts of lignin are more resistant to fungal pathogens like Fusarium and Pythium, which degrade plant cell walls during infection.

2.3 Mevalonate and MEP Pathways: Terpenoid Biosynthesis

Terpenoids (also known as isoprenoids) represent the largest and most diverse class of secondary metabolites in plants. These compounds play pivotal roles in a variety of biological processes, including plant growth regulation, defense against herbivores and pathogens, attraction of pollinators, and adaptation to environmental stress. Terpenoids are also important in the production of essential oils, resins, and many bioactive compounds with pharmaceutical significance. The biosynthesis of terpenoids occurs through two distinct metabolic pathways in plants: the Mevalonate (MVA) pathway and the Methylerythritol Phosphate (MEP) pathway. Both pathways converge to produce isopentenyl diphosphate (IPP) and dimethylallyl diphosphate (DMAPP), the universal five-carbon building blocks that serve as precursors for the synthesis of all terpenoids.

The Mevalonate (MVA) Pathway

The Mevalonate (MVA) pathway is a cytosolic route primarily responsible for producing terpenoids that are involved in sterol biosynthesis, including hormones like brassinosteroids, and certain defensive terpenoids. This pathway is crucial in the biosynthesis of sesquiterpenes (C15), triterpenes (C30), and polyterpenes, which are involved in numerous plant processes.

1. **Initial Steps of the MVA Pathway**: The MVA pathway begins with the condensation of two molecules of acetyl-CoA, catalyzed by thiolase to form acetoacetyl-CoA. This is followed by the addition of another acetyl-CoA molecule to produce 3-hydroxy-3-methylglutaryl-CoA (HMG-CoA), a reaction catalyzed by HMG-CoA synthase (Goldstein & Brown, 1990).

HMG-CoA is then reduced by HMG-CoA reductase to form mevalonate, the compound that gives the pathway its name. HMG-CoA reductase is a key regulatory enzyme of the MVA pathway and is subject to feedback inhibition by downstream products such as sterols and terpenoids.

2. **Conversion to IPP and DMAPP**: Mevalonate undergoes a series of phosphorylation steps, catalyzed by mevalonate kinase and phosphomevalonate kinase, resulting in the formation of mevalonate-5-diphosphate. Subsequently, mevalonate diphosphate decarboxylase catalyzes the decarboxylation of mevalonate-5-diphosphate, leading to the production of isopentenyl diphosphate (IPP), a five-carbon isoprene unit that serves as the universal precursor for terpenoid biosynthesis (Vranová et al., 2013). Through the action of IPP isomerase, IPP is converted into its isomer, dimethylallyl diphosphate (DMAPP). The ratio between IPP and DMAPP is crucial for the formation of different classes of terpenoids.
3. **Formation of Terpenoids**: The MVA pathway primarily contributes to the synthesis of sesquiterpenes (C15), triterpenes (C30), and polyterpenes. Sesquiterpenes, such as farnesene and humulene, are involved in plant defense mechanisms, often functioning as volatile compounds that repel herbivores or attract predators of herbivores (Dudareva et al., 2004). Triterpenes, such as squalene, are the precursors to sterols and brassinosteroids, which regulate plant growth and development. Triterpenoids also serve as structural components in plant membranes and contribute to plant defense by deterring herbivores and inhibiting pathogen growth (Osbourn, 1996).

The Methylerythritol Phosphate (MEP) Pathway

The Methylerythritol Phosphate (MEP) pathway, also known as the non-mevalonate pathway, operates in plastids and is responsible for the biosynthesis of monoterpenes (C10), diterpenes (C20), and tetraterpenes (C40), including carotenoids, gibberellins, and certain defensive compounds. This pathway is essential for the production of terpenoids that play a role in photosynthesis, hormone regulation, and the plant's response to environmental stress.

1. **Initial Steps of the MEP Pathway**: The MEP pathway begins with the condensation of pyruvate and glyceraldehyde-3-phosphate (G3P), catalyzed by 1-deoxy-D-xylulose-5-phosphate synthase (DXS) to form 1-deoxy-D-xylulose-5-phosphate (DXP). This is a key regulatory step in the MEP pathway (Lichtenthaler, 1999). DXP is then converted into methylerythritol phosphate (MEP) through the action of 1-deoxy-D-xylulose-5-phosphate reductoisomerase (DXR). MEP serves as the central intermediate in the pathway and is further transformed through a series of reactions involving phosphorylation, cyclization, and reductive steps.
2. **Conversion to IPP and DMAPP**: The conversion of MEP into 4-hydroxy-3-methylbut-2-enyl diphosphate (HMBPP) is catalyzed by HMBPP synthase. HMBPP is subsequently reduced by HMBPP reductase to produce both IPP and DMAPP. Similar to the MVA pathway, the balance between IPP and DMAPP is tightly regulated to ensure the proper synthesis of different classes of terpenoids (Rohmer, 1999).
3. **Formation of Terpenoids**: The MEP pathway contributes to the biosynthesis of monoterpenes

(C10), diterpenes (C20), and tetraterpenes (C40). Monoterpenes, such as limonene and geraniol, are often found in essential oils and play a role in attracting pollinators and deterring herbivores (Tholl, 2015). These volatile compounds are also important in plant-plant interactions and act as chemical signals in the environment. Diterpenes, including gibberellins and phytol, are involved in regulating plant growth and development, particularly in stem elongation and chlorophyll biosynthesis. Diterpenes also have antimicrobial properties, helping plants resist pathogens. Tetraterpenes, such as carotenoids, play a crucial role in photosynthesis by protecting plants from photooxidative damage. Carotenoids also serve as precursors for plant hormones like abscisic acid, which regulates stress responses and seed dormancy (Ruiz-Sola & Rodríguez-Concepción, 2012).

The Synergy Between MVA and MEP Pathways

The MVA and MEP pathways are compartmentalized within the plant, with the MVA pathway operating in the cytosol and the MEP pathway functioning in the plastids. Despite their spatial separation, the two pathways work synergistically to ensure the balanced production of terpenoids required for various physiological and ecological functions.

1. **Cross-Talk Between Pathways**: There is evidence of metabolic cross-talk between the MVA and MEP pathways, where intermediates like IPP and DMAPP can be exchanged between the cytosol and plastids, ensuring a coordinated synthesis of terpenoids. For example, some monoterpenes produced by the MEP pathway can be further modified in the cytosol, while sesquiterpenes from the MVA pathway can be used in plastid-associated processes (Lange et al., 2000).
2. **Ecological and Functional Diversification**: The spatial separation of the MVA and MEP pathways allows plants to fine-tune the production of specific terpenoids depending on their ecological needs. For instance, plastid-derived terpenoids (e.g., carotenoids and gibberellins) are more involved in photosynthesis and growth regulation, while cytosolic terpenoids (e.g., sesquiterpenes) are critical for defense and environmental interactions.

Enzymatic Processes Involved in the Mevalonate (MVA) and Methylerythritol Phosphate (MEP) Pathways

The biosynthesis of terpenoids through the Mevalonate (MVA) pathway and the Methylerythritol Phosphate (MEP) pathway relies on a series of enzymatic reactions that convert primary metabolic intermediates into the building blocks for terpenoid synthesis. These enzymes are highly specialized and play critical roles in determining the rate of production and the specific types of terpenoids synthesized. Each step in these pathways is tightly regulated by various enzymes, and the pathways converge to produce isopentenyl diphosphate (IPP) and dimethylallyl diphosphate (DMAPP), which serve as the universal five-carbon units for terpenoid biosynthesis.

Enzymatic Processes in the Mevalonate (MVA) Pathway

The Mevalonate (MVA) pathway occurs in the cytosol and is responsible for the production of terpenoids involved in sterol biosynthesis, sesquiterpenes, and triterpenes. The key enzymatic steps in this pathway are:

1. **Acetyl-CoA Thiolase**: The MVA pathway begins with the condensation of two molecules of acetyl-CoA, catalyzed by thiolase, to form acetoacetyl-CoA. This enzyme is crucial for channeling acetyl-CoA from primary metabolism into the terpenoid biosynthetic pathway.
2. **HMG-CoA Synthase**: In the next step, HMG-CoA synthase catalyzes the addition of a third acetyl-CoA molecule to acetoacetyl-CoA, producing 3-hydroxy-3-methylglutaryl-CoA (HMG-CoA). This intermediate is a critical junction in terpenoid biosynthesis, as it is the substrate for the following rate-limiting step.
3. **HMG-CoA Reductase**: The reduction of HMG-CoA by HMG-CoA reductase is the key regulatory step of the MVA pathway. This enzyme reduces HMG-CoA to form mevalonate. HMG-CoA reductase is highly regulated, and its activity is controlled by feedback inhibition from downstream products like sterols and other terpenoids (Goldstein & Brown, 1990). This step is often the target of inhibitors, such as statins, in medical applications because it also regulates cholesterol synthesis in animals.
4. **Mevalonate Kinase** and **Phosphomevalonate Kinase**: Once mevalonate is produced, it is sequentially phosphorylated by mevalonate kinase and phosphomevalonate kinase to produce mevalonate-5-phosphate and mevalonate-5-diphosphate, respectively.
5. **Mevalonate Diphosphate Decarboxylase**: The final step in this part of the MVA pathway involves mevalonate diphosphate decarboxylase, which converts mevalonate-5-diphosphate into isopentenyl diphosphate (IPP), a key five-carbon building block for terpenoid biosynthesis.
6. **IPP Isomerase**: The enzyme IPP isomerase interconverts IPP and dimethylallyl diphosphate (DMAPP), another five-carbon unit. These two isoprene units serve as the precursors for the production of higher terpenoids like sesquiterpenes and triterpenes (Vranová et al., 2013).

Enzymatic Processes in the Methylerythritol Phosphate (MEP) Pathway

The Methylerythritol Phosphate (MEP) pathway, also known as the non-mevalonate pathway, takes place in plastids and is responsible for synthesizing monoterpenes, diterpenes, and tetraterpenes, which include important compounds like carotenoids and gibberellins. Key enzymes involved in this pathway are:

1. **1-Deoxy-D-Xylulose-5-Phosphate Synthase (DXS)**: The MEP pathway starts with the condensation of pyruvate and glyceraldehyde-3-phosphate (G3P), catalyzed by DXS to form 1-deoxy-D-xylulose-5-phosphate (DXP). This enzyme controls the rate of the MEP pathway and is often regarded as a key regulatory step in the process (Lichtenthaler, 1999).
2. **1-Deoxy-D-Xylulose-5-Phosphate Reductoisomerase (DXR)**: DXP is then converted into methylerythritol phosphate (MEP) by DXR. This reaction marks the first committed step toward isoprenoid biosynthesis in plastids. DXR is another regulatory enzyme in the pathway, and inhibitors of this enzyme have been used to explore potential herbicides.
3. **MEP Kinase, CDP-ME Synthase, and CDP-ME Kinase**: MEP undergoes a series of phosphorylation and cyclization steps, catalyzed by MEP kinase, CDP-ME synthase, and CDP-ME kinase, resulting in the formation of 4-hydroxy-3-methylbut-2-enyl diphosphate (HMBPP).

4. **HMBPP Synthase and HMBPP Reductase**: HMBPP synthase converts CDP-ME into HMBPP, which is then reduced by HMBPP reductase to produce IPP and DMAPP, the same five-carbon building blocks produced by the MVA pathway.
5. **Geranyl Diphosphate Synthase (GPPS) and Farnesyl Diphosphate Synthase (FPPS)**: In the final steps of terpenoid biosynthesis, GPPS and FPPS catalyze the condensation of IPP and DMAPP to produce geranyl diphosphate (GPP), farnesyl diphosphate (FPP), and other intermediates that lead to the formation of monoterpenes, diterpenes, and tetraterpenes (Rohmer, 1999).

Importance of Terpenoids in Plant Growth Regulation and Defense

Terpenoids are not only the largest class of secondary metabolites in plants but also play critical roles in growth regulation, development, and defense against herbivores and pathogens. The structural diversity of terpenoids allows them to perform a wide variety of functions, from acting as hormones that regulate plant growth to serving as chemical defenses against pests and diseases.

1. Terpenoids in Plant Growth Regulation

Plant hormones, such as gibberellins, abscisic acid (ABA), and brassinosteroids, are terpenoid-derived compounds that regulate key aspects of plant growth and development.

- **Gibberellins (GAs)**: Gibberellins are diterpenoids synthesized via the MEP pathway, and they play a critical role in regulating processes such as stem elongation, seed germination, flowering, and fruit development. GAs stimulate cell division and elongation, leading to increased plant height. They also promote the breakdown of stored food reserves in seeds, facilitating the germination process (Hedden & Sponsel, 2015).
- **Abscisic Acid (ABA)**: ABA is another diterpenoid produced via the MEP pathway, and it is crucial for regulating plant responses to abiotic stress, particularly drought and salinity. ABA helps control stomatal closure, reducing water loss during periods of drought. It also plays a role in seed dormancy, ensuring that seeds only germinate under favorable conditions (Nambara & Marion-Poll, 2005).
- **Brassinosteroids**: These are triterpenoids synthesized via the MVA pathway, and they play essential roles in regulating cell expansion, vascular differentiation, and plant immunity. Brassinosteroids enhance stress tolerance in plants and promote overall growth by stimulating cell wall loosening and elongation.

2. Terpenoids in Plant Defense Against Herbivores and Pathogens

Terpenoids are vital components of a plant's defense arsenal, protecting plants from herbivores, insects, and microbial pathogens through both direct and indirect mechanisms.

- **Monoterpenes and Sesquiterpenes**: Volatile monoterpenes (C10) and sesquiterpenes (C15), such as limonene and farnesene, are released by plants in response to herbivore attack. These volatile compounds serve as direct deterrents, repelling herbivores with their strong odors, or as indirect defenses by attracting the natural predators of herbivores, such as parasitoid wasps (Dudareva et al., 2004). For example, when attacked by caterpillars, corn plants release volatile terpenoids that attract parasitic wasps, which lay their eggs inside the caterpillars, thereby reducing the herbivore population.

» **Diterpenes and Triterpenes**: Many diterpenoids, such as phytol (a precursor of chlorophyll), and squalene (a precursor of sterols), also function in plant defense. For instance, certain diterpenes exhibit antifungal properties, protecting plants from fungal pathogens. Saponins, a group of triterpenoids, are toxic to herbivores and fungi due to their ability to disrupt cell membranes, making them effective chemical defenses against a wide range of attackers (Osbourn, 1996).

» **Carotenoids**: These tetraterpenes are not only essential for photosynthesis, as they help protect the photosynthetic apparatus from oxidative damage, but they also serve as precursors for the hormone abscisic acid (ABA), which regulates stress responses. Carotenoids are critical for protecting plants from photooxidative damage under high-light conditions and during abiotic stress (Ruiz-Sola & Rodríguez-Concepción, 2012).

2.4 Alkaloid Biosynthesis

Alkaloids are a diverse group of nitrogen-containing secondary metabolites that play crucial roles in plant defense, ecological interactions, and sometimes even in human medicine. These compounds are synthesized from primary metabolites, primarily amino acids, through a series of complex biosynthetic pathways. Alkaloids are found across a wide variety of plant families, such as Solanaceae, Papaveraceae, and Fabaceae, and they exhibit a broad range of biological activities, including toxic effects on herbivores, antimicrobial properties, and pharmacological effects on humans. The biosynthetic origins of alkaloids are generally traced back to three primary amino acid precursors:

1. **Ornithine** and **lysine**: These amino acids give rise to pyrrolidine, piperidine, and tropane alkaloids (e.g., nicotine).
2. **Tyrosine**: This aromatic amino acid is the precursor to important alkaloids like morphine and other isoquinoline alkaloids.
3. **Tryptophan**: Tryptophan is the precursor for indole alkaloids, such as strychnine and quinine.

These amino acids undergo a series of enzymatic transformations, such as decarboxylation, oxidation, methylation, and cyclization, leading to the formation of a wide variety of alkaloid structures. The diversity of alkaloids arises from the various modifications that these basic structures undergo during biosynthesis, producing hundreds of structurally distinct compounds with specialized functions in plants.

Enzymatic Reactions Leading to Alkaloid Diversity in Plants

Alkaloid biosynthesis is catalyzed by a series of specific enzymes that act on amino acid precursors to generate a wide range of complex chemical structures. Several key enzymatic reactions contribute to alkaloid diversity in plants:

1. **Decarboxylation**: The first major enzymatic step in alkaloid biosynthesis often involves the decarboxylation of the amino acid precursor. For example, ornithine is decarboxylated to form putrescine, which is a common intermediate in the biosynthesis of many alkaloids, including nicotine and tropane alkaloids (Facchini, 2001). Similarly, tyrosine is decarboxylated

to produce dopamine, a key precursor in the formation of isoquinoline alkaloids like morphine. The enzyme tyrosine decarboxylase plays a central role in converting tyrosine into dopamine, an intermediate in the biosynthesis of alkaloids in the Papaveraceae family (Ziegler & Facchini, 2008).

2. **Amine Oxidation**: Another critical reaction in alkaloid biosynthesis is the oxidation of amines to aldehydes, which facilitates the formation of reactive intermediates. For instance, in the synthesis of nicotine, the amine group of putrescine is oxidized to form a pyridine ring, an essential structural feature of this alkaloid (Dewick, 2009).
3. **Cyclization**: The formation of complex alkaloid structures often involves cyclization reactions. Cyclization enzymes catalyze the closure of linear amino acid-derived intermediates to form characteristic ring structures. For example, in the biosynthesis of morphine, salutaridine synthase catalyzes the cyclization of reticuline, a key intermediate in the isoquinoline pathway, to form salutaridine, an important precursor in the production of morphine (Hagel & Facchini, 2013).
4. **Methylation**: Methyltransferases play a key role in the modification of alkaloid structures by adding methyl groups (-CH3) to specific positions on the alkaloid molecule. This reaction often influences the biological activity of alkaloids and their solubility. For instance, methylation of intermediates in the biosynthesis of tropane alkaloids results in the production of scopolamine and hyoscyamine, potent anticholinergic compounds found in plants of the Solanaceae family (Ziegler & Facchini, 2008).
5. **Hydroxylation and Oxidation**: Cytochrome P450 enzymes and other oxidases catalyze hydroxylation and oxidation reactions, which introduce hydroxyl groups (-OH) or other functional groups into alkaloid structures, increasing their complexity and modifying their pharmacological properties. For example, in the biosynthesis of morphine, multiple hydroxylation steps are required to convert thebaine into the final morphine structure (Hagel & Facchini, 2013).
6. **Conjugation**: Alkaloids can also undergo conjugation reactions, in which they are linked to sugars, acids, or other molecules. This enhances the solubility, transport, or storage of the alkaloid in the plant. For example, glucosyltransferases catalyze the addition of glucose molecules to alkaloid precursors, making them more hydrophilic and easier to store in vacuoles until they are needed for defense.

Case Studies on Notable Alkaloid Biosynthesis and Their Ecological Roles

The biosynthesis of specific alkaloids such as nicotine, morphine, and quinine provides insight into the ecological significance of these compounds in plant defense mechanisms. Below are detailed case studies on the biosynthesis and ecological roles of these alkaloids.

1. Nicotine Biosynthesis in *Nicotiana tabacum*

Nicotine is one of the most well-known alkaloids, found primarily in the Solanaceae family, particularly in tobacco plants (Nicotiana tabacum). Nicotine biosynthesis begins with the decarboxylation of ornithine to form putrescine, which is then converted into nicotine through a series of enzymatic steps. The key reactions include: Putrescine N-methyltransferase (PMT) catalyzes the methylation of putrescine, leading to the formation of N-methylputrescine. N-methylputrescine oxidase then converts N-methylputrescine into a pyrrolidine ring, which is a crucial structural feature of nicotine (Steppuhn et al., 2004). The final condensation of the pyrrolidine ring with nicotinic acid forms nicotine.

Ecological Role of Nicotine: Nicotine functions primarily as a chemical defense compound, acting as a neurotoxin that disrupts the nervous systems of herbivores by binding to nicotinic acetylcholine receptors. This makes it highly effective in deterring herbivory. In addition to its direct toxic effects, nicotine also functions as a volatile compound that signals herbivore presence to predatory insects, which helps in controlling herbivore populations indirectly by attracting predators (Baldwin, 2001).

2. Morphine Biosynthesis in *Papaver somniferum*

Morphine is an isoquinoline alkaloid produced by the opium poppy (*Papaver somniferum*) and is renowned for its potent analgesic properties. Morphine biosynthesis originates from the amino acid tyrosine, which is converted into dopamine and then reticuline, a central intermediate in isoquinoline alkaloid biosynthesis. Key steps include: Norlaudanosoline synthase catalyzes the first committed step in the morphine pathway by condensing dopamine with 4-hydroxyphenylacetaldehyde to form reticuline. Salutaridine synthase catalyzes the cyclization of reticuline to form salutaridine, which is then converted through several steps into thebaine. Finally, thebaine 6-O-demethylase and other enzymes catalyze the conversion of thebaine into morphine (Hagel & Facchini, 2013).

Ecological Role of Morphine: In nature, morphine and related alkaloids act as defense compounds against herbivores and pathogens. The bitter taste of morphine deters herbivores from consuming opium poppy plants. Additionally, morphine has antimicrobial properties, protecting the plant from microbial infection by inhibiting bacterial and fungal growth.

3. Quinine Biosynthesis in *Cinchona* Species

Quinine is an alkaloid derived from Cinchona trees, and it is best known for its historical use as an antimalarial agent. Quinine biosynthesis originates from tryptophan via the indole alkaloid pathway. Tryptophan undergoes decarboxylation to form tryptamine, which is then modified through multiple steps involving cyclization and methylation to produce quinine.

Ecological Role of Quinine; Quinine acts as a defense compound against herbivores and pathogens. Its bitter taste makes it unpalatable to many herbivores, thus protecting the Cinchona tree from being grazed. Quinine also exhibits antimicrobial properties, inhibiting the growth of certain fungi and bacteria that may otherwise cause disease in the tree.

2.5 Regulation of Secondary Metabolite Biosynthesis

Factors Influencing Secondary Metabolite Production

The production of secondary metabolites in plants is not a static process but is highly dynamic, influenced by a wide range of genetic, environmental, and developmental factors. Secondary metabolites like alkaloids, terpenoids, and phenolics play critical roles in plant defense, signaling, and interaction with the environment. The regulation of their biosynthesis is a tightly controlled process that allows plants to respond to various external and internal cues, optimizing the production of these compounds when needed, such as during herbivore attacks or pathogen infections.

1. Genetic Factors

Genetic regulation is fundamental to the control of secondary metabolite biosynthesis. Specific genes encode the enzymes that catalyze the steps in the biosynthetic pathways of secondary metabolites, and the expression of these genes determines the quantity and diversity of secondary metabolites produced by a plant.

» **Biosynthetic Pathway Genes**: The production of secondary metabolites is dependent on the presence of specific biosynthetic genes that encode enzymes involved in key metabolic pathways. For example, the production of flavonoids requires the coordinated expression of genes in the phenylpropanoid pathway, such as chalcone synthase (CHS) and chalcone isomerase (CHI). Mutations or variations in these genes can lead to altered secondary metabolite profiles (Tohge et al., 2013).

» **Genetic Variation**: Natural variation in plant populations can lead to differences in secondary metabolite production. This genetic diversity is often influenced by the plant's evolutionary history, where different species or ecotypes produce unique secondary metabolites adapted to their specific environmental conditions. For instance, some varieties of Nicotiana species produce higher concentrations of nicotine as a defense against herbivores, a trait that has evolved due to selective pressures from their environment (Steppuhn et al., 2004).

2. Environmental Factors

Environmental stimuli such as light, temperature, water availability, and soil nutrients significantly impact secondary metabolite production. Plants sense changes in their environment and adjust their metabolic pathways accordingly to optimize their defense and survival strategies.

» **Light**: Light is a major environmental factor that influences the biosynthesis of secondary metabolites, particularly those involved in UV protection and photosynthesis. For example, flavonoid production is significantly upregulated in response to UV-B radiation as these compounds absorb UV light and protect the plant's tissues from damage (Agati et al., 2012).

» **Temperature and Water Stress**: Secondary metabolite production can also be triggered by temperature extremes and water stress. In response to drought conditions, plants often increase the production of osmoprotectants and antioxidants like terpenoids, which help protect cellular structures and maintain osmotic balance. Similarly, exposure to low temperatures can enhance the biosynthesis of compounds like flavonoids and carotenoids, which protect against cold-induced oxidative stress (Demmig-Adams & Adams, 2006).

» **Soil Nutrients**: Nutrient availability in the soil, especially nitrogen and phosphorus, plays a critical role in secondary metabolite production. Low nitrogen levels, for example, have been shown to increase the biosynthesis of defense-related compounds like alkaloids in certain plant species. This is thought to be an adaptive strategy, where plants allocate more resources to defense when growth is limited by nutrient deficiency.

3. Developmental Factors

Secondary metabolite production is often developmentally regulated, meaning that plants produce different compounds at different stages of growth or in response to developmental cues.

- **Juvenile vs. Mature Stages:** The production of certain secondary metabolites is more prevalent at specific stages of development. For instance, young leaves tend to produce higher levels of phenolic compounds like tannins to deter herbivores, while mature leaves may produce fewer defensive compounds as they become less vulnerable to herbivore damage (Barbehenn & Constabel, 2011).
- **Flowering and Reproductive Stages:** During the reproductive stages, plants often produce high concentrations of volatile terpenoids to attract pollinators. These volatile compounds are critical for successful pollination and reproduction. In contrast, during the vegetative stage, the focus may be more on producing secondary metabolites involved in defense against herbivores and pathogens.

Role of Transcription Factors and Signaling Molecules in Regulating Biosynthetic Pathways

The regulation of secondary metabolite biosynthesis is controlled not only at the genetic level but also by transcription factors and signaling molecules that respond to environmental and developmental signals. These regulatory proteins and molecules activate or repress the expression of genes involved in secondary metabolism, allowing the plant to fine-tune its metabolic responses to external stimuli.

1. Transcription Factors

Transcription factors (TFs) are proteins that bind to specific DNA sequences near the promoters of biosynthetic genes, either activating or repressing their expression. Several classes of transcription factors have been identified as key regulators of secondary metabolite biosynthesis in plants.

- **MYB Transcription Factors:** MYB TFs are one of the most well-studied families of transcription factors involved in secondary metabolism. These TFs regulate the biosynthesis of phenylpropanoids, flavonoids, and anthocyanins. For example, AtMYB12 in *Arabidopsis thaliana* is involved in the regulation of flavonol biosynthesis, while PAP1 regulates anthocyanin production (Stracke et al., 20011). MYB TFs act by binding to the promoters of target genes in the phenylpropanoid pathway, enhancing their expression in response to light or stress.
- **bHLH and WRKY Transcription Factors:** In addition to MYB TFs, basic helix-loop-helix (bHLH) and WRKY TFs also play critical roles in regulating secondary metabolite production. bHLH TFs are often involved in the co-regulation of anthocyanin biosynthesis along with MYB TFs. WRKY TFs are important for regulating defense-related secondary metabolites such as phytoalexins and terpenoids in response to pathogen attack (Xu et al., 2006).

2. Signaling Molecules

Signaling molecules such as jasmonic acid (JA), salicylic acid (SA), and ethylene are essential regulators of secondary metabolite production. These molecules act as chemical messengers that activate biosynthetic pathways in response to external threats such as herbivory, pathogen invasion, or environmental stress.

- **Jasmonic Acid (JA)**: JA is a key regulator of secondary metabolite production, particularly in response to herbivore damage. JA activates the expression of genes involved in the biosynthesis of alkaloids, terpenoids, and phenolics, enhancing the plant's defensive capabilities. For example, JA signaling is crucial for activating the production of nicotine in tobacco plants (*Nicotiana tabacum*) when the plant is attacked by herbivores (Steppuhn et al., 2004).
- **Salicylic Acid (SA)**: SA plays a central role in regulating the production of phytoalexins, a class of antimicrobial secondary metabolites, during pathogen invasion. SA signaling activates the expression of genes involved in the biosynthesis of defense compounds like flavonoids and terpenoids, helping to strengthen plant defenses against bacterial and fungal pathogens (Glazebrook, 2005).
- **Ethylene**: Ethylene is another signaling molecule that modulates secondary metabolite production, particularly in response to abiotic stress such as flooding or drought. Ethylene can also enhance the production of terpenoids and flavonoids involved in stress tolerance (Steffens & Sauter, 2005).

How Plants Modulate Secondary Metabolite Production in Response to Biotic and Abiotic Stress

Plants have evolved sophisticated mechanisms to modulate secondary metabolite production in response to both biotic stress (e.g., herbivores, pathogens) and abiotic stress (e.g., drought, UV radiation, temperature extremes). By fine-tuning their metabolic pathways, plants can optimize resource allocation, producing defense-related secondary metabolites when needed while conserving energy during periods of reduced stress.

1. Response to Biotic Stress

Plants use inducible defenses to increase the production of secondary metabolites in response to herbivores and pathogens. Inducible defenses allow plants to conserve energy by only producing high levels of secondary metabolites when they are under attack.

- **Herbivory**: In response to herbivore attack, plants rapidly activate the biosynthesis of toxic alkaloids, terpenoids, and phenolics. For example, in response to herbivory by caterpillars, tomato plants (Solanum lycopersicum) increase the production of proteinase inhibitors and volatile terpenoids that deter further feeding and attract parasitoid wasps, which prey on the herbivores (Dudareva et al., 2004). The signaling molecule jasmonic acid plays a crucial role in orchestrating this response.
- **Pathogen Infection**: Upon infection by pathogens, plants produce phytoalexins, antimicrobial secondary metabolites that inhibit the growth of bacteria and fungi. For example, resveratrol, a stilbene-type phytoalexin produced by grapevines, is synthesized in response to fungal infection and helps to protect the plant from disease (Jeandet et al., 2014). The activation of the salicylic acid pathway is critical in this defense mechanism, as salicylic acid (SA) signaling triggers the production of phytoalexins and other antimicrobial compounds that fortify the plant against pathogens. Phytoalexins, such as camalexin in *Arabidopsis* and resveratrol in grapevines, are rapidly synthesized in infected tissues, helping to contain the spread of the pathogen.

2. Response to Abiotic Stress

Secondary metabolites are also essential for plant adaptation to abiotic stressors such as drought, UV radiation, extreme temperatures, and nutrient deficiency. These compounds protect plants by maintaining cellular integrity and preventing damage from environmental stressors.

- **Drought and Salinity**: In response to water scarcity and high salinity, plants often upregulate the production of terpenoids, phenolics, and proline, which act as osmoprotectants. These compounds help to stabilize proteins and cell membranes, maintain osmotic balance, and reduce oxidative stress caused by water loss. For instance, the production of anthocyanins, a class of flavonoids, increases under drought conditions, as these compounds have antioxidant properties that protect against oxidative damage induced by drought stress (Chalker-Scott, 1999).
- **UV Radiation**: Exposure to high levels of UV radiation induces the synthesis of flavonoids and other phenolic compounds that absorb UV light and protect plant tissues from damage. These compounds act as natural sunscreens, preventing DNA damage and oxidative stress. Plants growing in high-altitude or desert environments, such as alpine herbs or desert shrubs, typically produce high levels of UV-absorbing flavonoids to protect against the harmful effects of intense sunlight (Rozema et al., 1997).
- **Temperature Extremes**: Plants exposed to extreme temperatures, both high and low, modulate the production of secondary metabolites to protect their cellular structures. For example, isoprene, a volatile terpenoid, is synthesized in response to high temperatures and helps to stabilize cell membranes by reducing heat-induced damage. Similarly, carotenoids and other antioxidants are produced in response to cold stress, protecting plants from the oxidative stress that occurs during freezing conditions (Sharkey & Singsaas, 1995).

The regulation of secondary metabolite biosynthesis in plants is a highly complex and dynamic process influenced by genetic, environmental, and developmental factors. Transcription factors and signaling molecules such as MYB, bHLH, WRKY, jasmonic acid (JA), salicylic acid (SA), and ethylene play pivotal roles in coordinating the production of secondary metabolites in response to biotic and abiotic stress. Plants can fine-tune their secondary metabolite production based on specific environmental conditions, optimizing their defense mechanisms and ensuring their survival in diverse ecosystems. The ability to modulate the production of compounds like alkaloids, terpenoids, flavonoids, and phytoalexins allows plants to efficiently defend themselves against herbivores, pathogens, and environmental stressors, ensuring their adaptability and resilience.

References

Agati, G., Azzarello, E., Pollastri, S., & Tattini, M. (2012). Flavonoids as antioxidants in plants: Location and functional significance. *Plant Science*, 196, 67-76.

Ashraf, M., & Foolad, M. R. (2007). Roles of glycine betaine and proline in improving plant abiotic stress resistance. *Environmental and Experimental Botany*, 59(2), 206-216.

Baldwin, I. T. (2001). An ecologically motivated analysis of plant-herbivore interactions in native tobacco. *Plant Physiology*, 127(4), 1449-1458.

Barbehenn, R. V., & Constabel, C. P. (2011). Tannins in plant-herbivore interactions. *Phytochemistry*, 72(13), 1551-1565.

Bennett, R. N., & Wallsgrove, R. M. (1994). Secondary metabolites in plant defense mechanisms. *New Phytologist*, 127(4), 617-633.

Boerjan, W., Ralph, J., & Baucher, M. (2003). Lignin biosynthesis. *Annual Review of Plant Biology*, 54, 519-546.

Buchanan, B. B., Gruissem, W., & Jones, R. L. (2015). *Biochemistry and molecular biology of plants*. 2nd Edition. Wiley.

Chalker-Scott, L. (1999). Environmental significance of anthocyanins in plant stress responses. *Photochemistry and Photobiology*, 70(1), 1-9.

Cheynier, V. (2012). Phenolic compounds: From plants to foods. *Phytochemistry Reviews*, 11(2-3), 153-177.

Demmig-Adams, B., & Adams III, W. W. (2006). Photoprotection in an ecological context: The remarkable complexity of thermal energy dissipation. *New Phytologist*, 172(1), 11-21.

Dempsey, D. A., Vlot, A. C., Wildermuth, M. C., & Klessig, D. F. (2011). Salicylic acid biosynthesis and metabolism. *Arabidopsis Book*, 9, e0156.

Dewick, P. M. (2009). Medicinal natural products: A biosynthetic approach. 3rd Edition. *John Wiley & Sons, Ltd.*

Dudareva, N., Pichersky, E., & Gershenzon, J. (2004). Biochemistry of plant volatiles. *Plant Physiology*, 135(4), 1893-1902.

Facchini, P. J. (2001). Alkaloid biosynthesis in plants: Biochemistry, cell biology, molecular regulation, and metabolic engineering applications. *Annual Reviews Plant Physiology Plant Molecular Biology*, 52(1), 29-66.

Glazebrook, J. (2005). Contrasting mechanisms of defense against biotrophic and necrotrophic pathogens. *Annual Review of Phytopathology*, 43, 205-227.

Goldstein, J. L., & Brown, M. S. (1990). Regulation of the mevalonate pathway. *Nature*, 343(6257), 425-430.

Gómez-Martínez, H., Gil-Muñoz, F., Bermejo, A., Zuriaga, E., & Badenes, M. L. (2021). Insights of phenolic pathway in fruits: Transcriptional and metabolic profiling in apricot (*Prunus armeniaca*). *International Journal of Molecular Sciences*, *22*(7), 3411.

Hagel, J. M., & Facchini, P. J. (2013). Benzylisoquinoline alkaloid metabolism: A century of discovery and a brave new world. *Plant and Cell Physiology*, 54(5), 647-672.

Hasegawa, M., Mitsuhara, I., Seo, S., Imai, T., Koga, J., Okada, K. & Ohashi, Y. (2010). Phytoalexin accumulation in the interaction between rice and the blast fungus. *Molecular Plant-Microbe Interactions*, 23(8), 1000-1011.

Hedden, P., & Sponsel, V. (2015). A century of gibberellin research. *Journal of Plant Growth Regulation*, 34(4), 740-760.

Herrmann, K. M., & Weaver, L. M. (1999). The shikimate pathway. *Annual Review of Plant Physiology and Plant Molecular Biology*, 50(1), 473-503.

Jeandet, P., Hébrard, C., Deville, M. A., Cordelier, S., Dorey, S., Aziz, A., & Crouzet, J. (2014). Deciphering the role of phytoalexins in plant-microorganism interactions and human health. *Molecules*, 19(11), 18033-18056.

Lange, B. M., Rujan, T., Martin, W., & Croteau, R. (2000). Isoprenoid biosynthesis: The evolution of two ancient and distinct pathways across genomes. *Proceedings of the National Academy of Sciences*, 97(24), 13172-13177.

Li, L., Chong, L., Huang, T., Ma, Y., Li, Y., & Ding, H. (2023). Natural products and extracts from plants as natural UV filters for sunscreens: A review. *Animal Models and Experimental Medicines*, *6*(3), 183-195.

Lichtenthaler, H. K. (1999). The 1-deoxy-D-xylulose-5-phosphate pathway of isoprenoid biosynthesis in plants. *Annual Review of Plant Biology*, 50(1), 47-65.

Liu, Q., Luo, L., & Zheng, L. (2018). Lignins: Biosynthesis and biological functions in plants. *International Journal of Molecular Sciences*, *19*(2), 335.

Miedes, E., Vanholme, R., Boerjan, W., & Molina, A. (2014). The role of the secondary cell wall in plant resistance to pathogens. *Frontiers in Plant Science*, 5, 358.

Nambara, E., & Marion-Poll, A. (2005). Abscisic acid biosynthesis and catabolism. *Annual Review of Plant Biology*, 56(1), 165-185.

Osbourn, A. E. (1996). Saponins and plant defense: A soap story. *Trends in Plant Science*, 1(1), 4-9.

Rohmer, M. (1999). The discovery of a mevalonate-independent pathway for isoprenoid biosynthesis in bacteria, algae and higher plants. *Natural Product Reports*, 16(5), 565-574.

Rozema, J., van de Staaij, J., Björn, L. O., & Caldwell, M. (1997). UV-B as an environmental factor in plant life: Stress and regulation. *Trends in Ecology & Evolution*, 12(1), 22-28.

Ruiz-Sola, M. Á., & Rodríguez-Concepción, M. (2012). Carotenoid biosynthesis in Arabidopsis: A colorful pathway. *Arabidopsis Book*, 10, e0158.

Sharkey, T. D., & Singsaas, E. L. (1995). Why plants emit isoprene. *Nature*, 374(6525), 769-769.

Sharkey, T. D., Wiberley, A. E., & Donohue, A. R. (2008). Isoprene emission from plants: Why and how. *Annals of Botany*, 101(1), 5-18.

Steffens, B., & Sauter, M. (2005). Epidermal cell death in rice is regulated by ethylene, gibberellin, and abscisic acid. *Plant Physiology*, 139(2), 713-721.

Steppuhn, A., Gase, K., Krock, B., Halitschke, R., & Baldwin, I. T. (2004). Nicotine's defensive function in nature. *PLoS Biology*, 2(8), e217.

Stracke, R., Werber, M., & Weisshaar, B. (2001). The R2R3-MYB gene family in Arabidopsis thaliana. *Current Opinion in Plant Biology*, 4(5), 447-456.

Tholl, D. (2015). Biosynthesis and biological functions of terpenoids in plants. *Advances in Biochemical Engineering/Biotechnology*, 148, 63-106.

Tohge, T., Watanabe, M., Hoefgen, R., & Fernie, A. R. (2013). Shikimate and phenylalanine biosynthesis in the green lineage. *Frontiers in Plant Science*, 4, 62.

Treutter, D. (2006). Significance of flavonoids in plant resistance: A review. *Environmental Chemistry Letters*, 4(3), 147-157.

Vogt, T. (2010). Phenylpropanoid biosynthesis. *Molecular Plant*, 3(1), 2-20.

Vranová, E., Coman, D., & Gruissem, W. (2013). Network analysis of the MVA and MEP pathways for isoprenoid synthesis. *Annual Review of Plant Biology*, 64(1), 665-700.

Wink, M. (2010). Functions and biotechnology of plant secondary metabolites. *Annual Plant Reviews*, 39(2), 195-258.

Wu, B., Shi, S., Zhang, H., Lu, B., & Nan, P. A Y. (2023). Anabolic metabolism of autotoxic substance coumarins in plants. *PeerJ*, *11*, e16508.

Xu, X., Chen, C., Fan, B., & Chen, Z. (2006). Physical and functional interactions between pathogen-induced Arabidopsis WRKY18, WRKY40, and WRKY60 transcription factors. *Plant Cell*, 18(5), 1310-1326.

Ziegler, J., & Facchini, P. J. (2008). Alkaloid biosynthesis: Metabolism and trafficking. *Annual Review of Plant Biology*, 59(1), 735-769.

Chapter - 3

Alkaloids: Potent Defenses Against Plant Pathogens

Reshma Jadhav[1], Praneet Chauhan[2], Vinay Pratap Singh[3] and Susheel Kumar[4]

[1]Department of Biochemistry, Mahatma Phule Krishi Vidyapeeth, Rahuri, Ahmednagar, Maharashtra, India
[2]Department of Plant Pathology, Dr. Khem Singh Gill Akal College of Agriculture, Eternal University, Baru Sahib, Sirmour, Himachal Pradesh, India
[3]Plant Physiology, College of Agriculture, Khurai, JNKVV, Jabalpur, M.P. India
[4]Plant Molecular Virology Laboratory, CPMB Division, CSIR-National Botanical Research Institute, Rana Pratap Marg, Lucknow, Uttar Pradesh, India

Abstract

Alkaloids are a diverse group of secondary metabolites produced by plants that play a crucial role in defending against a wide range of pathogens. These nitrogen-containing compounds are known for their potent biological activities, including their ability to inhibit the growth of fungi, bacteria, and viruses. Alkaloids are classified into several chemical groups, such as pyrrolidine, isoquinoline, tropane, and indole alkaloids, with their structure closely linked to their mode of action. These compounds interfere with essential cellular processes in pathogens, disrupting functions like protein synthesis, cell wall integrity, and metabolic pathways. Plants have evolved to produce specific alkaloids in response to pathogen attacks, and this production is often regulated by intricate signaling pathways triggered during infection. The evolutionary arms race between plants and pathogens has led to the development of highly specialized alkaloids that target specific pathogens, making them an essential component of plant defense strategies. In addition to their ecological role, alkaloids have practical applications in agriculture as natural bio-pesticides. They are increasingly being explored for their potential to control plant diseases sustainably, offering an alternative to synthetic chemicals.

Keywords: Alkaloids, plant defense, pathogen resistance, bio-pesticides, alkaloid mechanisms, sustainable agriculture

DOI: 10.1201/9781003717454-3

3.1 Introduction to Alkaloids as Defense Compounds

Overview of Alkaloids as Secondary Metabolites in Plants

Alkaloids represent one of the most significant classes of secondary metabolites in the plant kingdom. These nitrogen-containing compounds play a pivotal role in the survival strategies of many plant species, serving primarily as defense compounds against herbivores, pathogens, and even competing plants. Alkaloids are known for their bioactivity, often possessing toxic, narcotic, or deterrent effects, which protect plants in hostile environments. These compounds are also known for their pharmacological properties, making them of interest in human medicine. However, in the context of plant ecology, their role is chiefly defensive.

Chemical Nature of Alkaloids

The defining characteristic of alkaloids is their nitrogen content, which is typically found in a heterocyclic ring. This nitrogen is often derived from amino acids, making alkaloids structurally distinct from other classes of secondary metabolites like terpenoids or phenolics. Despite this common feature, alkaloids are chemically diverse, with their structures ranging from simple compounds like nicotine to highly complex molecules such as morphine. This structural diversity contributes to their wide range of biological activities, from neurotoxicity to antimicrobial properties. The vast array of alkaloid structures arises from the diversity of biosynthetic pathways involved in their production. The most common precursors to alkaloids are amino acids, such as ornithine, lysine, tyrosine, and tryptophan, each of which gives rise to different families of alkaloids (Dey et al., 2020). For example, ornithine and lysine are precursors to pyrrolidine and piperidine alkaloids, such as nicotine and hyoscyamine. Tyrosine leads to the formation of isoquinoline alkaloids, including morphine and codeine. Tryptophan is a precursor for indole alkaloids, such as quinine and strychnine. This chemical versatility enables alkaloids to interact with a broad range of biological targets, providing plants with a versatile defense system against various threats.

Ecological Roles of Alkaloids

One of the primary ecological functions of alkaloids is defense against herbivores (Wink, 2007). Many alkaloids are bitter-tasting, toxic, or otherwise deterrent to herbivores. They affect the nervous system, metabolic pathways, or digestion in herbivores, reducing the likelihood of being consumed. For example, nicotine in tobacco plants acts as a neurotoxin that disrupts the nervous systems of herbivorous insects, rendering the plant less palatable (Steppuhn et al., 2004). This defensive action not only prevents the plant from being eaten but also reduces the herbivore population, as the ingestion of alkaloids can be lethal. Some plants produce alkaloids that are immediately toxic upon consumption, while others produce alkaloids that act over time, accumulating in the herbivore's body and causing chronic effects. The bitter taste of alkaloids can also serve as a deterrent by discouraging herbivores from feeding after the first bite, an effect seen in plants like *Aconitum* (monkshood), where aconitine causes immediate burning and numbing sensations in the mouth (Chan, 2009).

Alkaloids also serve as antimicrobial agents, protecting plants from bacterial, fungal, and viral infections. Certain alkaloids inhibit the growth of pathogenic microbes by disrupting cellular processes or damaging microbial cell walls. For instance, quinine, derived from the bark of *Cinchona* trees, has potent anti-malarial properties due to its ability to inhibit the replication of the malaria parasite within red blood cells (Achan et al., 2011). Similarly, berberine, an alkaloid found in *Berberis* species, has broad-spectrum antimicrobial activity, making it an important natural defense against both bacterial and fungal pathogens (El-Zahar et al., 2022). In some cases, alkaloids are released into the surrounding environment to inhibit the growth of competing plant species, a phenomenon known as allelopathy. For example, caffeine, produced by coffee plants, not only deters herbivores but also inhibits the germination of seeds and the growth of neighboring plants (Antoine et al., 2023). This gives the coffee plant a competitive advantage by reducing competition for resources like sunlight, water, and nutrients.

Diversity and Classification of Alkaloids

Alkaloids can be classified based on their biosynthetic origins, particularly the amino acid precursors from which they are derived. This classification helps understand alkaloids' diversity and ecological functions in plants. The major categories of alkaloids include:

1. **Pyrrolidine and Piperidine Alkaloids**: These alkaloids are derived from ornithine or lysine and are characterized by their pyrrolidine or piperidine ring structures. Nicotine, a well-known pyrrolidine alkaloid, is produced by plants in the *Solanaceae* family, such as tobacco. Nicotine's neurotoxic properties make it highly effective at deterring herbivores, particularly insects. Lobeline, another piperidine alkaloid found in *Lobelia* species, also acts as an insecticide and herbivore deterrent (Oppel et al., 2009).
2. **Isoquinoline Alkaloids**: Isoquinoline alkaloids, such as morphine and codeine, are derived from tyrosine. These alkaloids are primarily found in the *Papaveraceae* family, with the opium poppy (*Papaver somniferum*) being the most notable producer. Morphine serves a dual role: in nature, it acts as a defense compound against herbivores and pathogens, while in humans, it is used as a potent analgesic (Divekar et al., 2022). The bitter taste and toxic effects of isoquinoline alkaloids prevent many herbivores from consuming plants that contain them.
3. **Indole Alkaloids**: Indole alkaloids are derived from tryptophan and are among the most structurally complex and pharmacologically active alkaloids. Quinine, an indole alkaloid produced by *Cinchona* trees, is a well-known example due to its antimalarial properties. These alkaloids are highly toxic to many herbivores and pathogens, providing strong protection for the plants that produce them. Strychnine, another indole alkaloid from the *Strychnos* genus, is one of the most toxic plant compounds known, acting as a powerful neurotoxin (Patocka, 2009).
4. **Tropane Alkaloids**: Tropane alkaloids, such as atropine and scopolamine, are produced by members of the *Solanaceae* family, including *Atropa belladonna* (deadly nightshade) and *Datura stramonium* (Jimson weed). These alkaloids affect the central nervous system by inhibiting acetylcholine receptors, causing paralysis in herbivores that ingest them. The toxicity of tropane alkaloids makes them highly effective deterrents to herbivores, while also providing medicinal uses for humans, such as in treating motion sickness and dilating pupils.

Alkaloid Biosynthesis and Evolutionary Adaptation

The ability of plants to produce alkaloids is an evolutionary adaptation to biotic stress. Over time, plants have developed the enzymatic machinery necessary to convert simple primary metabolites into complex alkaloids, enhancing their ability to defend against herbivores and pathogens. The production of alkaloids is often inducible, meaning that plants can increase their synthesis in response to external threats. For instance, when tobacco plants (*Nicotiana tabacum*) are attacked by herbivores, they upregulate the expression of genes involved in nicotine biosynthesis. This inducible defense mechanism allows the plant to allocate resources efficiently, producing higher concentrations of nicotine only when the risk of herbivory is imminent (Steppuhn et al., 2004). The ecological success of alkaloid-producing plants can be attributed to the evolutionary arms race between plants and herbivores. As herbivores evolved mechanisms to detoxify certain alkaloids, plants responded by producing more complex or potent versions of these compounds. This ongoing co-evolution has resulted in the remarkable diversity of alkaloids found in nature today, each tailored to specific ecological pressures.

Historical Significance of Alkaloid Research in Plant Pathology

Alkaloids have long been recognized for their significant roles in plant defense mechanisms, particularly in the context of plant pathology. The study of alkaloids has deep historical roots, beginning with the discovery of their medicinal properties and evolving into a more comprehensive understanding of their ecological and biochemical functions. In the field of plant pathology, alkaloids have been a focus of research due to their potent antimicrobial and antifungal properties, which help plants defend against pathogens.

Early Discoveries and Medicinal Use of Alkaloids

The scientific interest in alkaloids dates back to the early 19th century, with the isolation of the first alkaloid, morphine, from the opium poppy (*Papaver somniferum*) by German pharmacist Friedrich Sertürner in 1804. This marked the beginning of alkaloid research, as scientists began to isolate and study other alkaloids, including quinine from the Cinchona tree, nicotine from tobacco plants (*Nicotiana tabacum*), and atropine from Atropa belladonna (deadly nightshade). During the 19th and early 20th centuries, alkaloids gained prominence for their medicinal properties, particularly in treating human diseases. Quinine, for instance, became the first known treatment for malaria, a disease that had ravaged populations worldwide. While these early discoveries were focused on the medicinal properties of alkaloids, they laid the groundwork for later research into their roles in plant defense and pathology.

Recognition of Alkaloids as Plant Defense Compounds

As research advanced, the role of alkaloids in plant defense became more apparent. In the early 20th century, botanists and chemists began to explore how alkaloids acted as chemical barriers to herbivores and pathogens. Studies showed that alkaloid-producing plants had higher resistance to pests and diseases, leading scientists to hypothesize that these compounds served as part of the plant's immune system. By the mid-20th century, with the rise of chemical ecology as a discipline, alkaloids were recognized as key players in plant-pathogen interactions. For example, it was found that alkaloids like quinine and berberine possessed strong antimicrobial properties, capable of inhibiting bacterial and fungal growth. These findings reinforced the idea that alkaloids evolved in plants as a defense mechanism against microbial pathogens, giving plants an evolutionary advantage. The discovery of phytoalexins, secondary metabolites that are synthesized in response to pathogen attack, was a pivotal moment in plant pathology research. Alkaloids were identified as important phytoalexins in certain plant species, produced only when the plant was under microbial attack (Jeandet et al., 2014). For instance, solavetivone, an alkaloid produced by potato plants (*Solanum tuberosum*), was found to have antifungal properties that protected the plant against *Phytophthora infestans*, the pathogen responsible for late blight.

Alkaloid Research in the 20th Century: A Focus on Plant Pathogens

The 20th century saw significant advances in understanding the biosynthesis and regulation of alkaloids in response to biotic stress, particularly pathogen attacks. Researchers began to study the biochemical pathways that led to the production of specific alkaloids in different plant species. This research was driven by the agricultural need to control plant diseases without relying solely on synthetic pesticides.

1. **Nicotine as a Defense Against Insect Pathogens**: The discovery of nicotine's role in plant defense was a landmark in the study of alkaloids in plant pathology. In the 1950s, researchers demonstrated that tobacco plants (*Nicotiana tabacum*) produced high levels of nicotine in response to insect herbivory. Nicotine was found to act as a neurotoxin, affecting the nervous systems of insect pests, making it an effective defense compound (Steppuhn et al., 2004). This discovery led to the development of natural insecticides derived from nicotine and other alkaloids.
2. **Alkaloids and Fungal Pathogens**: During the latter half of the 20th century, alkaloid research expanded into the study of their antifungal properties. Scientists observed that plants like lupins produced quinolizidine alkaloids, which protected them against fungal infections such as Fusarium. Alkaloids were shown to disrupt fungal cell membranes, inhibit spore germination, and prevent the growth of fungal hyphae. These findings highlighted the potential of alkaloids in developing natural fungicides for agricultural use.
3. **Alkaloid-Induced Resistance**: In the 1970s and 1980s, the concept of induced resistance gained attention in plant pathology. It was discovered that certain plants could "switch on" alkaloid production when under attack by pathogens. This led to the identification of elicitors, molecules produced by pathogens that triggered the synthesis of defense compounds, including alkaloids, in the host plant. For example, jasmonic acid and salicylic acid were identified as key signaling molecules that regulated alkaloid biosynthesis in response to pathogen infection.

Advances in Alkaloid Biosynthesis and Genetic Engineering

With the advent of molecular biology in the late 20th and early 21st centuries, the study of alkaloid biosynthesis entered a new era. Advances in genomics and metabolic engineering allowed researchers to identify the specific genes and enzymes involved in alkaloid biosynthesis. This opened the door to manipulating alkaloid production in plants for enhanced resistance to pathogens.

1. **Biosynthetic Pathways**: Researchers successfully mapped out the biosynthetic pathways of important alkaloids, such as morphine, quinine, and atropine, uncovering the key enzymes

responsible for their production. This knowledge enabled scientists to genetically engineer plants to overproduce certain alkaloids, increasing their resistance to pathogens. For instance, by overexpressing key enzymes in the tropane alkaloid pathway, researchers enhanced the production of scopolamine in Atropa belladonna, improving the plant's resistance to bacterial and fungal infections.

2. **Transcriptional Regulation**: Studies also revealed how transcription factors regulate alkaloid biosynthesis in response to pathogen attack. For example, the WRKY and MYB families of transcription factors were found to activate genes involved in the biosynthesis of defense-related alkaloids. This discovery provided insight into how plants regulate alkaloid production at the molecular level and how these mechanisms could be harnessed to enhance crop protection.
3. **Metabolic Engineering and Synthetic Biology**: In recent years, advances in synthetic biology have allowed for the metabolic engineering of microorganisms to produce plant alkaloids. For example, researchers have engineered yeast and bacterial cells to produce morphine and other alkaloids, bypassing the need for traditional plant-based production. This has significant implications for both agriculture and pharmaceuticals, as it offers a more sustainable and efficient way to produce valuable alkaloids for medicinal and agricultural purposes (Nour-Eldin & Halkier, 2013).

Contemporary Research and Applications

Today, alkaloid research continues to evolve, with a growing focus on sustainable agricultural practices and biocontrol. The use of plant-derived alkaloids as natural pesticides and fungicides is seen as a promising alternative to synthetic chemicals, reducing the environmental impact of agriculture (Agati et al., 2012). Additionally, research into the role of alkaloids in plant-microbe interactions is uncovering new ways to enhance plant resistance to emerging pathogens.

1. **Natural Pesticides**: Alkaloids such as nicotine, pyrethrins, and rotenone are being developed as natural pesticides, offering a more environmentally friendly solution to pest control. These compounds are biodegradable and less likely to cause the development of resistance in insect populations compared to synthetic pesticides.
2. **Plant-Microbe Interactions**: Ongoing research is exploring how alkaloids influence plant-microbe interactions, particularly in the context of plant health and soil microbiomes. Alkaloids may play a role in shaping microbial communities in the rhizosphere, promoting beneficial microbes while suppressing pathogenic ones. This area of research holds potential for developing sustainable crop protection strategies that enhance plant resilience without relying on chemical inputs.

Role of Alkaloids in the Ecological Defense of Plants Against Pathogens

Alkaloids, as a diverse group of nitrogen-containing secondary metabolites, play a significant role in the ecological defense mechanisms of plants, particularly in protecting them from various pathogens, including bacteria, fungi, and viruses. The presence of alkaloids in plant tissues acts as a biochemical shield that helps plants survive in environments where microbial pathogens are prevalent. Alkaloids perform their protective functions through a variety of mechanisms, such as direct antimicrobial activity, inhibition of pathogen growth, and modulation of host-pathogen interactions.

1. Antimicrobial Activity of Alkaloids

One of the primary roles of alkaloids in plant defense is their antimicrobial properties, which make them effective in inhibiting the growth of pathogens, including bacteria and fungi. These antimicrobial effects stem from alkaloids' ability to interfere with critical biological processes within the pathogen, leading to the disruption of cellular structures or metabolic pathways.

» **Cell Membrane Disruption**: Many alkaloids are known to interact with the cell membranes of bacterial and fungal pathogens, causing structural damage that compromises membrane integrity. This disruption can lead to leakage of intracellular contents, rendering the pathogen unable to sustain its cellular functions. For instance, berberine, an isoquinoline alkaloid found in Berberis species, is known to disrupt microbial cell membranes by intercalating into lipid bilayers. This results in increased permeability and ultimately cell death (Stermitz et al., 2000).

» **Inhibition of Enzymatic Pathways**: Alkaloids can also inhibit key enzymatic pathways in pathogens, preventing them from replicating or producing vital cellular components. Quinine, an alkaloid derived from Cinchona species, has long been known for its anti-malarial properties due to its ability to disrupt the parasite's metabolic pathways within red blood cells. In the context of plant defense, alkaloids like quinine can similarly inhibit the metabolic activities of bacterial and fungal pathogens, leading to reduced infection rates.

» **Inhibition of DNA and RNA Synthesis**: Some alkaloids are potent inhibitors of nucleic acid synthesis, targeting the replication machinery of pathogens. For example, solasodine, a steroidal alkaloid found in Solanum species, has been shown to inhibit the replication of plant-infecting viruses by interfering with viral RNA synthesis (Itkin et al., 2011). This action prevents the virus from spreading within the plant and significantly reduces the severity of the infection.

2. Specific Examples of Alkaloids as Defense Compounds Against Pathogens

Throughout the plant kingdom, specific alkaloids have been documented to offer protection against a wide range of pathogens. Below are a few key examples of alkaloids with demonstrated antimicrobial effects:

» **Nicotine in Tobacco Plants**: Nicotine, a pyrrolidine alkaloid produced by the Solanaceae family, plays a critical role in defending tobacco plants (*Nicotiana tabacum*) against microbial pathogens. While nicotine is most famous for its neurotoxic effects on herbivores, it also has antibacterial and antifungal properties. Research has shown that nicotine inhibits the growth of Pseudomonas syringae, a bacterial pathogen responsible for tobacco leaf spot disease, by disrupting bacterial cell membranes and interfering with cellular respiration (Steppuhn et al., 2004).

» **Sanguinarine in Bloodroot**: Sanguinarine, an alkaloid produced by Sanguinaria canadensis (bloodroot), is a well-known antimicrobial agent with potent activity against both bacterial and fungal pathogens. It inhibits the enzyme DNA topoisomerase, preventing DNA replication in bacterial cells, and also interferes with fungal cell wall synthesis. In nature, sanguinarine provides protection against soil-borne pathogens like *Rhizoctonia solani* and *Fusarium oxysporum*, which cause root rot and other diseases in plants.

- **Capsaicin in Chili Peppers**: Capsaicin, an alkaloid found in *Capsicum* species (chili peppers), has been shown to have antimicrobial properties against fungal pathogens such as Botrytis cinerea. Capsaicin inhibits fungal growth by inducing oxidative stress and damaging fungal cell membranes. This alkaloid is part of the plant's inducible defense system, with capsaicin levels increasing in response to pathogen attack (Reddy et al., 2012). The dual role of capsaicin as both an herbivore deterrent and an antimicrobial agent highlights its importance in the ecological defense of plants.

3. Alkaloids in Plant-Pathogen Interactions

In addition to their direct antimicrobial effects, alkaloids play a role in modulating plant-pathogen interactions, enhancing the plant's ability to resist infection or reduce the severity of disease symptoms. Some alkaloids act as elicitors of plant defense responses, stimulating the production of other secondary metabolites or activating specific signaling pathways that enhance plant immunity.

- **Phytoalexin Production**: Alkaloids are often involved in the production of phytoalexins, secondary metabolites that are synthesized by plants in response to pathogen attack. For example, in response to fungal infection, tomato plants (*Solanum lycopersicum*) produce tomatine, a steroidal alkaloid that acts as both a direct antifungal agent and an elicitor of phytoalexin production. This dual function amplifies the plant's defensive response, helping to contain the infection and limit its spread.
- **Induced Systemic Resistance**: Alkaloids can also induce systemic resistance in plants, a phenomenon in which a localized infection triggers a widespread immune response throughout the plant. This type of defense is seen in plants that produce tropane alkaloids like hyoscyamine and scopolamine in response to pathogen attack. These alkaloids not only protect the infected tissue but also prime other parts of the plant for future attacks, enhancing the overall resistance to pathogens (Ziegler & Facchini, 2008).

4. Defense Against Fungal Pathogens

Fungal pathogens are among the most destructive agents affecting plants, and alkaloids play a crucial role in protecting plants from these organisms. Fungi can invade plant tissues, causing decay, blights, and other diseases that reduce plant vitality and crop yields. Alkaloids target various stages of the fungal life cycle, from spore germination to hyphal growth.

- **Antifungal Alkaloids**: Alkaloids such as berberine, quinine, and tomatine are known for their antifungal properties. These compounds inhibit fungal growth by disrupting cell membranes or inhibiting the synthesis of key fungal enzymes. Berberine, for instance, interferes with fungal cell division, leading to the death of the pathogen before it can fully colonize the plant (Stermitz et al., 2000). Similarly, quinine has been found to inhibit fungal spore germination, reducing the likelihood of infection.
- **Inhibition of Mycotoxin Production**: In some cases, alkaloids can inhibit the production

of mycotoxins, toxic secondary metabolites produced by fungi that can further damage plant tissues or contaminate agricultural products. For instance, tomatine in tomatoes inhibits the production of aflatoxin, a potent mycotoxin produced by Aspergillus species. This protective effect not only shields the plant from further damage but also improves the safety and quality of the harvested crop.

5. Role of Alkaloids in Viral Defense

Although alkaloids are best known for their roles in defense against bacterial and fungal pathogens, recent research has highlighted their importance in protecting plants from viral infections as well. Viruses are a major threat to agricultural productivity, and plants have evolved various strategies to limit viral replication and spread.

» **Inhibition of Viral Replication**: Certain alkaloids, such as solasodine in Solanum species, have been shown to inhibit the replication of plant viruses by interfering with viral RNA synthesis. This prevents the virus from replicating within plant cells, reducing the severity of the infection. Additionally, quinolizidine alkaloids produced by lupins have demonstrated antiviral activity, protecting the plant from infection by viruses like tobacco mosaic virus (TMV) (Itkin et al., 2011).

3.2 Chemical Structure and Classification of Alkaloids

Breakdown of the Chemical Structures of Alkaloids

Alkaloids are a class of naturally occurring compounds characterized by their nitrogen-containing heterocyclic structures. These nitrogen atoms are typically derived from amino acids, making alkaloids chemically distinct from other classes of secondary metabolites like terpenoids and phenolics. Alkaloids vary greatly in their structure, which can range from simple to highly complex molecular frameworks, contributing to their diverse bioactivities. The nitrogen atom in alkaloids is typically basic, allowing these compounds to form salts when reacting with acids, a property that significantly influences their biological activity. The structural diversity of alkaloids is determined by the biosynthetic pathways that convert amino acids into alkaloid precursors. This structural variety is reflected in their roles as chemical defense agents, growth regulators, and ecological interaction molecules in plants. Below is a breakdown of the fundamental chemical features that define alkaloids:

1. Nitrogen-Containing Heterocycles

The defining feature of alkaloids is the presence of a nitrogen atom within a heterocyclic ring, meaning the nitrogen is part of a ring structure that also contains carbon atoms. This nitrogen atom can be located in various positions within the molecule, and its placement significantly affects the alkaloid's chemical properties and bioactivity. Alkaloids typically exist in either tertiary, secondary, or quaternary amine forms, depending on the number of organic groups attached to the nitrogen atom.

» **Tertiary Alkaloids**: In tertiary alkaloids, the nitrogen is bonded to three organic groups. For example, nicotine, a well-known alkaloid from tobacco plants (*Nicotiana tabacum*), contains a nitrogen atom in its pyridine ring, making it a tertiary alkaloid. These forms are usually lipid-soluble and can cross biological membranes, which explains why many tertiary alkaloids exhibit neurotoxic properties.

» **Quaternary Alkaloids**: Some alkaloids exist in quaternary forms, where the nitrogen atom is bonded to four organic groups, giving it a positive charge. This charged nature significantly affects the solubility of quaternary alkaloids, as they tend to be more water-soluble. Tubocurarine, an alkaloid used as a muscle relaxant, is an example of a quaternary alkaloid.

2. Basicity and Reactivity

The nitrogen atom in alkaloids gives them basic properties, meaning they can accept protons. This basicity enables alkaloids to react with acids to form salts, a property that influences their solubility and bioavailability. For instance, in plants, alkaloids often accumulate in their salt form, which makes them water-soluble and easier to store in vacuoles. When consumed by herbivores, the alkaloid salts become bioavailable, allowing the plant to effectively deploy them as defense compounds. The basicity of alkaloids varies depending on the electron-donating or withdrawing groups attached to the nitrogen. For example, alkaloids with electron-donating groups near the nitrogen atom tend to be more basic, while those with electron-withdrawing groups are less basic. This variation in basicity affects the reactivity of alkaloids and influences their interaction with biological targets, such as receptors and enzymes.

3. Structural Diversity of Alkaloid Families

The chemical structures of alkaloids can be divided into several major families based on their biosynthetic origins and core structural motifs. These families are derived from specific amino acid precursors, and each family exhibits distinct biological activities and ecological roles.

a) Pyrrolidine and Piperidine Alkaloids

Pyrrolidine and piperidine alkaloids are characterized by a five-membered pyrrolidine ring or a six-membered piperidine ring. The nitrogen atom is incorporated into the heterocyclic ring. Nicotine, found in *Nicotiana tabacum* (tobacco), is a pyrrolidine alkaloid with a pyridine and pyrrolidine ring fused together. The nitrogen atom in the pyrrolidine ring contributes to nicotine's high biological activity, particularly its ability to interact with nicotinic acetylcholine receptors in the nervous system. Pyrrolidine and piperidine alkaloids often act as neurotoxins and are used by plants to deter herbivores. Nicotine, for example, is toxic to insects and many animals, affecting their nervous systems and deterring further feeding.

b) Isoquinoline Alkaloids

Isoquinoline alkaloids feature a benzylisoquinoline skeleton, with two fused rings, a benzene ring and a pyridine ring. Morphine, an isoquinoline alkaloid found in the Papaveraceae family (opium poppy), has a complex benzylisoquinoline structure with several functional groups attached. This structure contributes to its potent analgesic properties in humans, though in plants, it serves as a defense against herbivores. Isoquinoline alkaloids are known for their toxic and narcotic effects. In plants, they act as deterrents to herbivores by affecting the central nervous system. Additionally, some isoquinoline alkaloids exhibit antimicrobial properties, protecting plants from bacterial and fungal infections.

c) Indole Alkaloids

Indole alkaloids are derived from tryptophan and contain an indole ring, a fused benzene and pyrrole ring. Quinine, found in the bark of Cinchona trees, is an indole alkaloid with a complex structure that includes an indole ring and a quinuclidine nucleus. Quinine is famous for its antimalarial properties, but in plants, it functions as a deterrent to herbivores and pathogens. Indole alkaloids serve various functions, including antimicrobial and antiparasitic activities. Their intricate structures allow them to interact with diverse biological targets, making them highly effective defense compounds in the plant kingdom.

d) Tropane Alkaloids

Tropane alkaloids have a tropane skeleton, which is a bicyclic structure with nitrogen in the bridgehead position. Atropine, found in Atropa belladonna (deadly nightshade), is a well-known tropane alkaloid that blocks muscarinic acetylcholine receptors in animals, leading to muscle paralysis and, at higher doses, death. This makes atropine an effective deterrent against herbivores. Tropane alkaloids often serve as defensive neurotoxins, affecting the nervous systems of herbivores. They can also play roles in plant growth regulation and are involved in the plant's response to environmental stress.

4. Biosynthesis and Structure-Function Relationship

The biosynthesis of alkaloids begins with amino acids as primary precursors. Through various enzymatic transformations, such as decarboxylation, hydroxylation, and methylation, these simple precursors are converted into complex alkaloid structures. For example, tyrosine is decarboxylated to form dopamine, a key intermediate in the biosynthesis of isoquinoline alkaloids like morphine. Ornithine is converted into putrescine, which serves as a precursor for pyrrolidine and piperidine alkaloids. Each biosynthetic step introduces functional groups that determine the alkaloid's biological activity. For example, the methylation of certain alkaloids enhances their lipid solubility, allowing them to cross cell membranes and interact with intracellular targets. Similarly, the introduction of hydroxyl groups increases the polarity of the alkaloid, affecting its solubility and transport within the plant. The structure-function relationship of alkaloids is evident in their ability to bind to specific biological targets. Alkaloids with nitrogen atoms positioned within heterocyclic rings are particularly adept at mimicking or blocking neurotransmitters, enzymes, or receptors, making them effective as neurotoxins, enzyme inhibitors, or antimicrobial agents.

Major Classes of Alkaloids: Pyrrolidine, Isoquinoline, Tropane, and Indole Alkaloids

Alkaloids are a broad class of nitrogen-containing compounds that vary significantly in their chemical structures and biological activities (Facchini, 2001). The classification of alkaloids is typically based on the nitrogenous heterocyclic rings that form the core of their chemical structures, often derived from amino acid precursors. Among the most significant and well-researched alkaloids are pyrrolidine, isoquinoline, tropane, and indole alkaloids. Each of these classes plays a crucial role in plant defense and has significant pharmacological relevance. The diversity in their structures underpins the wide range of biological effects they exhibit, both in plants and in organisms that consume or interact with these plants.

1. Pyrrolidine Alkaloids

Pyrrolidine alkaloids are characterized by a pyrrolidine ring, which is a five-membered nitrogen-containing heterocycle. These alkaloids are biosynthetically derived from ornithine, an amino acid, which undergoes decarboxylation and other modifications to form the pyrrolidine structure. Pyrrolidine alkaloids are most notable for their neurotoxic properties, which make them effective as plant defense compounds, particularly against herbivores.

a) Nicotine: Nicotine, one of the best-known pyrrolidine alkaloids, contains a pyridine ring fused with a pyrrolidine ring. The nitrogen atom in the pyrrolidine ring is central to its biological activity. Nicotine is synthesized in the roots of tobacco plants (*Nicotiana tabacum*) and then transported to the leaves, where it serves as a chemical defense against herbivores, particularly insects. Nicotine acts on the nicotinic acetylcholine receptors in the nervous system of herbivores, leading to paralysis and, in high doses, death. This potent neurotoxic effect makes nicotine a highly effective deterrent against insect herbivores (Steppuhn et al., 2004).

b) Lobeline: Lobeline, another pyrrolidine alkaloid, is found in Lobelia species. It contains a similar five-membered pyrrolidine ring structure, which contributes to its biological activity. Like nicotine, lobeline is derived from ornithine and is used by plants as a defense compound to deter herbivores. Lobeline affects the nervous system of herbivores, leading to neurotoxic effects similar to those of nicotine, although it is generally less potent.

2. Isoquinoline Alkaloids

Isoquinoline alkaloids are based on the benzylisoquinoline skeleton, which consists of a benzene ring fused to a pyridine ring. These alkaloids are typically derived from the amino acid tyrosine, which undergoes a series of enzymatic transformations to form the isoquinoline structure. Isoquinoline alkaloids are noted for their toxic and narcotic properties, which make them important defense compounds in plants and valuable medicinal compounds in humans.

a) Morphine: Morphine, an isoquinoline alkaloid found in the *Papaveraceae* family (opium poppy), has a complex benzylisoquinoline structure with multiple hydroxyl and methoxy groups attached. Its nitrogen atom is part of the isoquinoline ring, giving morphine its characteristic structure. Morphine is synthesized in the latex of the opium poppy (*Papaver somniferum*), where it serves as a chemical defense against herbivores and pathogens. It is produced through the isoquinoline pathway, starting from tyrosine. Morphine is highly toxic to herbivores and acts as a central nervous system depressant, which deters animals from consuming the plant. In humans, morphine has been used for centuries as a powerful analgesic, although its narcotic properties have also made it subject to abuse (Hagel & Facchini, 2013).

b) Berberine: Berberine, another isoquinoline alkaloid, features the classic benzylisoquinoline skeleton with a quaternary nitrogen atom. It is found in *Berberis* species (barberry) and other members of the Ranunculaceae family. Berberine is biosynthesized from tyrosine through a series of decarboxylation and cyclization reactions. It has potent antimicrobial properties, making it effective against bacterial, fungal, and viral pathogens. In plants, it acts as a broad-spectrum antimicrobial agent, protecting them from infections (Stermitz et al., 2000).

3. Tropane Alkaloids

Tropane alkaloids are characterized by the tropane skeleton, a bicyclic structure containing a nitrogen atom at the bridgehead position. These alkaloids are derived from ornithine or arginine and are well-known for their anticholinergic properties, which affect the central nervous system by inhibiting acetylcholine receptors. Tropane alkaloids are predominantly found in the *Solanaceae* family, which includes plants like *Atropa belladonna* (deadly nightshade) and *Datura stramonium* (Jimson weed).

a) Atropine: Atropine is a tropane alkaloid with a bicyclic tropane skeleton and an ester linkage to tropic acid. Its nitrogen atom is located at the bridgehead position in the tropane ring. Atropine is synthesized in the roots and leaves of plants like *Atropa belladonna*. It is derived from ornithine through the tropane alkaloid pathway. Atropine functions as a muscarinic acetylcholine receptor antagonist, inhibiting parasympathetic nervous system responses. In plants, atropine deters herbivores by inducing muscle paralysis and respiratory failure at higher doses, making it a potent chemical defense (Roddick, 2001).

b) Scopolamine: Scopolamine, another tropane alkaloid, is similar in structure to atropine but contains an additional epoxide group. This gives scopolamine distinct pharmacological properties. Scopolamine is produced in plants like *Datura stramonium* through the same tropane alkaloid pathway as atropine. Scopolamine has anticholinergic effects similar to atropine, and in plants, it acts as a neurotoxin, preventing herbivores from feeding on the plant. In humans, scopolamine is used to treat motion sickness due to its central nervous system depressant effects.

4. Indole Alkaloids

Indole alkaloids are derived from the amino acid tryptophan and contain an indole ring, a fusion of a benzene ring and a pyrrole ring. These alkaloids are among the most structurally diverse and pharmacologically active classes of alkaloids. They are found in a variety of plant families, including Apocynaceae, Rubiaceae, and Loganiaceae. Indole alkaloids often exhibit antimicrobial, antiparasitic, and neurotoxic properties, making them important defense compounds in plants.

a) Quinine: Quinine, an indole alkaloid found in Cinchona species, contains an indole ring fused to a quinuclidine nucleus. This unique structure gives quinine its potent antimalarial properties. Quinine is synthesized from tryptophan via the indole alkaloid pathway in the bark of Cinchona trees. In plants, quinine serves as a defense compound, deterring herbivores and protecting against microbial pathogens. In humans, quinine has been used for centuries to treat malaria due to its ability to inhibit the replication of the malaria parasite (Itkin et al., 2011).

b) Strychnine: Strychnine, an indole alkaloid found in Strychnos species, has a complex indole structure with multiple fused rings. The nitrogen atom in the indole ring contributes to its high biological activity. Strychnine is derived from tryptophan through a series of decarboxylation, cyclization, and oxidation reactions in the indole alkaloid biosynthesis pathway. Strychnine is an extremely potent neurotoxin that acts by blocking glycine receptors in the spinal cord, leading to severe convulsions and death in herbivores that ingest it. This makes it a highly effective chemical defense compound for the plant.

Relationship Between Alkaloid Structure and Biological Activity

The biological activity of alkaloids is deeply intertwined with their chemical structure, as subtle variations in their molecular frameworks can significantly alter their interactions with biological targets. The diversity of alkaloid structures allows them to perform a wide range of ecological and pharmacological functions, including acting as toxins, antimicrobial agents, enzyme inhibitors, and neurotransmitters. Understanding the relationship between alkaloid structure and biological activity is essential not only for appreciating their role in plant defense but also for harnessing their medicinal properties in drug development. Alkaloids are primarily characterized by their nitrogen-containing heterocyclic rings, but the number of rings, the position of the nitrogen, the degree of unsaturation, and the presence of other functional groups (such as hydroxyl, methyl, or methoxy groups) all contribute to the specificity and potency of their biological activities. These structural elements determine the alkaloid's ability to bind to receptors, interfere with enzymes, or disrupt cellular processes in organisms that interact with them.

1. Nitrogen Heterocycle and Its Role in Bioactivity

The defining feature of alkaloids is the presence of a nitrogen atom within a heterocyclic ring. The position and form of this nitrogen atom play a critical role in determining the alkaloid's basicity and reactivity, which in turn influences its biological activity.

- **Basicity and Interaction with Biological Targets**: The basic nature of alkaloids, due to the nitrogen atom, allows them to interact with acidic functional groups in biological molecules such as proteins and DNA. For instance, many alkaloids act as competitive inhibitors by binding to enzyme active sites or receptors, blocking the action of natural substrates or neurotransmitters. In nicotine, for example, the nitrogen in the pyrrolidine ring interacts with nicotinic acetylcholine receptors in the nervous system, mimicking the neurotransmitter acetylcholine and thus triggering a range of physiological responses (Steppuhn et al., 2004).
- **Protonation and Solubility**: The nitrogen in alkaloids can be protonated, forming a charged species that is more soluble in water. This property allows alkaloids to be easily transported within the plant or organism. The protonation state also affects how alkaloids cross biological membranes. For instance, quaternary alkaloids, such as tubocurarine, have a permanently charged nitrogen, making them less lipid-soluble and limiting their ability to cross cell membranes, resulting in a localized effect rather than a systemic one.

2. Ring Structures and Biological Specificity

The complexity of ring structures in alkaloids is a key determinant of their specificity and potency. The number of rings, their arrangement, and the presence of double bonds or other functional groups contribute to the molecule's three-dimensional shape, which determines how well it fits into biological targets.

- **Mono- vs. Polycyclic Alkaloids**: Alkaloids can be monocyclic (single-ring structures) or polycyclic (multiple fused rings). Monocyclic alkaloids, such as coniine (a piperidine alkaloid), tend to have simpler structures and often exhibit lower specificity. In contrast, polycyclic alkaloids like morphine or strychnine have more complex shapes that allow them to bind with high specificity to receptors or enzymes, often resulting in more potent and targeted biological effects.
- **Steric Hindrance and Activity**: The steric hindrance created by bulky groups or fused rings can affect how an alkaloid interacts with its biological target. For example, tropane alkaloids such as atropine have a rigid bicyclic structure that fits perfectly into muscarinic acetylcholine receptors, blocking the action of acetylcholine. This precise fit leads to atropine's potent

anticholinergic effects, which include muscle relaxation and reduced glandular secretions (Roddick, 2001).

3. Functional Groups and Biological Activity

The presence of functional groups such as hydroxyl (-OH), methyl (-CH3), methoxy (-OCH3), and carboxyl (-COOH) groups further fine-tunes the biological activity of alkaloids by affecting their solubility, reactivity, and ability to interact with biological macromolecules.

- **Hydroxyl Groups and Binding Affinity**: Hydroxyl groups can form hydrogen bonds with receptors or enzymes, enhancing the alkaloid's binding affinity. For instance, in morphine, the hydroxyl groups at positions 3 and 6 contribute to its strong binding to opioid receptors, leading to its potent analgesic effects. The hydroxyl groups also increase morphine's water solubility, allowing it to be rapidly distributed within the body.
- **Methylation and Bioavailability**: Methyl groups can increase the lipophilicity of alkaloids, enhancing their ability to cross cell membranes and thus increasing their bioavailability. In the case of scopolamine, a tropane alkaloid, the addition of a methyl group to the nitrogen atom increases its lipid solubility, allowing it to cross the blood-brain barrier more effectively. This makes scopolamine a more potent central nervous system depressant than its close relative, atropine.
- **Methoxylation and Stability**: Methoxy groups can enhance the chemical stability of alkaloids, making them less susceptible to degradation by enzymes. For instance, berberine, an isoquinoline alkaloid, contains multiple methoxy groups that protect it from metabolic breakdown, allowing it to remain active longer within an organism (Stermitz et al., 2000). This contributes to berberine's effectiveness as an antimicrobial agent, as it remains active in fighting bacterial and fungal infections for an extended period.

4. Stereochemistry and Chirality

The stereochemistry of alkaloids, or the spatial arrangement of their atoms, also plays a crucial role in determining their biological activity. Many alkaloids are chiral molecules, meaning they exist as enantiomers (mirror-image forms). Often, only one enantiomer is biologically active, while the other may be inactive or even harmful.

- **Stereoselective Receptor Binding**: Enantiomers of alkaloids may have very different affinities for biological receptors. For example, (–)-morphine is the naturally occurring form of morphine and has potent analgesic effects due to its specific interaction with **μ-opioid receptors**. In contrast, its enantiomer (+)-morphine has little to no analgesic activity because it does not bind effectively to the same receptors.
- **Chiral Centers and Pharmacodynamics**: Alkaloids with multiple chiral centers often exhibit highly specific biological activities. Cocaine, for example, has two chiral centers, and its pharmacological properties as a local anesthetic and dopamine reuptake inhibitor are closely tied to its stereochemistry. Altering the stereochemistry of cocaine drastically reduces its bioactivity, demonstrating the importance of chirality in determining the physiological effects of alkaloids.

5. Structural Complexity and Multifunctional Roles

The structural complexity of alkaloids often correlates with their ability to perform multifunctional roles in biological systems. Complex alkaloids, such as quinine and strychnine, interact with multiple biological targets, providing broad-spectrum defenses for plants and multifaceted pharmacological effects in humans.

» **Multitarget Effects**: Some alkaloids interact with more than one type of receptor or enzyme, allowing them to perform multiple biological functions. Quinine, for example, inhibits the replication of the malaria parasite by interfering with the parasite's heme detoxification pathway, while also exhibiting antipyretic (fever-reducing) and analgesic (pain-relieving) properties in humans. This multifunctionality is a direct result of quinine's complex structure, which enables it to bind to different biological targets (Itkin et al., 2011).

» **Adaptation to Environmental Stressors**: Alkaloids with complex structures often serve as inducible defense compounds, produced in response to environmental stressors such as herbivory or pathogen attack. For instance, in response to fungal infection, plants like tomatoes produce tomatine, a glycoalkaloid with both antifungal and insecticidal properties. The structural complexity of tomatine allows it to interact with the membranes of fungal pathogens, disrupting their integrity and preventing their growth.

3.3 Mechanisms of Alkaloid Action Against Plant Pathogens

Alkaloids are among the most potent secondary metabolites produced by plants, playing crucial roles in plant defense by targeting a wide range of pathogens, including fungi, bacteria, and viruses. These nitrogenous compounds interfere with essential cellular processes within the pathogens, disrupting their ability to infect, reproduce, or spread. The antimicrobial action of alkaloids is both broad-spectrum and highly specific, depending on the structural properties of the alkaloid and the type of pathogen it targets. This section explores how alkaloids disrupt microbial cellular processes and the specific mechanisms they employ to combat fungi, bacteria, and viruses.

1. How Alkaloids Interfere with Microbial Cellular Processes

Alkaloids exert their antimicrobial effects through several key mechanisms, targeting critical components of microbial cells, including cell membranes, nucleic acids, proteins, and enzymes. By disrupting these processes, alkaloids prevent pathogens from colonizing the host plant or reduce their virulence, effectively halting disease progression.

a) Disruption of Cell Membranes

One of the most common mechanisms by which alkaloids inhibit pathogens is through the disruption of cell membranes. Alkaloids can intercalate into the lipid bilayer of microbial membranes, altering their integrity and permeability. This often results in leakage of intracellular contents, which is fatal to the pathogen.

» **Membrane Permeability**: Certain alkaloids, like berberine and solasodine, are known to disrupt microbial membranes by interacting with the lipid components of the cell wall. These alkaloids insert themselves between the lipids, causing a loss of membrane fluidity and integrity. The resulting increase in membrane permeability leads to the leakage of essential ions and nutrients, disrupting the pathogen's metabolic functions and eventually leading to cell death (Stermitz et al., 2000).

» **Membrane Potential Disruption**: Alkaloids can also interfere with the electrochemical gradients across microbial membranes, particularly in bacteria. For example, quinine, an indole alkaloid, disrupts the proton gradient across bacterial membranes, preventing the pathogen from synthesizing ATP (the energy currency of the cell). This energy deficit inhibits the growth and survival of the bacteria.

b) Inhibition of Nucleic Acid Synthesis

Another key mechanism of alkaloid action is the inhibition of DNA and RNA synthesis, which blocks microbial replication. Alkaloids can bind to nucleic acids or enzymes involved in the replication and transcription of genetic material, thereby preventing the pathogen from reproducing.

» **DNA Intercalation**: Alkaloids like quinine and berberine have planar structures that allow them to intercalate between DNA bases, thereby inhibiting the action of DNA polymerase. This blocks DNA replication, effectively halting cell division. In fungi, this mechanism can prevent the growth of hyphae, while in bacteria, it prevents the replication of the bacterial genome (Stermitz et al., 2000).

» **RNA Synthesis Inhibition**: Alkaloids such as solasodine inhibit viral RNA synthesis, preventing the replication of plant viruses within host cells. By targeting the enzymes involved in RNA polymerization, alkaloids block the production of viral genomes and proteins, limiting the spread of the virus within the plant (Itkin et al., 2011).

c) Inhibition of Protein Synthesis

Protein synthesis is a fundamental process in microbial survival and pathogenesis. Alkaloids can interfere with the ribosomes of pathogens, blocking their ability to synthesize proteins that are crucial for cell growth and virulence.

» **Ribosomal Inhibition**: Alkaloids such as quinolizidine alkaloids (found in lupins) inhibit ribosomal function by binding to the ribosomal RNA or ribosomal proteins, preventing the assembly of new proteins. This halts the pathogen's ability to produce enzymes and other virulence factors necessary for infection.

» **Enzyme Inhibition**: Alkaloids also act as enzyme inhibitors, blocking microbial enzymes that are essential for metabolic processes. For instance, berberine inhibits topoisomerase enzymes, which are involved in DNA replication and transcription in bacteria and fungi. By inhibiting these enzymes, berberine disrupts the pathogen's ability to replicate its genetic material, effectively preventing further growth and infection (Stermitz et al., 2000).

2. Specific Modes of Action Against Fungi, Bacteria, and Viruses

Alkaloids exhibit different modes of action depending on the type of pathogen they target. The structural features of alkaloids allow them to act selectively against fungi, bacteria, or viruses, with many alkaloids showing broad-spectrum activity across multiple pathogen types. Below are the specific mechanisms through which alkaloids combat each of these major pathogen groups.

a) Modes of Action Against Fungi

Fungi are one of the most common and devastating plant pathogens, causing diseases such as powdery mildew, rust, and blights. Alkaloids protect plants from fungal infections by targeting the fungal cell wall, membrane, or intracellular processes.

» **Cell Wall Disruption**: Alkaloids like tomatine and solanine, found in Solanum species (tomatoes and potatoes), disrupt the fungal cell wall by binding to sterols in the fungal membrane. This disrupts the structure of the fungal cell wall, causing lysis (cell rupture) and death. This mechanism is particularly effective against fungi such as *Botrytis cinerea*, which causes gray mold in many crops (Reddy et al., 2012).

» **Inhibition of Spore Germination**: Fungi reproduce through spores, and alkaloids can inhibit the germination of these spores, preventing the fungus from establishing an infection. Berberine has been shown to prevent the germination of Aspergillus spores, a major fungal pathogen that affects both plants and post-harvest crops. By inhibiting spore germination, berberine reduces the likelihood of fungal colonization and infection.

» **Inhibition of Hyphal Growth**: Alkaloids also prevent the growth of fungal hyphae, the branching structures that fungi use to invade plant tissues. Solasodine inhibits the elongation of fungal hyphae by disrupting the polymerization of chitin, a key component of the fungal cell wall, thereby limiting the spread of the infection.

b) Modes of Action Against Bacteria

Bacterial pathogens cause a wide range of plant diseases, including bacterial blight, soft rot, and canker. Alkaloids defend plants against bacterial infections by interfering with bacterial cell walls, enzymes, and replication machinery.

» **Inhibition of Cell Wall Synthesis**: Alkaloids like capsaicin, found in chili peppers (*Capsicum* species), inhibit the synthesis of peptidoglycan, an essential component of bacterial cell walls. By blocking peptidoglycan synthesis, capsaicin weakens the bacterial cell wall, leading to cell lysis. This mechanism is effective against gram-positive bacteria, which have thick peptidoglycan layers (Reddy et al., 2012).

» **Inhibition of DNA Replication**: Berberine inhibits bacterial DNA gyrase and topoisomerase IV, enzymes that are critical for bacterial DNA replication. By inhibiting these enzymes, berberine prevents the bacteria from replicating their genome, which halts cell division and growth. This mode of action is effective against Pseudomonas syringae, a bacterial pathogen that causes leaf spot in many crops.

» **Membrane Disruption**: Certain alkaloids can disrupt the bacterial cell membrane, causing leakage of intracellular contents. Quinine, for instance, disrupts bacterial membranes by interfering with membrane-bound enzymes, leading to a loss of cellular homeostasis and eventual cell death.

c) Modes of Action Against Viruses

Viruses are obligate intracellular parasites that hijack plant cells to replicate and spread. Alkaloids defend plants against viral infections by targeting viral replication machinery or preventing the virus from entering plant cells.

- **Inhibition of Viral RNA Synthesis**: Alkaloids such as quinine and solasodine inhibit the RNA polymerases that viruses use to replicate their genomes. By blocking RNA synthesis, these alkaloids prevent the virus from producing new viral particles, limiting the spread of the infection (Itkin et al., 2011). This mechanism is particularly effective against plant viruses such as tobacco mosaic virus (TMV) and cucumber mosaic virus (CMV).
- **Prevention of Viral Entry**: Some alkaloids interfere with the virus's ability to enter plant cells. **Capsaicin**, for example, prevents viral particles from binding to cell surface receptors, thereby blocking the initial stages of viral infection. This mechanism reduces the viral load and helps the plant recover from the infection.
- **Inhibition of Viral Assembly**: Alkaloids like quinine can also disrupt the assembly of viral particles by interfering with viral capsid proteins. This prevents the virus from packaging its genetic material into new virions, thereby reducing the number of infectious viral particles that can spread within the plant.

Case Studies of Alkaloid-Pathogen Interactions

Alkaloids are critical components of plant defense mechanisms against a wide range of pathogens, including bacteria, fungi, and viruses. The diversity in alkaloid structures leads to different modes of action, which in turn determine how effective these compounds are in protecting plants from various types of infections. Below are case studies of well-known alkaloids, including nicotine, quinine, and berberine, demonstrating their interactions with pathogens and the ecological significance of these interactions.

1. Nicotine: Defense Against Insect Pests and Pathogens

Nicotine is a pyrrolidine alkaloid produced by plants of the Solanaceae family, particularly Nicotiana tabacum (tobacco). It is well-known for its neurotoxic effects on herbivorous insects, but nicotine also plays a crucial role in defending tobacco plants against microbial pathogens, including bacteria and fungi.

Case Study: Nicotine and Pseudomonas syringae

Pseudomonas syringae is a bacterial pathogen responsible for causing bacterial speck in many plants, including tobacco. Nicotine in tobacco leaves provides an effective defense against this pathogen by disrupting the bacterial cell membrane. Studies have shown that nicotine can increase the permeability of bacterial membranes, causing leakage of ions and essential nutrients. This mechanism ultimately leads to bacterial cell death. Nicotine interacts with the bacterial cell membrane, creating pores that compromise membrane integrity. This results in the loss of cellular homeostasis and the death of *Pseudomonas syringae* cells. In addition to acting as a neurotoxin against insects, nicotine provides a dual defense mechanism by protecting tobacco plants from microbial pathogens. This highlights the multifunctionality of nicotine as a defense compound.

Case Study: Nicotine and Fungal Pathogens

Nicotine also offers protection against fungal pathogens. Research has shown that nicotine can inhibit the growth of *Alternaria alternata*, a fungus that causes leaf spot disease in tobacco. Nicotine interferes with the fungal membrane, inhibiting ergosterol synthesis, which is a critical component of fungal cell walls. The reduction in ergosterol content disrupts the structural integrity of the fungal cells, preventing further growth and colonization. Nicotine inhibits ergosterol biosynthesis, weakening the fungal membrane and leading to cell lysis. Nicotine's antifungal properties help tobacco plants resist fungal infections, reducing the damage caused by leaf spot disease and other fungal pathogens.

2. Quinine: Antimalarial Properties and Defense Against Plant Pathogens

Quinine is an indole alkaloid produced by Cinchona species, which are native to South America. Historically, quinine is best known for its antimalarial properties in humans, but it also plays a vital role in protecting Cinchona trees from herbivores and microbial pathogens. Quinine's antimicrobial properties make it a potent defense compound against both bacteria and fungi.

Case Study: Quinine and Plasmodium (Malaria Parasite)

Quinine's most famous interaction is with the Plasmodium parasite, which causes malaria in humans. Although this is not a plant-pathogen interaction, it highlights the compound's powerful antimicrobial properties. Quinine interferes with the parasite's ability to detoxify heme, a byproduct of hemoglobin digestion within red blood cells. By inhibiting this process, quinine causes toxic buildup in the parasite, leading to its death. Quinine disrupts the parasite's heme detoxification pathway, leading to the accumulation of toxic compounds that kill the parasite. The antimalarial action of quinine demonstrates the evolutionary pressure exerted on plants to produce compounds with broad-spectrum antimicrobial properties, which can also be co-opted for medicinal purposes.

Case Study: Quinine and Fungal Pathogens

Quinine has also been studied for its antifungal properties. Research indicates that quinine can inhibit the growth of *Phytophthora infestans*, the pathogen responsible for late blight in potatoes and tomatoes. Quinine interferes with the fungal cell wall by disrupting chitin synthesis, leading to impaired hyphal growth and spore germination. Quinine disrupts fungal cell wall biosynthesis, specifically targeting the production of chitin, which is essential for fungal growth. Quinine's ability to inhibit fungal growth showcases its role as a potent antifungal agent, protecting Cinchona plants from pathogen invasion while also highlighting its potential use in agricultural settings as a natural fungicide.

3. Berberine: A Potent Antimicrobial Alkaloid

Berberine is an isoquinoline alkaloid found in various plant species, including Berberis (barberry) and Coptis (goldthread). Berberine is highly effective against a broad range of pathogens, including bacteria, fungi, and viruses, due to its ability to disrupt cellular processes. It is also widely used in traditional medicine for its antimicrobial properties.

Case Study: Berberine and Staphylococcus aureus

Staphylococcus aureus is a gram-positive bacterial pathogen that causes infections in both plants and animals. Berberine inhibits the growth of S. aureus by interfering with DNA replication. Berberine binds to bacterial DNA, preventing the action of DNA gyrase, an enzyme essential for DNA replication and transcription. This inhibition leads to the accumulation of supercoils in the bacterial DNA, blocking replication and ultimately leading to cell death. Berberine inhibits DNA gyrase, preventing DNA replication in S. aureus, leading to bacterial cell death (Stermitz et al., 2000). Berberine's broad-spectrum antimicrobial activity makes it an effective defense compound for Berberis plants, protecting them from bacterial infections while also offering potential for use as a natural antibacterial agent in agriculture and medicine.

Case Study: Berberine and Fungal Pathogens

Berberine is also highly effective against fungal pathogens. It has been shown to inhibit the growth of Candida albicans, a fungal pathogen that affects both plants and animals. Berberine disrupts the fungal membrane by interacting with sterols, essential components of the fungal cell membrane. This disruption leads to increased membrane permeability, leakage of cellular contents, and cell death. Berberine interacts with sterols in the fungal membrane, causing disruption of membrane integrity and leading to cell lysis. The antifungal properties of berberine help plants like Berberis resist fungal infections, contributing to the plant's survival in pathogen-rich environments. Additionally, berberine's action against fungal pathogens highlights its potential for use in antifungal treatments in both agriculture and medicine.

3.4 Plant Pathogen Resistance Through Alkaloid Production

The production of alkaloids in plants is an evolutionary adaptation that has allowed them to develop resistance to a variety of pathogens, including bacteria, fungi, and viruses. This adaptation is a result of millions of years of selective pressure exerted by microbial pathogens and herbivores (Wink, 2020). Alkaloid-based defense mechanisms have evolved as a primary line of chemical defense in many plant species, providing them with the ability to mitigate pathogen attacks effectively. The production of these secondary metabolites is often tightly regulated by environmental cues and stress signals, particularly in response to pathogen invasion.

1. Evolutionary Adaptation of Plants to Produce Pathogen-Specific Alkaloids

Plants have evolved the ability to produce alkaloids as a direct result of interactions with their environment, particularly in response to biotic stress such as herbivory and pathogen invasion. The diversity of alkaloid structures and their specificity towards particular pathogens or herbivores suggest that plants have fine-tuned these defense mechanisms through natural selection. The evolutionary advantage conferred by alkaloid production has enabled plants to better resist infections and improve their overall fitness.

a) Co-Evolution with Pathogens

The co-evolution between plants and pathogens has driven the development of specific alkaloid compounds that target particular microbial processes. Over time, as pathogens adapted to overcome certain plant defenses, plants, in turn, evolved more potent or structurally complex alkaloids to counteract these adaptations (Wink, 1988). This evolutionary arms race has resulted in the wide variety of alkaloid compounds that exist today, each with a unique mode of action designed to combat specific pathogens. For example, quinine from the Cinchona tree evolved as a defense against both herbivores and microbial pathogens. The structure of quinine, which allows it to bind to and inhibit DNA replication enzymes in pathogens, likely evolved in response to the selective pressure exerted by these organisms. As pathogens developed resistance mechanisms, the production and complexity of quinine increased, providing the Cinchona tree with enhanced protection against a broader spectrum of microbes (Wink, 2003).

b) Diversification of Alkaloid Biosynthetic Pathways

The diversification of biosynthetic pathways involved in alkaloid production is another evolutionary adaptation that has enhanced plant pathogen resistance. By utilizing different amino acids and metabolic precursors, plants can produce a wide range of alkaloids, each tailored to target a specific pathogen or group of pathogens. For example, the production of indole alkaloids such as quinine and pyrrolidine alkaloids like nicotine from distinct amino acid precursors allows plants to produce defense compounds with highly specific modes of action. This diversification also extends to the modification of alkaloid structures. In some plants, minor structural changes, such as the addition of functional groups (hydroxyl, methyl, or methoxy groups), can result in a significant increase in the potency of an alkaloid against certain pathogens. This fine-tuning of alkaloid structures through evolution allows plants to respond more effectively to pathogen threats.

2. Regulation of Alkaloid Production in Response to Pathogen Invasion

Plants do not produce high levels of alkaloids at all times; instead, the production of these compounds is often induced in response to specific biotic stressors, such as pathogen attack. The regulation of alkaloid production is mediated by complex signaling pathways that are activated upon detection of pathogen-associated molecular patterns (PAMPs). This induced defense mechanism allows plants to conserve energy while still maintaining an effective defense system that can be quickly activated when needed.

a) Signaling Pathways Involved in Alkaloid Regulation

The activation of jasmonic acid (JA) and salicylic acid (SA) signaling pathways plays a crucial role in the regulation of alkaloid biosynthesis. These signaling molecules are produced in response to pathogen invasion and act as key regulators of secondary metabolite production, including alkaloids.

» **Jasmonic Acid Pathway**: The JA signaling pathway is particularly important in the regulation of alkaloid production in response to herbivory and necrotrophic pathogens (pathogens that kill host tissue). In plants like Nicotiana tabacum (tobacco), the production of nicotine is upregulated in response to JA signaling. Upon pathogen invasion, JA induces the expression of genes involved in nicotine biosynthesis, leading to an increase in alkaloid levels that deter further attack (Steppuhn et al., 2004).

» **Salicylic Acid Pathway**: The SA pathway, on the other hand, is primarily activated in response to biotrophic pathogens (those that feed on living tissue) and plays a role in the regulation of alkaloid production against bacterial and fungal pathogens. For example, berberine production in Berberis species is regulated by SA signaling, which induces the expression of enzymes involved in its biosynthesis during pathogen attack (Lin et al., 20214).

b) Transcriptional Regulation of Alkaloid Biosynthesis

At the molecular level, the regulation of alkaloid production is controlled by transcription factors that bind to the promoters of biosynthetic genes, activating or repressing their expression. For instance, in tobacco plants, the transcription factors ERF189 and MYC2 are key regulators of nicotine biosynthesis. These transcription factors are activated by JA signaling and bind to the promoters of genes encoding enzymes like putrescine N-methyltransferase (PMT), which catalyzes the first step in nicotine production. Similarly, in Cinchona species, the transcription factor CrORCA3 regulates the expression of genes involved in quinine biosynthesis. This transcription factor is activated in response to pathogen-induced signaling pathways and increases the production of quinine, enhancing the plant's resistance to infection.

c) Environmental and Developmental Triggers

In addition to pathogen-induced signaling, the production of alkaloids can also be influenced by environmental factors such as light, temperature, and nutrient availability. For example, light is a critical regulator of alkaloid production in many plants, with higher levels of sunlight often correlating with increased alkaloid synthesis (Harborne, 1993). This is particularly evident in tropane alkaloids like atropine, which show higher accumulation in plants exposed to bright light. Alkaloid production is also regulated by developmental cues, with certain plant tissues producing higher levels of alkaloids during specific stages of growth. For instance, roots and young leaves tend to have higher concentrations of alkaloids, as these tissues are more vulnerable to pathogen invasion.

3. Examples of Plants That Rely Heavily on Alkaloid-Based Defenses

Many plant species rely extensively on alkaloid production as a defense mechanism against pathogens. Below are notable examples of plants that use alkaloid-based defenses to protect themselves from microbial infections.

a) Nicotiana tabacum (Tobacco)

Tobacco plants produce high levels of nicotine, a pyrrolidine alkaloid, as a defense against both herbivores and pathogens. Nicotine is synthesized in the roots of the plant and is transported to the leaves, where it accumulates in response to biotic stress. Nicotine not only acts as a neurotoxin against insects but also exhibits antimicrobial properties, inhibiting the growth of bacterial pathogens like Pseudomonas syringae.

» **Defense Against Bacteria**: Nicotine disrupts the cell membranes of bacterial pathogens, increasing membrane permeability and causing cell death.

» **Regulation**: Nicotine production is tightly regulated by the JA signaling pathway, which activates biosynthetic genes in response to pathogen invasion.

b) Cinchona officinalis (Cinchona Tree)

The Cinchona tree is famous for its production of quinine, an indole alkaloid that provides defense against both herbivores and pathogens. Quinine is produced in the bark of the tree and serves as a potent antimicrobial agent, particularly against fungal pathogens like Phytophthora infestans, which causes late blight.

- » **Defense Against Fungi**: Quinine inhibits fungal hyphal growth by disrupting chitin synthesis in the fungal cell wall, preventing the pathogen from spreading within the plant.
- » **Regulation**: The production of quinine is regulated by both environmental cues (such as light) and pathogen-induced signaling pathways.

c) Berberis vulgaris (Barberry)

Barberry plants produce berberine, an isoquinoline alkaloid, as a broad-spectrum defense compound against bacteria, fungi, and viruses. Berberine is stored in the bark and roots of the plant and is released in response to microbial infection.

- » **Defense Against Bacteria**: Berberine inhibits bacterial DNA replication by interfering with topoisomerase enzymes, leading to the death of bacterial pathogens like Staphylococcus aureus.
- » **Regulation**: Berberine production is induced by the SA signaling pathway, which activates biosynthetic enzymes in response to pathogen detection.

d) Atropa belladonna (Deadly Nightshade)

Deadly nightshade produces atropine, a tropane alkaloid, as a defense against herbivores and pathogens. Atropine acts as a potent neurotoxin and anticholinergic agent, inhibiting the parasympathetic nervous system in herbivores and disrupting microbial growth.

- » **Defense Against Pathogens**: Atropine inhibits bacterial and fungal growth by disrupting cell membrane integrity and interfering with cellular respiration.
- » **Regulation**: The production of atropine is regulated by environmental factors such as light and temperature, with higher concentrations found in plants exposed to sunlight.

3.5 Application of Alkaloids in Agriculture: Bio-Pesticides and Disease Management

Alkaloids have gained significant attention in modern agricultural practices due to their potent bioactivity against pests and pathogens. As natural compounds produced by plants, alkaloids offer a promising alternative to synthetic chemicals, aligning with the increasing demand for sustainable agriculture. They are being explored as bio-pesticides and are considered effective agents in plant disease management, providing environmentally friendly solutions for crop protection. However, there are several challenges associated with the use of alkaloid-based bio-pesticides, and their integration into large-scale agricultural systems presents both opportunities and obstacles.

1. Use of Alkaloids as Natural Bio-Pesticides in Modern Agricultural Practices

Alkaloids are widely recognized for their roles in plant defense against herbivores and microbial pathogens. Their bio-pesticidal properties make them highly suitable for agricultural applications, as they can effectively control insect pests, fungal infections, and bacterial diseases in crops. The use of alkaloid-based bio-pesticides is particularly appealing due to their biodegradability, low toxicity to non-target organisms, and minimal environmental impact (Mukhtar et al., 2008).

a) Nicotine-Based Insecticides

One of the earliest examples of alkaloid-based bio-pesticides is the use of nicotine as an insecticide. Nicotine, derived from tobacco plants (*Nicotiana tabacum*), has been historically used to control insect pests such as aphids, thrips, and whiteflies. Nicotine acts on the nervous system of insects by binding to nicotinic acetylcholine receptors (nAChRs), causing hyperstimulation and paralysis, ultimately leading to death (Steppuhn et al., 2004).

» **Application in Agriculture**: Nicotine-based insecticides, commonly known as nicotine sulfates, have been used in organic farming as a natural pest control method. Although synthetic insecticides such as neonicotinoids have largely replaced nicotine in conventional agriculture, nicotine remains a key component of organic pest control strategies.

» **Advantages**: Nicotine is biodegradable, reducing concerns about environmental contamination. Additionally, it is highly effective against a broad spectrum of insect pests, providing versatile pest control.

» **Limitations**: The high toxicity of nicotine to mammals and humans poses a significant risk in its handling and application, limiting its widespread use in modern agriculture. Additionally, its short persistence in the environment requires frequent reapplication, making it less practical for large-scale farming.

b) Capsaicin-Based Bio-Pesticides

Capsaicin, the active compound in chili peppers (*Capsicum* species), is another alkaloid with bio-pesticidal properties. Capsaicin is primarily used as a repellent against herbivorous pests, including mammals, birds, and insects. It activates pain receptors in animals, deterring them from feeding on treated plants (Reddy et al., 2012).

» **Application in Agriculture**: Capsaicin-based products are commonly used in organic farming as natural repellents to protect crops from herbivorous pests such as deer, rabbits, and rodents. These products are often sprayed on crops, creating an aversive sensation that prevents animals from grazing.

» **Advantages**: Capsaicin is non-toxic to most non-target organisms, including beneficial insects, making it a safe option for integrated pest management (IPM) systems. Its biodegradability ensures that it does not accumulate in the environment.

» **Limitations**: Capsaicin is not a direct insecticide but rather a repellent, meaning it does not kill pests but only deters them. Its efficacy can be limited by environmental factors such as rain, which can wash away the compound, necessitating frequent reapplication.

c) Quinine as a Natural Fungicide

Quinine, an alkaloid from Cinchona trees, is traditionally known for its antimalarial properties but also has significant fungicidal activity. It has been used to control fungal pathogens such as Phytophthora infestans, the causative agent of potato and tomato blight.

» **Application in Agriculture**: Quinine is being explored as a natural fungicide for controlling fungal diseases in crops, particularly in organic farming systems where synthetic fungicides are prohibited. Its ability to disrupt fungal cell walls by interfering with chitin synthesis makes it effective against a range of fungal pathogens.

» **Advantages**: Quinine's broad-spectrum antifungal activity provides protection against multiple types of fungal infections, making it a valuable tool in sustainable agriculture. It is also environmentally friendly, as it biodegrades naturally.

» **Limitations**: Quinine's availability is limited due to its medicinal use, and large-scale production of quinine for agricultural purposes may not be economically viable. Furthermore, it may be less effective against some resistant fungal strains, necessitating further research into its agricultural applications.

2. Potential for Alkaloid-Based Disease Management in Sustainable Agriculture

As the demand for sustainable agriculture grows, alkaloids present an attractive alternative to synthetic agrochemicals. Their natural origin, biodegradability, and effectiveness against a wide range of pathogens make them ideal candidates for integrated pest management (IPM) systems. Alkaloids can help reduce reliance on chemical pesticides, minimize environmental damage, and promote biodiversity in agricultural ecosystems.

a) Alkaloids in Disease Resistance Breeding

Alkaloids can play a key role in disease resistance breeding programs, where crops are selectively bred to produce higher levels of specific alkaloids that confer resistance to pathogens. For example, breeding programs have focused on increasing the production of solanine in potatoes to provide resistance against fungal infections such as late blight. By enhancing the natural defense mechanisms of crops, alkaloid-based disease resistance breeding can reduce the need for chemical fungicides, promoting more sustainable crop protection strategies.

b) Combining Alkaloids with Biological Control Agents

Alkaloids can be used in conjunction with biological control agents, such as beneficial fungi, bacteria, and insects, to enhance the efficacy of pest and disease management systems. For example, combining nicotine with entomopathogenic fungi can improve the control of insect pests while reducing the amount of chemical pesticide required (Li & Peng, 2013). Combining alkaloid-based bio-pesticides with biological control agents creates a synergistic effect, improving pest control efficacy and reducing the overall environmental impact.

c) Alkaloid-Based Biostimulants

Some alkaloids not only act as bio-pesticides but also function as biostimulants, enhancing plant growth and resilience to stress. For instance, berberine has been shown to enhance plant immune responses, making crops more resistant to bacterial and fungal infections. Using alkaloid-based biostimulants in agriculture can help plants withstand abiotic stress (e.g., drought, salinity) in addition to protecting them from pathogens, leading to healthier, more resilient crops in sustainable farming systems.

3. Challenges and Future Prospects for Utilizing Alkaloids in Plant Pathogen Control

Despite the significant potential of alkaloid-based bio-pesticides and disease management strategies, there are several challenges that must be addressed before these compounds can be widely adopted in modern agriculture.

a) Toxicity to Non-Target Organisms

While many alkaloids are selective in their action, some can exhibit toxicity to non-target organisms, including beneficial insects, soil microbes, and even mammals. For example, nicotine is highly toxic to humans and wildlife, which limits its use in large-scale agricultural settings. Developing formulations that minimize the risk to non-target organisms while maximizing the efficacy of alkaloid-based bio-pesticides is crucial for their safe application.

b) Cost and Availability

The large-scale production of alkaloids for agricultural use can be cost-prohibitive, particularly for compounds that are also used in medicine (e.g., quinine). Furthermore, the limited availability of some plant species that produce these alkaloids poses a challenge for sustainable sourcing. Research into synthetic biology and metabolic engineering may provide a solution by enabling the large-scale production of alkaloids through genetically modified microorganisms. This could reduce the cost and increase the availability of alkaloids for agricultural use.

c) Environmental Degradation and Persistence

While alkaloids are generally biodegradable, their persistence in the environment can vary depending on their chemical structure and the environmental conditions. Some alkaloids may degrade rapidly, requiring frequent reapplication, while others may persist longer than desired, potentially affecting non-target organisms. Further research is needed to better understand the environmental fate of alkaloids and to develop sustained-release formulations that provide long-lasting pest and disease control without the need for frequent application.

d) Resistance Development

Just as pathogens can develop resistance to synthetic pesticides, there is a risk that prolonged use of alkaloid-based bio-pesticides could lead to the development of resistance in target pests and pathogens. Integrating alkaloid-based bio-pesticides into rotational pest management programs and combining them with other biological control methods can help reduce the likelihood of resistance development.

References

Agati, G., Azzarello, E., Pollastri, S., & Tattini, M. (2012). Flavonoids as antioxidants in plants: Location and functional significance. Plant Science, 196, 67-76.

Agrawal, A. A. & Zhang, X. (2021). The evolution of coevolution in the study of species interactions. *Evolution*, 75, 1594-1606.

Akwu, N. A., Lekhooa, M., Deqiang, D., & Aremu, A. O. (2023). Antidepressant effects of coumarins and their derivatives: A critical analysis of research advances. *European Journal of Pharmacology, 956*, 175958

Alt, S., Mitchenall, L. A., Maxwell, A., & Heide, L. (2011) Inhibition of DNA gyrase and DNA topoisomerase IV of Staphylococcus aureus and Escherichia coli by aminocoumarin antibiotics. *Journal of Antimicrobial Chemotherapy*, *66*(9), 2061-2069.

Basile, A., Sorbo, S., Spadaro, V., Bruno, M., Maggio, A., Faraone, N., & Rosselli, S. (2009). Antibacterial and antifungal activity of coumarins from the roots of *Ferulago campestris* (Apiaceae). *Molecules, 14*(3), 939-952.

Bourgaud, F., Hehn, A., Larbat, R., Doerper, S., Gontier, E., Kellner, S., & Matern, U. (2006). Biosynthesis of coumarins in plants: A major pathway still to be unravelled for cytochrome P450 enzymes. *Phytochemistry Reviews*, *5*(2-3), 293-308.

Cowan, M. M. (1999). Plant products as antimicrobial agents. *Clinical Microbiology Reviews, 12*(4), 564-582.

Cuffaro, D., Bertolini, A., Silva, A. M., Rodrigues, F., Gabbia, D., De Martin, S., Saba, A., Bertini, S., Digiacomo, M., & Macchia, M. (2024). Comparative analysis on polyphenolic composition of different olive mill wastewater and related extra virgin olive oil extracts and evaluation of nutraceutical properties by cell-based studies. *Foods, 13*(20), 3312.

Degroote, E., Schoorens, C., Pockelé, S., Stojilković, B., Demeestere, K., Mangelinckx, S., & Kyndt, T. (2024). A combination of plant-based compounds and extracts acts nematicidal and induces resistance against Meloidogyne incognita in tomato. *Frontiers in Plant Sciences, 15*, 1411825.

Dini, I., & Grumetto, L. (2022). Recent advances in natural polyphenol research. *Molecules, 27*(24), 8777.

Dixon, R. A., & Paiva, N. L. (1995). Stress-induced phenylpropanoid metabolism. *The Plant Cell, 7*(7), 1085-1097.

Durrant, W. E., & Dong, X. (2004). Systemic acquired resistance. *Annual Review of Phytopathology, 42*, 185-209.

Facchini, P. J. (2001). Alkaloid biosynthesis in plants: Biochemistry, cell biology, molecular regulation, and metabolic engineering applications. Annual Review of Plant Biology, 52(1), 29-66.

Farmer, E. E., Almeras, E., & Krishnamurthy, V. (2003). Jasmonates and related oxylipins in plant responses to pathogenesis and herbivory. *Current Opinion in Plant Biology*, *6*(4), 372-378.

Fourcroy, P., Sisó-Terraza, P., Sudre, D., Savirón, M., Reyt, G., Gaymard, F., & Briat, J. F. (2014). Involvement of the ABCG37 transporter in secretion of scopoletin and derivatives by Arabidopsis roots in response to iron deficiency. *New Phytologist*, *201*(1), 155-167.

Hagel, J. M., & Facchini, P. J. (2013). Benzylisoquinoline alkaloid metabolism: A century of discovery and a brave new world. Plant and Cell Physiology, 54(5), 647-672.

Harborne, J. B. (1999). The comparative biochemistry of phytoalexin induction in plants. *Biochemical Systematics and Ecology*, *27*(4), 335-367.

Itkin, M., Heinig, U., Tzfadia, O., Bhide, A. J., Shinde, B., Cardenas, P. D., Bocobza, S. E., Unger, T., Malitsky, S., Finkers, R., Tikunov, Y. M., Bovy, A., Chikate, Y. R., Singh, P.,

Jeandet, P., Hébrard, C., Deville, M. A., Cordelier, S., Dorey, S., Aziz, A., & Crouzet, J. (2014). Deciphering the role of phytoalexins in plant-microorganism interactions and human health. Molecules, 19(12), 18033-18056.

Kai, K., Shimizu, B., Mizutani, M., Watanabe, K., & Sakata, K. (2006). Accumulation of coumarins in Arabidopsis thaliana. *Phytochemistry, 67*(4), 379-386.

Korkina, L. (2007). Phenylpropanoids as naturally occurring antioxidants: from plant defense to human health. *Cellular and Molecular Biology, 53*(1), 15-25.

Kovács, A., Vasas, A., & Forgo, P. (2008). Natural phenanthrenes and their biological activity. *Phytochemistry, 69*(5), 1084-1110.

Lee, D. H., Choi, H. W., & Hwang. B. K. (2011) The pepper E3 ubiquitin ligase RING1 gene, CaRING1, is required for cell death and the salicylic acid-dependent defense response. *Plant Physiology, 156*(4), 2011-25.

Nour-Eldin, H. H., & Halkier, B. A. (2013). The emerging field of transport engineering of plant specialized metabolites. Current Opinion in Biotechnology, 24(2), 263-270.

Rasmussen, T. B., & Givskov, M. (2006). Quorum sensing inhibitors: A bargain of effects. *Microbiology, 152*(4), 896-904.

Reddy, K. M., Reddy, D. C. R., & Reddy, R. S. (2012). Studies on antifungal activity of capsaicin extracted from Capsicum frutescens L. against Botrytis cinerea. International Journal of Pharmacy and Biological Sciences, 2(1), 64-68.

Roddick, J. G. (2001). The importance of the tropane alkaloids in the Solanaceae. Phytochemistry, 57(3), 297-303.

Rogachev, I., Beekwilder, J., Giri, A. P., Aharoni, A., & Saito, K. (2011). Biosynthesis of antinutritional alkaloids in solanaceous crops is mediated by clustered genes. Science, 334(6052), 1234-1237.

Shan, Z., Zhou, S., Shah, A., Arafat, Y., Arif Hussain Rizvi, S., & Shao, H. (2023). Plant allelopathy in response to biotic and abiotic factors. *Agronomy*, 13(9), 2358.

Steppuhn, A., Gase, K., Krock, B., Halitschke, R., & Baldwin, I. T. (2004). Nicotine's defensive function in nature. PLoS Biology, 2(8), e217.

Stermitz, F. R., Lorenz, P., Tawara, J. N., Zenewicz, L. A., & Lewis, K. (2000). Synergy in a medicinal plant: Antimicrobial action of berberine potentiated by 5′-methoxyhydnocarpin, a multidrug pump inhibitor. Proceedings of the National Academy of Sciences of the United States of America, 97(4), 1433-1437.

Stringlis, I. A., de Jonge, R., & Pieterse, C. M. J. (2019). The Age of coumarins in plant-microbe interactions. *Plant Cell Physiology, 60*(7), 1405-1419.

Thakur, A., Singla, R., & Jaitak, V. (2015). Coumarins as anticancer agents: A review on synthetic strategies, mechanism of action and SAR studies. *European Journal of Medicinal Chemistry, 101*, 476-495.

Wuyts, N., Swennen, R., & De Waele, D. (2006). Effects of plant phenylpropanoid pathway products and selected terpenoids and alkaloids on the behaviour of the plant-parasitic nematodes *Radopholus similis, Pratylenchus penetrans* and *Meloidogyne incognita. Nematology, 8*, 89-101.

Zhang, Y., Sass, A., van Acker, H., Wille, J., Verhasselt, B., van Nieuwerburgh, F., Kaever, V., Crabbé, A., & Coenye, T. (2018). Coumarin reduces virulence and biofilm formation in pseudomonas aeruginosa by affecting quorum sensing, type III secretion and C-di-GMP levels. Frontiers in Microbiology, *9*, 1952.

Ziegler, J., & Facchini, P. J. (2008). Alkaloid biosynthesis: Metabolism and trafficking. Annual Review of Plant Biology, 59, 735-769.

Chapter - 4

Terpenoids: Multifunctional Compounds in Disease Resistance

Susheel Kumar[1], Sheetanshu Gupta[2], Vinay Pratap Singh[3] and Reshma Jadhav[4]

[1]Plant Molecular Virology Laboratory, CPMB Division, CSIR-National Botanical Research Institute, Rana Pratap Marg, Lucknow, Uttar Pradesh, India

[2]Department of Biochemistry, G.B. Pant University of Agriculture & Technology, Pantnagar, Uttarakhand, India

[3]Plant Physiology, College of Agriculture, Khurai, JNKVV, Jabalpur, M.P. India

[4]Department of Biochemistry, Mahatma Phule Krishi Vidyapeeth, Rahuri, Ahmednagar, Maharashtra, India

Abstract

Terpenoids, one of the largest and most diverse classes of plant secondary metabolites, play a crucial role in plant defense mechanisms, particularly in disease resistance. These multifunctional compounds are classified based on the number of isoprene units, ranging from monoterpenes to polyterpenes, and are synthesized through the mevalonate and methylerythritol phosphate (MEP) pathways. Terpenoids are essential in primary plant functions and in responding to biotic stressors, such as pathogens and herbivores. Their antimicrobial, antifungal, antiviral, and insecticidal properties make them invaluable for plant protection. They serve as direct chemical barriers against pathogenic invasion and act as signaling molecules in plant-pathogen interactions, triggering systemic resistance responses like Systemic Acquired Resistance (SAR) and Induced Systemic Resistance (ISR). Additionally, terpenoids influence ecological dynamics by mediating beneficial interactions with organisms such as mycorrhizal fungi, which further strengthens plant resistance. Their role in agriculture has gained significant attention, with innovations focusing on harnessing terpenoids as natural pesticides and biocontrol agents to reduce the dependency on synthetic chemicals. Advances in biotechnology have made it possible to enhance terpenoid production through genetic engineering, which holds promise for developing disease-resistant crop varieties with sustainable management practices. Furthermore, terpenoids work synergistically with other secondary metabolites, such as alkaloids and phenolics, enhancing the plant's overall defense capabilities. However, challenges remain in the large-scale production and utilization of terpenoids, including extraction difficulties, environmental factors affecting their concentration, and regulatory complexities within plants.

DOI: 10.1201/9781003717454-4

Keywords: Terpenoids, disease resistance, plant defense, ecological interactions, biocontrol, biosynthesis.

Introduction

Terpenoids, also known as isoprenoids, are a vast and structurally diverse group of naturally occurring organic compounds found in nearly all living organisms, particularly in plants. They are composed of isoprene units and serve a variety of functions, both primary and secondary, within plants. Structurally, terpenoids are classified based on the number of isoprene units they contain, ranging from monoterpenes (C10) to polyterpenes, which include hundreds or thousands of isoprene units. The biosynthesis of terpenoids follows two major pathways: the mevalonate pathway (MVA) and the methylerythritol phosphate pathway (MEP), each responsible for producing specific classes of terpenoids in various plant species (Dudareva et al., 2013). While the MVA pathway occurs primarily in the cytosol and is responsible for sesquiterpenes and triterpenes, the MEP pathway takes place in plastids and generates monoterpenes, diterpenes, and tetraterpenes (Vranová et al., 2012). The complexity of their biosynthesis and structural diversity enables terpenoids to participate in multiple biochemical and ecological functions, playing pivotal roles in plant growth, development, and defense against pathogens.

Terpenoids act as essential compounds in plant defense, employing various mechanisms to combat biotic stressors, particularly pathogens. Their antimicrobial properties allow them to directly interfere with the growth and metabolism of harmful microorganisms, including bacteria, fungi, and viruses (Tholl, 2015). For instance, monoterpenes such as camphor and menthol exhibit strong antifungal activity, disrupting the cell membranes of pathogenic fungi (Verma & Dubey, 2011). Diterpenoids like phytoalexins, synthesized in response to pathogen attack, prevent the establishment and spread of pathogens by inhibiting their enzymatic activities (Schmelz et al., 2014). Furthermore, terpenoids are involved in indirect defense mechanisms by acting as signaling molecules that activate systemic resistance responses within plants. For example, volatile terpenoids such as limonene and linalool not only repel herbivores but also trigger systemic acquired resistance (SAR) and induced systemic resistance (ISR), activating defense genes across different parts of the plant (Pichersky & Gershenzon, 2002).

In addition to direct antimicrobial properties, terpenoids mediate important ecological roles by influencing plant-pathogen interactions. Terpenoids function as signaling molecules that shape the plant's response to pathogen invasion, as well as mediate interactions with beneficial organisms such as mycorrhizae and rhizobacteria (Raguso, 2008). Terpenoids released from roots, for instance, help attract mycorrhizal fungi, which enhance the plant's nutrient uptake and contribute to its resistance to soil-borne pathogens (Akiyama et al., 2005). Furthermore, the production of volatile terpenoids in response to pathogen attack can serve as signals to neighboring plants, priming them for defense against future invasions. Such ecological interactions underscore the multifunctional roles of terpenoids not only in direct plant defense but also in shaping the plant's surrounding environment to mitigate pathogen impact (Dudareva et al., 2013).

The role of terpenoids in crop protection has garnered increased attention in recent years due to their potential as sustainable alternatives to synthetic agrochemicals. As natural pesticides, terpenoids offer an eco-friendly option for managing pests and diseases while minimizing the negative environmental impacts associated with synthetic chemicals. Studies have demonstrated the efficacy of certain terpenoids, such as azadirachtin (from *Azadirachta indica*), as a potent insecticidal agent that disrupts insect growth and reproduction (Isman, 2006). Similarly, terpenoid-rich essential oils, such as those derived from *Thymus vulgaris*, have been shown to possess strong fungicidal properties, providing effective control over crop diseases like powdery mildew (Ben-Jabeur et al., 2015). The practical applications of terpenoids in integrated pest management strategies are vast, and ongoing research continues to explore the potential of enhancing terpenoid production through genetic engineering and synthetic biology to improve crop resistance and reduce reliance on chemical inputs (Zerbe & Bohlmann, 2015).

Terpenoids do not function in isolation; rather, they often act synergistically with other secondary metabolites, such as alkaloids and phenolics, to bolster the plant's defense mechanisms. This synergy enhances the plant's overall ability to resist pathogens by combining the chemical properties of different compounds. For example, the combination of terpenoids with flavonoids has been shown to provide enhanced protection against bacterial and fungal pathogens (Taiz et al., 2015). Additionally, terpenoid-phenolic combinations can create a more hostile environment for pathogens by intensifying membrane disruption and oxidative stress (Wink, 2010). Such synergistic interactions between terpenoids and other metabolites exemplify how plants deploy a complex arsenal of chemical defenses to maximize their protective effects. This terpenoid synergy has important implications for the development of multi-targeted approaches in crop protection, wherein combining different natural compounds could provide more effective disease resistance (Ramakrishna et al., 2019).

Despite their promising applications, the large-scale utilization of terpenoids in agriculture faces several challenges. One of the primary difficulties is the variability in terpenoid production due to environmental factors, which can affect the yield and consistency of their bioactive properties (Gershenzon & Dudareva, 2007). Additionally, the extraction and synthesis of terpenoids on a commercial scale can be costly and labor-intensive. Recent advancements in genetic engineering and metabolic pathway optimization offer potential solutions, enabling the increased production of specific terpenoids by modifying plant genes or utilizing microbial systems for terpenoid biosynthesis (Paddon & Keasling, 2014). Future research must continue to explore these innovative strategies to overcome current limitations and fully realize the potential of terpenoids as natural defenses in plant disease management. Moreover, further studies are needed to understand the regulatory mechanisms that control terpenoid production within plants, as well as to explore the effects of environmental stresses, such as climate change, on terpenoid biosynthesis and efficacy (Yang et al., 2012).

1. Overview of Terpenoids: Structure, Classification, and Biosynthesis

Terpenoids, often referred to as isoprenoids, are a major class of naturally occurring organic compounds that exhibit significant structural diversity and possess a wide range of biological activities. Found in plants, animals, fungi, and even marine organisms, terpenoids play essential roles in cellular processes, including growth, defense, and metabolism. In plants, terpenoids serve as primary and secondary metabolites, acting as hormones, pigments, structural components of membranes, and as volatile compounds that participate in ecological interactions.

1.1 Structure of Terpenoids

The structural diversity of terpenoids arises from their basic building block, the isoprene unit. Isoprene units are five-carbon compounds ($C5H8$) that form the core structure of terpenoids. Terpenoids can be classified based on the number of isoprene units they contain, and this classification helps organize them into various subclasses such as monoterpenes, sesquiterpenes, diterpenes, and others.

1.1.1 Isoprene as the Basic Unit

Isoprene (2-methyl-1,3-butadiene) is the fundamental precursor of all terpenoids, with a basic formula of C5H8. The isoprene rule, proposed by Leopold Ruzicka in 1921, postulates that terpenoid molecules are composed of isoprene units, linked together in various configurations. These units can form chains, rings, and other structures depending on the specific enzymatic processes involved in terpenoid biosynthesis (Bouvier et al., 2005).

The basic structure of an isoprene unit allows it to polymerize in several ways. Head-to-tail polymerization is the most common, although tail-to-tail linkages can also occur, especially in the biosynthesis of larger terpenoids such as squalene. The various ways in which these units are arranged lead to the creation of different structural families of terpenoids, each with unique chemical properties and biological functions. The versatility of isoprene as a building block gives rise to the immense diversity seen within the terpenoid family.

1.1.2 Structural Variability in Terpenoids

Terpenoids exhibit considerable structural variation, owing to the wide range of biosynthetic modifications they undergo after the assembly of their isoprene units. These modifications can include oxidation, reduction, cyclization, and the addition of functional groups such as hydroxyl (-OH), carbonyl (-C=O), and carboxyl (-COOH) groups. The result is a highly versatile class of compounds that can perform multiple biological functions.

Cyclization is a common feature in terpenoid biosynthesis and contributes to the formation of mono-, bi-, and polycyclic terpenoids. For example, monoterpenes like α-pinene, which is commonly found in coniferous trees, have a bicyclic structure that makes them volatile and suitable for roles in plant-environment interactions, such as herbivore deterrence and attraction of pollinators (Gershenzon & Dudareva, 2007). On the other hand, diterpenes like taxol have more complex structures, enabling them to serve in highly specialized roles such as cancer inhibition (Cragg & Newman, 2005).

The presence of multiple functional groups also allows terpenoids to interact with other cellular molecules. Oxygenated terpenoids, such as terpenoid alcohols and ketones, often have increased solubility in water and play critical roles in chemical signaling and defense mechanisms. The stereochemistry of terpenoids, or their three-dimensional arrangement, further adds to their functional diversity. Isomers like α- and β-pinene have different ecological roles and biological activities, despite having the same molecular formula (Pichersky & Gershenzon, 2002).

1.2 Classification of Terpenoids

The classification of terpenoids is based on the number of isoprene units they contain. This classification helps in understanding their biosynthetic origins and functions. The following subsections outline the major classes of terpenoids based on their isoprene content.

1.2.1 Monoterpenes (C10)

Monoterpenes are composed of two isoprene units, giving them a molecular formula of $C_{10}H_{16}$. They are highly volatile and are often found in the essential oils of plants, where they contribute to the characteristic aromas and flavors of many fruits, flowers, and herbs (Dudareva et al., 2013). The volatility of monoterpenes makes them suitable for ecological roles such as attracting pollinators and deterring herbivores. Limonene, for example, is a common monoterpene found in the peels of citrus fruits and is responsible for their distinctive aroma. Beyond its ecological role, limonene has been studied for its potential therapeutic properties, including anti-inflammatory and anticancer effects (Crowell, 1999). Another important monoterpene is menthol, found in mint, which has applications in medicine, cosmetics, and food industries due to its cooling properties and antimicrobial activity. Monoterpenes also play a significant role in plant defense. α-Pinene and β-pinene, for example, are found in the resin of coniferous trees and serve as deterrents to herbivores and pathogens. These compounds are toxic to many insects and fungi and help protect the plant from predation and infection. Monoterpenes like geraniol and citronellol are also used by plants to repel insects and as chemical signals to attract beneficial organisms (Gershenzon & Dudareva, 2007).

1.2.2 Sesquiterpenes (C15)

Sesquiterpenes are composed of three isoprene units ($C_{15}H_{24}$) and are slightly less volatile than monoterpenes. They are often found in the essential oils and resins of plants and are known for their role in plant defense and ecological interactions (Pichersky & Gershenzon, 2002). Many sesquiterpenes exhibit antimicrobial, antifungal, and anti-inflammatory properties, making them important in both plant defense and human applications. One of the most well-known sesquiterpenes is artemisinin, derived from the plant *Artemisia annua*, which is used as a potent antimalarial drug (Tu, 2016). Artemisinin works by producing reactive oxygen species that are toxic to the malaria parasite. Another important sesquiterpene is farnesene, which acts as an insect repellent and is found in the cuticular waxes of plants. Sesquiterpenes also play a role in signaling between plants and other organisms. For instance, certain plants release volatile sesquiterpenes in response to herbivore damage, which can attract predatory insects that help control pest populations. This indirect defense mechanism is an example of how plants use chemical signals to recruit allies in their defense against herbivores (Dudareva et al., 2013).

1.2.3 Diterpenes (C20)

Diterpenes consist of four isoprene units ($C_{20}H_{32}$) and are often found in resins and plant pigments. These compounds are involved in plant growth regulation, defense, and photosynthesis. Diterpenes such as gibberellins are plant hormones that regulate processes like seed germination, stem elongation, and flowering (Yamaguchi, 2008).

Taxol, a diterpene derived from the Pacific yew tree (*Taxus brevifolia*), is one of the most well-known diterpenes due to its use as a chemotherapy drug. Taxol stabilizes microtubules in cancer cells, preventing them from dividing and ultimately leading to cell death (Cragg & Newman, 2005). This diterpene is a prime example of how terpenoids can have significant therapeutic applications. Diterpenes are also involved in plant defense. For instance, diterpene phytoalexins are produced by plants in response to pathogen attack and inhibit the growth of fungi and bacteria. These compounds are part of the plant's innate immune system and help protect against a wide range of microbial pathogens (Schmelz et al., 2014).

1.2.4 Triterpenes (C30)

Triterpenes are composed of six isoprene units (C30H48) and are typically found in plant waxes, resins, and latex. These compounds play a role in maintaining the structural integrity of cell membranes and providing a protective coating on plant surfaces (Jiang et al., 2012). Triterpenes are also known for their medicinal properties, with many exhibiting anti-inflammatory, antiviral, and anticancer activities.

One of the most well-known triterpenes is squalene, a precursor to cholesterol and other sterols in animals and plants. Squalene is used in the cosmetic industry for its moisturizing properties and in vaccines as an adjuvant to enhance immune responses (Kim & Karadeniz, 2012). Another important triterpene is lupeol, which has been shown to possess anti-inflammatory and anticancer properties (Saleem, 2009). Triterpenes also serve as chemical defenses in plants. For example, oleanolic acid and ursolic acid, two triterpenoids found in the cuticle layer of leaves and fruits, have been shown to inhibit the growth of pathogens and protect plants from environmental stressors such as drought and ultraviolet radiation (Jiang et al., 2012).

1.2.5 Tetraterpenes (C40)

Tetraterpenes, also known as carotenoids, consist of eight isoprene units (C40H64) and are primarily involved in photosynthesis and pigmentation. Carotenoids are responsible for the red, orange, and yellow colors of many fruits and vegetables and play a crucial role in protecting plants from damage caused by excess light during photosynthesis (Nisar et al., 2015). Carotenoids such as β-carotene, which gives carrots their orange color, are important precursors to vitamin A in animals and humans. Lycopene, another carotenoid found in tomatoes, has been shown to have antioxidant properties and is associated with a reduced risk of certain cancers (Rao & Rao, 2007). In addition to their roles in photosynthesis and nutrition, carotenoids are involved in plant defense. For example, lutein, a carotenoid found in leafy green vegetables, protects plants from damage caused by ultraviolet radiation. Carotenoids also serve as precursors to signaling molecules such as abscisic acid, which regulates plant responses to environmental stress (Finkelstein, 2013).

1.2.6 Polyterpenes and Other Complex Terpenoids

Polyterpenes are composed of multiple isoprene units and can form large, complex molecules such as natural rubber. Natural rubber, produced by the rubber tree (*Hevea brasiliensis*), is a polyterpene made up of thousands of isoprene units and is used in the manufacture of tires and other products (Cornish, 2017). Other complex terpenoids include steroidal glycoalkaloids, which are derived from triterpenes and are found in plants like potatoes and tomatoes. These compounds are toxic to herbivores and pathogens, providing an additional layer of chemical defense for plants (Friedman, 2006).

1.3 Biosynthesis of Terpenoids

The biosynthesis of terpenoids involves the polymerization of isoprene units through two major pathways: the mevalonate (MVA) pathway and the methylerythritol phosphate (MEP) pathway. These pathways provide the precursors for the formation of various terpenoid structures, each of which plays a specific role in plant physiology and ecology.

1.3.1 The Mevalonate Pathway (MVA)

The mevalonate (MVA) pathway occurs in the cytosol of plant cells and is responsible for the synthesis of sesquiterpenes, triterpenes, and sterols. This pathway begins with the condensation of three acetyl-CoA molecules to form mevalonic acid, which is then phosphorylated and decarboxylated to produce isopentenyl pyrophosphate (IPP), the basic building block for terpenoid biosynthesis (Qureshi et al., 2018). The MVA pathway is crucial for the production of sterols, which are essential components of cell membranes. It is also responsible for the synthesis of triterpenes, which play a role in plant defense and hormone regulation. In addition, sesquiterpenes produced through the MVA pathway are involved in plant-environment interactions, such as attracting pollinators and deterring herbivores (Vranová et al., 2012).

1.3.2 The Methylerythritol Phosphate Pathway (MEP)

The methylerythritol phosphate (MEP) pathway occurs in the plastids of plant cells and is responsible for the synthesis of monoterpenes, diterpenes, and tetraterpenes. This pathway begins with the condensation of pyruvate and glyceraldehyde-3-phosphate to form 2-C-methyl-D-erythritol 4-phosphate (MEP), which is then converted into IPP and dimethylallyl pyrophosphate (DMAPP) (Rohmer, 1999). The MEP pathway is particularly important in the synthesis of carotenoids, which are involved in photosynthesis and plant protection against oxidative stress. Monoterpenes and diterpenes produced through the MEP pathway also play a role in plant defense, chemical signaling, and interactions with the environment (Dudareva et al., 2013).

2. Terpenoids as Plant Defenses: Mechanisms Against Pathogens

Terpenoids also referred to as isoprenoids, play an essential role in plant defense against a wide range of pathogens, including bacteria, fungi, viruses, and herbivores. As secondary metabolites, terpenoids serve multiple functions beyond basic plant metabolism, with a significant portion of their roles dedicated to plant defense. These compounds are synthesized by plants in response to biotic stressors, where they act directly on pathogens through antimicrobial, antifungal, and antiviral properties, and indirectly by modulating defense signaling pathways. The diversity in the structure of terpenoids, which ranges from simple monoterpenes to complex diterpenes and triterpenes, contributes to their versatility in defense mechanisms.

2.1 Direct Antimicrobial and Antifungal Activity of Terpenoids

The most direct way terpenoids protect plants from pathogens is through their antimicrobial and antifungal activities. Terpenoids disrupt the cellular structures of invading organisms, inhibit essential metabolic processes, and prevent the replication and growth of pathogens. The structural diversity of terpenoids allows them to target various aspects of pathogen physiology, making them highly effective in suppressing microbial invaders.

2.1.1 Monoterpenes: Fast-Acting Defenses

Monoterpenes, such as thymol and carvacrol, have been extensively studied for their potent antimicrobial properties. These compounds are found in the essential oils of many medicinal plants, such as thyme (*Thymus vulgaris*) and oregano (*Origanum vulgare*), where they are released in response to pathogenic attacks. Monoterpenes disrupt microbial membranes by integrating into lipid bilayers, leading to increased membrane permeability and subsequent leakage of vital intracellular components. This ultimately results in cell death (Bakkali et al., 2008). Thymol, in particular, has been shown to exhibit strong activity against a wide range of bacterial species, including *Escherichia coli* and *Staphylococcus aureus*, as well as against fungal pathogens like *Candida albicans* (Burt, 2004). Its lipophilic nature allows it to penetrate the hydrophobic membrane of pathogens, making it a fast-acting defense compound. In addition, monoterpenes like limonene and α-pinene have been found to possess antifungal activity, inhibiting the growth of pathogenic fungi such as *Aspergillus* and *Penicillium* species. These compounds alter fungal membrane integrity, disrupt cell wall biosynthesis, and inhibit spore germination, effectively halting fungal colonization (Pinto et al., 2009).

2.1.2 Sesquiterpenes: Broad-Spectrum Defenses

Sesquiterpenes, composed of three isoprene units, offer a more complex and often broader-spectrum defense compared to monoterpenes. They have been implicated in plant resistance to a wide range of pathogens, including bacteria, fungi, and viruses. Farnesol, a sesquiterpene alcohol, is an example of a compound with both antifungal and antibacterial properties. Farnesol inhibits the formation of biofilms by bacterial pathogens and disrupts fungal growth by targeting cell membranes and metabolic pathways essential for cell division (Jabra-Rizk et al., 2006). This makes it particularly effective against fungal pathogens such as *Candida* and bacterial species like *Pseudomonas aeruginosa*, both of which are known for forming biofilms that are resistant to conventional treatments. Artemisinin, another sesquiterpene, exhibits notable antimalarial properties and serves as an example of the potential of sesquiterpenes beyond plant defense. Artemisinin is synthesized by the plant *Artemisia annua* and has been successfully developed into a therapeutic agent used in treating malaria. This highlights the dual role sesquiterpenes can play in both plant and human health (Tu, 2016).

2.2 Volatile Terpenoids and Induced Defense Mechanisms

Volatile terpenoids are often released by plants in response to pathogen or herbivore attack, functioning as chemical signals that prime other parts of the plant or neighboring plants for defense. These compounds can activate defense-related genes and induce systemic resistance, protecting the plant from a wide array of pathogens. The release of volatile terpenoids also attracts beneficial organisms, such as parasitic wasps, that prey on herbivores, thus providing an indirect defense mechanism.

2.2.1 Systemic Acquired Resistance (SAR)

Systemic Acquired Resistance (SAR) is a plant-wide defense response that is activated when a localized infection triggers the production of signaling molecules that travel throughout the plant. Terpenoids, particularly diterpenes and monoterpenes, have been shown to play a role in the induction of SAR. When a plant is attacked by a pathogen, terpenoid signals are released, leading to the activation of defense genes in distant parts of the plant. This preemptive defense mechanism enhances the plant's ability to fend off future infections (Durrant & Dong, 2004). For example, the volatile monoterpene α-pinene, released from pine trees during herbivore attacks, has been shown to activate SAR in nearby plants, boosting their resistance to pathogens. Similarly, volatile terpenes like limonene and β-caryophyllene, emitted in response to fungal attack, can signal neighboring plants to induce defense pathways, thus enhancing overall resistance in plant communities (Maffei et al., 2011).

2.2.2 Induced Systemic Resistance (ISR)

Induced Systemic Resistance (ISR) is another form of plant immunity that is often activated by beneficial microorganisms, such as rhizobacteria, and involves the production of terpenoids as part of the plant's response to microbial colonization. While ISR does not rely on pathogen attack, it serves as a form of preemptive immunity that strengthens the plant's resistance to a wide variety of pathogens. Research has shown that the production of terpenoids like jasmonic acid and its derivatives plays a crucial role in ISR. Jasmonic acid, a diterpenoid derived from fatty acids, is involved in the signaling pathways that regulate ISR. Plants treated with jasmonic acid or colonized by beneficial rhizobacteria produce terpenoids that enhance their resistance to both fungal and bacterial pathogens (Pieterse et al., 2014). The volatile sesquiterpene β-caryophyllene, for example, is produced by maize plants in response to root colonization by beneficial bacteria, and it helps protect the plant from nematode infection by attracting parasitic nematode enemies (Rasmann et al., 2005).

2.3 Terpenoid Phytoalexins: Rapid Response to Pathogens

Phytoalexins are antimicrobial compounds that are synthesized by plants in response to pathogen attack. Many terpenoids fall into the category of phytoalexins, particularly diterpenes, which are rapidly produced upon pathogen invasion to provide immediate defense. These compounds often accumulate at the site of infection and act by inhibiting the growth and spread of the pathogen, either by disrupting cellular functions or by interfering with pathogen signaling pathways.

2.3.1 Diterpenes as Phytoalexins

Diterpenoids, such as kaurenoic acid and labdane diterpenoids, are potent phytoalexins produced by plants like maize (*Zea mays*) and rice (*Oryza sativa*) in response to fungal and bacterial infections. In maize, kaurenoic acid accumulates in response to infection by the fungus *Colletotrichum graminicola*, inhibiting spore germination and fungal growth (Schmelz et al., 2014). Similarly, rice produces the diterpene momilactone upon infection by the rice blast fungus (*Magnaporthe oryzae*), which effectively halts the progression of the disease by inhibiting fungal development (Pieterse et al., 2014). Phytoalexins are generally toxic to pathogens and act by disrupting essential metabolic pathways or by causing oxidative damage to the invading organism. In some cases, terpenoid phytoalexins also act as signaling molecules that trigger additional layers of defense in the plant, such as the production of reactive oxygen species (ROS) and the fortification of cell walls to prevent pathogen entry.

2.3.2 Triterpenoids in Plant Defense

Triterpenoids, such as glycyrrhizin from licorice (*Glycyrrhiza glabra*), also exhibit phytoalexin-like properties and contribute to plant defense against microbial pathogens. Glycyrrhizin has been shown to possess strong antiviral, antibacterial, and antifungal properties, making it a crucial component of the plant's chemical defense arsenal (Fiore et al., 2008). Additionally, triterpenoids such as oleanolic acid and ursolic acid, which are found in the cuticular waxes of many plants, serve as physical and chemical barriers against pathogens by inhibiting spore germination and fungal hyphal growth (Jiang et al., 2012).

2.4 Terpenoid Synergy with Other Defense Mechanisms

Terpenoids do not act in isolation; they often work in concert with other defense mechanisms to provide a multi-layered approach to pathogen resistance. For example, terpenoids can interact synergistically with phenolic compounds, alkaloids, and flavonoids to enhance the plant's overall defense capacity. The combination of terpenoids with other secondary metabolites creates a hostile environment for pathogens, making it more difficult for them to overcome the plant's defenses.

2.4.1 Synergistic Effects with Phenolics

Phenolic compounds, such as flavonoids and tannins, are well-known for their antimicrobial properties. When combined with terpenoids, these compounds can enhance the antimicrobial efficacy of the defense response. Studies have shown that the combination of terpenoids like thymol and carvacrol with phenolics like quercetin results in a stronger inhibition of bacterial growth compared to either compound alone (Kamatou & Viljoen, 2010). This synergistic effect is particularly useful in controlling bacterial pathogens like *Escherichia coli* and *Salmonella* spp., which are known for developing resistance to single-compound treatments.

2.4.2 Terpenoid-Flavonoid Complexes in Defense

Flavonoids, a class of phenolic compounds, often form complexes with terpenoids, particularly in the outer tissues of plants such as leaves and fruits. These complexes serve as a physical and chemical barrier to pathogen invasion. The terpenoid-flavonoid interaction enhances the defense properties of both compounds, making it more difficult for pathogens to penetrate plant tissues or to neutralize the chemical defenses of the plant (Wink, 2010). For instance, the combination of flavonoids like rutin and terpenoids like menthol has been shown to inhibit the growth of fungal pathogens that attack fruit crops, thereby extending the shelf life of harvested fruits (Bakkali et al., 2008).

2.5 Future Directions in Terpenoid-Based Plant Defense Strategies

The study of terpenoids as plant defense compounds is a rapidly growing field with significant implications for sustainable agriculture. With increasing interest in reducing the reliance on synthetic chemical pesticides, terpenoids offer a natural alternative for controlling plant pathogens. Advances in genetic engineering and synthetic biology have enabled researchers to enhance the production of terpenoids in crop plants, making them more resistant to pathogens (Zerbe & Bohlmann, 2015).

2.5.1 Engineering Plants for Enhanced Terpenoid Production

Genetic engineering techniques, such as CRISPR-Cas9 and RNA interference (RNAi), have been used to manipulate the biosynthetic pathways responsible for terpenoid production. By overexpressing genes involved in the MVA and MEP pathways, researchers have been able to increase the levels of specific terpenoids, such as β-caryophyllene and limonene, in plants like maize and tomato. These engineered plants exhibit enhanced resistance to both herbivores and microbial pathogens, making them more resilient in agricultural settings (Gao et al., 2018).

2.5.2 Synthetic Biology Approaches

In addition to genetic engineering, synthetic biology approaches have been developed to produce terpenoids in microbial systems, such as yeast and bacteria. These microbial "factories" can be engineered to produce high-value terpenoids, such as artemisinin, on a large scale, providing a sustainable and cost-effective alternative to traditional plant extraction methods (Paddon & Keasling, 2014). The use of synthetic biology to produce terpenoids has significant potential for improving crop resistance to pathogens and for developing new therapeutic agents for human health.

3. Ecological Roles of Terpenoids in Plant-Pathogen Interactions

Terpenoids are among the most diverse and ecologically significant plant secondary metabolites, playing crucial roles in mediating interactions between plants and their environment. These interactions involve a wide range of ecological players, including herbivores, pathogens, beneficial microbes, and even neighboring plants. Terpenoids are central to the defense strategies that plants employ to mitigate attacks from pathogens, while also acting as signaling molecules that influence the behavior of other organisms within the ecosystem. The ecological roles of terpenoids in plant-pathogen interactions are multifaceted and include direct defense through antimicrobial activity, as well as indirect defense by recruiting beneficial organisms or by priming plant defenses. Additionally, terpenoids are involved in the plant's ability to signal to neighboring plants or the community at large, helping to establish resistance mechanisms across plant populations.

3.1 Direct Defense: Antimicrobial and Antifungal Roles of Terpenoids

Terpenoids serve as key defensive compounds in plants, providing a direct line of protection against microbial pathogens. These compounds are produced constitutively in some plant tissues or are synthesized de novo in response to pathogen attack. The antimicrobial and antifungal properties of terpenoids are well documented, with many terpenoids inhibiting the growth and reproduction of bacteria, fungi, and viruses.

3.1.1 Terpenoids in Preventing Pathogen Invasion

One of the primary functions of terpenoids is to create a chemical barrier that prevents pathogen invasion. Terpenoids, such as monoterpenes and sesquiterpenes, are often stored in specialized plant structures, such as trichomes, resin ducts, or oil glands, where they are ready to be released when a pathogen attempts to penetrate the plant tissue (Gershenzon & Dudareva, 2007). For example, in coniferous trees, the resin containing terpenoids such as α-pinene and β-pinene is produced in response to wounds or pathogen invasion. This resin acts as a chemical sealant that traps pathogens and prevents them from spreading further into the plant (Keeling & Bohlmann, 2006). In addition to physical entrapment, terpenoids also exhibit strong antimicrobial properties that actively inhibit pathogen growth. Monoterpenes like limonene, found in the essential oils of citrus plants, disrupt the cellular membranes of fungal and bacterial pathogens, leading to cell lysis and death (Pinto et al., 2009). Similarly, sesquiterpenes such as caryophyllene are known to inhibit the germination and growth of fungal pathogens like *Botrytis cinerea*, a common cause of gray mold in crops (Dudareva et al., 2006).

3.1.2 Antifungal Activities of Terpenoids

Fungal pathogens pose a significant threat to agricultural productivity, and plants have evolved to produce specific terpenoids that target these organisms. One well-known antifungal terpenoid is artemisinin, a sesquiterpene lactone produced by *Artemisia annua*. Artemisinin not only acts as a potent antifungal agent but also plays a broader role in plant defense against both pathogens and herbivores (Tu, 2016). Studies have shown that artemisinin disrupts the development of fungal cell walls and inhibits spore germination, effectively protecting the plant from fungal colonization (Li et al., 2007). Other terpenoids, such as diterpenes and triterpenes, also exhibit strong antifungal properties. For instance, the diterpene kaurene is produced in response to fungal infection and acts by inhibiting fungal growth and reproduction. Kaurene derivatives are known to accumulate at the site of infection, providing localized defense against a wide range of fungal pathogens, including *Fusarium* and *Colletotrichum* species (Schmelz et al., 2014). Similarly, triterpenoids like oleanolic acid and ursolic acid, which are commonly found in the waxy cuticle of leaves and fruits, have been shown to prevent fungal spore germination and hyphal growth, providing an additional layer of defense (Jiang et al., 2012).

3.2 Indirect Defense: Recruitment of Beneficial Organisms

In addition to their direct antimicrobial roles, terpenoids also function as important ecological signals that recruit beneficial organisms to assist in plant defense. Plants emit volatile terpenoids in response to herbivore or pathogen attack, which can attract natural enemies of herbivores, such as parasitic wasps, or beneficial microbes that enhance plant resistance to pathogens.

3.2.1 Terpenoid Signaling to Beneficial Insects

One of the most well-known examples of terpenoid-mediated indirect defense is the recruitment of parasitic wasps by maize plants. When maize roots are attacked by herbivorous insects such as the corn rootworm, the plant releases volatile terpenoids like β-caryophyllene into the soil. This compound serves as a chemical signal that attracts parasitic nematodes, which attack and kill the rootworm larvae, thus protecting the maize plant from further damage (Rasmann et al., 2005). This interaction demonstrates the ecological role of terpenoids in fostering mutualistic relationships between plants and beneficial organisms, effectively enhancing the plant's defense against herbivores and pathogens alike. Similarly, volatile terpenoids released by damaged leaves can attract predatory insects such as ladybugs or predatory mites, which help to control populations of herbivorous pests. The terpenoid compound linalool, for example, is emitted by several plant species in response to herbivore feeding and has been shown to attract parasitic wasps that lay their eggs inside herbivorous caterpillars (Degenhardt et al., 2003). These parasitic wasps help reduce the population of herbivores, indirectly protecting the plant from further damage and decreasing the likelihood of secondary infections by pathogens.

3.2.2 Terpenoid-Induced Recruitment of Beneficial Microbes

Terpenoids also play a role in recruiting beneficial microbes, such as mycorrhizal fungi and plant growth-promoting rhizobacteria (PGPR), which contribute to plant health and enhance resistance to pathogens. For example, certain sesquiterpenes are known to attract mycorrhizal fungi, which form symbiotic associations with plant roots. These fungi help plants by improving nutrient uptake and increasing resistance to soil-borne pathogens (Akiyama et al., 2005). The terpenoid signal strigolactone is released by plant roots to stimulate the growth and branching of mycorrhizal fungal hyphae, facilitating the colonization of plant roots and enhancing the plant's overall resistance to pathogens (Bouwmeester et al., 2007). In addition to mycorrhizal recruitment, volatile terpenoids like β-caryophyllene have been shown to attract beneficial rhizobacteria that induce systemic resistance in plants. These rhizobacteria colonize plant roots and trigger a systemic immune response throughout the plant, helping to protect against both above-ground and below-ground pathogens (Pieterse et al., 2014). This form of indirect defense illustrates how terpenoids not only deter pathogens directly but also play a key role in establishing beneficial microbial partnerships that enhance plant immunity.

3.3 Terpenoids as Signaling Molecules in Plant-Plant Interactions

In addition to mediating interactions between plants and beneficial organisms, terpenoids also serve as important signaling molecules in plant-plant communication. Plants that are under attack by pathogens or herbivores often release volatile terpenoids that can be detected by neighboring plants, priming them to activate their own defense mechanisms. This form of interplant signaling is an essential component of plant ecology, helping to coordinate defense strategies within plant communities.

3.3.1 Airborne Terpenoid Signals

Terpenoids such as methyl jasmonate and green leaf volatiles (GLVs) are released into the air by damaged plants and serve as signals to neighboring plants. When a plant is attacked by herbivores or pathogens, it emits these volatile terpenoids, which can be detected by nearby plants, triggering the activation of defense-related genes. This process, known as "priming," allows neighboring plants to enhance their resistance to potential attacks, even before they are directly affected by pathogens or herbivores (Farmer & Ryan, 1990). One example of airborne terpenoid signaling is the emission of GLVs by tomato plants in response to insect herbivory. These volatiles are detected by neighboring tomato plants, which then activate the expression of defense genes involved in the production of proteinase inhibitors, compounds that deter herbivores by interfering with their digestive enzymes (Arimura et al., 2009). In this way, terpenoid signals not only help the attacked plant but also contribute to the overall defense of the plant community.

3.3.2 Root-Exuded Terpenoids in Below-Ground Communication

Terpenoids also play a role in below-ground signaling between plants. Certain terpenoids are released from plant roots and act as chemical signals in the rhizosphere, influencing the behavior of neighboring plants and microbes. For instance, strigolactones, a class of terpenoids exuded by plant roots, are known to influence the germination of parasitic plant seeds and the growth of neighboring plants (Bouwmeester et al., 2007). In parasitic plants such as *Striga* spp., root-exuded strigolactones from host plants stimulate seed germination, allowing the parasite to locate its host and establish a connection. However, plants can also use strigolactones to suppress the growth of neighboring plants, reducing competition for resources like light, water, and nutrients.

3.4 Terpenoids in the Context of Ecosystem Health

The ecological significance of terpenoids extends beyond individual plant-pathogen interactions. Terpenoids play an essential role in maintaining ecosystem health by mediating interactions between plants, pathogens, herbivores, and beneficial organisms. Their roles in direct defense, signaling, and recruitment of beneficial partners contribute to the resilience of plant communities and the overall stability of ecosystems. In agricultural systems, understanding the ecological roles of terpenoids can lead to the development of more sustainable pest and pathogen management strategies, reducing the need for synthetic pesticides and promoting biodiversity. In natural ecosystems, the release of terpenoids into the environment can shape the composition of plant and microbial communities, influencing the diversity and dynamics of ecosystems. Terpenoid emissions from plants can affect the behavior and distribution of herbivores and pathogens, as well as the composition of microbial communities in the soil. This highlights the far-reaching ecological impact of terpenoids, which extends from the molecular level to ecosystem-wide processes.

3.5 Future Directions in Research on Terpenoid Ecological Roles

Research on the ecological roles of terpenoids is a rapidly growing field, with new discoveries continually emerging about how these compounds mediate interactions between plants and their environment. Future research will likely focus on understanding the complex networks of signaling pathways involving terpenoids, as well as the evolutionary adaptations that have shaped the diversity of terpenoid-mediated defenses in plants.

3.5.1 Genetic Engineering of Terpenoid Pathways

One promising area of research involves the genetic manipulation of terpenoid biosynthetic pathways to enhance plant resistance to pathogens and herbivores. Advances in biotechnology have made it possible to increase the production of specific terpenoids in crop plants, thereby improving their resilience to biotic stressors. For example, by overexpressing genes involved in the production of monoterpenes and sesquiterpenes, researchers have been able to develop plants with enhanced resistance to fungal pathogens and insect pests (Dudareva et al., 2013).

3.5.2 Impacts of Climate Change on Terpenoid-Mediated Interactions

Another important avenue of research is the study of how climate change affects terpenoid-mediated plant-pathogen interactions. Changes in temperature, atmospheric CO2 levels, and precipitation patterns are likely to influence the production and emission of terpenoids by plants, with potential consequences for plant-pathogen dynamics. Understanding how these changes will affect the ecological roles of terpenoids will be crucial for predicting the future resilience of plant communities in the face of global environmental change (Vickers et al., 2009).

4. Terpenoids in Crop Protection: Applications and Innovations

Terpenoids play a crucial role in the agricultural sector, particularly in the protection of crops from pests, pathogens, and environmental stressors. These naturally occurring compounds are part of plants' innate defense mechanisms and have been harnessed for use in crop protection, offering environmentally friendly alternatives to synthetic pesticides and chemical control methods. The increasing awareness of the negative environmental and health impacts of chemical pesticides has driven the search for safer, more sustainable agricultural practices. As a result, terpenoids have gained considerable attention for their application in integrated pest management (IPM), biocontrol strategies, and the development of resistant crop varieties through genetic engineering.

4.1 Terpenoids as Natural Pesticides

One of the most significant applications of terpenoids in crop protection is their use as natural pesticides. Unlike synthetic pesticides, which often pose risks to human health and the environment, terpenoids offer a safer, biodegradable alternative. Terpenoids are effective against a wide range of agricultural pests, including insects, fungi, bacteria, and nematodes, making them versatile tools in pest and pathogen management.

4.1.1 Insecticidal Properties of Terpenoids

Terpenoids have been widely studied for their insecticidal properties. Many plants produce volatile terpenoids that act as direct deterrents to insect herbivores by repelling them or by inhibiting their growth and reproduction. For instance, limonene, a monoterpene found in citrus fruits, is a potent insect repellent that has been used to manage pests such as aphids, whiteflies, and mosquitoes (Isman, 2006). Limonene works by disrupting the nervous systems of insects, leading to paralysis and eventual death. Its effectiveness against a broad spectrum of insect pests, combined with its low toxicity to humans, has made limonene a valuable component of natural insecticides used in organic farming. Another example is azadirachtin, a terpenoid derived from the neem tree (*Azadirachta indica*). Azadirachtin has strong insecticidal properties and functions as a growth regulator and feeding inhibitor for a variety of insect pests, including locusts, caterpillars, and beetles (Mordue & Nisbet, 2000). Azadirachtin interferes with the hormonal systems of insects, preventing molting and reproduction, and has been widely adopted as a natural pesticide in both conventional and organic farming systems. Its broad-spectrum activity, low environmental persistence, and minimal impact on non-target organisms make azadirachtin an important tool in sustainable agriculture.

4.1.2 Antifungal and Antibacterial Applications

In addition to their insecticidal properties, many terpenoids exhibit strong antifungal and antibacterial activity, making them valuable for controlling plant pathogens. Essential oils rich in terpenoids, such as thymol, carvacrol, and eugenol, have been shown to inhibit the growth of a variety of plant pathogens, including *Botrytis cinerea*, *Fusarium oxysporum*, and *Pseudomonas syringae* (Pinto et al., 2009). These terpenoids disrupt the cell membranes of fungi and bacteria, leading to cell lysis and death. Their use as natural fungicides has gained popularity in the management of post-harvest diseases, particularly in fruit and vegetable crops. Thymol, a monoterpene found in thyme (*Thymus vulgaris*), is particularly effective against fungal pathogens and has been used to control fungal diseases such as powdery mildew and gray mold in crops like grapes, strawberries, and tomatoes (Ben-Jabeur et al., 2015). Thymol's antifungal action works by disrupting the fungal cell wall and inhibiting spore germination. Carvacrol, another monoterpene found in oregano (*Origanum vulgare*), has similar antifungal and antibacterial properties and is often used in conjunction with thymol in biopesticide formulations.

4.1.3 Terpenoids as Nematocides

Nematodes, particularly root-knot nematodes (*Meloidogyne spp.*), pose a significant threat to many crops by damaging root systems and reducing nutrient uptake. Terpenoids are effective in managing nematode infestations. For instance, the sesquiterpene β-caryophyllene, emitted by maize roots, has been shown to attract entomopathogenic nematodes, which prey on root-feeding insect larvae (Rasmann et al., 2005). This form of indirect defense not only protects the plant from herbivores but also helps reduce nematode populations. Additionally, the essential oils of certain plants, such as marigold (*Tagetes spp.*), are rich in terpenoids like α-terthienyl, which exhibit nematocidal activity.

These oils are often used in crop rotation systems to suppress nematode populations in the soil. The use of terpenoid-based nematocides offers a sustainable alternative to chemical nematocides, which are often harmful to beneficial soil organisms and can lead to long-term soil degradation.

4.2 Terpenoids in Biocontrol Strategies

Terpenoids play a central role in biological control (biocontrol) strategies, where natural enemies of pests and pathogens are employed to manage agricultural threats. Biocontrol is a cornerstone of integrated pest management (IPM), and terpenoids contribute to the success of these strategies by attracting beneficial organisms that help control pest populations.

4.2.1 Terpenoid-Mediated Attraction of Natural Enemies

Many plants release volatile terpenoids in response to herbivore feeding, which attracts natural enemies such as predatory insects and parasitic wasps. This ecological interaction is crucial for reducing herbivore populations and preventing crop damage. One well-studied example is the release of β-caryophyllene by maize roots in response to insect herbivory. β-caryophyllene attracts entomopathogenic nematodes, which attack and kill root-feeding insect larvae, providing an indirect defense for the plant (Rasmann et al., 2005). Another example is the emission of the monoterpene linalool by plants such as cotton and cucumber. Linalool is released in response to caterpillar feeding and attracts parasitic wasps that lay their eggs inside the caterpillars. These parasitic wasps help control caterpillar populations, thereby reducing the damage caused to the crops (Degenhardt et al., 2003). This terpenoid-mediated tritrophic interaction demonstrates how plants can use natural chemical cues to recruit beneficial organisms as part of their defense strategy.

4.2.2 Terpenoids in Fungal Biocontrol

Terpenoids are also involved in fungal biocontrol strategies, where beneficial fungi are used to control plant pathogens. Certain terpenoids produced by plants or fungal biocontrol agents inhibit the growth of pathogenic fungi and promote the colonization of beneficial fungi. For example, the triterpenoid oleanolic acid, found in olive leaves, has been shown to inhibit the growth of pathogenic fungi such as *Verticillium dahliae*, while promoting the growth of beneficial fungi like *Trichoderma harzianum* (Benitez et al., 2004). *Trichoderma* species are commonly used in biocontrol programs to suppress soil-borne pathogens and improve plant health by enhancing nutrient uptake and stimulating plant growth. Similarly, volatile terpenoids produced by beneficial fungi, such as *Trichoderma* and *Beauveria bassiana*, can inhibit the growth of plant pathogens. These beneficial fungi colonize the root systems of plants, creating a protective barrier that prevents the establishment of pathogenic fungi. The use of terpenoid-producing biocontrol agents is an effective strategy for reducing the reliance on chemical fungicides and promoting sustainable crop protection.

4.3 Innovations in Genetic Engineering for Terpenoid Production

Advancements in genetic engineering have opened new avenues for enhancing the production of terpenoids in crop plants, leading to improved resistance against pests and pathogens. By manipulating the biosynthetic pathways responsible for terpenoid production, researchers have been able to increase the levels of specific terpenoids in crops, enhancing their defense capabilities.

4.3.1 Overexpression of Terpenoid Biosynthetic Genes

One approach to increasing terpenoid production in crops is the overexpression of key enzymes involved in terpenoid biosynthesis, such as terpene synthases. Terpene synthases are responsible for the conversion of precursor molecules into a wide variety of terpenoids. By introducing genes encoding for terpene synthases into crop plants, researchers have been able to boost the production of terpenoids that confer resistance to pests and pathogens. For instance, overexpression of the gene encoding for β-caryophyllene synthase in maize has been shown to increase the production of β-caryophyllene in maize roots, enhancing the plant's ability to attract entomopathogenic nematodes and protect against root-feeding pests (Gao et al., 2018). Similarly, the overexpression of genes involved in the biosynthesis of monoterpenes like limonene and menthol has been explored in crops such as tomato and tobacco, with promising results in terms of increased resistance to insect herbivores and fungal pathogens (Aharoni et al., 2003).

4.3.2 Metabolic Engineering for Terpenoid Pathway Optimization

In addition to overexpression strategies, metabolic engineering techniques have been developed to optimize the entire terpenoid biosynthetic pathway in crops. This involves modifying multiple steps in the pathway to enhance the flux of precursor molecules toward the production of specific terpenoids. For example, the metabolic engineering of the methylerythritol phosphate (MEP) pathway in tomatoes has been used to increase the production of geraniol, a monoterpene with strong antifungal properties (Kumar et al., 2016). Metabolic engineering approaches also involve the introduction of regulatory elements, such as transcription factors, that control the expression of terpenoid biosynthetic genes. By manipulating these regulatory elements, researchers can fine-tune the production of terpenoids in response to environmental cues or pathogen attack, creating crops that are more resilient to biotic stress.

4.3.3 Synthetic Biology Approaches for Terpenoid Production

The emerging field of synthetic biology has provided new tools for the large-scale production of terpenoids in microbial systems. By engineering yeast and bacteria to produce plant terpenoids, researchers have developed microbial "factories" that can produce terpenoids in a cost-effective and sustainable manner. This approach has been particularly successful in the production of high-value terpenoids, such as artemisinin, which is used as an antimalarial drug (Paddon & Keasling, 2014). Synthetic biology approaches are also being applied to develop terpenoid-based biopesticides. For instance, by engineering microbial systems to produce insecticidal terpenoids like limonene and caryophyllene, researchers have created microbial strains that can be applied to crops as biological pesticides, offering a sustainable alternative to chemical pesticides.

4.4 Future Directions in Terpenoid-Based Crop Protection

The future of terpenoid-based crop protection lies in the continued exploration of their ecological roles, the development of new biocontrol strategies, and the advancement of genetic engineering and synthetic biology techniques. The use of terpenoids in crop protection is aligned with the goals of sustainable agriculture, reducing the reliance on chemical inputs and promoting biodiversity.

4.4.1 Expanding the Use of Terpenoids in IPM

Integrated Pest Management (IPM) is a holistic approach to crop protection that combines biological, cultural, and chemical control methods to manage pest populations. Terpenoids are ideally suited for use in IPM programs due to their natural origin, low toxicity, and multifunctional properties. Expanding the use of terpenoid-based biopesticides and biocontrol agents in IPM systems will help reduce the environmental impact of agriculture while maintaining crop yields.

4.4.2 Breeding for Enhanced Terpenoid Production

The development of crop varieties with enhanced terpenoid production is another promising avenue for future research. Traditional breeding techniques, combined with modern genetic engineering methods, can be used to select crops that naturally produce higher levels of terpenoids, improving their resistance to pests and pathogens. Breeding programs focused on enhancing the production of specific terpenoids, such as monoterpenes and sesquiterpenes, will contribute to the development of more resilient crop varieties.

5. Terpenoid Synergy with Other Plant Metabolites in Disease Management

Terpenoids are one of the most structurally diverse classes of secondary metabolites in plants, playing vital roles in defense against pathogens, herbivores, and environmental stressors. However, their effectiveness in disease management is often enhanced when they work synergistically with other plant metabolites. These interactions can lead to the development of stronger and more effective defense mechanisms that improve plant resistance to a wide range of pathogens. The combined action of terpenoids with phenolics, alkaloids, flavonoids, and other secondary metabolites can enhance the antimicrobial properties of plants, strengthen the plant's structural defenses, and modulate various signaling pathways involved in disease resistance.

5.1 Synergy Between Terpenoids and Phenolic Compounds

Phenolic compounds are another major class of plant secondary metabolites, known for their potent antioxidant, antimicrobial, and anti-inflammatory properties. The combination of terpenoids with phenolic compounds often results in synergistic effects that enhance the overall defense response of plants to pathogens. Phenolics, such as tannins, flavonoids, and lignins, are widely distributed in plants and play a crucial role in protecting plants from biotic stress.

5.1.1 Terpenoid-Phenolic Complexes in Plant Defense

Terpenoids and phenolics often work together to reinforce plant cell walls, making it more difficult for pathogens to invade. Phenolic compounds like lignins are essential for strengthening plant cell walls, while terpenoids provide an additional chemical barrier by disrupting the cell membranes of invading pathogens. In many cases, terpenoids act as signaling molecules that activate the biosynthesis of phenolic compounds in response to pathogen attack, creating a multi-layered defense system. For example, the interaction between terpenoids like thymol and phenolics such as quercetin has been shown to enhance the antimicrobial efficacy against bacterial pathogens like *Escherichia coli* and *Salmonella* spp. (Burt, 2004). The combined action of these compounds leads to greater disruption of bacterial cell membranes, increasing permeability and resulting in bacterial cell death. This synergy makes the plant's defense system more robust and less susceptible to pathogen invasion.

5.1.2 Enhanced Antimicrobial Properties Through Terpenoid-Phenolic Synergy

The synergy between terpenoids and phenolic compounds also amplifies their antimicrobial properties. For instance, the monoterpene thymol, when combined with phenolic acids like ferulic acid, exhibits significantly enhanced antifungal activity against pathogens such as *Fusarium oxysporum* and *Botrytis cinerea* (Tian et al., 2011). These pathogens are responsible for causing wilt diseases and fruit rots in a wide range of crops, including tomatoes, strawberries, and grapes. The terpenoid-phenolic combination disrupts fungal cell walls and interferes with fungal metabolism, reducing spore germination and hyphal growth. Similarly, flavonoids such as quercetin and kaempferol, when combined with terpenoids like menthol, show enhanced antibacterial properties, making them more effective against a broad spectrum of bacterial pathogens (Kamatou & Viljoen, 2010). This synergistic relationship not only enhances the plant's natural defenses but also opens new avenues for developing plant-based antimicrobial treatments for use in agriculture and medicine.

5.2 Synergy Between Terpenoids and Alkaloids

Alkaloids, another group of plant secondary metabolites, are known for their potent biological activity, including their role in plant defense against pathogens and herbivores. When terpenoids interact with alkaloids, the resulting synergy can lead to more effective disease management by enhancing the toxicity of these compounds against pathogens and pests.

5.2.1 Terpenoid-Alkaloid Interactions in Plant Immunity

Alkaloids like nicotine, morphine, and caffeine have long been recognized for their toxic effects on herbivores and pathogens. When combined with terpenoids, their antimicrobial properties are significantly enhanced. For example, studies have shown that the combination of terpenoids like limonene and alkaloids like nicotine increases the toxic effects against insect pests and microbial pathogens (Wink, 2010). This synergy enhances the plant's ability to repel herbivores while simultaneously preventing infection by bacterial or fungal pathogens. In many cases, terpenoids act as carriers or facilitators for alkaloid activity, increasing the permeability of microbial cell membranes and allowing alkaloids to penetrate and disrupt cellular processes more efficiently. This interaction also triggers the upregulation of genes involved in the plant's immune response, leading to a more rapid and robust defense against pathogen invasion.

5.2.2 Terpenoid-Alkaloid Synergy in Pathogen Resistance

The synergistic effects of terpenoids and alkaloids are particularly useful in the management of soil-borne pathogens, which are difficult to control using conventional chemical treatments. For instance, the combination of sesquiterpenes with alkaloids like berberine has been shown to inhibit the growth of *Phytophthora infestans*, a major pathogen responsible for potato late blight (Shah et al., 2005). This synergy not only reduces the virulence of the pathogen but also enhances the plant's systemic resistance to future infections. Moreover, terpenoids and alkaloids together can protect against viral pathogens. Some alkaloids, like caffeine, have been shown to inhibit the replication of plant viruses, while terpenoids enhance the penetration and systemic distribution of these alkaloids within plant tissues (Sallé et al., 2017). This interaction creates a more comprehensive defense mechanism that can prevent the spread of viral infections within crops.

5.3 Synergy Between Terpenoids and Flavonoids

Flavonoids, a subclass of phenolic compounds, are highly regarded for their antioxidant and antimicrobial properties. Their interaction with terpenoids leads to a synergistic effect that enhances plant resistance to various pathogens. Flavonoids play an essential role in regulating plant growth, development, and defense responses, and when combined with terpenoids, their antimicrobial activity becomes significantly more potent.

5.3.1 Terpenoid-Flavonoid Complexes in Antimicrobial Defense

Flavonoids such as kaempferol and quercetin, when combined with terpenoids like menthol or eucalyptol, show enhanced antimicrobial properties against bacterial and fungal pathogens. These flavonoid-terpenoid complexes are often found in the outer tissues of plants, such as leaves, stems, and fruits, where they act as both physical and chemical barriers to pathogen invasion (Bakkali et al., 2008). The combined action of these metabolites inhibits microbial colonization by disrupting cell walls and interfering with metabolic processes essential for pathogen survival. For example, the synergy between the monoterpene menthol and the flavonoid quercetin has been demonstrated to inhibit the growth of fungal pathogens responsible for fruit rot, such as *Penicillium* and *Aspergillus* species. These terpenoid-flavonoid complexes prevent spore germination and fungal hyphal development, effectively reducing the incidence of post-harvest diseases in fruits like apples, oranges, and grapes (Bakkali et al., 2008).

5.3.2 Flavonoid-Terpenoid Synergy in Antioxidant Defense

Flavonoids are also known for their antioxidant properties, which help plants manage oxidative stress caused by pathogen attacks, UV radiation, and environmental pollutants. The combination of terpenoids and flavonoids creates a more robust antioxidant defense system, protecting plant tissues from oxidative damage while simultaneously enhancing antimicrobial activity. For instance, the flavonoid kaempferol, when combined with the diterpene ginkgolide from *Ginkgo biloba*, has been shown to significantly reduce oxidative stress in plants exposed to environmental pollutants and pathogen-induced oxidative damage (Shah et al., 2005). This synergistic relationship not only boosts the plant's antioxidant capacity but also strengthens its defense against bacterial and fungal pathogens.

5.4 Terpenoids and Other Secondary Metabolites: Synergistic Defense Networks

Terpenoids also interact with other secondary metabolites, such as saponins, glucosinolates, and tannins, creating complex defense networks that enhance plant resilience to pathogens. These interactions are often highly specific and vary depending on the plant species, the type of pathogen, and environmental conditions.

5.4.1 Terpenoid-Saponin Synergy in Cell Membrane Disruption

Saponins are glycosides that exhibit strong antimicrobial and antifungal properties by disrupting the cell membranes of pathogens. When combined with terpenoids, such as triterpenoids like oleanolic acid, the antimicrobial efficacy of saponins is significantly enhanced (Jiang et al., 2012). Terpenoids increase the permeability of microbial cell membranes, allowing saponins to penetrate and disrupt essential cellular functions, leading to cell death. For example, the combination of saponins with oleanolic acid has been shown to effectively control fungal pathogens in crops like tomatoes and cucumbers. This synergistic interaction also helps prevent the development of fungal resistance, as the combined action of terpenoids and saponins targets multiple pathways within the pathogen, making it more difficult for the pathogen to adapt.

5.4.2 Terpenoids and Glucosinolates in Pest and Pathogen Management

Glucosinolates are sulfur-containing compounds found in cruciferous plants like broccoli, cabbage, and mustard. When combined with terpenoids, such as monoterpenes and sesquiterpenes, glucosinolates create a potent defense system against herbivores and microbial pathogens. The breakdown products of glucosinolates, such as isothiocyanates, work in conjunction with terpenoids to inhibit the growth of bacterial and fungal pathogens, while also deterring herbivores through their bitter taste and toxic effects (Fahey et al., 2001). The interaction between terpenoids and glucosinolates is particularly effective in managing soil-borne pathogens, such as *Rhizoctonia solani* and *Pythium* species, which cause root rot in crops like carrots and radishes. The combination of terpenoids and glucosinolates reduces the need for synthetic pesticides, making it a valuable tool in organic and sustainable farming practices.

5.5 Applications of Terpenoid Synergy in Sustainable Agriculture

The synergistic interactions between terpenoids and other plant metabolites offer significant potential for the development of new crop protection strategies that are both effective and environmentally friendly. By harnessing the natural defense mechanisms of plants, farmers can reduce their reliance on synthetic pesticides and fungicides, promoting more sustainable agricultural practices.

5.5.1 Terpenoid-Based Biopesticides and Biofungicides

The development of terpenoid-based biopesticides and biofungicides is a promising area of research in sustainable agriculture. These natural products, when combined with other plant metabolites, offer a potent alternative to chemical pesticides. For example, biopesticides containing combinations of terpenoids, flavonoids, and alkaloids have been shown to effectively control insect pests and fungal pathogens in crops like corn, soybeans, and wheat (Burt, 2004). Moreover, these natural products are biodegradable and pose minimal risk to non-target organisms, including beneficial insects, pollinators, and soil microbes. As a result, terpenoid-based biopesticides are becoming increasingly popular in organic farming systems, where the use of synthetic chemicals is restricted.

5.5.2 Enhancing Crop Resilience Through Breeding and Genetic Engineering

The synergistic effects of terpenoids with other plant metabolites can also be harnessed through traditional breeding and genetic engineering approaches. By selecting for crop varieties that naturally produce higher levels of terpenoids, flavonoids, and alkaloids, plant breeders can develop crops that are more resistant to pests and diseases. In addition, advances in genetic engineering have made it possible to manipulate the biosynthetic pathways of terpenoids and other metabolites, creating plants that produce higher levels of these compounds in response to pathogen attack (Aharoni et al., 2003). This approach has already been successfully applied to crops like tomatoes, tobacco, and maize, where the overexpression of terpenoid and flavonoid biosynthetic genes has led to increased resistance to fungal and bacterial pathogens (Gao et al., 2018). As research in this area continues to advance, the potential for creating crops that are more resilient to environmental stressors and biotic threats will expand, contributing to more sustainable and productive agricultural systems.

Conclusion

Terpenoids represent one of the largest and most diverse classes of natural compounds, playing a crucial role in the survival and protection of plants. Their structural complexity and diversity are reflected in their classification into monoterpenes, sesquiterpenes, diterpenes, and other categories, each defined by the number of isoprene units they contain. These compounds are synthesized through two major biosynthetic pathways: the mevalonate (MVA) pathway and the methylerythritol phosphate (MEP) pathway. The structural variations of terpenoids contribute to their versatile biological activities, making them essential components of plant defense systems and vital to crop protection strategies.

The role of terpenoids in plant defense mechanisms is profound, offering plants a powerful arsenal against pathogens, herbivores, and environmental stressors. Terpenoids function through direct toxicity to pathogens, disruption of microbial growth, and interference with pathogen signaling processes. They are also involved in the induction of systemic acquired resistance (SAR) in plants, a long-term defense response that strengthens the plant's ability to combat a variety of pathogens. The bioactivity of terpenoids is often tailored to specific threats, providing plants with a dynamic and adaptable defense system. Ecologically, terpenoids serve as key mediators in plant-pathogen interactions, influencing both plant health and ecological balance. These compounds play a vital role in plant communication with their environment, acting as chemical signals to attract beneficial organisms such as pollinators and natural enemies of herbivores. Additionally, terpenoids can inhibit pathogen colonization by altering the microbial community around the plant. Their ecological significance extends beyond direct pathogen defense, influencing plant interactions with insects, microbes, and neighboring plants in complex ecosystems.

The application of terpenoids in crop protection has gained significant attention due to their natural origin and environmentally friendly properties. Terpenoids have been utilized in innovative crop protection strategies, including biopesticides, plant growth regulators, and elicitors of plant defense responses. The development of synthetic biology techniques has further expanded the potential of terpenoids in agriculture, allowing for the production of specific terpenoids in greater quantities and their incorporation into integrated pest management programs. Terpenoid-based biopesticides, for instance, offer a promising alternative to chemical pesticides, reducing the risk of environmental contamination and the development of resistance in pests.

Moreover, the synergy between terpenoids and other plant metabolites enhances their efficacy in disease management. Terpenoids often interact with alkaloids, phenolics, and other secondary metabolites, resulting in additive or synergistic effects that improve plant defense mechanisms. These interactions can enhance the potency of plant defenses against pathogens, leading to more effective control of diseases. The ability of terpenoids to work in concert with other compounds highlights the complexity and sophistication of plant chemical defenses and opens new avenues for developing more potent crop protection strategies. Despite the promising potential of terpenoids in plant defense and crop protection, there remain several challenges in their research and utilization. The structural diversity of terpenoids complicates their study, as does the variability in their biosynthesis and regulation across different plant species. Additionally, the scalability of terpenoid production for commercial applications poses technical challenges, as many terpenoids are produced in low quantities and are difficult to extract. Advances in genetic engineering and synthetic biology hold promise for overcoming these limitations, but more research is needed to optimize these techniques and ensure their cost-effectiveness.

Looking ahead, the future of terpenoid research and utilization in agriculture and plant defense appears bright. The ongoing development of bioengineering technologies, such as CRISPR and metabolic pathway engineering, offers the potential to enhance terpenoid production and tailor these compounds for specific agricultural applications. Moreover, as global agricultural systems face increasing pressure from climate change and evolving pest populations, the demand for sustainable, natural crop protection solutions like terpenoids is expected to grow. Continued research into the ecological roles of terpenoids, their interactions with other metabolites, and their applications in modern agriculture will be crucial in harnessing their full potential for sustainable crop protection.

References

Aharoni, A., Giri, A. P., Verstappen, F. W. A., Bertea, C. M., Sevenier, R., Sun, Z., Jongsma, M. A., Schwab, W., & Bouwmeester, H. J. (2003). Terpenoid metabolism in wild-type and transgenic Arabidopsis plants. *The Plant Cell, 15*(12), 2866-2884. https://doi.org/10.1105/tpc.015883

Akiyama, K., Matsuzaki, K., & Hayashi, H. (2005). Plant sesquiterpenes induce hyphal branching in arbuscular mycorrhizal fungi. *Nature, 435*(7043), 824-827. https://doi.org/10.1038/nature03608

Arimura, G., Ozawa, R., Shimoda, T., Nishioka, T., Boland, W., & Takabayashi, J. (2009). Herbivory-induced volatiles elicit defence genes in lima bean leaves. *Nature, 406*(6795), 512-515. https://doi.org/10.1038/35020072

Bakkali, F., Averbeck, S., Averbeck, D., & Idaomar, M. (2008). Biological effects of essential oils – A review. *Food and Chemical Toxicology, 46*(2), 446-475. https://doi.org/10.1016/j.fct.2007.09.106

Benitez, T., Rincon, A. M., Limon, M. C., & Codon, A. C. (2004). Biocontrol mechanisms of *Trichoderma* strains. *International Microbiology, 7*(4), 249-260.

Ben-Jabeur, M., Ghabri, E., Myriam, M. K., & Hamada, W. (2015). Thyme essential oil as a defense tool against postharvest fungal diseases of citrus fruit. *Journal of Applied Microbiology, 119*(4), 1013-1022. https://doi.org/10.1111/jam.12905

Bouvier, F., Rahier, A., & Camara, B. (2005). Biogenesis, molecular regulation, and function of plant isoprenoids. *Progress in Lipid Research, 44*(6), 357-429. https://doi.org/10.1016/j.plipres.2005.09.002

Bouwmeester, H. J., Roux, C., Lopez-Raez, J. A., & Bécard, G. (2007). Rhizosphere communication of plants, parasitic plants, and AM fungi. *Trends in Plant Science, 12*(5), 224-230. https://doi.org/10.1016/j.tplants.2007.03.009

Burt, S. (2004). Essential oils: Their antibacterial properties and potential applications in foods – A review. *International Journal of Food Microbiology, 94*(3), 223-253. https://doi.org/10.1016/j.ijfoodmicro.2004.03.022

Cornish, K. (2017). Alternative natural rubber crops: Why should we care? *Technology and Innovation, 19*(4), 221-229. https://doi.org/10.21300/19.4.2017.221

Cragg, G. M., & Newman, D. J. (2005). Plants as a source of anti-cancer agents. *Journal of Ethnopharmacology, 100*(1-2), 72-79. https://doi.org/10.1016/j.jep.2005.05.011

Crowell, P. L. (1999). Prevention and therapy of cancer by dietary monoterpenes. *The Journal of Nutrition, 129*(3), 775S-778S. https://doi.org/10.1093/jn/129.3.775S

Degenhardt, J., Gershenzon, J., Baldwin, I. T., & Kessler, A. (2003). Attracting predatory arthropods using herbivore-induced plant volatiles: The future in pest control? *Journal of Plant Physiology, 146*(5), 711-717. https://doi.org/10.1093/jxb/erg083

Dudareva, N., Negre, F., Nagegowda, D. A., & Orlova, I. (2013). Plant volatiles: Recent advances and future perspectives. *Critical Reviews in Plant Sciences, 32*(5), 417-440. https://doi.org/10.1080/07352689.2013.755610

Dudareva, N., Pichersky, E., & Gershenzon, J. (2006). Biochemistry of plant volatiles. *Plant Physiology, 135*(4), 1893-1902. https://doi.org/10.1104/pp.104.049981

Durrant, W. E., & Dong, X. (2004). Systemic acquired resistance. *Annual Review of Phytopathology, 42*(1), 185-209. https://doi.org/10.1146/annurev.phyto.42.040803.140421

Fahey, J. W., Zalcmann, A. T., & Talalay, P. (2001). The chemical diversity and distribution of glucosinolates and isothiocyanates among plants. *Phytochemistry, 56*(1), 5-51. https://doi.org/10.1016/S0031-9422(00)00316-2

Farmer, E. E., & Ryan, C. A. (1990). Interplant communication: Airborne methyl jasmonate induces synthesis of proteinase inhibitors in plant leaves. *Proceedings of the National Academy of Sciences, 87*(19), 7713-7716. https://doi.org/10.1073/pnas.87.19.7713

Finkelstein, R. (2013). Abscisic acid synthesis and response. *The Arabidopsis Book, 11*, e0166. https://doi.org/10.1199/tab.0166

Fiore, C., Eisenhut, M., Krausse, R., Ragazzi, E., Pellati, D., Armanini, D., & Bielenberg, J. (2008). Antiviral effects of Glycyrrhiza species. *Phytotherapy Research, 22*(2), 141-148. https://doi.org/10.1002/ptr.2295

Friedman, M. (2006). Potato glycoalkaloids and metabolites: Roles in the plant and in the diet. *Journal of Agricultural and Food Chemistry, 54*(23), 8655-8681. https://doi.org/10.1021/jf061471t

Gao, Y., Honzatko, R. B., & Peters, R. J. (2018). Terpenoid synthase structures: A so far incomplete view of complex catalysis. *Natural Product Reports, 35*(5), 515-530. https://doi.org/10.1039/C8NP00014A

Gershenzon, J., & Dudareva, N. (2007). The function of terpene natural products in the natural world. *Nature Chemical Biology, 3*(7), 408-414. https://doi.org/10.1038/nchembio.2007.5

Isman, M. B. (2006). Botanical insecticides, deterrents, and repellents in modern agriculture and an increasingly regulated world. *Annual Review of Entomology, 51*, 45-66. https://doi.org/10.1146/annurev.ento.51.110104.151146

Jabra-Rizk, M. A., Shirtliff, M., James, C., & Meiller, T. F. (2006). Effect of farnesol on Staphylococcus aureus biofilm formation and antimicrobial susceptibility. *Antimicrobial Agents and Chemotherapy, 50*(4), 1463-1469. https://doi.org/10.1128/AAC.50.4.1463-1469.2006

Jiang, J. G., Zhu, L., & Zhang, Q. W. (2012). Pharmacological activities and clinical application of oleanolic acid and its derivatives: A mini review. *Archives of Pharmacal Research, 35*(1), 5-11. https://doi.org/10.1007/s12272-012-0117-y

Kamatou, G. P. P., & Viljoen, A. M. (2010). A review of the application and pharmacological properties of α-bisabolol and α-bisabolol-rich oils. *Journal of the American Oil Chemists' Society, 87*(1), 1-10. https://doi.org/10.1007/s11746-009-1483-3

Keeling, C. I., & Bohlmann, J. (2006). Diterpene resin acids in conifers. *Phytochemistry, 67*(22), 2415-2423. https://doi.org/10.1016/j.phytochem.2006.09.023

Kim, S. K., & Karadeniz, F. (2012). Biological importance and applications of squalene and squalane. *Advances in Food and Nutrition Research, 65*, 223-233. https://doi.org/10.1016/B978-0-12-416003-3.00014-7

Kumar, P., Lucini, L., Rouphael, Y., Cardarelli, M., Kalunke, R. M., & Colla, G. (2016). Insight into the role of grafting and arbuscular mycorrhiza on yield and metabolic profiling of greenhouse-grown tomato. *Journal of Agricultural and Food Chemistry, 63*(44), 10516-10528. https://doi.org/10.1021/acs.jafc.5b02868

Li, Y., Xu, Y., Jie, X., Liu, Y., Wu, Y., Chen, W., & Song, X. (2007). Artemisinin protects against hydrogen peroxide-induced injury in endothelial cells via the ERK1/2 pathway. *Biochemical and Biophysical Research Communications, 360*(4), 1087-1092. https://doi.org/10.1016/j.bbrc.2007.06.155

Maffei, M. E., Gertsch, J., & Appendino, G. (2011). Plant volatiles: Production, function and pharmacology. *Natural Product Reports, 28*(8), 1359-1380. https://doi.org/10.1039/c1np00021g

Mordue, A. J., & Nisbet, A. J. (2000). Azadirachtin from the neem tree *Azadirachta indica*: Its action against insects. *Anais da Sociedade Entomológica do Brasil, 29*(4), 615-632. https://doi.org/10.1590/S0301-80592000000400001

Nisar, N., Li, L., Lu, S., Khin, N. C., & Pogson, B. J. (2015). Carotenoid metabolism in plants. *Molecular Plant, 8*(1), 68-82. https://doi.org/10.1016/j.molp.2014.12.007

Paddon, C. J., & Keasling, J. D. (2014). Semi-synthetic artemisinin: A model for the use of synthetic biology in pharmaceutical development. *Nature Reviews Microbiology, 12*(5), 355-367. https://doi.org/10.1038/nrmicro3240

Pichersky, E., & Gershenzon, J. (2002). The formation and function of plant volatiles: Perfumes for pollinator attraction and defense. *Current Opinion in Plant Biology, 5*(3), 237-243. https://doi.org/10.1016/S1369-5266(02)00251-0

Pieterse, C. M., van der Does, D., Zamioudis, C., Leon-Reyes, A., & Van Wees, S. C. (2014). Hormonal modulation of plant immunity. *Annual Review of Cell and Developmental Biology, 28*(1), 489-521. https://doi.org/10.1146/annurev-cellbio-100913-013125

Pieterse, C. M., Zamioudis, C., Berendsen, R. L., Weller, D. M., Van Wees, S. C., & Bakker, P. A. (2014). Induced systemic resistance by beneficial microbes. *Annual Review of Phytopathology, 52*(1), 347-375. https://doi.org/10.1146/annurev-phyto-082712-102340

Pinto, E., Pina-Vaz, C., Salgueiro, L., Gonçalves, M. J., Costa-de-Oliveira, S., Cavaleiro, C., Palmeira,

A., Rodrigues, A. G., & Martinez-de-Oliveira, J. (2009). Antifungal activity of the essential oil of Thymus vulgaris on Candida, Aspergillus and dermatophyte species. *Journal of Medical Microbiology, 58*(11), 1454-1462. https://doi.org/10.1099/jmm.0.010538-0

Qureshi, N., Porter, J. W., & Peffley, D. M. (2018). The mevalonate pathway in plants. In D. Neidleman & A. I. Laskin (Eds.), *Advances in Applied Microbiology* (Vol. 32, pp. 217-243). Academic Press. https://doi.org/10.1016/S0065-2164(08)70122-1

Raguso, R. A. (2008). Wake up and smell the roses: The ecology and evolution of floral scent. *Annual Review of Ecology, Evolution, and Systematics, 39*, 549-569.

Ramakrishna, A., Ravishankar, G. A., & Srivastava, S. (2019). Secondary metabolites and plant defense responses to pathogenic microorganisms. In *Plant Biotechnology and Agriculture* (pp. 515-533). Academic Press.

Rao, A. V., & Rao, L. G. (2007). Carotenoids and human health. *Pharmacological Research, 55*(3), 207-216. https://doi.org/10.1016/j.phrs.2007.01.012

Rasmann, S., Köllner, T. G., Degenhardt, J., Hiltpold, I., Toepfer, S., Kuhlmann, U., Gershenzon, J., & Turlings, T. C. (2005). Recruitment of entomopathogenic nematodes by insect-damaged maize roots. *Nature, 434*(7034), 732-737. https://doi.org/10.1038/nature03451

Rohmer, M. (1999). The discovery of a mevalonate-independent pathway for isoprenoid biosynthesis in bacteria, algae, and higher plants. *Natural Product Reports, 16*(5), 565-574. https://doi.org/10.1039/A809199E

Saleem, M. (2009). Lupeol, a novel anti-inflammatory and anti-cancer dietary triterpene. *Cancer Letters, 285*(2), 109-115. https://doi.org/10.1016/j.canlet.2009.04.033

Schmelz, E. A., Kaplan, F., Huffaker, A., & Dafoe, N. J. (2014). Diterpenoid phytoalexins accumulate and enhance resistance in maize. *Plant Physiology, 166*(3), 1562-1572. https://doi.org/10.1104/pp.114.244251

Schmelz, E. A., Kaplan, F., Huffaker, A., & Dafoe, N. J. (2014). Diterpenoid phytoalexins accumulate and enhance resistance in maize. *Plant Physiology, 166*(3), 1562-1572.

Shah, A., Mahmood, S., & Rahman, S. (2005). In vitro antifungal activity of selected Pakistani medicinal plants. *Journal of Medicinal Plants Research, 15*(5), 101-106. https://doi.org/10.4314/jmpr.v15i5

Tholl, D. (2015). Terpene synthases and the regulation, diversity, and biological roles of terpene metabolism. *Current Opinion in Plant Biology, 25*, 55-61.

Tu, Y. (2016). Artemisinin: A gift from traditional Chinese medicine to the world (Nobel Lecture). *Angewandte Chemie International Edition, 55*(35), 10210-10226. https://doi.org/10.1002/anie.201601967

Verma, R. S., & Dubey, N. K. (2011). Prospects of botanical pesticides in sustainable agriculture. *Current Science, 100*(9), 1246-1250.

Vickers, C. E., Gershenzon, J., Lerdau, M. T., & Loreto, F. (2009). A unified mechanism of action for volatile isoprenoids in plant abiotic stress. *Nature Chemical Biology, 5*(5), 283-291. https://doi.org/10.1038/nchembio.158

Vranová, E., Coman, D., & Gruissem, W. (2012). Structure and dynamics of the isoprenoid pathway network. *Molecular Plant, 5*(2), 318-333. https://doi.org/10.1093/mp/sss015

Wink, M. (2010). Functions of plant secondary metabolites and their exploitation in biotechnology. *Annual Plant Reviews*, *39*, 20-30.

Yamaguchi, S. (2008). Gibberellin metabolism and its regulation. *Annual Review of Plant Biology, 59*(1), 225-251. https://doi.org/10.1146/annurev.arplant.59.032607.092804

Yang, L., Wen, K. S., Ruan, X., Zhao, Y. X., Wei, F., & Wang, Q. (2018). Response of plant secondary metabolites to environmental factors. *Molecules*, *23*(4), 762.

Zerbe, P., & Bohlmann, J. (2015). Plant diterpene synthases: Exploring modularity and metabolic diversity for bioengineering. *Trends in Biotechnology*, *33*(8), 419-428.

Chapter - 5

Phenolics and Flavonoids: Antioxidant and Antimicrobial Roles in Plant Health

Zuby Gohar Ansari[1], Susheel Kumar[2], Vinay Pratap Singh[3] and Pratima Raypa[4]

[1]B.M. College of Agriculture, Jaswadi Road Khandwa, Madhya Pradesh, India
[2]Plant Molecular Virology Laboratory, CPMB Division, CSIR-National Botanical Research Institute, Rana Pratap Marg, Lucknow, Uttar Pradesh, India
[3]Plant Physiology, College of Agriculture, Khurai, JNKVV, Jabalpur, M.P. India
[4]Department of Biochemistry, H.N.B. Garhwal University, Srinagar, Uttarakhand, India

Abstract

Phenolics and flavonoids are crucial plant secondary metabolites that play a vital role in plant health, primarily through their antioxidant and antimicrobial properties. These compounds are widely distributed across various plant species and are synthesized via the shikimate and phenylpropanoid pathways, contributing to a vast array of structurally diverse molecules. Phenolics, which include phenolic acids and tannins, and flavonoids, such as flavonols, flavones, and anthocyanins, are integral to plant defense systems. Their antioxidant properties enable them to neutralize reactive oxygen species (ROS) generated during environmental stresses, thus protecting plants from oxidative damage. By scavenging free radicals, phenolics and flavonoids help maintain cellular integrity and support the plant's metabolic processes during biotic and abiotic stress conditions. In addition to their antioxidant roles, these compounds exhibit potent antimicrobial activities, targeting a wide range of plant pathogens, including bacteria, fungi, and viruses. The antimicrobial mechanisms of phenolics and flavonoids include disruption of microbial cell membranes, inhibition of essential microbial enzymes, and interference with pathogen quorum sensing, which is crucial for coordinated pathogenicity. Furthermore, phenolics and flavonoids contribute to strengthening the plant's physical barriers, such as cell walls, making them more resistant to pathogen invasion. The synergy between their antioxidant and antimicrobial functions makes them essential components in enhancing plant resilience to environmental challenges. In agriculture, these compounds are gaining attention as natural alternatives to synthetic chemicals for sustainable crop protection. Their ability to mitigate oxidative stress and inhibit pathogen growth offers the potential

DOI: 10.1201/9781003717454-5

for developing eco-friendly pest and disease management strategies. Current research also explores the application of phenolics and flavonoids in bioengineering efforts aimed at increasing crop resistance to diseases and environmental stressors. Despite the growing interest in their use, challenges remain in the large-scale production and efficient application of these metabolites in agriculture. Advances in biotechnology, such as metabolic engineering, offer promising avenues to overcome these limitations and optimize the use of phenolics and flavonoids for plant health and sustainable crop protection. Their dual role in antioxidant defense and antimicrobial action underpins their significance in maintaining plant health and ensuring resilience against stress, making them valuable tools in both natural plant ecosystems and agricultural systems. These properties place phenolics and flavonoids at the forefront of innovative solutions for environmentally sustainable agricultural practices, positioning them as critical players in disease management and crop protection.

Keywords: Phenolics, flavonoids, antioxidants, antimicrobial, plant health, sustainable agriculture.

Introduction

Phenolics and flavonoids are among the most important classes of plant secondary metabolites that contribute significantly to plant growth, development, and defense mechanisms. These compounds are widely distributed in the plant kingdom and are known for their roles in protecting plants from various biotic and abiotic stressors. Phenolics, including phenolic acids, tannins, lignins, and flavonoids, a subclass of phenolics, exhibit structural diversity that enables them to participate in multiple biochemical interactions, offering plants a robust defense system (Lattanzio et al., 2006). They serve as antioxidants, antimicrobial agents, and signaling molecules, playing vital roles in maintaining plant health and resilience.

The biosynthesis of phenolics and flavonoids occurs through complex enzymatic pathways derived from the shikimate, phenylpropanoid, and polyketide pathways. These pathways lead to the production of structurally diverse compounds, each with specific biological functions. Phenolics are derived from aromatic amino acids such as phenylalanine and tyrosine, which are converted into a wide array of phenolic acids and polyphenols (Cheynier, 2012). Flavonoids, characterized by their C6-C3-C6 carbon skeleton, are synthesized through the flavonoid biosynthetic pathway, starting with the conversion of phenylalanine to naringenin (Pourcel et al., 2007). This structural diversity allows flavonoids and phenolics to participate in plant defense mechanisms against pathogens, herbivores, and environmental stressors.

Phenolics and flavonoids play a critical role in mitigating oxidative stress within plants. Reactive oxygen species (ROS) are produced in response to various environmental factors such as drought, UV radiation, and pathogen attack, which can lead to oxidative damage and impair plant growth. Both phenolics and flavonoids act as antioxidants, scavenging free radicals and reducing oxidative stress in plant cells (Gill & Tuteja, 2010). Flavonoids, in particular, are known for their ability to chelate metal ions, inhibit ROS formation, and enhance the plant's defense response (Agati et al., 2012). This antioxidant activity not only protects plants from cellular damage but also contributes to their overall stress tolerance. In addition to their antioxidant functions, phenolics and flavonoids exhibit strong antimicrobial properties, making them vital components of plant defense against pathogens. These compounds disrupt microbial membranes, inhibit enzyme activity, and interfere with the replication of pathogens, providing an effective chemical barrier against infection (Cowan, 1999). For example, flavonoids such as quercetin and kaempferol have been shown to inhibit bacterial growth and reduce fungal spore germination, making them potent antimicrobial agents (Dixon & Paiva, 1995). Furthermore, phenolics such as tannins and lignins strengthen plant cell walls, providing a physical barrier that reduces pathogen entry.

Phenolics and flavonoids also play crucial roles in the plant's response to environmental stress. In conditions of drought, salinity, or nutrient deficiency, these compounds accumulate in plant tissues, helping to mitigate the harmful effects of stress. Phenolics, such as lignins, contribute to cell wall rigidity, which helps plants maintain structural integrity under stress conditions (Moura et al., 2010). Flavonoids, on the other hand, act as UV protectants by absorbing harmful radiation and preventing cellular damage (Agati et al., 2012). The ability of these compounds to modulate plant responses to environmental stressors highlights their importance in enhancing plant resilience. In agricultural applications, phenolics and flavonoids are being explored for their potential to improve crop resistance to pests and diseases. Their natural antioxidant and antimicrobial properties make them valuable for developing eco-friendly, plant-based treatments that reduce reliance on chemical pesticides and enhance plant health (Zaynab et al., 2018). The synergistic effects of phenolics and flavonoids with other plant metabolites are also being studied to understand how these compounds can be harnessed to improve crop protection strategies in sustainable agriculture (Lattanzio et al., 2006). As research into phenolics and flavonoids continues to expand, it is becoming increasingly clear that these compounds play a fundamental role in plant health and defense. Their ability to function as antioxidants, antimicrobial agents, and modulators of plant stress responses underscores their significance in both natural plant ecosystems and agricultural systems.

1. Biosynthesis and Structural Diversity of Phenolics and Flavonoids

Phenolics and flavonoids are a large and diverse group of plant secondary metabolites that are synthesized through intricate biochemical pathways. These compounds play essential roles in plant growth, development, and defense mechanisms. The structural diversity of phenolics and flavonoids, arising from their complex biosynthetic pathways, contributes to their wide range of biological activities, including antioxidant, antimicrobial, and signaling functions.

1.1 Biosynthesis of Phenolics

The biosynthesis of phenolics is primarily derived from the shikimate and phenylpropanoid pathways, which convert simple aromatic amino acids, such as phenylalanine and tyrosine, into a wide array of phenolic compounds. The shikimate pathway generates aromatic amino acids, which are then used in the phenylpropanoid pathway to produce basic phenolic structures (Vogt, 2010). The phenylpropanoid pathway is responsible for synthesizing various classes of phenolics, including phenolic acids, coumarins, tannins, and lignins.

1.1.1 Shikimate Pathway

The shikimate pathway is central to the biosynthesis of aromatic amino acids in plants, which serve as precursors for many secondary metabolites, including phenolics. This pathway starts with the condensation of phosphoenolpyruvate (PEP) and erythrose-4-phosphate (E4P), leading to the formation of chorismate, a key intermediate in the production of phenylalanine, tyrosine, and tryptophan (Herrmann & Weaver, 1999). Phenylalanine, in particular, plays a critical role in the phenylpropanoid pathway, where it is converted into cinnamic acid by the action of phenylalanine ammonia-lyase (PAL).

1.1.2 Phenylpropanoid Pathway

The phenylpropanoid pathway is the primary route for the synthesis of phenolic compounds in plants. After the deamination of phenylalanine to cinnamic acid by PAL, a series of hydroxylation, methylation, and glycosylation reactions occurs, leading to the production of various phenolics. Cinnamic acid is converted into p-coumaric acid, which serves as a precursor for a wide range of phenolic compounds, including flavonoids, stilbenes, and lignins (Metsämuuronen & Sirén, 2019). The structural diversity of phenolics arises from modifications such as hydroxylation, methylation, and glycosylation at different positions on the aromatic ring. For example, hydroxycinnamic acids, such as ferulic acid and caffeic acid, differ in their degree of hydroxylation and methoxylation, leading to variations in their biological activities. These compounds are involved in plant cell wall strengthening, protection against ultraviolet (UV) radiation, and defense against pathogens.

1.2 Biosynthesis of Flavonoids

Flavonoids, a subclass of phenolics, are synthesized through the flavonoid biosynthetic pathway, which branches off from the phenylpropanoid pathway. The key precursor for flavonoid biosynthesis is p-coumaroyl-CoA, which combines with three molecules of malonyl-CoA to form chalcone, the central intermediate in the flavonoid pathway. The enzyme chalcone synthase (CHS) catalyzes this reaction, marking the first committed step in flavonoid biosynthesis (Winkel-Shirley, 2001).

1.2.1 Flavonoid Pathway

After the formation of chalcone, a series of enzymatic reactions, including cyclization, hydroxylation, and methylation, lead to the production of various classes of flavonoids. These include flavonols, flavones, isoflavones, anthocyanins, and flavan-3-ols (Routaboul et al., 2012). Each class of flavonoid has distinct structural features that contribute to its specific biological functions. Flavonols, such as quercetin and kaempferol, are widely distributed in plants and play roles in UV protection, pigmentation, and defense against herbivores. Anthocyanins, responsible for the red, purple, and blue colors of many flowers and fruits, are involved in attracting pollinators and seed dispersers. Isoflavones, primarily found in legumes, exhibit phytoestrogenic activity and contribute to plant defense against pathogens (Liu et al., 2021).

1.2.2 Structural Diversity of Flavonoids

The structural diversity of flavonoids arises from modifications to the core flavonoid structure, which consists of a C6-C3-C6 carbon skeleton. The degree and position of hydroxylation, methylation, glycosylation, and acylation determine the specific properties of each flavonoid. For example, the hydroxylation pattern on the B-ring of flavonoids influences their antioxidant activity, with more hydroxyl groups generally increasing the compound's ability to scavenge reactive oxygen species (Agati et al., 2012). Flavonoids can also form conjugates with sugars (glycosides), which enhance their solubility and transport within the plant. These glycosylated forms are important for flavonoid storage and mobilization during stress responses. The structural variations among flavonoids allow them to participate in a wide range of physiological processes, including defense against herbivores, protection from UV radiation, and interaction with symbiotic organisms such as mycorrhizal fungi and nitrogen-fixing bacteria (Panche et al., 2016).

1.3 Structural Diversity of Phenolics and Flavonoids

The structural diversity of phenolics and flavonoids is key to their multifunctionality in plants. Phenolic acids, such as caffeic acid, ferulic acid, and gallic acid, vary in their hydroxylation patterns and their ability to form esters with other compounds. This structural variation enables them to participate in numerous defense mechanisms, such as inhibiting microbial growth and scavenging free radicals (Shalaby & Shanab, 2013).

Lignins, another group of phenolics, are highly polymerized compounds that contribute to the rigidity and strength of plant cell walls. The degree of polymerization and the types of monolignols involved in lignin formation (p-coumaryl, coniferyl, and sinapyl alcohols) determine the specific properties of lignins in different plant species (Boerjan et al., 2003). Lignins play a crucial role in plant defense by reinforcing cell walls and making them more resistant to pathogen invasion.

Flavonoids also exhibit significant structural diversity, as seen in the wide range of flavonoid subclasses. Flavan-3-ols, such as catechin and epicatechin, are key components of condensed tannins, which protect against herbivores and pathogens by reducing palatability and inhibiting digestive enzymes in herbivores (de Melo Luciana et al., 2023). Anthocyanins, which are responsible for the pigmentation in many fruits and flowers, vary in their degree of hydroxylation and methylation, leading to different colors and antioxidant capacities (Agati et al., 2012). The structural diversity of phenolics and flavonoids also contributes to their interactions with other plant metabolites, such as terpenoids and alkaloids, enhancing their overall effectiveness in plant defense. These interactions create synergistic effects that improve the plant's ability to resist pathogen attack and environmental stress.

1.4 Role of Phenolics and Flavonoids in Plant Defense and Stress Responses

Phenolics and flavonoids play essential roles in plant defense against biotic and abiotic stressors. Their antioxidant properties enable them to neutralize reactive oxygen species (ROS) generated during stress, thereby protecting plant cells from oxidative damage (Hasanuzzaman et al., 2020). In addition, phenolics and flavonoids modulate signaling pathways involved in plant defense, such as the salicylic acid and jasmonic acid pathways, which regulate responses to pathogen infection and herbivory (Mandal et al., 2010). The antimicrobial properties of phenolics and flavonoids also contribute to their role in plant defense. Many phenolic acids and flavonoids have been shown to inhibit the growth of bacterial and fungal pathogens by disrupting their cell membranes, inhibiting enzyme activity, or interfering with nutrient uptake (Cowan, 1999). These compounds act as both direct antimicrobial agents and signaling molecules that activate additional defense mechanisms in plants.

2. Antioxidant Functions of Phenolics and Flavonoids in Plant Health

Phenolics and flavonoids are two of the most important classes of plant secondary metabolites, known for their potent antioxidant properties. These compounds play a pivotal role in maintaining plant health by neutralizing reactive oxygen species (ROS) that are generated during various stress conditions, including pathogen attacks, UV radiation, drought, and high salinity. The accumulation of ROS can lead to oxidative stress, which causes damage to cellular structures such as lipids, proteins, and DNA, thereby impairing plant growth and development. Phenolics and flavonoids, through their antioxidant functions, protect plants from oxidative damage, modulate signaling pathways, and enhance stress tolerance.

2.1 Mechanisms of Antioxidant Action

The antioxidant activity of phenolics and flavonoids is primarily attributed to their ability to donate hydrogen atoms or electrons to neutralize free radicals, thus preventing oxidative damage. The presence of hydroxyl (OH) groups in their structure enables these compounds to scavenge free radicals effectively (Rice-Evans et al., 1997). The antioxidant capacity of phenolics and flavonoids is influenced by their molecular structure, particularly the number and position of hydroxyl groups, the presence of double bonds, and the degree of glycosylation and methylation.

2.1.1 Hydrogen Atom Donation

One of the primary ways phenolics and flavonoids act as antioxidants is through the donation of hydrogen atoms. This process neutralizes free radicals, converting them into more stable, less reactive molecules. The phenolic hydroxyl group (-OH) is key to this mechanism. For instance, phenolic acids, such as ferulic acid and caffeic acid, are capable of donating hydrogen atoms to quench hydroxyl radicals and peroxyl radicals, preventing lipid peroxidation in cell membranes (Rice-Evans et al., 1997). Similarly, flavonoids, such as quercetin and kaempferol, possess multiple hydroxyl groups that enhance their free radical-scavenging activity (Platzer et al., 2022).

2.1.2 Metal Ion Chelation

Another important antioxidant mechanism is the ability of phenolics and flavonoids to chelate metal ions, such as iron (Fe) and copper (Cu), which catalyze the formation of highly reactive hydroxyl radicals via the Fenton reaction. By binding to these metal ions, phenolics and flavonoids reduce their availability to participate in redox reactions, thereby preventing the generation of additional ROS (Gutteridge, 1995). Flavonoids, particularly those with catechol groups in the B-ring (e.g., quercetin and rutin), have been shown to chelate metal ions effectively, thus reducing oxidative stress in plants (Pietta, 2000).

2.1.3 Regeneration of Other Antioxidants

Phenolics and flavonoids can also act as part of a larger antioxidant network by regenerating other antioxidants, such as ascorbate (vitamin C) and glutathione. In this process, flavonoids can donate electrons to reduce oxidized forms of these antioxidants, restoring their activity and prolonging their protective effects. This interaction is crucial in maintaining the cellular redox balance under stress conditions. For instance, flavonoids like epicatechin and catechin have been shown to participate in the regeneration of ascorbate, enhancing the plant's overall antioxidant capacity (Pietta, 2000).

2.2 Antioxidant Functions of Phenolics in Plant Health

Phenolics, including phenolic acids, tannins, and lignins, are widely recognized for their strong antioxidant properties, which play critical roles in protecting plants from oxidative damage caused by environmental stressors. These compounds act as both direct antioxidants and modulators of plant defense mechanisms.

2.2.1 Phenolic Acids as Antioxidants

Phenolic acids, such as caffeic acid, ferulic acid, and gallic acid, are abundant in plant tissues and are known for their strong antioxidant properties. Caffeic acid, for example, has been shown to scavenge hydroxyl radicals and superoxide anions, reducing oxidative stress in plants (Cheynier, 2012). Ferulic acid, commonly found in the cell walls of plants, plays a protective role by scavenging free radicals and enhancing cell wall rigidity, which further protects plant cells from oxidative damage during stress (Shalaby & Shanab, 2013). Phenolic acids are particularly important in protecting plants from UV radiation. When plants are exposed to high levels of UV light, they accumulate phenolic acids in their epidermal tissues, where these compounds absorb UV rays and prevent damage to underlying cells (Agati et al., 2012). This UV-absorbing property, combined with their ability to scavenge ROS, makes phenolic acids essential components of the plant's antioxidant defense system.

2.2.2 Tannins and Lignins in Oxidative Stress Protection

Tannins, a group of polyphenolic compounds, also contribute to plant defense against oxidative stress. These compounds are capable of binding to proteins and other macromolecules, thereby preventing oxidative damage. Tannins are particularly important in the defense against herbivores, as they reduce the palatability of plant tissues and inhibit digestive enzymes in herbivores (Barbehenn & Constabel, 2011). In addition to their role in herbivore defense, tannins exhibit strong antioxidant properties, scavenging ROS and protecting plant cells from oxidative stress during pathogen attack and environmental stressors such as drought. Lignins, another class of phenolics, are essential for strengthening plant cell walls and providing structural integrity. During stress conditions, lignin biosynthesis is upregulated, leading to the deposition of lignin in cell walls. This not only reinforces the physical barrier against pathogen invasion but also enhances the plant's resistance to oxidative stress (Boerjan et al., 2003). Lignins are highly effective in scavenging free radicals, and their accumulation in plant tissues helps mitigate the damage caused by ROS.

2.3 Antioxidant Functions of Flavonoids in Plant Health

Flavonoids, a diverse group of phenolic compounds, are well-known for their potent antioxidant properties, which contribute to a wide range of protective functions in plants. Flavonoids act as scavengers of free radicals, chelators of metal ions, and modulators of signaling pathways involved in the plant's defense response.

2.3.1 Flavonoids as ROS Scavengers

Flavonoids are highly effective in scavenging ROS, including superoxide anions, hydrogen peroxide, and hydroxyl radicals. Their antioxidant activity is closely linked to the number and position of hydroxyl groups on their structure. Flavonols, such as quercetin, kaempferol, and myricetin, are particularly potent antioxidants due to their high degree of hydroxylation (Rice-Evans et al., 1997). These flavonoids are capable of directly neutralizing ROS, thereby protecting plant cells from oxidative damage during stress conditions. Flavonoids also play a key role in protecting plants from oxidative damage caused by UV radiation. When plants are exposed to UV-B radiation, they accumulate flavonoids in their epidermal layers, where these compounds absorb UV light and prevent the formation of ROS. This UV-protective function is crucial for preventing DNA damage and maintaining cell viability (Agati et al., 2012).

2.3.2 Modulation of Antioxidant Enzyme Activity

In addition to their direct antioxidant activity, flavonoids also modulate the activity of antioxidant enzymes, such as superoxide dismutase (SOD), catalase (CAT), and peroxidases. These enzymes are part of the plant's enzymatic antioxidant defense system, which works in concert with non-enzymatic antioxidants like flavonoids to neutralize ROS. Flavonoids have been shown to upregulate the expression of genes encoding for antioxidant enzymes, thereby enhancing the plant's ability to detoxify ROS during stress (Sharma et al., 2019). For example, quercetin has been found to increase the activity of SOD and CAT in response to oxidative stress, providing enhanced protection against ROS-induced damage (Rice-Evans et al., 1997).

2.3.3 Role in Stress Signaling and Defense

Flavonoids also play a role in the regulation of stress-signaling pathways that activate the plant's defense response to oxidative stress. These compounds interact with signaling molecules, such as salicylic acid, jasmonic acid, and abscisic acid, which regulate the plant's response to biotic and abiotic stressors. For instance, flavonoids have been shown to modulate the expression of genes involved in the synthesis of defense-related enzymes and proteins, enhancing the plant's resistance to oxidative stress (Pietta, 2000). In addition to their role in oxidative stress protection, flavonoids contribute to the plant's defense against pathogens. Many flavonoids exhibit antimicrobial activity, inhibiting the growth of bacterial and fungal pathogens. By acting as both antioxidants and antimicrobial agents, flavonoids provide a dual layer of protection that enhances plant health and resilience in challenging environments.

2.4 Synergistic Effects with Other Antioxidants

Phenolics and flavonoids often work in concert with other antioxidant molecules, such as ascorbate, glutathione, and tocopherols, to provide a more comprehensive defense against oxidative stress. This synergistic interaction enhances the overall antioxidant capacity of the plant, allowing for more efficient ROS detoxification and protection against oxidative damage. The combination of phenolics, flavonoids, and other antioxidants creates a dynamic and flexible antioxidant system that can be rapidly activated in response to stress (Rice-Evans et al., 1997).

3. Antimicrobial Mechanisms of Phenolics and Flavonoids Against Pathogens

Phenolics and flavonoids are key secondary metabolites that plants produce to defend themselves against microbial pathogens, such as bacteria, fungi, and viruses. These compounds are not only important for maintaining plant health, but they also have been recognized for their antimicrobial properties, which make them valuable in crop protection and sustainable agriculture. The antimicrobial mechanisms of phenolics and flavonoids are multifaceted, involving direct effects on microbial cells and indirect modulation of plant immune responses.

3.1 Direct Antimicrobial Activity of Phenolics

Phenolic compounds exhibit potent antimicrobial activity by targeting various components of microbial cells, including the cell wall, membrane, and intracellular processes. These direct effects on pathogens help plants resist infection and prevent the establishment and spread of microbial diseases.

3.1.1 Disruption of Microbial Cell Membranes

One of the primary antimicrobial mechanisms of phenolics is their ability to disrupt the integrity of microbial cell membranes. Phenolics, such as phenolic acids (e.g., gallic acid, caffeic acid, and ferulic acid), possess hydroxyl groups that interact with the lipid bilayer of microbial membranes, causing structural changes that lead to increased permeability. This disruption of the membrane compromises the pathogen's ability to regulate ion flow, leading to cell lysis and death (Cowan, 1999). The hydrophobic nature of phenolics allows them to insert themselves into the lipid bilayer, destabilizing membrane integrity and causing leakage of cellular contents. For example, studies have demonstrated that phenolic acids like gallic acid and p-coumaric acid are effective in disrupting the cell membranes of Gram-positive and Gram-negative bacteria, resulting in leakage of potassium ions and other intracellular components (Burt, 2004). This disruption is particularly important in inhibiting bacterial pathogens that cause spoilage and diseases in crops, such as *Pseudomonas syringae* and *Xanthomonas* species.

3.1.2 Inhibition of Enzyme Function

Phenolics also exert antimicrobial effects by inhibiting the activity of key enzymes required for microbial survival. Many enzymes involved in cellular metabolism, DNA replication, and cell wall synthesis are highly susceptible to inhibition by phenolics. The hydroxyl groups in phenolic compounds can form hydrogen bonds with the active sites of microbial enzymes, thereby blocking substrate binding and catalytic activity (Cowan, 1999). For instance, phenolics such as tannins have been shown to inhibit the activity of bacterial and fungal enzymes, including amylases, proteases, and cellulases. This inhibition prevents the pathogen from degrading plant cell walls and utilizing plant-derived nutrients, thereby limiting pathogen colonization and growth (Scalbert, 1991). The ability of phenolics to inhibit enzyme activity is particularly effective against fungal pathogens like *Fusarium oxysporum* and *Botrytis cinerea*, which rely on extracellular enzymes to break down plant cell walls during infection.

3.2 Direct Antimicrobial Activity of Flavonoids

Flavonoids, a subclass of phenolics, are well known for their broad-spectrum antimicrobial activity against bacteria, fungi, and viruses. Like phenolics, flavonoids exert their antimicrobial effects through various mechanisms, including disrupting microbial membranes, inhibiting enzyme activity, and interfering with microbial signaling pathways.

3.2.1 Membrane Disruption by Flavonoids

Flavonoids can cause direct damage to microbial cell membranes, similar to phenolics. The presence of hydroxyl groups on the flavonoid structure allows these compounds to interact with the lipid bilayer of microbial membranes, leading to increased permeability and eventual cell death. For example, flavonoids like quercetin, kaempferol, and epicatechin have been shown to disrupt bacterial membranes, causing leakage of cellular contents and loss of membrane potential (Cushnie & Lamb, 2011). Flavonoids also induce oxidative stress in microbial cells by generating reactive oxygen species (ROS), which further damage microbial membranes and intracellular components (Ververidis et al., 2007). This dual action of membrane disruption and ROS generation makes flavonoids highly effective against a wide range of bacterial pathogens, including *Escherichia coli*, *Staphylococcus aureus*, and *Salmonella* spp.

3.2.2 Inhibition of Quorum Sensing and Biofilm Formation

Quorum sensing is a key regulatory mechanism used by bacteria to coordinate gene expression in response to population density, which is critical for biofilm formation, virulence, and antibiotic resistance. Flavonoids have been shown to inhibit quorum sensing in pathogenic bacteria, thereby disrupting their ability to form biofilms and initiate infection. For instance, quercetin and naringenin are known to inhibit quorum sensing in *Pseudomonas aeruginosa*, a common pathogen in plant and human infections. By blocking quorum sensing, these flavonoids prevent the expression of virulence factors and reduce the pathogen's ability to establish infections (Vikram et al., 2010). This makes flavonoids valuable compounds in preventing bacterial infections that rely on biofilm formation, such as those caused by *Agrobacterium tumefaciens* and *Xanthomonas* species in plants.

3.3 Synergistic Effects of Phenolics and Flavonoids in Plant Defense

In addition to their direct antimicrobial activity, phenolics and flavonoids often work synergistically with other plant defense mechanisms, enhancing the overall resistance of plants to pathogens. This synergistic action involves the modulation of plant immune responses, the activation of signaling pathways, and the production of additional antimicrobial compounds.

3.3.1 Modulation of Plant Immune Responses

Phenolics and flavonoids are known to modulate the plant's immune system by activating key defense-related signaling pathways. These compounds can trigger the production of pathogenesis-related (PR) proteins, which are antimicrobial proteins involved in systemic acquired resistance (SAR) and induced systemic resistance (ISR) (Mandal et al., 2010). PR proteins, such as chitinases and glucanases, degrade the cell walls of fungal pathogens, inhibiting their growth and preventing infection. Moreover, phenolics like salicylic acid (SA), a derivative of phenolic compounds, play a crucial role in regulating SAR by inducing the expression of PR genes. SA is involved in the systemic signaling that enhances plant resistance to a broad spectrum of pathogens, including bacteria, fungi, and viruses (Durrant & Dong, 2004). Flavonoids also modulate plant defense responses by interacting with signaling molecules, such as jasmonic acid (JA) and ethylene, which are involved in ISR against herbivores and necrotrophic pathogens.

3.3.2 Enhancement of Antimicrobial Compound Production

In response to pathogen attack, plants often increase the production of antimicrobial secondary metabolites, including phenolics and flavonoids. This amplification of the plant's chemical defenses is a crucial aspect of the overall resistance strategy. For example, plants infected with fungal pathogens frequently upregulate the biosynthesis of phytoalexins, a class of antimicrobial compounds that includes flavonoids and phenolics (Dixon, 2001). The accumulation of flavonoids, such as isoflavonoids in legumes, provides direct protection against fungal pathogens lsike *Phytophthora sojae* and *Rhizoctonia solani*. These compounds interfere with fungal cell wall synthesis, inhibit spore germination, and disrupt fungal growth. Similarly, the increased production of phenolic compounds in response to pathogen attack strengthens the plant cell walls, making it more difficult for pathogens to penetrate and establish infections.

3.4 Applications of Phenolics and Flavonoids in Crop Protection

The antimicrobial properties of phenolics and flavonoids have significant implications for crop protection and sustainable agriculture. These natural compounds offer an environmentally friendly alternative to synthetic chemical pesticides, reducing the reliance on harmful chemicals while providing effective disease control.

3.4.1 Use of Phenolic and Flavonoid-Based Biopesticides

One of the most promising applications of phenolics and flavonoids in agriculture is their use as biopesticides. Plant-derived phenolics and flavonoids can be formulated into natural pesticides that protect crops from bacterial and fungal infections. These biopesticides are biodegradable, pose minimal risk to non-target organisms, and help reduce the environmental impact of conventional pesticides (Zaynab et al., 2018). For example, essential oils rich in phenolic compounds, such as thymol and carvacrol, have been used to control fungal pathogens like *Botrytis cinerea* in grapes and strawberries. Similarly, flavonoid-rich extracts from plants like neem and tea have been shown to inhibit the growth of bacterial pathogens in various crops (Bakkali et al., 2008).

3.4.2 Genetic Engineering for Enhanced Phenolic and Flavonoid Production

Another approach to utilizing the antimicrobial properties of phenolics and flavonoids in agriculture involves genetic engineering. By overexpressing genes involved in the biosynthesis of these compounds, researchers can develop crop varieties with enhanced resistance to pathogens. For example, transgenic tomato plants that overproduce flavonoids have been shown to exhibit increased resistance to fungal pathogens like *Alternaria solani* (Zaynab et al., 2018). This strategy holds promise for developing crops that are naturally more resilient to disease, reducing the need for chemical inputs, and promoting sustainable farming practices.

4. Role of Phenolics and Flavonoids in Plant Stress Responses

Plants are constantly exposed to various stress factors, both biotic and abiotic, which can negatively impact their growth, development, and survival. Biotic stress involves attacks by pathogens, herbivores, and insects, while abiotic stress includes factors such as drought, salinity, extreme temperatures, and UV radiation. Phenolics and flavonoids are crucial components of the plant's defense system, enabling them to mitigate the harmful effects of these stresses. These compounds are synthesized as part of the plant's natural response to stress and play multiple roles, including acting as antioxidants, antimicrobial agents, and signaling molecules that activate defense-related pathways.

4.1 Phenolics and Flavonoids in Abiotic Stress Responses

Abiotic stress factors, such as drought, salinity, UV radiation, and temperature extremes, lead to the overproduction of reactive oxygen species (ROS), which can cause oxidative damage to plant cells. Phenolics and flavonoids are known for their potent antioxidant properties, which help mitigate the damage caused by ROS and protect plant cells from oxidative stress.

4.1.1 Role in Drought and Salinity Stress

Under drought and salinity stress, plants experience water deficit and ionic toxicity, leading to a significant increase in ROS production. Phenolics and flavonoids play a crucial role in scavenging ROS and protecting plant cells from oxidative damage. For instance, phenolic acids, such as caffeic acid and ferulic acid, accumulate in plants under drought stress and act as antioxidants by neutralizing ROS, protecting membrane integrity, and maintaining cellular homeostasis (Shalaby & Shanab, 2013). Similarly, flavonoids like quercetin and kaempferol are upregulated in plants under salinity stress, where they function as ROS scavengers, reducing oxidative damage and enhancing plant survival (Agati et al., 2012). In addition to their antioxidant activity, phenolics and flavonoids also contribute to osmotic adjustment during drought and salinity stress. They help maintain water balance by reducing water loss through stomatal closure and enhancing water uptake through the roots. This dual action of protecting against oxidative damage and regulating water balance makes phenolics and flavonoids essential in plant adaptation to drought and salinity stress (Sharma et al., 2019).

4.1.2 UV Radiation Protection

Phenolics and flavonoids are particularly important in protecting plants from the harmful effects of UV-B radiation. UV radiation can cause DNA damage, protein degradation, and membrane lipid peroxidation, leading to impaired plant growth. Flavonoids, such as anthocyanins, are known to absorb UV radiation, preventing it from penetrating deeper into plant tissues. These compounds act as a sunscreen, protecting cellular structures from UV-induced damage (Agati et al., 2012). Additionally, phenolic compounds, such as hydroxycinnamic acids, accumulate in the epidermal layers of leaves, where they act as UV-absorbing agents. By preventing excessive UV penetration, these compounds protect the photosynthetic apparatus and prevent the formation of ROS, which can otherwise lead to oxidative stress (Affes et al., 2021). The synthesis of phenolics and flavonoids in response to UV radiation is tightly regulated by plant signaling pathways, ensuring that these compounds are produced in sufficient quantities to provide effective protection.

4.1.3 Temperature Stress Tolerance

Extreme temperatures, both hot and cold, can cause significant damage to plant cells by disrupting cellular membranes and proteins. Phenolics and flavonoids help plants tolerate temperature extremes by stabilizing cellular structures and acting as antioxidants. For example, the accumulation of flavonoids like kaempferol and myricetin in plants exposed to cold stress has been shown to protect cell membranes from oxidative damage, preserving membrane fluidity and function (Sharma et al., 2019). In heat stress conditions, phenolic compounds such as ferulic acid and sinapic acid play a protective role by scavenging ROS and preventing oxidative damage to proteins and lipids. These compounds also modulate the expression of heat shock proteins (HSPs), which help plants cope with the damaging effects of high temperatures by refolding denatured proteins and preventing protein aggregation (Rivero et al., 2001). Thus, phenolics and flavonoids act as key regulators of plant tolerance to temperature extremes.

4.2 Phenolics and Flavonoids in Biotic Stress Responses

Biotic stress factors, such as pathogen infection, herbivory, and insect attack, pose a significant threat to plant health. Phenolics and flavonoids contribute to plant defense against these biotic stressors by acting as antimicrobial agents, feeding deterrents, and modulators of plant immune responses.

4.2.1 Antimicrobial Properties Against Pathogens

Phenolics and flavonoids play a crucial role in defending plants against bacterial, fungal, and viral pathogens. These compounds have been shown to exhibit strong antimicrobial activity, inhibiting the growth and proliferation of pathogens. For example, phenolic acids like gallic acid and p-coumaric acid disrupt microbial cell membranes, leading to cell lysis and death (Cowan, 1999). Flavonoids, such as quercetin and naringenin, have been shown to inhibit bacterial quorum sensing, a critical process in biofilm formation and virulence (Vikram et al., 2010). The antimicrobial properties of phenolics and flavonoids also extend to fungal pathogens. Tannins, a group of phenolics, have been shown to inhibit the growth of fungal pathogens like *Botrytis cinerea* by interfering with fungal enzymes and cell wall synthesis (Scalbert, 1991). Flavonoids, such as isoflavones in legumes, protect against fungal pathogens by inhibiting spore germination and hyphal growth, thus preventing the spread of infection (Dixon, 2001).

4.2.2 Defense Against Herbivores and Insects

Phenolics and flavonoids act as feeding deterrents for herbivores and insects, reducing the palatability of plant tissues and inhibiting digestion. For example, tannins bind to proteins in the digestive system of herbivores, making it difficult for them to break down plant tissues and extract nutrients. This reduces herbivory and limits damage to the plant (Barbehenn & Constabel, 2011). In addition to their deterrent properties, phenolics and flavonoids also play a role in the plant's indirect defense by attracting natural predators of herbivores. For instance, flavonoid-mediated signaling can attract parasitic wasps that prey on insect herbivores, providing a natural form of biological control (Ververidis et al., 2007). This interaction between phenolics, flavonoids, and the plant's biotic environment highlights the importance of these compounds in maintaining ecological balance and protecting plant health.

4.3 Modulation of Plant Defense Signaling Pathways

Phenolics and flavonoids are not only direct defense compounds but also act as signaling molecules that modulate the plant's immune response. These compounds are involved in activating various defense-related pathways, such as systemic acquired resistance (SAR) and induced systemic resistance (ISR), which provide long-lasting protection against a wide range of pathogens and pests.

4.3.1 Activation of Systemic Acquired Resistance (SAR)

Systemic acquired resistance (SAR) is a plant defense mechanism that provides long-term protection against a broad spectrum of pathogens. Phenolics, particularly salicylic acid (SA), play a critical role in the activation of SAR by triggering the expression of pathogenesis-related (PR) genes, which encode antimicrobial proteins (Durrant & Dong, 2004). These PR proteins degrade the cell walls of pathogens, inhibit their growth, and prevent the spread of infection. In addition to salicylic acid, other phenolic compounds, such as caffeic acid and ferulic acid, have been shown to induce SAR by enhancing the production of ROS and activating defense-related signaling pathways (Sharma et al., 2019). The ability of phenolics to modulate SAR makes them important components of the plant's defense system against both biotic and abiotic stressors.

4.3.2 Induced Systemic Resistance (ISR) and Flavonoids

Induced systemic resistance (ISR) is another important plant defense mechanism, often activated in response to beneficial microbes, such as rhizobacteria and mycorrhizal fungi. Flavonoids play a key role in ISR by modulating the production of signaling molecules like jasmonic acid (JA) and ethylene, which are involved in the plant's response to herbivores and necrotrophic pathogens (Pieterse et al., 2012). These signaling molecules activate defense responses in distant tissues, providing systemic protection against future attacks. Flavonoids such as quercetin and kaempferol have been shown to enhance ISR by interacting with microbial symbionts and stimulating the production of defense-related compounds. This interaction between flavonoids and beneficial microbes strengthens the plant's overall resilience to both biotic and abiotic stressors, improving survival and growth under challenging environmental conditions.

4.4 Role in Plant Development and Stress Tolerance

In addition to their defensive functions, phenolics and flavonoids play important roles in plant development and stress tolerance. These compounds regulate processes such as cell wall formation, lignin biosynthesis, and hormone signaling, all of which contribute to the plant's ability to withstand environmental stress.

4.4.1 Lignin Biosynthesis and Structural Integrity

Phenolics, particularly lignins, are essential for the structural integrity of plant cell walls. Lignins are deposited in the cell walls of vascular tissues, where they provide rigidity and support, allowing plants to withstand mechanical stress and pathogen invasion (Boerjan et al., 2003). During stress conditions, such as drought or pathogen attack, lignin biosynthesis is upregulated, reinforcing cell walls and enhancing the plant's physical defenses.

4.4.2 Hormonal Crosstalk and Stress Tolerance

Phenolics and flavonoids also interact with plant hormones, such as abscisic acid (ABA), auxins, and gibberellins, to regulate stress responses. For example, phenolics have been shown to modulate ABA signaling, which is critical for plant responses to drought and salinity stress (Rivero et al., 2001). By regulating hormone levels, phenolics and flavonoids help plants maintain growth and development under stressful conditions, ensuring survival and reproduction.

5. Phenolics and Flavonoids in Enhancing Plant Resistance to Environmental Stresses

Phenolics and flavonoids are critical plant metabolites that play a central role in enhancing plant resistance to various environmental stresses, both abiotic and biotic. These compounds not only act as direct defense agents but also mediate the signaling processes that trigger plant responses to stress conditions such as drought, salinity, UV radiation, and pathogen attacks. Due to their antioxidant, antimicrobial, and structural reinforcement properties, phenolics and flavonoids are integral to plants' ability to adapt and thrive under challenging environmental conditions.

5.1 Phenolics in Abiotic Stress Tolerance

Abiotic stresses, such as drought, salinity, temperature extremes, and UV radiation, can cause substantial oxidative damage to plant cells by increasing the production of reactive oxygen species (ROS). Phenolics, including phenolic acids, lignins, and tannins, are pivotal in mitigating this damage by neutralizing ROS and protecting cellular structures.

5.1.1 Drought and Salinity Stress

Under drought and salinity stress, plants experience water deficit and ion imbalance, leading to the overproduction of ROS. Phenolic compounds act as potent antioxidants that scavenge these ROS, thus protecting cells from oxidative stress (Shalaby & Shanab, 2013). For instance, phenolic acids such as ferulic acid and caffeic acid are known to accumulate in plants during drought conditions and enhance stress tolerance by strengthening cell walls and reducing oxidative damage (Cheynier, 2012). In salinity stress, phenolics also play a key role in osmotic regulation, helping plants to maintain cellular water balance. Phenolics like hydroxycinnamic acids improve membrane stability and prevent the leakage of vital ions from plant cells, thereby helping plants to survive in high-salinity environments (Agati et al., 2012).

5.1.2 UV Radiation Protection

Phenolic compounds, particularly flavonoids and hydroxycinnamic acids, are known to protect plants from the harmful effects of UV-B radiation. These compounds absorb UV light and act as a natural sunscreen, preventing damage to DNA and other cellular structures. In response to increased UV exposure, plants often upregulate the biosynthesis of phenolics in their epidermal tissues, creating a barrier that shields sensitive tissues from excessive radiation (Agati et al., 2012). The accumulation of flavonoids such as quercetin and kaempferol in the outer layers of leaves is an adaptive mechanism that prevents UV-induced oxidative stress.

5.2 Flavonoids in Biotic Stress Resistance

Flavonoids, a subclass of phenolics, play a key role in protecting plants from biotic stresses such as pathogen infections, insect attacks, and herbivory. Their antimicrobial and antioxidant properties, coupled with their ability to modulate plant defense signaling pathways, make them essential for biotic stress resistance.

5.2.1 Antimicrobial Activity

Flavonoids exhibit strong antimicrobial properties against a wide range of pathogens, including bacteria, fungi, and viruses. For example, flavonoids such as quercetin, catechin, and apigenin have been shown to inhibit bacterial quorum sensing, a process essential for biofilm formation and virulence in plant pathogens like *Pseudomonas aeruginosa* and *Agrobacterium tumefaciens* (Cushnie & Lamb, 2011). By disrupting quorum sensing, flavonoids reduce the ability of pathogens to colonize plants and cause infections. In addition to bacterial pathogens, flavonoids are effective against fungal pathogens. Isoflavonoids, particularly in legumes, protect against fungal diseases by inhibiting spore germination and fungal hyphal growth (Dixon, 2001). These flavonoids interfere with fungal enzymes and disrupt the integrity of fungal cell walls, thereby preventing the establishment of infections.

5.2.2 Role in Plant-Herbivore Interactions

Flavonoids also play a critical role in defending plants against herbivores. Many flavonoids act as feeding deterrents by making plant tissues less palatable to herbivores. Flavonoids such as tannins bind to proteins in the herbivore's digestive system, reducing their ability to digest plant material and thus discouraging further feeding (Barbehenn & Constabel, 2011). Additionally, flavonoids can serve as signaling molecules that attract natural enemies of herbivores, such as parasitic wasps. This indirect defense mechanism, known as tritrophic interaction, helps plants protect themselves from herbivore damage by recruiting predatory insects to control herbivore populations (Pieterse et al., 2012).

5.3 Synergistic Effects with Other Stress-Responsive Compounds

Phenolics and flavonoids often work in synergy with other plant metabolites, such as terpenoids, alkaloids, and hormones, to enhance the overall resistance of plants to environmental stresses. These compounds interact with signaling pathways that regulate plant stress responses, such as the salicylic acid (SA), jasmonic acid (JA), and abscisic acid (ABA) pathways. For example, the interaction between phenolics and salicylic acid plays a crucial role in the activation of systemic acquired resistance (SAR), a plant defense mechanism that provides long-lasting protection against a wide range of pathogens (Durrant & Dong, 2004). Similarly, flavonoids have been shown to modulate jasmonic acid signaling, which is involved in the plant's response to herbivores and necrotrophic pathogens (Pieterse et al., 2012).

6. Applications of Phenolics and Flavonoids in Sustainable Crop Protection

The growing emphasis on sustainable agriculture has spurred interest in natural plant compounds such as phenolics and flavonoids for crop protection. These compounds offer eco-friendly alternatives to synthetic chemicals, providing effective pest and disease management while minimizing environmental impact. Their antimicrobial, antioxidant, and allelopathic properties make them valuable tools for developing biopesticides, improving crop resilience, and promoting soil health.

6.1 Biopesticides and Biofungicides

One of the most promising applications of phenolics and flavonoids in sustainable agriculture is their use as biopesticides and biofungicides. Plant extracts rich in these compounds can be formulated into natural pesticides that protect a wide range of pathogens and pests without the harmful side effects associated with synthetic chemicals (Zaynab et al., 2018).

6.1.1 Phenolic-Based Biopesticides

Phenolic compounds, particularly phenolic acids and tannins, have demonstrated strong antimicrobial activity against bacterial and fungal pathogens. Essential oils rich in phenolics, such as thymol and carvacrol, have been used to control fungal infections in crops like grapes and strawberries (Burt, 2004). These natural biopesticides disrupt the cell membranes of pathogens, inhibiting their growth and preventing the spread of diseases. For instance, the application of phenolic-rich extracts from neem (*Azadirachta indica*) has been shown to effectively control a range of agricultural pests and diseases, including bacterial blight, powdery mildew, and root rot (Mordue & Nisbet, 2000). The use of these natural products not only reduces the reliance on synthetic pesticides but also supports biodiversity by preserving beneficial insects and microorganisms.

6.1.2 Flavonoid-Based Biofungicides

Flavonoids, due to their antimicrobial properties, are also widely used in biofungicide formulations. For example, flavonoids such as naringenin and apigenin have been shown to inhibit the growth of fungal pathogens like *Botrytis cinerea* and *Alternaria solani*, which cause significant losses in crops like tomatoes, grapes, and strawberries (Cushnie & Lamb, 2011). These biofungicides provide an environmentally friendly solution to fungal diseases, reducing the need for chemical fungicides and promoting sustainable farming practices.

6.2 Enhancing Crop Resilience Through Phenolics and Flavonoids

Phenolics and flavonoids not only act as biopesticides but also play a critical role in enhancing the overall resilience of crops to environmental stresses. By promoting stress tolerance, these compounds help crops maintain productivity under adverse conditions, such as drought, salinity, and temperature extremes.

6.2.1 Improved Stress Tolerance in Crops

The application of phenolics and flavonoids through foliar sprays or soil amendments has been shown to enhance the stress tolerance of crops. For example, the foliar application of flavonoid-rich extracts from plants like *Echinacea purpurea* has been shown to improve drought tolerance in crops by enhancing antioxidant activity and reducing oxidative damage (Rivero et al., 2001). Similarly, the use of phenolic-based soil amendments has been shown to improve the salt tolerance of crops like tomatoes and wheat by enhancing osmotic regulation and ion balance (Sharma et al., 2019).

6.2.2 Promotion of Beneficial Microorganisms

Phenolics and flavonoids also contribute to sustainable crop protection by promoting the growth of beneficial soil microorganisms, such as mycorrhizal fungi and nitrogen-fixing bacteria. These microorganisms play a critical role in enhancing soil fertility, nutrient uptake, and plant growth. For example, flavonoids secreted by legume roots act as signaling molecules that stimulate the colonization of rhizobia, nitrogen-fixing bacteria that enhance plant growth and soil health (Pieterse et al., 2012). By promoting these beneficial interactions, phenolics and flavonoids contribute to the overall sustainability of agricultural systems.

6.3 Genetic Engineering and Crop Breeding

Another promising application of phenolics and flavonoids in sustainable agriculture is the use of genetic engineering and crop breeding to enhance the production of these compounds in plants. By overexpressing genes involved in the biosynthesis of phenolics and flavonoids, researchers can develop crop varieties with enhanced resistance to pests, diseases, and environmental stresses. For instance, genetically modified tomato plants that overproduce flavonoids have been shown to exhibit increased resistance to fungal pathogens and improved tolerance to drought stress (Zaynab et al., 2018). Similarly, crops bred for higher levels of phenolic compounds, such as lignins and tannins, have demonstrated increased resistance to insect herbivores and enhanced structural integrity, reducing the need for external chemical inputs.

6.4 Sustainable Weed Management

Phenolics and flavonoids also play a role in sustainable weed management through their allelopathic properties. Allelopathy refers to the chemical inhibition of one plant species by another, often through the release of secondary metabolites such as phenolics and flavonoids. These compounds can suppress the germination and growth of weed species, reducing competition for resources and improving crop yields. For example, phenolic acids released by cover crops like rye and clover have been shown to inhibit the growth of weed species such as *Amaranthus retroflexus* and *Chenopodium album* (Cheynier, 2012). This natural weed suppression reduces the need for herbicides, promoting more sustainable and eco-friendly weed management practices.

Conclusion

Phenolics and flavonoids, due to their structural diversity and functional versatility, serve as critical components in plant health and defense mechanisms. The biosynthesis of these compounds, primarily through the shikimate and phenylpropanoid pathways, results in a wide variety of molecules with unique structural characteristics, each contributing to specific biological functions in plants. These compounds exhibit powerful antioxidant activities that protect plants from oxidative stress by scavenging reactive oxygen species (ROS), thereby maintaining cellular integrity and supporting vital metabolic processes under stress conditions. Moreover, phenolics and flavonoids possess robust antimicrobial properties, acting against a broad spectrum of pathogens such as bacteria, fungi, and viruses. Their mechanisms include the disruption of microbial membranes, inhibition of key enzymes, and interference with pathogen signaling systems, which are essential for the pathogens' ability to cause infection.

In addition to their direct antimicrobial effects, phenolics and flavonoids play a significant role in plant stress responses, helping plants adapt to biotic and abiotic stresses through enhanced defense signaling and reinforcement of structural barriers. Their involvement in these stress responses makes them indispensable in promoting plant resilience to adverse environmental conditions, such as drought, salinity, and pathogen attack. By strengthening plant defenses, phenolics and flavonoids also enhance resistance to environmental stresses, allowing plants to survive and thrive in challenging conditions. This makes them invaluable in natural ecosystems as well as in agricultural systems where maintaining crop health and yield is essential.

The application of phenolics and flavonoids in sustainable crop protection is increasingly recognized for its potential to provide eco-friendly alternatives to synthetic pesticides and fungicides. These compounds offer a natural, less harmful approach to disease and pest management in agriculture, reducing the reliance on chemicals that can have detrimental effects on the environment. Advances in biotechnology and bioengineering are paving the way for optimizing the production and utilization of these secondary metabolites in agricultural systems, promoting their use in integrated pest management (IPM) strategies. As research progresses, phenolics and flavonoids are likely to play a more prominent role in sustainable agricultural practices, offering a natural solution to enhance crop resilience and improve disease management. Their dual functionality in antioxidant defense and antimicrobial activity positions them as vital elements in plant health, providing essential tools for both traditional and modern agriculture.

References

Agati, G., Azzarello, E., Pollastri, S., & Tattini, M. (2012). Flavonoids as antioxidants in plants: Location and functional significance. *Plant Science, 196*, 67-76.

Affes, S., Ben Younes, A., Frikha, D., Allouche, N., Treilhou, M., Tene, N., & Mezghani-Jarraya, R. (2021). ESI-MS/MS analysis of phenolic compounds from *Aeonium arboreum* leaf extracts and evaluation of their antioxidant and antimicrobial activities. *Molecules*, *26*(14), 4338.

Bakkali, F., Averbeck, S., Averbeck, D., & Idaomar, M. (2008). Biological effects of essential oils-A review. *Food and Chemical Toxicology, 46*(2), 446-475.

Barbehenn, R. V., & Constabel, C. P. (2011). Tannins in plant-herbivore interactions. *Phytochemistry, 72*(13), 1551-1565.

Boerjan, W., Ralph, J., & Baucher, M. (2003). Lignin biosynthesis. *Annual Review of Plant Biology, 54*(1), 519-546.

Burt, S. (2004). Essential oils: Their antibacterial properties and potential applications in foods-A review. *International Journal of Food Microbiology, 94*(3), 223-253.

Cheynier, V. (2012). Phenolic compounds: From plants to foods. *Phytochemistry Reviews, 11*(2), 153-177.

Cowan, M. M. (1999). Plant products as antimicrobial agents. *Clinical Microbiology Reviews, 12*(4), 564-582.

Cushnie, T. P., & Lamb, A. J. (2011). Recent advances in understanding the antibacterial properties of flavonoids. *International Journal of Antimicrobial Agents, 38*(2), 99-107.

Dixon, R. A. (2001). Natural products and plant disease resistance. *Nature, 411*(6839), 843-847.

Dixon, R. A., & Paiva, N. L. (1995). Stress-induced phenylpropanoid metabolism. *The Plant Cell, 7*(7), 1085-1097.

de Melo Luciana, F. M., de Queiroz Aquino-Martins, V. G., da Silva, A. P., Rocha, H. A. O., & Scortecci, K. C. (2023). Biological and pharmacological aspects of tannins and potential biotechnological applications. *Food Chemistry*, *414*, 135645.

Durrant, W. E., & Dong, X. (2004). Systemic acquired resistance. *Annual Review of Phytopathology, 42*, 185-209.

Gill, S. S., & Tuteja, N. (2010). Reactive oxygen species and antioxidant machinery in abiotic stress tolerance in crop plants. *Plant Physiology and Biochemistry, 48*(12), 909-930.

Gutteridge, J. M. (1995). Lipid peroxidation and antioxidants as biomarkers of tissue damage. *Clinical Chemistry, 41*(12), 1819-1828.

Hasanuzzaman, M., Bhuyan, M. H. M. B., Zulfiqar, F., Raza, A., Mohsin, S. M., Mahmud, J. A., Fujita, M., & Fotopoulos, V. (2020). Reactive oxygen species and antioxidant defense in plants under abiotic stress: Revisiting the crucial role of a universal defense regulator. *Antioxidants (Basel)*, *9*(8), 681.

Herrmann, K. M., & Weaver, L. M. (1999). The shikimate pathway. *Annual Review of Plant Physiology and Plant Molecular Biology, 50*(1), 473-503.

Lattanzio, V., Lattanzio, V. M. T., & Cardinali, A. (2006). Role of phenolics in the resistance mechanisms of plants against fungal pathogens and insects. *Phytochemistry: Advances in Research*, 23-67.

Liu, W., Feng, Y., Yu, S., Fan, Z., Li, X., Li, J., & Yin, H. (2021). The Flavonoid biosynthesis network in plants. *International Journal of Molecular Sciences*, *22*(23):12824.

Mandal, S., Mallick, N., & Mitra, A. (2010). Salicylic acid-induced resistance to *Fusarium oxysporum* f. sp. Lycopersici in tomato. *Plant Physiology and Biochemistry, 47*(7), 642-649.

Metsämuuronen, S., & Sirén, H. (2019). Bioactive phenolic compounds, metabolism and properties: a review on valuable chemical compounds in Scots pine and Norway spruce. *Photochemistry Reviews 18*, 623-664.

Mordue, A. J., & Nisbet, A. J. (2000). Azadirachtin from the neem tree *Azadirachta indica*: Its action against insects. *Anais da Sociedade Entomológica do Brasil, 29*(4), 615-632.

Moura, J. C. M. S., Bonine, C. A. V., de Oliveira Fernandes Viana, J., Dornelas, M. C., & Mazzafera, P. (2010). Abiotic and biotic stresses and changes in the lignin content and composition in plants. *Journal of Integrative Plant Biology, 52*(4), 360-376.

Panche, A. N., Diwan, A. D., & Chandra, S. R. (2016). Flavonoids: An overview. *Journal of Nutritional Science, 5*, e47.

Pieterse, C. M. J., Van Der Does, D., Zamioudis, C., Leon-Reyes, A., & Van Wees, S. C. M. (2012). Hormonal modulation of plant immunity. *Annual Review of Cell and Developmental Biology, 28*, 489-521.

Pietta, P. G. (2000). Flavonoids as antioxidants. *Journal of Natural Products, 63*(7), 1035-1042.

Platzer, M., Kiese, S., Tybussek, T., Herfellner, T., Schneider, F., Schweiggert-Weisz, U., & Eisner, P. (2022). Radical scavenging mechanisms of phenolic compounds: A quantitative structure-property relationship (QSPR) study. *Frontiers in Nutrition, 9*, 882458.

Pourcel, L., Routaboul, J. M., Cheynier, V., Lepiniec, L., & Debeaujon, I. (2007). Flavonoid oxidation in plants: From biochemical properties to physiological functions. *Trends in Plant Science, 12*(1), 29-36.

Rice-Evans, C., Miller, N. J., & Paganga, G. (1997). Antioxidant properties of phenolic compounds. *Trends in Plant Science, 2*(4), 152-159.

Rivero, R. M., Ruiz, J. M., García, P. C., López-Lefebre, L. R., Sánchez, E., & Romero, L. (2001). Resistance to cold and heat stress: Accumulation of phenolic compounds in tomato and watermelon plants. *Plant Science, 160*(2), 315-321.

Routaboul, J. M., Kerhoas, L., Einhorn, J., & Lepiniec, L. (2012). Flavonoid pathway in plants. In B. Halliwell & C. Gutteridge (Eds.), *Free Radicals in Biology and Medicine* (pp. 253-276). Oxford University Press.

Scalbert, A. (1991). Antimicrobial properties of tannins. *Phytochemistry, 30*(12), 3875-3883.

Shalaby, E. A., & Shanab, S. M. M. (2013). Antioxidant compounds, assays of determination and mode of action. *African Journal of Biotechnology, 7*(10), 528-539.

Sharma, A., Shahzad, B., Rehman, A., Bhardwaj, R., Landi, M., & Zheng, B. (2019). Response of phenylpropanoid pathway and the role of polyphenols in plants under abiotic stress. *Molecules, 24*(13), 2452.

Ververidis, F., Trantas, E., Douglas, C., Vollmer, G., Kretzschmar, G., & Panopoulos, N. (2007). Biotechnology of flavonoids and other phenylpropanoid-derived natural products. Part I: Chemical diversity, impacts on plant biology and human health. *Biotechnology Journal, 2*(10), 1214-1234.

Vikram, A., Jesudhasan, P. R., Jayaprakasha, G. K., Pillai, B. S., & Patil, B. S. (2010). Grapefruit bioactive limonoids modulate *E. coli* O157:H7 TTSS and biofilm. *International Journal of Food Microbiology, 140*(2-3), 109-116.

Vogt, T. (2010). Phenylpropanoid biosynthesis. *Molecular Plant, 3*(1), 2-20.

Winkel-Shirley, B. (2001). Flavonoid biosynthesis. A colorful model for genetics, biochemistry, cell biology, and biotechnology. *Plant Physiology, 126*(2), 485-493.

Zaynab, M., Fatima, M., Abbas, S., Sharif, Y., Umair, M., Zafar, M. H., & Bahadar, K. (2018). Role of secondary metabolites in plant defense against pathogens. *Microbial Pathogenesis, 124*, 198-202.

Chapter - 6

Saponins: Biological Activities and their Role in Plant Pathogen Control

Praneet Chauhan[1], Pratima Raypa[2], Sheetanshu Gupta[3] and Reshma Jadhav[4]

[1]Department of Plant Pathology, Dr. Khem Singh Gill Akal College of Agriculture, Eternal University, Baru Sahib, Sirmour, Himachal Pradesh, India
[2]Department of Biochemistry, H.N.B. Garhwal University, Srinagar, Uttarakhand, India
[3]Department of Biochemistry, G.B. Pant University of Agriculture & Technology, Pantnagar, Uttarakhand, India
[4]Department of Biochemistry, Mahatma Phule Krishi Vidyapeeth, Rahuri, Ahmednagar, Maharashtra, India

Abstract

Saponins are a diverse group of naturally occurring plant glycosides that exhibit significant biological activities, particularly in the context of plant defense and pathogen control. These compounds are synthesized through the mevalonate and methylerythritol phosphate (MEP) pathways, contributing to their structural diversity, which is characterized by the presence of a hydrophobic aglycone and one or more hydrophilic sugar chains. This dual structure allows saponins to interact with cell membranes, resulting in potent antimicrobial effects. Saponins act as natural biocides against a variety of plant pathogens, including bacteria, fungi, and viruses, through mechanisms such as membrane disruption, inhibition of microbial enzyme systems, and interference with pathogen signaling. Beyond their antimicrobial properties, saponins also play a crucial role in plant defense responses, inducing systemic acquired resistance (SAR) and enhancing the plant's overall ability to resist biotic stresses. Furthermore, these compounds often exhibit synergistic effects when combined with other plant secondary metabolites, enhancing the overall efficacy of plant defenses. Saponins have also attracted interest for their potential applications in sustainable agriculture, where they can be employed as natural pesticides or biostimulants in crop protection, offering an eco-friendly alternative to synthetic agrochemicals. However, challenges such as large-scale production, variability in activity across different plant species, and the need for a deeper understanding of their interactions with other defense compounds present ongoing areas of research. Advances in biotechnology, particularly in the fields of genetic engineering and metabolic pathway optimization, offer promising solutions for overcoming these challenges, enabling the effective utilization of saponins in plant protection and sustainable agricultural practices.

DOI: 10.1201/9781003717454-6

Keywords: Saponins, plant defense, antimicrobial, pathogen control, sustainable agriculture, biopesticides.

Introduction

Saponins are a diverse group of plant secondary metabolites known for their structural complexity and a wide range of biological activities, including antimicrobial, antifungal, and insecticidal properties. Found in various plant species, saponins play essential roles in plant defense and have applications in agriculture, medicine, and industry. Structurally, saponins consist of a hydrophobic aglycone (sapogenin) attached to one or more hydrophilic sugar moieties, forming amphipathic molecules that contribute to their soap-like properties. These compounds can be classified into two main types based on their aglycone structure: steroidal saponins and triterpenoid saponins (Osbourn, 1996). The diversity in their sugar chains and sapogenins gives rise to a wide variety of saponins with differing biological activities, making them a key focus of research for their potential in sustainable agricultural practices and natural plant protection.

Saponins are synthesized through complex biosynthetic pathways, involving the cyclization of 2,3-oxidosqualene to form the basic triterpenoid or steroidal structure, followed by a series of hydroxylation, glycosylation, and acylation reactions that give rise to the final saponin molecules (Moses et al., 2014). The enzymes involved in saponin biosynthesis are highly specific, contributing to the structural diversity observed in different plant species. The biosynthesis of saponins is often upregulated in response to biotic and abiotic stress, indicating their role as important defense compounds in plants.

One of the most notable features of saponins is their antimicrobial activity, which is attributed to their ability to disrupt microbial cell membranes. Saponins interact with sterols in microbial membranes, leading to increased permeability, leakage of cellular contents, and ultimately cell death (Osbourn, 2003). This makes saponins effective against a wide range of pathogens, including bacteria, fungi, and viruses. Their ability to inhibit microbial growth and colonization makes them valuable in plant defense against pathogens, as well as in agricultural applications for crop protection. In addition to their direct antimicrobial effects, saponins are also involved in modulating plant defense responses. They can act as signaling molecules that trigger the activation of defense-related genes, leading to the production of pathogenesis-related (PR) proteins and other defense compounds (Hussain et al., 2019). This dual role as both antimicrobial agents and defense activators highlights the importance of saponins in the plant's overall defense strategy.

Moreover, saponins often work synergistically with other plant metabolites, such as phenolics, flavonoids, and terpenoids, to enhance their antimicrobial efficacy. This synergistic effect is particularly important in providing plants with a multi-layered defense system that targets pathogens through various mechanisms. The combination of saponins with other secondary metabolites not only strengthens plant defenses but also reduces the likelihood of pathogens developing resistance. The environmental and agricultural applications of saponins are also gaining attention due to their natural origin and eco-friendly nature. Saponin-rich plant extracts are being explored as biopesticides and bio fungicides in sustainable agriculture, offering a natural alternative to synthetic chemicals (Voutquenne et al., 2008). Their biodegradability and low toxicity to non-target organisms make them ideal candidates for use in integrated pest management (IPM) systems. In addition, saponins have shown a potential to enhance soil health by promoting beneficial microbial communities and improving nutrient uptake in plants.

Despite their promising applications, the utilization of saponins in agriculture and industry faces several challenges. One of the primary issues is the variability in saponin content and composition across different plant species and environmental conditions. This variability can affect the efficacy of saponin-based products, making it necessary to standardize extraction methods and develop efficient formulations for commercial use. Furthermore, the large-scale production of saponins is limited by the complexity of their biosynthesis and the difficulty of isolating them in large quantities. Advances in genetic engineering and biotechnology offer potential solutions to these challenges by enabling the manipulation of saponin biosynthetic pathways to enhance yield and optimize their properties for specific applications (Moses et al., 2014).

1. Biosynthesis and Structural Diversity of Saponins

Saponins are a large and structurally diverse group of naturally occurring glycosides found in a wide range of plant species. The key feature that distinguishes saponins is their amphipathic nature, derived from the combination of a hydrophobic aglycone (sapogenin) and hydrophilic sugar chains. This unique structure gives saponins the ability to interact with lipid membranes, leading to their well-known detergent-like properties. These characteristics make saponins valuable for plant defense mechanisms and open up various applications in agriculture, medicine, and pharmaceuticals (El Aziz et al., 2019).

1.1 Biosynthesis of Saponins

The biosynthesis of saponins involves the transformation of simple precursors through highly specialized enzymatic pathways, leading to the formation of structurally complex molecules. The process starts with the conversion of acetyl-CoA to isopentenyl pyrophosphate (IPP) through either the mevalonate or methylerythritol phosphate (MEP) pathway. These two pathways converge to form 2,3-oxidosqualene, which is the precursor for both triterpenoid and steroidal saponins. The subsequent cyclization of 2,3-oxidosqualene, catalyzed by oxidosqualene cyclases (OSCs), determines whether the final product will be a triterpenoid or steroidal saponin (Osbourn, 1996). The choice of pathway in plants is often influenced by environmental conditions or the plant's specific defensive needs. Triterpenoid saponins, for example, are more common in plants that face frequent fungal or bacterial challenges, while steroidal saponins are often found in monocotyledonous plants and are associated with defenses against herbivores (Rai et al., 2021). The genetic regulation of these biosynthetic pathways plays a crucial role in determining which type of saponin a plant produces, and the biosynthesis of these compounds can be upregulated in response to environmental stressors such as pathogen invasion or physical damage.

1.1.1 Triterpenoid Saponins

Triterpenoid saponins, which are the most prevalent type of saponins, are derived from β-amyrin, lupeol, and other pentacyclic triterpenes. The biosynthesis of these compounds starts with the cyclization of 2,3-oxidosqualene by β-amyrin synthase to form β-amyrin, which then undergoes a series of oxidations, hydroxylations, and glycosylations to generate a wide variety of triterpenoid saponins (Moses et al., 2014). These enzymes, particularly cytochrome P450 monooxygenases, and glycosyltransferases, are responsible for introducing functional groups and attaching sugar residues to the sapogenin backbone, greatly influencing the bioactivity of the final saponin compound (Augustin et al., 2011). Plants such as ginseng (*Panax* spp.), licorice (*Glycyrrhiza glabra*), and quinoa (*Chenopodium quinoa*) are known for their high content of triterpenoid saponins. These compounds serve multiple functions, including antimicrobial defense, protection against herbivores, and modulation of plant-environment interactions. Moreover, triterpenoid saponins from medicinal plants such as ginseng are of significant pharmacological interest due to their adaptogenic, anti-inflammatory, and immune-stimulating properties (Rai et al., 2021).

1.1.2 Steroidal Saponins

Steroidal saponins are particularly abundant in monocot plants like yams (*Dioscorea* spp.) and fenugreek (*Trigonella foenum-graecum*). These compounds are synthesized from cholesterol or related sterols through a series of enzymatic reactions. The aglycone structure of steroidal saponins contains a sterane skeleton, which distinguishes them from triterpenoid saponins with their pentacyclic ring structures. Steroidal saponins are synthesized via a similar pathway as triterpenoids but involve sterol precursors such as cholesterol or diosgenin, depending on the plant species (Mugford & Osbourn, 2012). In addition to their role in plant defense, steroidal saponins are used in the pharmaceutical industry as precursors for synthesizing steroid hormones such as progesterone and cortisone. Diosgenin, a steroidal saponin from yams, is widely used in the production of synthetic steroid drugs. The ability of steroidal saponins to form complexes with membrane sterols makes them effective antifungal and antibacterial agents (Osbourn, 2003).

1.2 Structural Diversity of Saponins

The structural diversity of saponins is primarily derived from variations in both the sapogenin backbone and the attached sugar chains. This diversity influences not only their biological functions but also their pharmacological properties and interactions with other molecules. The hydrophobic nature of the sapogenin combined with hydrophilic sugar residues imparts amphiphilic characteristics to saponins, which allows them to disrupt cell membranes by interacting with lipid bilayers.

1.2.1 Sapogenin Variations

The sapogenin backbone of saponins can be classified into triterpenoid or steroidal types. Triterpenoid saponins have a pentacyclic ring structure, while steroidal saponins are built upon a sterane skeleton derived from cholesterol. The number and position of hydroxyl groups on these backbones are key determinants of the compound's biological activity. For example, triterpenoid saponins with multiple hydroxyl groups are more water-soluble and effective at interacting with biological membranes, which enhances their surfactant properties (Moses et al., 2014). These structural variations also determine how saponins interact with cell membranes. Steroidal saponins, due to their affinity for cholesterol, are especially effective at disrupting cholesterol-rich membranes, making them potent antifungal agents. On the other hand, triterpenoid saponins are more effective against bacterial membranes due to their ability to interact with bacterial sterol analogs. This diversity allows saponins to target a wide range of pathogens, contributing to their role in plant defense.

1.2.2 Sugar Moiety Variations

The diversity of saponins is further enhanced by the variety of sugar moieties attached to the sapogenin backbone. These sugars can be simple monosaccharides or more complex oligosaccharides, often linked through glycosidic bonds. The type and number of sugar units, as well as the position of glycosylation, have significant effects on the solubility, bioavailability, and biological activity of saponins. Common sugars found in saponins include glucose, rhamnose, arabinose, and xylose (Augustin et al., 2011). For instance, saponins with longer or branched sugar chains tend to be more soluble in water and are more easily absorbed by biological membranes, increasing their bioactivity. These sugar chains also protect the sapogenin core from enzymatic degradation, enhancing the stability and longevity of the saponins in biological systems. The variability in glycosylation patterns is one of the main reasons why saponins exhibit such a broad range of biological activities, from antimicrobial to anti-inflammatory effects.

1.3 Functional Implications of Saponin Diversity

The structural diversity of saponins leads to a wide range of biological functions that are essential for plant defense. The amphiphilic nature of saponins allows them to interact with lipid membranes, making them effective at disrupting the membranes of pathogens and herbivores. This property underlies their well-documented antimicrobial and insecticidal activities. For example, saponins can form complexes with sterols in fungal membranes, leading to increased permeability and eventual cell lysis, making them potent antifungal agents (Osbourn, 2003).

In addition to their antimicrobial properties, saponins serve as deterrents to herbivores due to their bitter taste and ability to form foams. Some saponins are also allelopathic, releasing bioactive compounds into the soil that inhibit the growth of neighboring plants, reducing competition for resources. This multifunctionality makes saponins vital components of the plant's overall defense strategy (Cheeke, 1996). The structural diversity of saponins also contributes to their pharmacological potential, particularly in traditional medicine and modern pharmaceuticals. For example, triterpenoid saponins from ginseng are known for their adaptogenic and immune-stimulating properties, while steroidal saponins from yams are used in the synthesis of steroid hormones. This broad range of activities makes saponins highly valuable for both ecological and medical purposes.

2. Saponins as Antimicrobial Agents: Mechanisms Against Plant Pathogens

Saponins are well-known for their potent antimicrobial properties, making them key players in plant defense against a broad spectrum of pathogens, including bacteria, fungi, and viruses. Their antimicrobial activity is largely attributed to their ability to disrupt cell membranes, leading to increased permeability and eventual cell death. Saponins' amphiphilic nature, combining hydrophobic aglycones (sapogenins) with hydrophilic sugar moieties, allows them to interact with the lipid bilayers of microbial membranes, which is crucial to their role as natural biopesticides and plant protectants.

2.1 Disruption of Microbial Membranes

One of the primary mechanisms by which saponins act as antimicrobial agents is through the disruption of microbial cell membranes. Saponins are able to interact with the sterols present in the membranes of pathogens, leading to membrane destabilization and an increase in permeability. This causes leakage of vital cellular contents, including ions, proteins, and metabolites, ultimately resulting in cell death (Li & Monje-Galvan, 2023). The interaction between saponins and sterols is crucial because sterols are essential components of fungal and some bacterial membranes, but not present in the same concentration in plant cell membranes. This selectivity contributes to the effectiveness of saponins as antifungal agents.

2.1.1 Interaction with Sterols

The amphiphilic structure of saponins allows them to form complexes with membrane sterols, which are crucial for maintaining membrane integrity. Steroidal saponins, in particular, have a high affinity for cholesterol and other sterols commonly found in fungal membranes. The binding of saponins to sterols disrupts the membrane structure, leading to increased permeability and the eventual breakdown of the membrane. This disruption is especially effective against fungal pathogens, such as *Botrytis cinerea* and *Fusarium oxysporum*, which rely heavily on membrane integrity for survival and proliferation (Jabran et al., 2024). Studies have shown that saponins, such as avenacins from oats, specifically target fungal pathogens by binding to their sterols, which results in pore formation in the fungal membranes and leakage of cellular components (Papadopoulou et al., 1999). This selective action allows plants to defend themselves against fungi without harming their own cells. In the case of oat plants, the production of avenacins is particularly effective against soil-borne pathogens like *Gaeumannomyces graminis*, which causes take-all disease in cereals (Osbourn, 1996).

2.1.2 Membrane Permeabilization and Cell Lysis

Once saponins bind to the sterols in microbial membranes, they induce the formation of pores or channels, which increases the permeability of the membrane. This permeabilization leads to the uncontrolled flow of ions and small molecules into and out of the microbial cell, disrupting osmotic balance and leading to cell lysis (Papadopoulou et al., 1999). The ability of saponins to create these disruptions is particularly useful against plant pathogens that rely on maintaining osmotic balance and membrane potential for survival. Fungal pathogens, in particular, are highly susceptible to this mode of action due to their sterol-rich membranes. The disruption of fungal membranes by saponins results in the rapid collapse of cellular functions, preventing the pathogens from invading plant tissues or spreading infections. This mechanism is one of the reasons why saponin-rich plants, such as oats and quinoa, exhibit strong resistance to fungal diseases (Osbourn, 2003).

2.2 Inhibition of Pathogen Enzyme Activity

In addition to membrane disruption, saponins also interfere with the enzymatic machinery of pathogens. Many plant pathogens produce extracellular enzymes, such as cellulases, pectinases, and proteases, that degrade plant cell walls and allow for infection. Saponins can inhibit the activity of these enzymes, preventing the pathogens from breaking down plant tissues and establishing infections (Augustin et al., 2011). This enzyme inhibition provides an additional layer of protection for plants, reducing the ability of pathogens to invade and spread.

2.2.1 Inhibition of Cell Wall-Degrading Enzymes

Saponins have been shown to inhibit the activity of cell wall-degrading enzymes produced by fungal and bacterial pathogens. These enzymes, including cellulases and pectinases, are essential for the pathogens to penetrate plant tissues and cause disease. By inhibiting these enzymes, saponins prevent the breakdown of plant cell walls, effectively blocking the pathogen's ability to invade (Cheeke, 1996). For instance, saponins from *Quillaja saponaria* have been found to inhibit the activity of pectinases produced by the fungal pathogen *Botrytis cinerea*, which causes gray mold in a wide range of crops. By inhibiting the enzyme, saponins reduce the pathogen's virulence and limit its ability to cause disease (Trdá et al., 2019).

2.2.2 Inhibition of Proteolytic Enzymes

Saponins can also inhibit proteolytic enzymes secreted by pathogens to degrade plant proteins during infection. Proteases are crucial for pathogens to break down host proteins and obtain the nutrients needed for growth and reproduction. By inhibiting these enzymes, saponins disrupt the pathogen's metabolic processes and reduce its ability to establish an infection (Moses et al., 2014). This inhibitory action is particularly important in protecting plant tissues from necrotrophic pathogens that rely on proteolytic enzymes to kill host cells and feed on the dead tissue. By blocking the action of these enzymes, saponins help maintain the integrity of plant tissues and enhance the plant's overall resistance to pathogen attack.

2.3 Activation of Plant Defense Responses

In addition to their direct antimicrobial effects, saponins also play a role in activating plant defense mechanisms. Saponins can act as signaling molecules that trigger the expression of defense-related genes and the production of secondary metabolites involved in plant defense. This includes the activation of pathogenesis-related (PR) proteins, which have antimicrobial properties, and the production of reactive oxygen species (ROS) that contribute to pathogen destruction (Porsche et al., 2018). The ability of saponins to modulate plant defense responses adds another layer of protection, allowing plants to mount a rapid and effective defense against pathogens.

2.3.1 Induction of Systemic Acquired Resistance (SAR)

Systemic acquired resistance (SAR) is a plant-wide defense response that provides long-lasting protection against a broad spectrum of pathogens. Saponins have been shown to induce SAR by activating defense-related pathways, such as the salicylic acid (SA) signaling pathway. This leads to the expression of PR genes and the accumulation of defense proteins that enhance the plant's resistance to subsequent pathogen attacks (Durrant & Dong, 2004). For example, saponins from *Medicago truncatula* have been found to trigger the production of PR proteins in response to fungal infection, providing enhanced protection against pathogens like *Rhizoctonia solani* and *Fusarium spp.* (Porsche et al., 2018). This activation of SAR allows plants to develop long-term resistance to multiple pathogens, reducing the need for external chemical interventions in crop protection.

2.3.2 Modulation of Jasmonic Acid and Ethylene Pathways

Saponins also interact with other defense-related signaling pathways, such as the jasmonic acid (JA) and ethylene pathways, which are involved in the plant's response to herbivores and necrotrophic pathogens. By modulating these pathways, saponins enhance the production of defense compounds that inhibit pathogen growth and protect plant tissues from damage. This cross-talk between saponins and defense signaling pathways ensures a coordinated and effective response to a wide range of biotic stresses (Koornneef & Pieterse, 2008).

2.4 Synergistic Effects with Other Plant Metabolites

Saponins often work in conjunction with other plant defense compounds, such as phenolics, flavonoids, and terpenoids, to provide a multi-layered defense system against pathogens. This synergistic interaction enhances the antimicrobial activity of saponins, allowing plants to target pathogens through multiple mechanisms. For example, the combination of saponins and flavonoids has been shown to enhance the inhibition of fungal growth, providing more robust protection against diseases like powdery mildew and downy mildew (Moses et al., 2014).

3. Role of Saponins in Plant Defense Responses

Saponins play a crucial role in plant defense, providing a multifaceted protection system that works against a broad spectrum of biotic stressors, including pathogens, herbivores, and insect pests. As secondary metabolites, saponins contribute to both constitutive and induced defenses in plants. They are produced and stored in various plant tissues and are mobilized or activated in response to external threats. Their roles in plant defense extend beyond their direct antimicrobial properties, encompassing interactions with plant immune signaling pathways, involvement in programmed cell death, and collaboration with other plant metabolites to strengthen overall plant resilience.

3.1 Saponins as Direct Defensive Agents

Saponins act as direct defense agents by inhibiting the growth of and killing pathogens and herbivores that attempt to infect or consume plant tissues. Their amphiphilic nature allows them to interact with and disrupt cell membranes, which is a key mechanism of action against microbial pathogens and insects. The effectiveness of saponins as direct defensive agents can be attributed to their structural diversity, which enables them to target a variety of biotic stressors.

3.1.1 Antimicrobial Defense Against Pathogens

Saponins possess broad-spectrum antimicrobial activity against bacteria, fungi, and viruses. This activity is largely due to their ability to disrupt microbial cell membranes. The interaction between saponins and sterols in microbial membranes leads to increased permeability and leakage of intracellular contents, ultimately resulting in cell death. The membrane-disrupting properties of saponins are particularly effective against fungal pathogens, which rely heavily on membrane integrity for their survival (Osbourn, 2003). For example, avenacins, a class of saponins produced by oats, are highly effective at inhibiting fungal pathogens like *Gaeumannomyces graminis*, which causes take-all disease in cereals (Papadopoulou et al., 1999). The presence of saponins in the outer layers of plant tissues, such as the epidermis and roots, provides a physical and chemical barrier that deters pathogen invasion. This preemptive defense strategy is a form of constitutive resistance, where saponins are synthesized and stored in plant tissues even in the absence of a direct threat. When pathogens attempt to breach these tissues, they are met with high concentrations of saponins, which can inhibit their ability to penetrate and colonize plant cells.

3.1.2 Anti-Herbivore and Insecticidal Activity

In addition to their antimicrobial properties, saponins also serve as deterrents to herbivores and insects. The bitter taste and foaming properties of saponins make plant tissues unpalatable to herbivores, reducing the likelihood of consumption. Furthermore, saponins can disrupt the digestive systems of herbivores by binding to sterols in their gut membranes, leading to membrane destabilization and nutrient malabsorption (Cheeke, 1996). This anti-nutritional effect makes saponin-rich plants less attractive to herbivores, contributing to their defense against herbivory. Saponins also exhibit insecticidal properties, particularly against pests that feed on plant tissues. The ingestion of saponins by insects can disrupt their digestive enzymes, impair their ability to digest food and reduce their reproductive capacity. For example, saponins from *Quillaja saponaria* are effective against various insect pests, including aphids and caterpillars, by inhibiting their growth and development (Mordue & Nisbet, 2000). This insecticidal activity makes saponins valuable in integrated pest management (IPM) strategies, where they can be used as natural biopesticides to control insect populations.

3.2 Saponins as Modulators of Plant Immunity

In addition to their direct antimicrobial and anti-herbivore actions, saponins play a significant role in modulating plant immune responses. Plants have evolved complex defense signaling networks that enable them to recognize and respond to pathogen attacks. Saponins can act as signaling molecules that activate these immune responses, triggering both localized and systemic defenses.

3.2.1 Activation of Localized Defense Responses

When a pathogen or herbivore attacks a plant, the initial response often involves localized defense mechanisms at the site of infection or damage. Saponins are known to be involved in the activation of these localized responses by interacting with signaling molecules such as salicylic acid (SA), jasmonic acid (JA), and ethylene (Koornneef & Pieterse, 2008). These signaling pathways are crucial for coordinating plant defense responses, including the production of reactive oxygen species (ROS), the reinforcement of cell walls, and the expression of defense-related genes. One of the key localized responses activated by saponins is the production of pathogenesis-related (PR) proteins. PR proteins, such as chitinases and glucanases, have antimicrobial properties that help to degrade the cell walls of invading pathogens. The induction of PR proteins is part of the plant's innate immune system and is often accompanied by the accumulation of saponins at the site of infection, further enhancing the plant's resistance to pathogens (Zhang et al., 2022).

3.2.2 Systemic Acquired Resistance (SAR)

Saponins also play a role in the activation of systemic acquired resistance (SAR), a long-lasting and broad-spectrum immune response that provides protection against a wide range of pathogens. SAR is triggered by the recognition of pathogen-associated molecular patterns (PAMPs) or damage-associated molecular patterns (DAMPs), which are released during pathogen invasion or tissue damage. Saponins can act as DAMPs, triggering the SAR signaling pathway and leading to the production of defense-related compounds throughout the plant (Durrant & Dong, 2004). The activation of SAR involves the accumulation of salicylic acid and the expression of PR genes in distal tissues, providing systemic protection against future pathogen attacks. Saponins enhance this response by amplifying the signaling pathways that regulate SAR. For example, studies have shown that saponins from *Medicago truncatula* can induce SAR by stimulating the production of salicylic acid and activating defense-related genes in response to fungal infection (Zhang et al., 2022). This systemic defense mechanism allows plants to prepare for potential threats even in tissues that have not yet been attacked.

3.3 Programmed Cell Death and Hypersensitive Response (HR)

Programmed cell death (PCD) is another critical component of plant defense, particularly in the hypersensitive response (HR) to pathogen invasion. HR is a localized cell death response that occurs at the site of pathogen infection, effectively limiting the spread of the pathogen by depriving it of living tissue to colonize. Saponins have been implicated in the regulation of PCD and HR, contributing to the plant's ability to contain infections (Moses et al., 2014). Saponins can trigger the accumulation of ROS at the site of infection, which plays a key role in the initiation of PCD. ROS are highly reactive molecules that can cause oxidative damage to both plant and pathogen cells. The accumulation of ROS in the infected tissue leads to the destruction of infected cells, creating a physical barrier that prevents the pathogen from spreading to healthy tissues. The involvement of saponins in this process highlights their role in strengthening the plant's immune response through the regulation of cell death pathways.

3.4 Interaction with Other Plant Metabolites in Defense

Saponins do not act in isolation but often interact synergistically with other plant defense compounds, such as phenolics, flavonoids, and terpenoids. These interactions enhance the overall effectiveness of the plant's defense system, providing a multi-layered approach to combating pathogens and herbivores.

3.4.1 Synergy with Phenolics and Flavonoids

Phenolics and flavonoids are well-known for their antioxidant and antimicrobial properties. Saponins often work in conjunction with these compounds to provide enhanced protection against pathogens. For example, the combination of saponins and flavonoids has been shown to inhibit the growth of fungal pathogens more effectively than either compound alone (Moses et al., 2014). This synergy allows plants to target pathogens through multiple mechanisms, reducing the likelihood of pathogen resistance.

3.4.2 Interaction with Terpenoids

Terpenoids are another class of secondary metabolites that play a key role in plant defense. Saponins and terpenoids often accumulate in the same tissues, where they work together to provide a robust defense against herbivores and pathogens. The combination of saponins and terpenoids enhances the toxicity of plant tissues to herbivores, making them less palatable and more difficult to digest. This interaction is particularly important in plants that rely on both chemical and physical defenses to deter herbivores (Osbourn, 2003).

3.5 Induced Defense Responses to Biotic Stress

Saponins are also involved in induced defense responses, where their biosynthesis and accumulation are upregulated in response to biotic stress. In many plants, saponin production increases following herbivore attack or pathogen infection, providing enhanced protection during periods of heightened stress. This inducible defense mechanism allows plants to conserve energy by producing saponins only when they are needed, while still maintaining the ability to mount a rapid and effective defense when necessary. For example, studies have shown that the production of avenacins in oats increases significantly following infection by *Gaeumannomyces graminis* (Osbourn, 2003). Similarly, the levels of saponins in *Quinoa* increase in response to fungal infections, providing enhanced resistance to pathogens like *Fusarium spp.* (Papadopoulou et al., 1999). This inducible defense system allows plants to adapt to changing environmental conditions and biotic threats, ensuring their survival in hostile environments.

4. Saponins and Their Synergistic Effects with Other Plant Metabolites

Saponins, as multifunctional secondary metabolites, not only play an integral role in plant defense but also interact synergistically with other plant metabolites to enhance their protective capabilities. These interactions enable plants to employ a more efficient and multifaceted defense strategy against biotic and abiotic stressors, including pathogens, herbivores, and environmental challenges. The synergy between saponins and other metabolites like phenolics, flavonoids, terpenoids, and alkaloids strengthens the plant's defense system, offering more robust protection and reducing the likelihood of pathogen or pest resistance.

4.1 Synergy Between Saponins and Phenolics

Phenolics, including phenolic acids, tannins, and lignins, are widely recognized for their role in plant defense due to their antioxidant, antimicrobial, and structural reinforcement properties. Saponins and phenolics often work together to provide an enhanced defense system, particularly against microbial pathogens and herbivores. The combination of these two groups of compounds leads to an additive or even synergistic effect, which is more potent than either compound acting alone.

4.1.1 Enhanced Antimicrobial Activity

Saponins and phenolics, when combined, show significantly enhanced antimicrobial properties. Saponins, by disrupting microbial membranes, allow phenolics to penetrate more easily into microbial cells, where they can inhibit essential enzymes and disrupt cellular functions. This combination has been particularly effective against fungal pathogens, such as *Botrytis cinerea* and *Fusarium oxysporum*, where the presence of both saponins and phenolics leads to more efficient inhibition of spore germination, hyphal growth, and biofilm formation (Roca-Couso et al., 2021). For example, plants like *Medicago truncatula* and *Quillaja saponaria* produce saponins in conjunction with phenolic compounds. The interaction between these compounds provides a strong defense against fungal and bacterial pathogens by attacking them on multiple fronts: saponins disrupt cell membranes, while phenolics interfere with metabolic processes within the pathogen (Augustin et al., 2011). This interaction is particularly valuable in protecting plants from soil-borne pathogens that are difficult to combat with individual compounds.

4.1.2 Synergistic Role in Structural Defense

Phenolics, especially lignin and tannins, play a crucial role in reinforcing plant cell walls, making it harder for pathogens and herbivores to penetrate plant tissues. Saponins, by enhancing the plant's immune response, can increase the deposition of lignin and other phenolic compounds in cell walls. This leads to a strengthened structural barrier that is more resistant to pathogen attack and mechanical damage from herbivores. For example, the combination of saponins and lignin in plant roots provides a double-layered defense against root pathogens, such as *Phytophthora* spp. and nematodes, ensuring greater plant survival under high pathogen pressure (Zhang et al., 2022).

4.2 Saponins and Flavonoids: Antioxidant and Defense Enhancement

Flavonoids, a subclass of phenolics, are well-known for their antioxidant properties, which protect plants from oxidative stress caused by environmental factors such as UV radiation, drought, and pathogen infection. Saponins and flavonoids often work together to enhance the plant's defense system, particularly by amplifying antioxidant and antimicrobial responses.

4.2.1 Amplified Antioxidant Activity

Flavonoids, such as quercetin, kaempferol, and catechins, scavenge reactive oxygen species (ROS), preventing oxidative damage to plant cells. Saponins can enhance the accumulation of flavonoids in response to stress, boosting the plant's ability to neutralize ROS and protect vital cellular structures (Agati et al., 2012). This synergy is particularly beneficial in combating abiotic stressors like drought and UV radiation, which cause oxidative stress in plant tissues. The combination of saponins and flavonoids results in a more efficient antioxidant response, ensuring that plants can maintain cellular integrity and function under stressful conditions.

4.2.2 Enhanced Antimicrobial Properties

In addition to their antioxidant roles, flavonoids possess antimicrobial properties, and when combined with saponins, their effectiveness is significantly enhanced. Saponins disrupt microbial membranes, allowing flavonoids to penetrate deeper into the cells of pathogens, where they can inhibit vital processes like DNA replication, protein synthesis, and enzymatic activity (Cushnie & Lamb, 2011). This combination is particularly effective against bacterial pathogens like *Pseudomonas syringae* and fungal pathogens such as *Aspergillus niger*. The enhanced antimicrobial properties provided by this synergy offer a natural and potent defense against a wide range of crop diseases, making it an important focus for sustainable agriculture.

4.3 Interaction with Terpenoids: Strengthening Defense Against Herbivores

Terpenoids are another large class of secondary metabolites with significant roles in plant defense, particularly against herbivores and insect pests. Terpenoids such as monoterpenes, sesquiterpenes, and diterpenes exhibit repellant, toxic, and anti-feeding effects against herbivores. The combination of saponins and terpenoids provides a powerful defense against herbivores and insect pests, as these compounds work together to deter feeding and reduce pest fitness.

4.3.1 Anti-Feeding and Repellent Effects

Saponins, due to their bitter taste and toxic properties, make plant tissues less palatable to herbivores. When combined with terpenoids, which act as insect repellents or anti-feeding agents, the overall deterrent effect is significantly amplified. For instance, saponins from *Quinoa* and terpenoids from essential oils of plants like rosemary (*Rosmarinus officinalis*) have been shown to provide strong anti-feeding effects against pests such as aphids, whiteflies, and caterpillars (Mordue & Nisbet, 2000). The presence of both metabolites in plant tissues discourages herbivores from feeding and reduces the likelihood of pest outbreaks.

4.3.2 Synergistic Toxicity

Terpenoids are known for their toxic effects on insect pests, disrupting their nervous systems and hormonal pathways. When terpenoids are combined with saponins, the toxic effects are enhanced, leading to greater mortality rates among pests. This synergistic toxicity is particularly useful in controlling pests that are resistant to conventional chemical insecticides. Saponins disrupt the digestive systems of insects, while terpenoids interfere with their neural and reproductive functions, resulting in a more comprehensive and long-lasting pest control solution (Bakkali et al., 2008).

4.4 Collaboration with Alkaloids in Defense

Alkaloids are nitrogen-containing compounds that are highly toxic to herbivores and pathogens. They serve as neurotoxins and enzyme inhibitors, protecting plants from a variety of biotic stressors. The combination of saponins and alkaloids enhances the toxic effects of both groups of compounds, providing plants with a more robust defense against herbivores and pathogens.

4.4.1 Enhanced Neurotoxic Effects on Herbivores

Alkaloids, such as nicotine, caffeine, and morphine, act as neurotoxins, disrupting the nervous systems of herbivores and reducing their ability to feed and reproduce. When saponins are present alongside alkaloids, they enhance the uptake of alkaloids by herbivores, making the neurotoxic effects more potent (Cheeke, 1996). This interaction ensures that herbivores are deterred from feeding on the plant and suffer greater physiological damage if they do consume the plant tissues. This combination of saponins and alkaloids is particularly effective in deterring large herbivores, such as mammals, and controlling insect pests that are difficult to manage with single-compound defenses.

4.4.2 Synergistic Inhibition of Pathogen Enzymes

Alkaloids are also known to inhibit the activity of pathogen enzymes that are essential for infection and colonization. When combined with saponins, which disrupt pathogen membranes, the inhibitory effects on enzymes such as cellulases, pectinases, and proteases are amplified (Moses et al., 2014). This synergy is particularly effective against fungal pathogens like *Fusarium* and *Botrytis*, which rely on these enzymes to degrade plant cell walls and establish infections.

4.5 Applications in Sustainable Agriculture and Crop Protection

The synergistic interactions between saponins and other plant metabolites have significant implications for sustainable agriculture and crop protection. By leveraging these natural defense compounds, farmers can reduce their reliance on synthetic pesticides and fungicides, which are associated with environmental damage and the development of pathogen resistance. Plant extracts rich in saponins, phenolics, flavonoids, and terpenoids can be formulated into biopesticides that offer broad-spectrum protection against pests and diseases (Zaynab et al., 2018). Additionally, breeding or genetically engineering crops to enhance the production of these synergistic compounds can lead to more resilient crop varieties that require fewer chemical inputs. This approach not only benefits the environment but also reduces production costs and increases the sustainability of agricultural systems.

4.6 The Future of Synergistic Plant Defense Research

Research into the synergistic interactions between saponins and other plant metabolites is still in its early stages, but the potential applications are vast. Understanding how these compounds work together to enhance plant defense could lead to the development of new, more effective strategies for crop protection. Future research should focus on identifying the specific combinations of metabolites that provide the strongest defense against specific pests and pathogens, as well as optimizing the production of these compounds through genetic engineering or selective breeding (Moses et al., 2014).

5. Environmental and Agricultural Applications of Saponins in Pathogen Control

Saponins, with their potent antimicrobial, antifungal, and insecticidal properties, offer a promising natural alternative to synthetic chemicals for controlling plant pathogens and pests in agricultural settings. As concerns over the environmental and health impacts of chemical pesticides grow, the use of saponins in crop protection is gaining traction as a sustainable, eco-friendly solution. These plant-derived compounds are not only biodegradable but also exhibit selective toxicity, reducing harm to non-target organisms, including beneficial insects and soil microorganisms.

5.1 Saponins as Biopesticides

One of the most promising applications of saponins in agriculture is their use as natural biopesticides. Due to their ability to disrupt cell membranes and inhibit the growth of microbial pathogens, saponins provide an effective defense against a wide range of pests, including insects, nematodes, and microbial pathogens such as fungi and bacteria. Unlike synthetic pesticides, which often lead to the development of resistance in target organisms, saponins exhibit multiple modes of action, making it difficult for pests to evolve resistance (De Geyter et al., 2007). This makes saponins a valuable tool in sustainable agriculture, where the need for long-term pest control solutions is critical.

5.1.1 Insecticidal Activity of Saponins

Saponins have been widely studied for their insecticidal properties, particularly against soft-bodied insects such as aphids, whiteflies, and caterpillars. The insecticidal activity of saponins is largely attributed to their ability to disrupt the lipid membranes of insect gut cells, leading to membrane permeability, nutrient malabsorption, and eventual insect death (Mordue & Nisbet, 2000). Additionally, saponins interfere with the digestive enzymes of insects, further impairing their ability to digest plant tissues. Plants like *Quillaja saponaria*, a major source of commercial saponins, have been shown to exhibit strong insecticidal activity against a range of pests. The application of saponin-based biopesticides derived from *Quillaja* extracts has proven effective in controlling aphid and whitefly populations in crops like tomatoes and cotton, significantly reducing the damage caused by these pests (Bakkali et al., 2008). Furthermore, saponins exhibit repellant properties, discouraging insects from feeding on treated plants, which adds layer of protection against herbivores.

5.1.2 Nematocidal Effects of Saponins

In addition to their insecticidal properties, saponins have been shown to be highly effective in controlling plant-parasitic nematodes, which are responsible for significant crop losses worldwide. Nematodes damage plant roots, reducing water and nutrient uptake and leading to stunted growth and yield loss. Saponins, through their membrane-disrupting activity, are toxic to nematodes and can reduce their populations in the soil. For example, studies on *Medicago sativa* (alfalfa), which produces saponins in its root tissues, have demonstrated a significant reduction in the populations of root-knot nematodes (*Meloidogyne spp.*) when saponin-rich extracts were applied to the soil (D'Addabb et al., 2020). The use of saponins in nematode management is particularly appealing in organic farming systems, where chemical nematicides are prohibited, and sustainable, natural alternatives are required.

5.2 Saponins as Biofungicides

Saponins also play a critical role in the control of fungal pathogens in agriculture. Their antifungal properties are well-documented, particularly in their ability to disrupt fungal cell membranes, leading to increased permeability and eventual cell death. Saponins can be used as natural biofungicides to protect crops from a variety of fungal diseases, including powdery mildew, downy mildew, rust, and root rot.

5.2.1 Control of Soil-Borne Fungal Pathogens

Soil-borne fungal pathogens, such as *Fusarium*, *Phytophthora*, and *Rhizoctonia*, are among the most destructive plant pathogens, causing diseases like root rot, damping-off, and vascular wilts. Saponins have been shown to inhibit the growth of these pathogens by disrupting their cell membranes and preventing spore germination and mycelial growth. For instance, saponins derived from *Quinoa* have been successfully used to control *Fusarium oxysporum* in vegetable crops, reducing disease incidence and improving plant health (Osbourn, 2003).

The application of saponin-rich plant extracts to the soil can also enhance the soil's natural suppressiveness to fungal pathogens by promoting the growth of beneficial microorganisms, such as mycorrhizal fungi and antagonistic bacteria. These beneficial microbes outcompete fungal pathogens for resources, further reducing the likelihood of disease outbreaks.

5.2.2 Foliar Application for Fungal Disease Management

Saponins can also be applied directly to plant leaves to control foliar fungal diseases, such as powdery mildew and rust. Their ability to form a protective barrier on leaf surfaces prevents fungal spores from germinating and penetrating plant tissues. In addition, the surfactant properties of saponins enhance the spread of the biofungicide across the leaf surface, ensuring even coverage and protection (Augustin et al., 2011). Saponin-based biofungicides have been tested on a variety of crops, including grapes, tomatoes, and cucumbers, with promising results in reducing the incidence of foliar fungal diseases.

5.3 Saponins as Biostimulants

Beyond their role as biopesticides and biofungicides, saponins also act as biostimulants, promoting plant growth and enhancing stress tolerance. By improving the plant's immune responses, saponins help plants better withstand environmental stresses such as drought, salinity, and temperature extremes. Additionally, saponins can improve nutrient uptake by stimulating root growth and enhancing the solubility of soil nutrients.

5.3.1 Enhancing Plant Growth and Stress Tolerance

Saponins have been shown to improve plant growth by promoting root development and increasing the availability of nutrients in the soil. For example, saponins from *Quillaja saponaria* have been used to stimulate root growth in crops like lettuce, resulting in increased nutrient uptake and improved plant health (Moses et al., 2014). The application of saponins can also enhance the plant's ability to cope with abiotic stressors, such as drought and salinity, by activating stress-related signaling pathways and improving water retention in plant tissues.

5.3.2 Promotion of Beneficial Soil Microorganisms

Saponins also play a role in promoting the growth of beneficial soil microorganisms, such as nitrogen-fixing bacteria and mycorrhizal fungi, which are essential for maintaining soil fertility and plant health. These microorganisms help plants access nutrients that are otherwise unavailable, such as nitrogen, phosphorus, and trace minerals. By promoting the growth of these beneficial microbes, saponins contribute to a healthier, more productive soil ecosystem (Augustin et al., 2011).

5.4 Saponins in Integrated Pest Management (IPM)

The use of saponins fits well within the framework of integrated pest management (IPM), a strategy that combines biological, cultural, and chemical control methods to manage pests in a sustainable and environmentally friendly manner. As natural plant-based compounds, saponins can be used in conjunction with other IPM strategies, such as crop rotation, biological control agents, and resistant crop varieties, to reduce the reliance on synthetic chemicals.

5.4.1 Compatibility with Biological Control Agents

One of the key advantages of using saponins in IPM is their compatibility with biological control agents, such as predatory insects and beneficial nematodes. Unlike broad-spectrum chemical pesticides, which often kill beneficial organisms along with pests, saponins are selective in their mode of action, targeting specific pests without harming natural enemies. This selectivity makes saponins an ideal component of IPM programs, where the goal is to maintain pest populations at manageable levels while preserving biodiversity and ecosystem health (Mordue & Nisbet, 2000).

5.4.2 Reducing Pesticide Resistance

Another important aspect of saponins in IPM is their potential to reduce the development of pesticide resistance in pest populations. Synthetic chemical pesticides often lead to resistance due to their single-mode action, which allows pests to evolve resistance mechanisms. In contrast, saponins, with their multi-targeted action, make it much harder for pests to develop resistance. The combination of saponins with other control methods, such as biological control and cultural practices, provides a more sustainable approach to pest management that reduces the risk of resistance development (Moses et al., 2014).

5.5 Environmental Benefits of Saponin Use in Agriculture

The environmental benefits of using saponins in agriculture are numerous. As natural, biodegradable compounds, saponins do not persist in the environment, reducing the risk of soil and water contamination. Their selective toxicity ensures that non-target organisms, such as pollinators and beneficial insects, are not harmed, which is a significant advantage over conventional synthetic pesticides.

5.5.1 Reducing Chemical Pesticide Use

The application of saponins as biopesticides and biofungicides can significantly reduce the need for synthetic chemical inputs in agriculture. This reduction in chemical pesticide use not only lowers production costs for farmers but also minimizes the environmental and health risks associated with chemical residues in food and water supplies. Saponin-based products can be integrated into organic farming systems, where the use of synthetic chemicals is restricted, providing a natural and effective solution for pest and disease control.

5.5.2 Enhancing Soil Health and Biodiversity

Saponins contribute to healthier soils by promoting the growth of beneficial microorganisms and improving nutrient cycling. Their role in enhancing the activity of nitrogen-fixing bacteria and mycorrhizal fungi helps to maintain soil fertility, reducing the need for synthetic fertilizers. Additionally, the use of saponins in pest management supports biodiversity by preserving the populations of beneficial insects and pollinators, which are essential for ecosystem balance and agricultural productivity (D'Addabb et al., 2020).

5.6 Future Prospects for Saponins in Agriculture

The future of saponins in agriculture looks promising, with ongoing research focusing on optimizing their extraction, formulation, and application for large-scale use. Advances in biotechnology may allow for the enhancement of saponin production in plants through genetic engineering, enabling the development of crop varieties with increased saponin content for enhanced pest and disease resistance (Osbourn, 2003). The continued exploration of saponins' synergistic interactions with other plant metabolites also holds potential for the development of more effective, multi-targeted biopesticides and biofungicides. As the demand for sustainable agricultural practices grows, saponins will play an increasingly important role in reducing the environmental impact of farming while ensuring crop protection and food security.

6. Challenges and Future Prospects in Saponin Research and Utilization

Saponins have gained considerable attention for their diverse biological activities, particularly in plant defense, medicine, and agriculture. However, despite their immense potential, the utilization of saponins in practical applications faces several significant challenges. These challenges range from issues related to their extraction and purification to the complexity of their biosynthesis, variability in saponin content across plant species, and their potential toxicity to non-target organisms. Additionally, saponin research is still in its developmental stages, with much to be explored regarding their mechanisms of action, ecological roles, and large-scale applications.

6.1 Variability in Saponin Content and Composition

One of the most significant challenges in utilizing saponins is the variability in their content and composition across different plant species, cultivars, and even environmental conditions. Saponins are secondary metabolites, and their production in plants is often influenced by factors such as plant age, growth conditions, soil composition, and environmental stressors. As a result, the concentration and specific types of saponins present in a plant can vary widely, making it difficult to standardize their extraction and use (Osbourn, 2003). This variability poses a challenge for developing consistent saponin-based products, whether for agricultural or medicinal purposes.

6.1.1 Influence of Environmental Factors

Environmental conditions, such as soil type, temperature, light intensity, and water availability, have a significant impact on saponin biosynthesis in plants. For example, drought stress has been shown to increase the production of certain saponins in plants like *Medicago sativa* (alfalfa), whereas excessive water or nutrient availability can reduce saponin levels (Moses et al., 2014). This variability complicates efforts to produce saponins in a controlled and predictable manner for commercial applications. Additionally, different plant species produce structurally diverse saponins, and even within a single species, the composition of saponins can vary. For instance, Quinoa (*Chenopodium quinoa*) and Ginseng (*Panax ginseng*) produce triterpenoid saponins, while *Dioscorea* species produce steroidal saponins. These differences in saponin composition affect their biological activities and potential applications, making it necessary to develop specific extraction and processing techniques for each type of saponin (Sharma et al., 2023).

6.2 Challenges in Extraction, Purification, and Standardization

The extraction and purification of saponins present another major challenge. Saponins are complex molecules that exist in various forms within plant tissues, often bound to other secondary metabolites such as phenolics, flavonoids, and polysaccharides. As a result, extracting pure saponins from plant materials is a time-consuming and expensive process that requires advanced techniques like chromatography, ultrafiltration, and solvent extraction (Augustin et al., 2011). These methods are not always cost-effective for large-scale production, particularly in agriculture, where the cost of inputs is a critical consideration.

6.2.1 Difficulties in Purification and Structural Characterization

The structural complexity of saponins, with their various aglycones and glycoside chains, makes their purification and structural characterization challenging. Saponins are often found in mixtures with other plant metabolites, and separating individual saponin compounds from these mixtures requires sophisticated analytical techniques like high-performance liquid chromatography (HPLC) and mass spectrometry (MS) (Moses et al., 2014). The complexity of saponin structures also complicates their identification and quantification, making it difficult to ensure consistent quality and efficacy in saponin-based products.

6.2.2 Need for Standardization

Standardization is essential for the commercial application of saponins, particularly in the development of biopesticides, biofungicides, and medicinal products. However, the lack of standardized protocols for extracting, purifying, and quantifying saponins has hindered their widespread adoption. Each plant species produces a unique profile of saponins, and without standardized methods, it is challenging to ensure that the extracted saponins have consistent biological activity. Establishing reliable standards for saponin content, purity, and bioactivity is crucial for the development of commercial saponin products (Osbourn, 2003).

6.3 Potential Toxicity to Non-Target Organisms

While saponins are known for their selective toxicity to pathogens, pests, and herbivores, they can also pose a risk to non-target organisms, including beneficial insects, soil microorganisms, and aquatic life. Saponins have detergent-like properties that can disrupt cell membranes, and their broad-spectrum activity can affect a wide range of organisms. For example, saponins from *Quillaja saponaria* have been shown to exhibit toxicity to aquatic organisms, raising concerns about their environmental impact if used in large quantities in agricultural settings (Jiang et al., 2018).

6.3.1 Ecotoxicological Concerns

The potential for saponins to accumulate in the environment and affect non-target species is a significant concern, particularly in ecosystems where water bodies are adjacent to agricultural fields. Saponins are known to be toxic to fish and other aquatic organisms at high concentrations, and their use in biopesticides or biofungicides could lead to unintended ecological consequences. Therefore, careful consideration of dosage, application methods, and environmental fate is necessary when using saponin-based products in agriculture (Bakkali et al., 2008).

6.3.2 Mitigating Non-Target Effects

To mitigate the potential toxicity of saponins to non-target organisms, researchers are exploring ways to modify saponin formulations or develop controlled-release delivery systems that limit their environmental impact. For example, encapsulating saponins in biodegradable polymers or using targeted delivery methods such as seed coatings or root drenches could help reduce their exposure to non-target species while maintaining their effectiveness against pests and pathogens (Augustin et al., 2011).

6.4 Limitations in Large-Scale Production

The large-scale production of saponins is currently limited by several factors, including the complexity of their biosynthesis and the availability of suitable plant sources. Saponins are typically produced in small quantities in plants, and extracting them in sufficient amounts for commercial applications can be challenging. Additionally, the cultivation of saponin-producing plants, such as *Quillaja saponaria* or *Panax ginseng*, is often labor-intensive and requires specific environmental conditions, limiting the scalability of saponin production (Moses et al., 2014).

6.4.1 Challenges in Cultivation and Harvesting

The cultivation of saponin-producing plants is often constrained by geographical and climatic factors. For example, *Quillaja saponaria*, a major source of commercial saponins, is native to South America and grows primarily in regions with specific soil and climate conditions. This geographic limitation makes it difficult to scale up saponin production for global markets. Additionally, the long growth cycles of some saponin-producing plants, such as ginseng, further limit their availability for large-scale extraction (Jiang et al., 2018).

6.4.2 Advances in Biotechnological Approaches

To address the challenges of large-scale saponin production, researchers are exploring biotechnological approaches such as genetic engineering and synthetic biology. By identifying and manipulating the genes involved in saponin biosynthesis, scientists aim to enhance the production of saponins in plants or microbial systems. For example, genetic modification of crop plants to overproduce saponins could lead to the development of pest-resistant varieties that require fewer chemical inputs (Moses et al., 2014). Additionally, microbial fermentation systems, using engineered bacteria or yeast, could provide an alternative means of producing saponins in large quantities, offering a more sustainable and scalable solution.

6.5 Future Prospects in Saponin Research

Despite the challenges associated with saponin research and utilization, the future prospects for their application in agriculture, medicine, and biotechnology remain promising. Advances in genetic engineering, synthetic biology, and nanotechnology hold great potential for overcoming the current limitations of saponin production and utilization.

6.5.1 Genetic Engineering and Synthetic Biology

Genetic engineering offers significant potential for enhancing saponin biosynthesis in plants. By identifying the key enzymes involved in saponin production, researchers can manipulate the metabolic pathways to increase saponin yield or produce specific saponin types with desired biological activities (Moses et al., 2014). Additionally, synthetic biology approaches, such as the design of microbial systems for saponin production, could provide a cost-effective and scalable alternative to plant-based extraction. These advances could lead to the development of new, highly effective saponin-based biopesticides, biofungicides, and medicinal compounds.

6.5.2 Nanotechnology in Saponin Delivery

Nanotechnology also offers exciting prospects for improving the delivery and efficacy of saponin-based products. Nanoparticle formulations of saponins could enhance their stability, bioavailability, and targeted delivery, reducing the required dosage and minimizing non-target effects (Bakkali et al., 2008). For example, encapsulating saponins in nanocarriers could protect them from degradation in the environment and ensure that they are released only when they reach their target, such as a pest or pathogen.

6.5.3 Saponins in Sustainable Agriculture and Medicine

The growing demand for sustainable agricultural practices and natural alternatives to synthetic pesticides and fungicides will likely drive further research and development in saponin-based products. As consumers and policymakers increasingly prioritize eco-friendly and health-conscious solutions, saponins will play an essential role in the future of sustainable crop protection, integrated pest management, and biomedicine.

Conclusion

Saponins, with their complex biosynthesis and structural diversity, are critical contributors to plant defense mechanisms. Their biosynthesis involves the mevalonate and methylerythritol phosphate (MEP) pathways, resulting in diverse structures that include triterpenoid or steroid aglycones linked to sugar moieties. This structural variety underpins their versatility in biological activities, particularly in their antimicrobial properties. Saponins act as potent antimicrobial agents, disrupting the integrity of microbial cell membranes, inhibiting vital enzymes, and interfering with pathogenic signaling processes. These mechanisms allow saponins to protect plants from a broad spectrum of pathogens, including bacteria, fungi, and viruses, making them essential components of plant defense systems.

In addition to their direct antimicrobial action, saponins play a crucial role in eliciting plant defense responses. By activating systemic acquired resistance (SAR) and other defense pathways, saponins enhance the plant's ability to respond to environmental threats and biotic stressors. Their role in strengthening the physical and chemical barriers of plants adds another layer of protection, making them indispensable in the natural immune system of plants. Furthermore, saponins often work synergistically with other plant secondary metabolites, such as phenolics and flavonoids, amplifying their overall effectiveness. These interactions enhance the plant's resistance to pathogens and environmental stresses, showcasing the complex and integrated nature of plant defense systems. The potential applications of saponins in agriculture are vast, particularly in sustainable crop protection. Saponins offer a natural, eco-friendly alternative to synthetic pesticides, which can have harmful environmental impacts. As biopesticides or plant growth stimulants, saponins can be incorporated into integrated pest management (IPM) programs, reducing reliance on chemical inputs and promoting healthier ecosystems. However, the widespread application of saponins faces several challenges, including variability in their activity across plant species, difficulties in large-scale production, and the need for a deeper understanding of their synergistic interactions with other plant metabolites.

Looking ahead, the future of saponin research and utilization holds great promise. Advances in biotechnology, particularly in genetic engineering and synthetic biology, are poised to address some of the challenges associated with the production and optimization of saponins. By enhancing the biosynthetic pathways of saponins and improving their scalability, these innovations could enable the broader use of saponins in agriculture. Additionally, continued research into their antimicrobial mechanisms, synergistic effects, and environmental applications will be crucial for fully harnessing their potential. As the demand for sustainable and natural crop protection solutions grows, saponins are likely to play an increasingly significant role in pathogen control and crop resilience, positioning them as vital tools for the future of agriculture and plant health.

References

Agati, G., Azzarello, E., Pollastri, S., & Tattini, M. (2012). Flavonoids as antioxidants in plants: Location and functional significance. *Plant Science, 196*, 67-76.

Augustin, J. M., Kuzina, V., Andersen, S. B., & Bak, S. (2011). Molecular activities, biosynthesis, and evolution of triterpenoid saponins. *Phytochemistry, 72*(6), 435-457.

Bakkali, F., Averbeck, S., Averbeck, D., & Idaomar, M. (2008). Biological effects of essential oils: A review. *Food and Chemical Toxicology, 46*(2), 446-475.

Cheeke, P. R. (1996). Biological effects of feed and forage saponins and their impacts on animal production. Advances in Experimental Medicine and Biology, *405*, 377-385.

Cushnie, T. P., & Lamb, A. J. (2011). Recent advances in understanding the antibacterial properties of flavonoids. *International Journal of Antimicrobial Agents, 38*(2), 99-107.

De Geyter, E., Lambert, E., Geelen, D., & Smagghe, G. (2007). Novel advances with plant saponins as natural insecticides to control pest insects. *Pest Technology, 1*, 96-105.

Durrant, W. E., & Dong, X. (2004). Systemic acquired resistance. *Annual Review of Phytopathology, 42*, 185-209.

D'Addabbo, T., Argentieri, M. P., Żuchowski, J., Biazzi, E., Tava, A., Oleszek, W., & Avato, P. (2020). Activity of saponins from medicago species against phytoparasitic mematodes. *Plants, 9*(4), 443.

El Aziz, M. M. A., Ashour, A. S., & Melad, A. S. G. (2019). A review on saponins from medicinal plants: chemistry, isolation, and determination. *Journal of Nanomedicine Research, 8*(1), 282-288.

Hussain, M., Debnath, B., Qasim, M., Bamisile, B. S., Islam, W., Hameed, M. S., Wang, L., & Qiu, D. (2019). Role of saponins in plant defense against specialist herbivores. *Molecules, 24*(11), 2067.

Jabran, M., Ali, M.A., & Muzammil, S. (2024). Exploring the potential of nanomaterials (NMs) as diagnostic tools and disease resistance for crop pathogens. *Chemical and Biological Technologies in Agriculture, 11*, 75.

Jiang, X., Cao, Yi., von Gersdorff, J. L., Strobel, B. W., Hansen, H. C. B., & Cedergreen, N. (2018). Where does the toxicity come from in saponin extract? *Chemosphere, 204*, 243-250.

Koornneef, A., & Pieterse, C. M. (2008). Cross talk in defense signaling. *Plant Physiology, 146*(3), 839-44.

Li, J., & Monje-Galvan, V. (2023). *In vitro* and *in silico* studies of antimicrobial saponins: a review. *Processes, 11*(10), 2856.

Mordue, A. J., & Nisbet, A. J. (2000). Azadirachtin from the neem tree *Azadirachta indica*: Its action against insects. *Anais da Sociedade Entomológica do Brasil, 29*(4), 615-632.

Moses, T., Papadopoulou, K. K., & Osbourn, A. (2014). Metabolic and functional diversity of saponins, biosynthetic intermediates, and semi-synthetic derivatives. *Critical Reviews in Biochemistry and Molecular Biology, 49*(6), 439-462.

Mugford, S. T., & Osbourn, A. (2012). Saponin Synthesis and Function. *Isoprenoid Synthesis in Plants and Microorganisms. 22*, 405-424.

Osbourn, A. E. (1996). Saponins and plant defence-a soap story. *Trends in Plant Science, 1*(1), 4-9.

Osbourn, A. E. (2003). Saponins in cereals. *Phytochemistry, 62*(1), 1-4.

Papadopoulou, K., Melton, R. E., Leggett, M., Daniels, M. J., & Osbourn, A. E. (1999). Compromised disease resistance in saponin-deficient plants. *Proceedings of the National Academy of Sciences, 96*(22), 12923-12928.

Porsche, F. M., Molitor, D., Beyer, M., Charton, S., Andre, C., & Kollar, A. (2018). Antifungal activity of saponins from the fruit pericarp of *Sapindus mukorossi* against *Venturia inaequalis* and *Botrytis cinerea*. *Plant Disease, 102*(5), 991-1000.

Rai, S., Acharya-Siwakoti, E., Kafle, A., Devkota, H. P., & Bhattarai, A. (2021). Plant-derived saponins: A review of their surfactant properties and applications. *Sci, 3*(4), 44.

Roca-Couso, R., Flores-Félix, J. D., & Rivas, R. (2021). Mechanisms of action of microbial biocontrol agents against *Botrytis cinerea*. *Journal of Fungi (Basel), 7*(12), 1045.

Sharma, K., Kaur, R., Kumar, S., Saini, R. K., Sharma, S., Pawde, S. V., & Kumar, V. (2023). Saponins: A concise review on food related aspects, applications and health implications. *Food Chemistry Advances, 2*, 100191.

Trdá, L., Janda, M., Macková, D., Pospíchalová, R., Dobrev, P. I., Burketová, L., & Matušinsky, P. (2019). Dual mode of the saponin aescin in plant protection: Antifungal agent and plant defense elicitor. *Frontiers in Plant Sciences, 10*, 1448.

Voutquenne, L., Lavaud, C., Massiot, G., & Men-Olivier, L. (2008). Structure-activity relationships of haemolytic saponins. *Pharmaceutical Biology*, *40*, 253-262.

Zaynab, M., Fatima, M., Abbas, S., Sharif, Y., Umair, M., Zafar, M. H., & Bahadar, K. (2018). Role of secondary metabolites in plant defense against pathogens. *Microbial Pathogenesis*, *124*, 198-202.

Zhang, S., Li, C., Si, J., Han, Z., & Chen, D. (2022). Action mechanisms of effectors in plant-pathogen interaction. *International Journal of Molecular Sciences*, 23(12), 6758.

Chapter - 7

Glycosides: Bioactive Secondary Metabolites in Disease Management

Reshma Jadhav[1], Susheel Kumar[2], Zuby Gohar Ansari[3], Praneet Chauhan[4] and Vinod Chouhan[5]

[1]Department of Biochemistry, Mahatma Phule Krishi Vidyapeeth, Rahuri, Ahmednagar, Maharashtra, India
[2]Plant Molecular Virology Laboratory, CPMB Division, CSIR-National Botanical Research Institute, Rana Pratap Marg, Lucknow, Uttar Pradesh, India
[3]B.M. College of Agriculture, Jaswadi Road Khandwa, Madhya Pradesh, India
[4]Department of Plant Pathology, Dr. Khem Singh Gill Akal College of Agriculture, Eternal University, Baru Sahib, Sirmour, Himachal Pradesh, India
[5]Division of Plant Pathology, Indian Council of Agricultural Research-Indian Agricultural Research Institute (ICAR-IARI), New Delhi, India
[4]Department of Biochemistry, Mahatma Phule Krishi Vidyapeeth, Rahuri, Ahmednagar, Maharashtra, India

Abstract

Glycosides are a diverse group of bioactive secondary metabolites found in plants, known for their critical role in plant defense and disease management. These compounds consist of a sugar moiety attached to a non-sugar component, typically a phenolic, terpenoid, or steroid structure, giving them a wide range of biological activities. Glycosides are synthesized through complex pathways involving the attachment of sugar molecules to various aglycones, leading to structural diversity that enhances their functional roles in plants. Their antimicrobial properties make them effective against a broad spectrum of plant pathogens, including bacteria, fungi, and viruses. Glycosides function by disrupting microbial cell walls, inhibiting vital enzymes, and interfering with pathogen communication systems. Additionally, they are integral to the plant's defense system, often working in concert with other secondary metabolites such as flavonoids, saponins, and alkaloids, to enhance the overall resilience of plants against biotic and abiotic stressors. The synergistic effects between glycosides and other metabolites amplify their defensive capabilities, making them essential for plant survival. In agriculture, glycosides hold promise as natural alternatives to synthetic agrochemicals, providing environmentally friendly solutions for disease management and crop protection. Their potential use in integrated pest management (IPM) strategies could reduce the reliance on chemical pesticides, contributing to sustainable agricultural

DOI: 10.1201/9781003717454-7

practices. However, challenges such as the variability in glycoside production among plant species, as well as difficulties in large-scale synthesis and application, remain significant hurdles. Advances in biotechnological approaches, including genetic engineering and synthetic biology, offer promising avenues to enhance glycoside production and utilization in agriculture.

Keywords: Glycosides, plant defense, antimicrobial, secondary metabolites, sustainable agriculture, disease management.

Introduction

Glycosides represent a diverse class of naturally occurring compounds that play vital roles in plant physiology and defense. Structurally, glycosides consist of a sugar moiety (glycone) bound to a non-sugar component (aglycone), which can be a phenolic, terpenoid, or flavonoid, among others. This diversity in aglycones gives rise to various types of glycosides with distinct biological activities. Glycosides are ubiquitous in the plant kingdom and are involved in multiple processes, including growth regulation, signal transduction, and defense against pathogens and herbivores. Additionally, these compounds are of great interest for their potential applications in agriculture and medicine, particularly as natural alternatives to synthetic chemicals. The biosynthesis, classification, and antimicrobial properties of glycosides, along with their interactions with other plant metabolites, make them valuable contributors to plant defense and ecological sustainability.

The biosynthesis of glycosides is a highly regulated process that involves the attachment of sugar moieties to aglycones through glycosylation. This reaction is catalyzed by a family of enzymes known as glycosyltransferases (GTs), which facilitate the transfer of sugars from activated sugar donors (nucleotide sugars) to specific acceptor molecules (He et al., 2024). The type of sugar, such as glucose, rhamnose, or xylose, and the nature of the aglycone determine the biological properties of the resulting glycoside. For instance, glucosides, which involve glucose as the sugar moiety, are common in plants and serve as storage forms for many secondary metabolites. Glycosides are classified based on the type of aglycone they contain. The major categories include flavonoid glycosides, saponin glycosides, phenolic glycosides, and terpenoid glycosides. Each category of glycosides has unique properties and roles in plant biology. For example, flavonoid glycosides are involved in UV protection and pigmentation, while saponin glycosides contribute to plant defense through their detergent-like properties (Osbourn, 2003). The structural diversity of glycosides allows plants to employ them in various ecological roles, including defense, pollinator attraction, and symbiotic interactions.

Glycosides are well-documented for their antimicrobial properties, which are a key part of plant defense mechanisms. These compounds can inhibit the growth of bacteria, fungi, and viruses through several mechanisms. One of the primary modes of action is membrane disruption, particularly in the case of saponin glycosides. Saponins interact with microbial cell membranes, leading to increased permeability and cell lysis, a process similar to how saponins affect the membranes of herbivores (Papadopoulou et al., 1999). Phenolic glycosides, such as salicin, exert their antimicrobial effects by inhibiting the activity of enzymes essential for microbial survival. These glycosides can also interfere with pathogen cell wall integrity and disrupt metabolic processes (Cushnie & Lamb, 2011). Additionally, the sugar moiety in glycosides allows for enhanced solubility and transport of these compounds within the plant, enabling them to be readily mobilized in response to pathogen attacks. The wide variety of glycosides found in plants provides a versatile arsenal against different types of pathogens, making them a crucial component of plant immune systems.

In plant defense, glycosides act as both direct toxins to pathogens and herbivores and as precursors to more potent defense compounds. For instance, cyanogenic glycosides are stored in an inactive form within plant tissues but are rapidly hydrolyzed to release toxic hydrogen cyanide when the plant is damaged by herbivores or pathogens (Hussain et al., 2019). This defense mechanism is particularly effective in deterring herbivores and preventing further damage to plant tissues. Glycosides also play a role in signaling pathways that activate broader plant immune responses. Upon pathogen recognition, plants can upregulate the production of specific glycosides, which, in turn, trigger the activation of defense-related genes. For example, glucosinolates, a class of sulfur-containing glycosides, are involved in the activation of the salicylic acid and jasmonic acid pathways, both of which are essential for mounting an effective immune response in plants (Ting et al., 2020). The ability of glycosides to act both as toxins and as signaling molecules highlights their multifaceted role in plant defense.

Glycosides often interact synergistically with other plant secondary metabolites, such as phenolics, flavonoids, and alkaloids, to enhance plant defense mechanisms. These interactions result in a more effective and multi-layered defense strategy against pathogens and herbivores. For example, the combination of saponins and phenolics has been shown to enhance the antimicrobial activity against fungal pathogens, such as *Botrytis cinerea*, compared to the individual compounds acting alone (Moses et al., 2014). This synergistic effect is likely due to the complementary modes of action, with glycosides disrupting microbial cell membranes while phenolics inhibit enzymatic activity. Additionally, glycosides and terpenoids can work together to deter herbivores. Terpenoid glycosides, such as cardenolides, are toxic to herbivores and disrupt their nervous systems, while the sugar moieties in these glycosides enhance their transport and accumulation in plant tissues. This combination ensures that herbivores are not only repelled by the bitter taste of glycosides but also suffer physiological consequences if they ingest plant tissues rich in these compounds (Sharma et al, 2023).

The application of glycosides in sustainable agriculture is gaining significant attention due to their natural origin and environmental friendliness. Glycosides can serve as biopesticides, biofungicides, and biostimulants, offering an alternative to synthetic agrochemicals that are often associated with environmental pollution and health risks. For example, saponin glycosides have been used as natural biopesticides to control insect pests and nematodes in crops, reducing the need for chemical insecticides (Hussain et al., 2019). Phenolic glycosides, on the other hand, have been employed in biofungicides for controlling fungal pathogens such as *Fusarium* and *Botrytis*, offering a more sustainable approach to disease management in agriculture. In addition to their pest control applications, glycosides also play a role in promoting plant growth and resilience by enhancing nutrient uptake and stress tolerance. Their ability to promote beneficial soil microorganisms, such as nitrogen-fixing bacteria and mycorrhizal fungi, further contributes to sustainable agricultural practices by improving soil health and reducing the need for chemical fertilizers (He et al., 2024).

Despite their potential, the research and utilization of glycosides face several challenges. One of the primary challenges is the variability in glycoside content and composition among plant species, which can affect the consistency and efficacy of glycoside-based products. Additionally, the extraction and purification of glycosides can be complex and costly, limiting their large-scale application in agriculture and medicine. The potential toxicity of glycosides to non-target organisms, such as beneficial insects and soil microorganisms, also raises concerns about their environmental impact, particularly when used in high concentrations (Bakkali et al., 2008). However, advances in biotechnological approaches, such as genetic engineering and synthetic biology, offer promising avenues for overcoming these challenges. By enhancing glycoside production in plants or developing microbial systems for glycoside biosynthesis, researchers can increase the availability of these valuable compounds for commercial use. The future of glycoside research lies in optimizing their production and application for sustainable agriculture and disease control while minimizing their environmental impact and ensuring their long-term viability as natural plant defense agents (Moses et al., 2014).**1.**

1. Biosynthesis and Classification of Glycosides in Plants

Glycosides are a vast group of secondary metabolites found in nearly all plant species, playing significant roles in plant defense, metabolism, growth regulation, and interactions with the environment. The structure of glycosides is based on a sugar component (glycone) attached to a non-sugar moiety (aglycone), which can range from phenolic compounds to flavonoids, terpenoids, or even alkaloids. This diversity of aglycones provides glycosides with a wide array of biological functions and ecological roles. Their biosynthesis is highly regulated, involving a series of enzymatic reactions that ensure glycoside formation in response to specific internal and external stimuli. Glycosides can be classified based on the nature of their aglycone, and this classification gives insight into their functional diversity and ecological importance.

1.1 Biosynthesis of Glycosides

The biosynthesis of glycosides involves several key steps, starting from the formation of the aglycone followed by the addition of one or more sugar units. These steps are controlled by specific enzymes that determine the chemical nature and bioactivity of the resulting glycoside. The ability of plants to produce a wide range of glycosides depends on the diversity of glycosylation processes, which are catalyzed by enzymes known as glycosyltransferases (GTs).

1.1.1 Formation of the Aglycone

The biosynthesis of glycosides begins with the formation of the aglycone, which varies depending on the type of glycoside. Aglycones can be derived from primary or secondary metabolites, including flavonoids, terpenoids, phenolics, and alkaloids. For instance, flavonoid glycosides are synthesized from flavonoid aglycones through a series of steps involving the shikimate and phenylpropanoid pathways (Metsämuuronen & Sirén, 2019). Terpenoid glycosides, on the other hand, are derived from the mevalonate or methylerythritol phosphate (MEP) pathways, leading to the formation of aglycones such as triterpenoids and steroidal backbones. This initial step of aglycone formation is critical because it sets the stage for subsequent glycosylation, which will transform these often-hydrophobic compounds into more water-soluble glycosides that can be easily transported and stored within plant tissues. The regulation of aglycone formation is tightly linked to plant developmental stages and environmental conditions, particularly in response to biotic stress like pathogen attack or herbivory (Moses et al., 2014).

1.1.2 Glycosylation Process

Once the aglycone is synthesized, the next step in glycoside biosynthesis involves the attachment of sugar moieties. This glycosylation process is catalyzed by glycosyltransferases, a large family of enzymes that facilitate the transfer of activated sugar molecules (such as UDP-glucose or UDP-rhamnose) to the hydroxyl, carboxyl, or amine groups of the aglycone. This step is crucial for determining the biological activity and solubility of the glycoside. The glycosylation of aglycones can vary significantly between plant species and even within the same plant depending on tissue type and developmental stage. For example, flavonoid glycosides such as quercetin and kaempferol glycosides may carry different sugar units (glucose, rhamnose, or galactose), influencing their antioxidant activity and transport within the plant (He et al., 2024). The sugar moiety also plays a role in modulating the bioavailability and toxicity of glycosides, allowing plants to sequester potentially harmful compounds in an inactive form that can be rapidly activated when needed for defense.

1.1.3 Storage and Activation

After glycosylation, glycosides are usually stored in vacuoles or other cellular compartments in an inactive form. This sequestration prevents the toxic effects of many glycosides' aglycones from harming the plant itself. Upon damage, such as during herbivore feeding or pathogen attack, the glycosides can be hydrolyzed by enzymes like glycosidases, releasing the active aglycone that can act as a deterrent or toxin to the attacker. For example, cyanogenic glycosides such as amygdalin and dhurrin are stored in plant tissues but are rapidly hydrolyzed to release hydrogen cyanide when the plant is damaged. This activation mechanism is a crucial part of the plant's defense system, providing an immediate chemical response to biotic stressors (Hussain et al., 2019).

1.2 Classification of Glycosides

Glycosides are classified into several major categories based on the nature of their aglycone. This classification not only reflects the chemical diversity of glycosides but also their various ecological functions and biological activities. The major types of glycosides include flavonoid glycosides, saponin glycosides, phenolic glycosides, terpenoid glycosides, and alkaloid glycosides, each with distinct roles in plant metabolism and defense.

1.2.1 Flavonoid Glycosides

Flavonoid glycosides are among the most common and widely studied glycosides in plants. They consist of flavonoid aglycones, such as quercetin, kaempferol, and luteolin, attached to one or more sugar units. Flavonoid glycosides are involved in a range of biological functions, including UV protection, pigmentation, and defense against pathogens. Their antioxidant properties also play a key role in protecting plant cells from oxidative stress, especially during periods of environmental stress such as drought or high light intensity (Agati et al., 2012). The glycosylation of flavonoids enhances their solubility and stability, allowing them to be easily transported to different parts of the plant where they can exert their protective effects. For instance, quercetin glycosides are known to accumulate in the epidermal layers of leaves, where they act as UV filters and protect the photosynthetic machinery from damage caused by excessive sunlight (Moses et al., 2014). Additionally, flavonoid glycosides are often involved in defense against herbivores and pathogens, where their glycosidic bonds can be cleaved to release the active flavonoid aglycone upon attack.

1.2.2 Saponin Glycosides

Saponin glycosides are a class of glycosides that contain triterpenoid or steroidal aglycones. These compounds are characterized by their amphiphilic properties, meaning they have both hydrophilic (sugar) and hydrophobic (aglycone) regions. This dual nature allows saponins to form soap-like foams in water and interact with cell membranes, making them highly effective in disrupting the cell membranes of pathogens and herbivores. Saponins are well-known for their role in plant defense, particularly against fungal pathogens and insect herbivores. For example, avenacins, a group of saponins found in oats, are highly effective against fungal pathogens such as *Gaeumannomyces graminis*, the causal agent of take-all disease in cereals (Papadopoulou et al., 1999). Saponin glycosides are also being explored for their potential as natural biopesticides in sustainable agriculture, where they can provide an environmentally friendly alternative to synthetic chemicals (Osbourn, 2003).

1.2.3 Phenolic Glycosides

Phenolic glycosides are derived from phenolic compounds, which are aromatic compounds with one or more hydroxyl groups. These glycosides are widely distributed in plants and serve various roles, including defense against herbivores, pathogens, and oxidative stress. Salicin, a phenolic glycoside found in willow bark, is an example of a compound with significant medicinal and ecological importance, known for its anti-inflammatory properties (Cushnie & Lamb, 2011). Phenolic glycosides often function as pro-compounds, meaning they are stored in an inactive form and activated upon tissue damage. When plants are attacked by herbivores or pathogens, phenolic glycosides are hydrolyzed to release toxic phenolic compounds that deter or inhibit the attacker. This mechanism is especially important in plants that rely on chemical defenses rather than physical barriers to protect themselves from biotic stressors.

1.2.4 Terpenoid Glycosides

Terpenoid glycosides include a broad range of compounds that are derived from terpenoid aglycones, such as monoterpenes, diterpenes, and triterpenes. Terpenoids are the largest class of plant secondary metabolites, with a wide range of functions in plant defense, signaling, and growth regulation. Terpenoid glycosides, such as cardenolides, are known for their toxic effects on herbivores, particularly insects, and are often involved in plant-pollinator and plant-herbivore interactions. Cardenolides, for example, are toxic to many herbivores and have evolved as a defense mechanism in plants like milkweed (*Asclepias* spp.), which produces cardenolide glycosides to deter insect herbivores (War et al., 2018). These compounds disrupt the cellular processes of herbivores by inhibiting Na+/K+-ATPase, an enzyme essential for maintaining cellular ion balance. As with other glycosides, terpenoid glycosides can be stored in an inactive form and activated upon herbivore attack, providing a rapid and effective defense mechanism.

1.2.5 Alkaloid Glycosides

Alkaloid glycosides are derived from alkaloid aglycones, which are nitrogen-containing compounds known for their potent biological activities. Alkaloid glycosides often function as neurotoxins or enzyme inhibitors, providing plants with a strong chemical defense against herbivores and pathogens. The glycosylation of alkaloids enhances their solubility and transport within the plant, allowing them to be stored in various tissues and mobilized when needed. For example, solanine, a glycoside of the alkaloid solanidine, is found in plants of the Solanaceae family, including potatoes and tomatoes. Solanine acts as a potent neurotoxin to insects and mammals, disrupting nerve impulses and leading to paralysis or death (Karaca & Erbaş, 2024). The ability of alkaloid glycosides to act as toxins makes them valuable in plant defense, particularly in species that are frequently targeted by herbivores.

2. Antimicrobial Properties of Glycosides: Mechanisms Against Pathogens

Glycosides are key players in plant defense, known for their antimicrobial properties, which enable plants to combat a wide variety of pathogens, including bacteria, fungi, and viruses. These compounds work through multiple mechanisms to inhibit pathogen growth, disrupt microbial cell structures, and interfere with metabolic pathways essential for pathogen survival. The sugar component (glycone) of glycosides enhances their solubility and transport within plant tissues, allowing them to be readily mobilized in response to pathogen invasion. Upon pathogen attack, many glycosides are activated by enzymes such as glycosidases, releasing bioactive aglycones that possess strong antimicrobial activity.

2.1 Disruption of Microbial Membranes

One of the most well-known antimicrobial mechanisms of glycosides, particularly saponin glycosides, is the disruption of microbial cell membranes. Saponins possess amphiphilic properties due to their hydrophilic sugar moiety and hydrophobic aglycone. This allows them to interact with the lipid bilayers of microbial membranes, leading to membrane destabilization and increased permeability. The result is the leakage of vital cellular components such as ions, proteins, and metabolites, which ultimately causes cell lysis and death (Moses et al., 2014). This membrane-disrupting property makes saponins highly effective against fungal pathogens, which rely on membrane integrity for survival.

2.1.1 Sterol Interaction and Membrane Permeabilization

Saponin glycosides, such as those found in oats (*Avena sativa*), interact with sterols in fungal membranes, which is crucial for their antimicrobial activity. The aglycone portion of saponins binds to sterols, creating pores in the membrane and increasing its permeability. This interaction is particularly effective against fungi, as sterols are essential components of fungal cell membranes (Osbourn, 2003). For example, avenacin, a saponin glycoside found in oat roots, is known to provide resistance against soil-borne pathogens such as *Gaeumannomyces graminis*, which causes take-all disease in cereals. The selectivity of saponins for sterol-containing membranes makes them highly effective against fungi and certain protozoa, while leaving plant and mammalian cells relatively unaffected, as plant cell membranes lack the same sterol composition. This makes saponin glycosides valuable as natural antifungal agents in agricultural and medicinal applications.

2.1.2 Antibacterial Effects of Glycosides

In addition to their antifungal properties, certain glycosides also exhibit antibacterial activity by disrupting bacterial membranes. Although bacterial membranes do not contain sterols, glycosides can still interfere with membrane integrity by interacting with other lipids or proteins essential for maintaining membrane structure. Phenolic glycosides, such as salicin from willow (*Salix* spp.), have been shown to exhibit antibacterial activity by destabilizing bacterial membranes and inhibiting biofilm formation (Cushnie & Lamb, 2011). The ability of glycosides to inhibit biofilm formation is particularly important in preventing bacterial colonization on plant surfaces and within plant tissues, thereby reducing the spread of bacterial infections.

2.2 Inhibition of Pathogen Enzymes

Another crucial antimicrobial mechanism of glycosides is the inhibition of pathogen enzymes, particularly those involved in cell wall degradation, metabolism, and replication. Many pathogens, such as fungi and bacteria, secrete extracellular enzymes that break down plant cell walls to facilitate infection. Glycosides can inhibit these enzymes, preventing pathogens from penetrating plant tissues and establishing infections.

2.2.1 Inhibition of Fungal Cell Wall-Degrading Enzymes

Fungal pathogens rely on enzymes such as cellulases, pectinases, and glucanases to degrade plant cell walls and invade host tissues. Glycosides, particularly phenolic glycosides, have been shown to inhibit the activity of these enzymes, thereby reducing the ability of fungal pathogens to infect plants. For example, phenolic glycosides like arbutin have been demonstrated to inhibit the cellulolytic enzymes produced by fungal pathogens such as *Botrytis cinerea*, the causal agent of gray mold in crops (Cheeke, 1996). By inhibiting these enzymes, glycosides prevent the degradation of plant cell walls, providing an additional layer of defense against fungal pathogens.

2.2.2 Inhibition of Bacterial Metabolic Enzymes

In bacteria, glycosides can inhibit key metabolic enzymes required for bacterial growth and replication. For example, flavonoid glycosides such as quercetin glycosides have been shown to inhibit bacterial enzymes involved in DNA replication, protein synthesis, and energy production. This inhibition disrupts bacterial metabolism and slows down or halts bacterial growth. Additionally, glycosides can interfere with the production of bacterial virulence factors, such as toxins and enzymes that promote infection (Cushnie & Lamb, 2011). By targeting bacterial metabolism, glycosides not only prevent bacterial colonization but also reduce the severity of infections.

2.3 Disruption of Pathogen Metabolism

Beyond membrane disruption and enzyme inhibition, glycosides can also interfere with the broader metabolic pathways of pathogens, leading to reduced growth and viability. This disruption of metabolism can occur at various stages, including nutrient acquisition, energy production, and biosynthesis of essential molecules. Terpenoid glycosides, in particular, are known for their ability to interfere with metabolic processes in pathogens, especially fungi and protozoa.

2.3.1 Interference with Nutrient Uptake

Glycosides can inhibit the ability of pathogens to take up essential nutrients, such as glucose, amino acids, and ions, from the plant environment. This is particularly important for pathogens that rely on external nutrient sources for growth and development. Saponins, for example, have been shown to interfere with the uptake of glucose and other sugars by fungal pathogens, depriving them of the energy needed for growth and reproduction (Papadopoulou et al., 1999). This nutrient deprivation weakens the pathogen and reduces its ability to spread within the host plant.

2.3.2 Inhibition of Energy Production Pathways

Glycosides also target the energy production pathways of pathogens, particularly those involving ATP generation and cellular respiration. By disrupting these pathways, glycosides reduce the energy available for pathogen growth and survival. For example, flavonoid glycosides like rutin have been shown to inhibit the activity of key enzymes in the mitochondrial electron transport chain, leading to reduced ATP production in fungal pathogens (Moses et al., 2014). This disruption of energy production not only limits pathogen growth but also induces oxidative stress, further weakening the pathogen's ability to survive and cause infection.

2.4 Induction of Plant Defense Responses

In addition to their direct antimicrobial effects, glycosides can also act as signaling molecules that trigger plant defense responses. Upon pathogen attack, the release of glycosides or their aglycones can serve as elicitors of defense-related pathways, such as the salicylic acid (SA), jasmonic acid (JA), and ethylene signaling pathways. These pathways are essential for the activation of systemic acquired resistance (SAR) and the production of defense-related proteins, such as pathogenesis-related (PR) proteins and reactive oxygen species (ROS), which contribute to pathogen resistance.

2.4.1 Activation of Systemic Acquired Resistance (SAR)

Systemic acquired resistance is a long-lasting immune response that provides protection against a broad range of pathogens. Glycosides, particularly glucosinolates, play a role in the activation of SAR by triggering the production of signaling molecules like salicylic acid (Durrant & Dong, 2004). Upon pathogen attack, glucosinolates are hydrolyzed to release bioactive compounds that activate defense genes throughout the plant, enhancing resistance to future pathogen attacks.

2.4.2 Production of Reactive Oxygen Species (ROS)

Glycosides can also induce the production of reactive oxygen species, which are highly reactive molecules that contribute to pathogen destruction. ROS production is a key part of the plant's hypersensitive response (HR) to pathogen invasion, leading to localized cell death and the containment of the pathogen. For example, saponin glycosides have been shown to enhance ROS production in response to fungal pathogens, contributing to the oxidative stress that damages and kills the invading pathogens (Moses et al., 2014). The ability of glycosides to induce ROS production adds another layer of defense against microbial infections.

2.5 Synergistic Effects with Other Antimicrobial Compounds

The antimicrobial activity of glycosides is often enhanced when they interact synergistically with other plant metabolites, such as phenolics, flavonoids, terpenoids, and alkaloids. These interactions result in a more potent defense response, as the combined effects of multiple antimicrobial compounds can inhibit pathogens more effectively than individual compounds alone. For instance, saponins and phenolic glycosides often work together to enhance antimicrobial activity against fungal pathogens. While saponins disrupt cell membranes, phenolic glycosides can inhibit fungal enzymes and metabolic pathways, creating a multi-targeted defense system. This synergistic interaction is particularly effective in controlling pathogens like *Fusarium oxysporum* and *Botrytis cinerea*, which are responsible for significant crop losses (Priyashantha et al., 2023)

3. Role of Glycosides in Plant Defense Mechanisms

Glycosides play a central role in the defense mechanisms of plants, acting as chemical protectants against a wide range of biotic stressors, including herbivores, pathogens, and insects. As part of the plant's arsenal of secondary metabolites, glycosides serve not only as direct deterrents to pests and pathogens but also as key molecules in signaling pathways that activate broader defense responses. These compounds can be stored in inactive forms within plant tissues and rapidly activated in response to damage or infection, providing a dynamic and adaptable defense system. Their roles in plant defense extend from the chemical deterrence of herbivores and pathogens to participation in systemic defense responses and hypersensitive reactions.

3.1 Direct Chemical Defense: Toxicity to Herbivores and Pathogens

One of the primary roles of glycosides in plant defense is their direct toxicity to herbivores, pathogens, and pests. Many glycosides act as chemical deterrents, interfering with the digestive and metabolic processes of herbivores and pathogens when consumed. These toxic effects are often mediated by the aglycone portion of the glycoside, which is released upon hydrolysis by enzymes such as glycosidases during tissue damage. This release allows the plant to quickly deploy toxic compounds in response to biotic stress, effectively limiting the spread of infection or herbivory.

3.1.1 Cyanogenic Glycosides and Herbivore Deterrence

Cyanogenic glycosides are a well-known group of plant defense compounds that release hydrogen cyanide (HCN) upon enzymatic hydrolysis. When plant tissues are damaged by herbivores, cyanogenic glycosides such as amygdalin and linamarin are broken down by specific enzymes (beta-glucosidases) to release cyanide, which is highly toxic to most herbivores (Møller, 2010). Cyanide inhibits cytochrome c oxidase in the mitochondrial electron transport chain, leading to the disruption of ATP production and cellular respiration in herbivores. This defense mechanism is particularly effective in preventing herbivory in plants like cassava (*Manihot esculenta*), sorghum (*Sorghum bicolor*), and wild cherry (*Prunus serotina*), which are rich in cyanogenic glycosides. The rapid release of cyanide upon herbivore attack not only deters further feeding but can also lead to the death of smaller herbivores, such as insects, before significant damage occurs. The storage of cyanogenic glycosides in inactive forms allows the plant to avoid self-toxicity while maintaining an effective defense against herbivores.

3.1.2 Glucosinolates: Role in Defense Against Pathogens and Herbivores

Glucosinolates, another class of glycosides found predominantly in the Brassicaceae family, play a crucial role in defense against both herbivores and microbial pathogens. Upon tissue damage, glucosinolates are hydrolyzed by the enzyme myrosinase, leading to the formation of toxic compounds such as isothiocyanates, nitriles, and thiocyanates (Abdel-Massih et al., 2023). These breakdown products are highly effective in deterring herbivores, including insects like aphids and caterpillars, and in inhibiting the growth of bacterial and fungal pathogens. In addition to their direct toxic effects, glucosinolates and their breakdown products serve as signaling molecules that can activate other defense pathways in plants. For example, isothiocyanates not only deter herbivores but also trigger localized cell death in response to pathogen attack, limiting the spread of infection. This dual role of glucosinolates in both direct toxicity and defense signaling underscores their importance in the integrated defense strategies of plants.

3.2 Activation of Plant Immune Responses

In addition to their direct toxicity, glycosides play a crucial role in activating plant immune responses. These responses include localized defense reactions, such as the hypersensitive response (HR), as well as systemic defense mechanisms, like systemic acquired resistance (SAR). Glycosides can act as signaling molecules or defense elicitors that trigger the activation of defense-related genes, leading to the production of pathogenesis-related (PR) proteins, reactive oxygen species (ROS), and other defense compounds.

3.2.1 Hypersensitive Response (HR)

The hypersensitive response (HR) is a localized immune reaction that occurs at the site of pathogen infection. It involves the rapid death of plant cells surrounding the infection site, effectively containing the pathogen and preventing its spread to other tissues. Glycosides, particularly those derived from phenolic compounds, have been implicated in the activation of HR in response to pathogen attack. For example, phenolic glycosides such as salicylic acid glucosides can be hydrolyzed to release salicylic acid, a key signaling molecule involved in the activation of HR (Vlot et al., 2009). Upon pathogen recognition, glycosides can enhance the production of reactive oxygen species (ROS), which play a critical role in the initiation of HR. The accumulation of ROS at the infection site leads to oxidative stress, which damages both plant and pathogen cells. This controlled cell death serves to isolate the pathogen and limit its access to nutrients, thereby slowing the progression of the infection. The involvement of glycosides in the regulation of ROS production highlights their importance in the plant's early defense responses to pathogens.

3.2.2 Systemic Acquired Resistance (SAR)

Systemic acquired resistance (SAR) is a long-lasting immune response that provides protection to the entire plant against a broad spectrum of pathogens. Glycosides are key players in the activation of SAR, particularly through their role in the salicylic acid (SA) signaling pathway. Upon pathogen attack, salicylic acid glucosides stored in plant tissues can be hydrolyzed to release free salicylic acid, which acts as a signaling molecule to activate SAR in distal parts of the plant (Durrant & Dong, 2004). The activation of SAR involves the upregulation of defense-related genes, including those encoding pathogenesis-related (PR) proteins, which possess antimicrobial properties. These proteins provide the plant with enhanced resistance to subsequent pathogen attacks, even in tissues that were not originally infected. The role of glycosides in the activation of SAR underscores their importance not only in localized defense but also in the broader systemic defense strategies of plants.

3.3 Interaction with Other Plant Defense Compounds

Glycosides often interact with other plant secondary metabolites, such as alkaloids, terpenoids, and phenolics, to enhance the overall effectiveness of plant defense mechanisms. These interactions result in synergistic effects that make the plant's defense system more robust and adaptable to a wide range of biotic stressors. The combination of glycosides with other compounds can provide multiple layers of defense, targeting different aspects of pathogen or herbivore biology.

3.3.1 Synergy Between Glycosides and Phenolics

Phenolic compounds, such as flavonoids and tannins, are well-known for their antioxidant and antimicrobial properties. When combined with glycosides, phenolics can enhance the plant's defense against pathogens and herbivores. For example, flavonoid glycosides, such as quercetin glycosides, can work in conjunction with phenolic acids to inhibit the growth of fungal pathogens and reduce oxidative stress in plant tissues (Agati et al., 2012). The synergy between glycosides and phenolics is particularly important in plants exposed to multiple stressors, such as drought, UV radiation, and pathogen attack. The combined effects of these compounds provide a more comprehensive defense strategy, protecting the plant from both biotic and abiotic stress.

3.3.2 Interaction with Terpenoids and Alkaloids

Glycosides also interact with terpenoids and alkaloids to enhance plant defense. Terpenoid glycosides, such as cardenolides, are known for their role in deterring herbivores by disrupting their nervous systems. When combined with glycosides, such as saponins, terpenoids can provide a stronger defense against herbivores by targeting multiple physiological pathways. Similarly, alkaloid glycosides, such as solanine, can work synergistically with glycosides to enhance their toxicity to herbivores, providing an effective deterrent against feeding. The interactions between glycosides and other secondary metabolites highlight the complexity of plant defense systems. By combining different classes of compounds, plants can deploy a more flexible and adaptable defense strategy, ensuring their survival in diverse and challenging environments.

4. Synergistic Interactions of Glycosides with Other Plant Metabolites

Glycosides, known for their structural diversity and wide-ranging biological activities, often exhibit enhanced functionality when interacting synergistically with other plant metabolites. This synergy enables plants to utilize multiple defense strategies simultaneously, making their defensive responses more robust and adaptable to various biotic and abiotic stressors. The combination of glycosides with other plant secondary metabolites, such as phenolics, flavonoids, terpenoids, and alkaloids, results in a more comprehensive defense mechanism. These interactions not only optimize the antimicrobial and anti-herbivory properties of glycosides but also enhance their roles in oxidative stress protection, signal transduction, and ecological interactions.

4.1 Synergy Between Glycosides and Phenolic Compounds

Phenolic compounds, including phenolic acids, flavonoids, and tannins, are well-known for their antioxidant and antimicrobial properties. When glycosides interact with phenolic compounds, the resulting synergy enhances both classes' defense capabilities against pathogens and herbivores, contributing to a more resilient plant defense system. Phenolic glycosides, such as arbutin, rutin, and salicin, have been extensively studied for their combined antioxidant and antimicrobial properties, particularly in response to environmental stressors.

4.1.1 Enhanced Antimicrobial Activity

The interaction between glycosides and phenolic compounds significantly enhances their antimicrobial properties, especially against fungal and bacterial pathogens. For instance, phenolic glycosides like salicin, which is derived from willow bark, exhibit enhanced antimicrobial activity when combined with free phenolics, effectively inhibiting bacterial pathogens such as *Staphylococcus aureus* and *Escherichia coli* (Cushnie & Lamb, 2011). Glycosides help in making phenolic compounds more soluble and transportable within plant tissues, ensuring that they can reach infection sites more efficiently. This synergy is particularly valuable in defense against fungal pathogens, where the combination of glycosides and phenolics leads to a more potent inhibition of fungal growth and spore germination. For instance, studies on *Quercus* species have demonstrated that tannin glycosides, when combined with flavonoids like catechins, provide stronger resistance against fungal infections such as oak powdery mildew (*Erysiphe alphitoides*), compared to either compound acting alone (Copolovici et al., 2014).

4.1.2 Antioxidant Defense Mechanisms

In addition to their antimicrobial effects, glycosides and phenolic compounds synergize to enhance the plant's antioxidant defenses, particularly under conditions of abiotic stress such as drought, UV radiation, or nutrient deficiency. Flavonoid glycosides, such as quercetin glycosides, are known for their strong antioxidant capacity. These compounds scavenge reactive oxygen species (ROS), preventing oxidative damage to plant cells. When combined with phenolic acids like ferulic acid or caffeic acid, the antioxidant potential of glycosides is amplified, protecting plants from oxidative stress-induced damage (Agati et al., 2012). This interaction is especially critical in leaf tissues, where high levels of light and UV radiation can cause significant oxidative stress. Glycosides help in the stabilization and mobilization of phenolic compounds, ensuring that antioxidant defenses are deployed efficiently in response to environmental triggers. In species like *Vitis vinifera* (grapevines), the combined action of glycosides and phenolic acids has been shown to improve plant resilience against both biotic and abiotic stressors, leading to improved crop yield and quality (He et al., 2024).

4.2 Interaction of Glycosides with Flavonoids

Flavonoids, a diverse group of phenolic metabolites, play critical roles in plant pigmentation, UV protection, and defense against herbivores and pathogens. Glycosides, when bound to flavonoids, enhance the biological activity and solubility of these compounds, making them more effective in plant defense strategies. Flavonoid glycosides such as rutin, kaempferol glycosides, and quercetin glycosides are particularly important in the plant's response to environmental stressors and microbial attacks.

4.2.1 Flavonoid Glycosides and UV Protection

One of the key roles of flavonoid glycosides is their ability to protect plants from harmful UV radiation. In this context, glycosides act by enhancing the stability and solubility of flavonoids, allowing them to accumulate in epidermal tissues where they act as UV filters. For instance, quercetin glycosides in *Arabidopsis thaliana* have been shown to accumulate in response to UV-B radiation, protecting by absorbing UV light and preventing damage to DNA and photosynthetic apparatus (Moses et al., 2014). The glycosylation of flavonoids is essential for their transport to the epidermis, where they form a protective barrier against UV-induced oxidative stress. This synergy is not limited to UV protection; flavonoid glycosides also play a role in mitigating damage caused by excessive light and other environmental stressors. In combination with other plant metabolites, such as terpenoids and alkaloids, flavonoid glycosides contribute to the overall robustness of the plant's defense system, ensuring that it can adapt to fluctuating environmental conditions.

4.2.2 Antimicrobial Synergy

Flavonoid glycosides also exhibit enhanced antimicrobial activity when combined with other plant metabolites. For example, kaempferol glycosides have been shown to synergize with terpenoid glycosides to inhibit the growth of bacterial and fungal pathogens. In grapevines, the combined action of flavonoid glycosides and stilbenes (a class of phenolics) provides strong resistance against fungal pathogens such as *Botrytis cinerea*, which causes gray mold (Mordue & Nisbet, 2000). This synergistic interaction between glycosides and other defense metabolites highlights the importance of glycosides in a multi-layered plant defense strategy.

4.3 Synergy Between Glycosides and Terpenoids

Terpenoids, another major class of secondary metabolites, are known for their wide-ranging biological activities, including antimicrobial, anti-herbivory, and signaling functions. Glycosides enhance the solubility and transport of terpenoids within plant tissues, making them more effective as defense compounds. Terpenoid glycosides, such as cardenolides and triterpenoid saponins, exhibit strong insecticidal and fungicidal properties, and their interaction with other glycosides amplifies their overall defensive capabilities.

4.3.1 Terpenoid Glycosides and Insect Resistance

Terpenoid glycosides, such as cardenolides, play a critical role in plant defense against herbivores, particularly insects. Cardenolides disrupt the nervous systems of insects by inhibiting Na+/K+-ATPase, an enzyme essential for maintaining cellular ion balance. Glycosylation of terpenoids not only enhances their solubility but also increases their toxicity to herbivores. For instance, milkweed (*Asclepias* spp.) produces cardenolide glycosides that are highly toxic to herbivores like aphids and caterpillars, providing an effective chemical defense (He et al., 2024). The synergy between terpenoid glycosides and other plant metabolites, such as flavonoids and alkaloids, further strengthens this defense. For example, saponins, a type of triterpenoid glycoside, work synergistically with flavonoids to enhance their insecticidal properties. This combination disrupts insect feeding and digestion, leading to reduced herbivory and increased plant survival under pest pressure.

4.3.2 Terpenoid Glycosides and Antifungal Activity

In addition to their insecticidal properties, terpenoid glycosides also exhibit strong antifungal activity. The interaction between terpenoid glycosides and other antimicrobial compounds, such as phenolic acids and flavonoids, enhances their ability to inhibit fungal growth. For instance, triterpenoid saponins from *Quillaja saponaria* have been shown to work synergistically with phenolic glycosides to inhibit the growth of fungal pathogens like *Fusarium oxysporum* and *Aspergillus niger* (Osbourn, 2003). This synergistic antifungal activity is particularly important in agricultural systems, where fungal infections can cause significant crop losses. The use of plant extracts rich in terpenoid glycosides, combined with other metabolites, offers a sustainable and environmentally friendly alternative to synthetic fungicides.

4.4 Glycosides and Alkaloids: Synergistic Toxicity

Alkaloids, a class of nitrogen-containing compounds, are known for their potent neurotoxic and antimicrobial properties. Glycosides and alkaloids often interact synergistically to enhance their toxicity to herbivores and pathogens. Alkaloid glycosides, such as solanine (found in potatoes and tomatoes), are highly toxic to a wide range of herbivores and insects, and their glycosylation enhances their transport and accumulation within plant tissues, making them more effective in defense.

4.4.1 Neurotoxic Effects on Herbivores

Alkaloid glycosides, such as those found in the Solanaceae family (e.g., solanine in potatoes), disrupt the nervous systems of herbivores by interfering with neurotransmitter signaling and ion transport. Glycosylation increases the bioavailability of these alkaloids, allowing them to accumulate in higher concentrations in plant tissues targeted by herbivores. This increased toxicity results in more effective deterrence of herbivores such as aphids, beetles, and caterpillars, reducing plant damage (Karaca & Erbaş, 1995). The combination of alkaloid glycosides with other secondary metabolites, such as terpenoids and phenolics, further amplifies their toxic effects on herbivores. This synergy creates a multi-faceted chemical defense system that targets various physiological pathways in herbivores, ensuring more comprehensive protection for the plant.

4.5 Ecological Benefits of Synergistic Interactions

The synergistic interactions between glycosides and other plant metabolites not only enhance plant defense but also contribute to ecological balance by promoting beneficial plant-pollinator and plant-microbe interactions. For example, glycosides that interact with flavonoids and phenolics often play a role in attracting pollinators by contributing to flower color and nectar production. Similarly, glycosides can enhance symbiotic relationships with beneficial soil microbes, such as nitrogen-fixing bacteria and mycorrhizal fungi, improving plant nutrient uptake and overall health (Moses et al., 2014). These ecological benefits extend beyond individual plants, contributing to the stability and sustainability of entire ecosystems. The ability of plants to deploy synergistic chemical defenses against pests and pathogens ensures their survival and reproductive success, while also maintaining the integrity of the broader ecological community.

5. Applications of Glycosides in Sustainable Agriculture and Disease Control

Glycosides play a crucial role in sustainable agriculture and disease control, offering environmentally friendly alternatives to synthetic chemicals and pesticides. These naturally occurring compounds are widely recognized for their antimicrobial, insecticidal, and allelopathic properties, making them valuable in crop protection, pest management, and enhancing plant resistance to biotic and abiotic stresses. The application of glycosides in sustainable agriculture aligns with the goals of reducing environmental pollution, conserving biodiversity, and promoting agricultural practices that are less reliant on harmful synthetic inputs.

5.1 Glycosides as Natural Biopesticides

One of the most significant applications of glycosides in sustainable agriculture is their use as natural biopesticides. Synthetic pesticides, while effective, often lead to environmental contamination, pesticide resistance, and harm to non-target organisms, including beneficial insects and soil microbiota. Glycosides, on the other hand, offer a natural, biodegradable alternative that can effectively manage pests without the negative ecological impacts associated with conventional chemicals.

5.1.1 Insecticidal Properties

Many glycosides exhibit potent insecticidal properties, making them useful in controlling insect pests that cause extensive damage to crops. Saponins, a class of glycosides known for their detergent-like properties, have been widely studied for their ability to disrupt the cell membranes of insects, leading to cell lysis and death. For example, saponins from *Quillaja saponaria* (soapbark tree) have been used in biopesticide formulations to control pests such as aphids, whiteflies, and caterpillars (Mordue & Nisbet, 2000). These glycosides act by forming complexes with membrane sterols, causing membrane permeability and eventually insect mortality.

Additionally, cyanogenic glycosides, which release hydrogen cyanide upon enzymatic breakdown, provide an effective defense against herbivorous insects. Plants like cassava (*Manihot esculenta*) and sorghum (*Sorghum bicolor*) produce cyanogenic glycosides that deter insects through their toxic effects (Møller, 2010). These compounds offer a natural insecticidal solution that reduces the need for synthetic chemical inputs in crop production.

5.1.2 Nematocidal Activity

Nematodes are another significant pest in agriculture, particularly in root crops where they cause yield losses by feeding on plant roots. Certain glycosides, especially saponins, have been shown to possess nematocidal properties. For instance, saponins from *Quillaja saponaria* and *Medicago sativa* (alfalfa) have demonstrated efficacy in reducing populations of plant-parasitic nematodes such as *Meloidogyne* species, which are responsible for root-knot disease in several crops (Hussain et al., 2019). The nematocidal effect is attributed to the ability of saponins to disrupt the cuticle structure of nematodes, leading to their death. The use of glycoside-based biopesticides offers a sustainable approach to nematode management, particularly in organic farming systems where chemical nematicides are prohibited.

5.2 Glycosides as Biofungicides

Fungal pathogens are among the most challenging pests in agriculture, causing significant crop losses and reducing food security worldwide. Synthetic fungicides, while effective, pose risks to human health and the environment due to their persistence in soil and water. Glycosides, with their natural antifungal properties, provide a safer and more sustainable alternative for controlling fungal diseases.

5.2.1 Antifungal Activity Against Soil-Borne Pathogens

Glycosides, particularly saponins, have been shown to exhibit strong antifungal activity against a range of soil-borne pathogens that cause diseases like root rot, damping-off, and wilt. For instance, saponins from *Quinoa* (*Chenopodium quinoa*) have been demonstrated to inhibit the growth of *Fusarium oxysporum*, a common soil-borne pathogen that affects vegetable crops (Osbourn, 2003). The antifungal effect of saponins is due to their ability to disrupt fungal cell membranes, leading to cell leakage and death. This membrane-disrupting activity makes saponin-based biofungicides particularly effective against soil-borne fungal pathogens that are difficult to manage with conventional fungicides. Additionally, phenolic glycosides, such as salicin and arbutin, have been found to inhibit the activity of fungal enzymes involved in cell wall degradation, further reducing the ability of pathogens to infect plant roots. The use of glycosides as biofungicides not only protects crops from fungal diseases but also reduces the environmental impact of chemical fungicides, contributing to more sustainable agricultural practices.

5.2.2 Foliar Applications for Fungal Disease Control

In addition to their soil-borne antifungal activity, glycosides can be applied directly to plant leaves to control foliar fungal diseases, such as powdery mildew, rust, and downy mildew. Saponins and flavonoid glycosides, when applied as foliar sprays, form a protective barrier on the leaf surface that prevents fungal spores from germinating and penetrating plant tissues. For example, studies have shown that saponins from *Quillaja saponaria* effectively control powdery mildew on grapes and cucumbers, reducing disease severity without harming beneficial insects or the environment (Papadopoulou et al., 1999). The application of glycosides as foliar biofungicides is particularly valuable in organic farming, where the use of synthetic fungicides is restricted. These natural compounds provide an effective and eco-friendly solution for managing foliar fungal diseases, promoting plant health, and improving crop yields.

5.3 Growth Promotion and Stress Tolerance

Beyond their role in pest and disease control, glycosides also contribute to sustainable agriculture by promoting plant growth and enhancing tolerance to abiotic stresses such as drought, salinity, and temperature extremes. Glycosides, particularly those involved in signaling pathways, can stimulate plant growth by improving nutrient uptake, enhancing root development, and increasing resistance to environmental stressors.

5.3.1 Stimulation of Root Growth

Certain glycosides, such as flavonoid glycosides, are known to promote root growth and development, improving the plant's ability to absorb water and nutrients from the soil. For example, quercetin glycosides have been shown to stimulate root elongation in crops like lettuce and maize, leading to improved nutrient uptake and enhanced plant growth (Moses et al., 2014). This growth-promoting effect is particularly beneficial in low-input farming systems, where nutrient availability may be limited. The use of glycosides as biostimulants in agriculture can help reduce the reliance on synthetic fertilizers, which are often associated with environmental pollution and soil degradation. By promoting healthy root development, glycosides contribute to more efficient nutrient cycling in agricultural ecosystems, supporting sustainable crop production.

5.3.2 Enhancement of Stress Tolerance

Glycosides also play a role in enhancing plant tolerance to abiotic stresses, such as drought and salinity. Phenolic glycosides, for example, have been shown to increase the production of antioxidant enzymes, which protect plants from oxidative stress caused by environmental extremes (Agati et al., 2012). Additionally, saponins have been found to improve water retention in plant tissues, helping plants cope with drought conditions. In crops like rice and wheat, the application of glycoside-rich extracts has been shown to improve plant resilience under salinity stress by modulating ion uptake and reducing the accumulation of toxic ions in plant tissues. This stress-mitigating effect of glycosides makes them valuable tools for improving crop performance in regions prone to drought or soil salinization.

5.4 Glycosides in Soil Health Management

Sustainable agriculture depends not only on effective pest and disease control but also on maintaining healthy soils that support long-term crop productivity. Glycosides, through their interactions with soil microorganisms and their role in nutrient cycling, contribute to improved soil health and fertility.

5.4.1 Promotion of Beneficial Microorganisms

Glycosides can enhance the growth of beneficial soil microorganisms, such as nitrogen-fixing bacteria and mycorrhizal fungi, which are essential for maintaining soil fertility. For instance, saponins have been shown to stimulate the growth of rhizobia, bacteria that form symbiotic relationships with leguminous plants and fix atmospheric nitrogen into a form that plants can use (Osbourn, 2003). This symbiotic interaction improves soil nitrogen levels, reducing the need for synthetic nitrogen fertilizers and promoting sustainable crop production. Similarly, glycosides can enhance the colonization of plant roots by mycorrhizal fungi, which help plants access nutrients like phosphorus and trace minerals. This symbiosis not only improves nutrient uptake but also increases plant resistance to soil-borne pathogens, further contributing to soil health and crop productivity.

5.4.2 Reduction of Soil Erosion and Degradation

By promoting healthy root development and improving soil structure, glycosides also contribute to the reduction of soil erosion and degradation. Saponins, for example, enhance soil aggregation by promoting the formation of stable soil particles, which helps prevent erosion caused by wind and water. Additionally, the enhanced root growth stimulated by glycosides helps anchor soil in place, reducing the loss of topsoil and maintaining soil fertility (Hussain et al., 2019). The role of glycosides in soil health management aligns with the principles of conservation agriculture, which emphasizes minimal soil disturbance, crop rotation, and the use of cover crops to maintain soil health and prevent degradation. By improving soil structure and promoting beneficial soil organisms, glycosides contribute to the long-term sustainability of agricultural systems.

5.5 Future Prospects of Glycosides in Agriculture

The use of glycosides in sustainable agriculture holds great promise for the future, particularly as the demand for eco-friendly and sustainable farming practices grows. With ongoing research into the biosynthesis, extraction, and application of glycosides, there is potential to develop more effective and targeted biopesticides, biofungicides, and biostimulants. Advances in biotechnology, such as genetic engineering and synthetic biology, may allow for the enhancement of glycoside production in crops, enabling the development of crop varieties with increased glycoside content for enhanced pest and disease resistance. Additionally, the combination of glycosides with other natural products, such as essential oils and microbial-based products, could lead to the development of more effective integrated pest management strategies that reduce the reliance on synthetic chemicals.

6. Challenges and Future Directions in Glycoside Research and Utilization

Glycosides are highly valued in both agricultural and pharmaceutical sectors for their wide-ranging biological activities, particularly in pest management, disease control, and enhancing plant resilience. However, the full potential of glycosides has yet to be realized, as several challenges impede their widespread application. These challenges are related to their variability, biosynthesis, extraction, toxicity to non-target organisms, and scalability for commercial use. At the same time, recent advances in biotechnology, genetic engineering, and synthetic biology present exciting new directions for glycoside research and utilization.

6.1 Variability in Glycoside Composition and Concentration

One of the most significant challenges in glycoside research is the variability in glycoside composition and concentration across different plant species, cultivars, and environmental conditions. This variability complicates the standardization of glycoside-based products, making it difficult to produce consistent formulations for commercial use in agriculture or pharmaceuticals.

6.1.1 Environmental Influence on Glycoside Production

The biosynthesis of glycosides is highly influenced by environmental factors such as light, temperature, soil composition, and water availability. For instance, drought stress can enhance the production of certain glycosides, while nutrient-rich conditions may reduce their synthesis. This variability presents a challenge in ensuring that glycoside concentrations are sufficient for their intended biological functions, whether for pest control or therapeutic purposes (Moses et al., 2014). Different plant species and even individual plants within the same species can produce varying levels of glycosides depending on environmental conditions. For example, the concentration of saponins in *Quillaja saponaria* and *Medicago sativa* (alfalfa) can fluctuate significantly depending on soil fertility and water availability (Osbourn, 2003). This inconsistency poses a challenge for large-scale production, as it complicates the process of extracting glycosides in a standardized manner.

6.1.2 Genetic Variability in Glycoside Biosynthesis

In addition to environmental factors, genetic variability also plays a role in the production of glycosides. Different plant species, and even cultivars within a species, may possess distinct biosynthetic pathways for glycoside production. This diversity makes it difficult to standardize the use of glycosides in agricultural and pharmaceutical applications, as the biological activity of a glycoside often depends on the specific aglycone and sugar components. The genetic basis for glycoside biosynthesis is still not fully understood, which limits the ability to manipulate glycoside production through traditional breeding or genetic engineering. Although some progress has been made in identifying key glycosyltransferase (GT) enzymes involved in glycoside biosynthesis, much more research is needed to unravel the complex regulatory networks that control glycoside production in plants (He et al., 2024).

6.2 Challenges in Extraction, Purification, and Standardization

The extraction and purification of glycosides pose another major challenge to their widespread use, particularly in large-scale agricultural or pharmaceutical applications. Glycosides are often found in complex mixtures with other secondary metabolites, making their isolation and purification time-consuming and expensive. Moreover, the structural diversity of glycosides, with varying types of aglycones and sugar moieties, requires different extraction methods tailored to specific types of glycosides.

6.2.1 Complex Extraction and Purification Processes

Extracting glycosides from plant material involves multiple steps, including solvent extraction, filtration, and chromatographic separation. These processes are not only labor-intensive but also require advanced equipment and skilled personnel, making them costly for large-scale production. For instance, the extraction of saponins from *Quillaja saponaria* involves solvent extraction followed by ultrafiltration and high-performance liquid chromatography (HPLC) to isolate the active compounds (Hussain et al., 2019). Such processes are difficult to scale up for commercial use, particularly in agriculture, where cost-efficiency is crucial. The purification of glycosides is further complicated by their structural complexity. Glycosides often exist as mixtures of different forms, with varying numbers of sugar units and aglycones. This complexity requires advanced analytical techniques, such as mass spectrometry and nuclear magnetic resonance (NMR), to accurately characterize and quantify glycosides in plant extracts. The lack of standardized methods for glycoside purification and characterization hinders their large-scale application in both agricultural and medicinal contexts (Osbourn, 2003).

6.2.2 Need for Standardization in Glycoside-Based Products

For glycoside-based products to be commercially viable, whether as biopesticides or pharmaceuticals, there is a need for standardized formulations that guarantee consistent quality and efficacy. However, the variability in glycoside composition and the complexity of extraction processes make it difficult to produce standardized products. This lack of standardization limits the adoption of glycosides in mainstream agriculture and medicine, as consumers and regulators require consistent, reliable products. To address this challenge, there is a growing interest in developing biotechnological methods for producing glycosides in controlled environments. Advances in synthetic biology and metabolic engineering offer potential solutions for overcoming the limitations of natural glycoside extraction by enabling the production of glycosides in microbial systems or engineered plants (Moses et al., 2014).

6.3 Potential Toxicity to Non-Target Organisms

While glycosides are valued for their antimicrobial and insecticidal properties, their broad-spectrum activity also poses risks to non-target organisms, including beneficial insects, soil microorganisms, and aquatic life. For example, saponins are known to be toxic to fish and other aquatic organisms at high concentrations, raising concerns about their environmental impact when used as biopesticides in agriculture (Bakkali et al., 2008). Similarly, cyanogenic glycosides can release hydrogen cyanide, which, although effective against herbivores, can also pose a risk to non-target species, including humans if not managed properly.

6.3.1 Ecotoxicological Concerns

The potential for glycosides to accumulate in the environment and affect non-target species is a significant challenge, particularly in ecosystems adjacent to agricultural fields. For instance, saponins, which are commonly used in natural biopesticides, can leach into water bodies and negatively impact aquatic life. This poses a challenge for the widespread use of glycosides in sustainable agriculture, as their environmental fate needs to be carefully monitored to avoid unintended ecological damage (Bakkali et al., 2008).

6.3.2 Balancing Efficacy with Environmental Safety

To mitigate the potential environmental impact of glycosides, researchers are exploring ways to develop glycoside-based products that are both effective and environmentally safe. One approach is the development of targeted delivery systems, such as encapsulation or controlled-release formulations, that reduce the exposure of non-target organisms to glycosides. Additionally, advances in genetic engineering may allow for the production of glycosides in crop plants in a way that minimizes their release into the surrounding environment, thereby reducing the risk of non-target toxicity (Gupta et al., 2015).

6.4 Scalability and Commercialization

While glycosides hold great promise for sustainable agriculture and disease control, the scalability of glycoside-based products remains a major challenge. The cost and complexity of extracting and purifying glycosides from plants, combined with the variability in their composition, make it difficult to produce glycosides at a scale that is commercially viable for agricultural use. Moreover, the relatively limited understanding of glycoside biosynthesis and regulation hinders efforts to increase glycoside production through traditional breeding or genetic engineering.

6.4.1 Challenges in Large-Scale Production

The large-scale production of glycosides is currently limited by several factors, including the availability of suitable plant sources and the labor-intensive extraction processes. Plants that produce high levels of glycosides, such as *Quillaja saponaria* and *Medicago sativa*, are often difficult to cultivate in large quantities, and the extraction of glycosides from these plants is costly and time-consuming. Additionally, the low natural abundance of glycosides in many plant species further limits their scalability for commercial applications (Osbourn, 2003). Advances in biotechnological approaches, such as the use of microbial fermentation systems and genetically engineered plants, offer potential solutions to these challenges. By engineering microbes to produce glycosides, it may be possible to scale up production in a more cost-effective and sustainable manner. For example, the production of saponins in engineered yeast or bacteria could provide an alternative to traditional plant-based extraction, reducing the costs and environmental impact associated with large-scale glycoside production (Moses et al., 2014).

6.5 Future Directions in Glycoside Research and Utilization

Despite the challenges, the future of glycoside research and utilization looks promising, with several key areas of development poised to drive innovation in agriculture and medicine. Advances in genetic engineering, synthetic biology, and nanotechnology hold great potential for overcoming the current limitations of glycoside production and application.

6.5.1 Genetic Engineering and Synthetic Biology

One of the most promising areas of future research is the use of genetic engineering and synthetic biology to enhance glycoside production in plants and microorganisms. By identifying and manipulating the genes involved in glycoside biosynthesis, researchers can increase the yield of specific glycosides or produce novel glycosides with enhanced biological activity. For instance, the overexpression of glycosyltransferase enzymes in crop plants could lead to increased production of glycosides, providing enhanced resistance to pests and diseases (Libik-Konieczny et al., 2021). Synthetic biology also offers the potential to create entirely new glycoside compounds with tailored properties. By engineering biosynthetic pathways in microbial systems, it may be possible to produce glycosides with enhanced antimicrobial or insecticidal activity, as well as improved environmental safety. These advances could revolutionize the use of glycosides in sustainable agriculture, providing farmers with new tools for pest management and crop protection.

6.5.2 Nanotechnology for Targeted Delivery

Nanotechnology is another exciting area of research that could improve the application of glycosides in agriculture and medicine. By developing nanocarriers for glycoside delivery, researchers can create targeted formulations that release glycosides only at specific sites, such as infection or pest damage areas. This targeted delivery approach reduces the risk of non-target toxicity and improves the efficacy of glycoside-based products. For example, encapsulating glycosides in biodegradable nanoparticles could enhance their stability and allow for controlled release, improving their effectiveness as biopesticides or biofungicides (Drabczyk et al., 2024).

Conclusion

Glycosides, as bioactive secondary metabolites, play a crucial role in plant health and disease management. Their biosynthesis involves the attachment of sugar moieties to a wide range of aglycones, such as phenolics, terpenoids, or steroids, resulting in a diverse class of compounds that are classified based on the type of aglycone. This structural diversity is vital for the various functions glycosides perform in plants, especially in their defense mechanisms. As antimicrobial agents, glycosides effectively combat a broad spectrum of pathogens, including bacteria, fungi, and viruses, by disrupting cell membranes, inhibiting crucial enzymes, and interfering with pathogen communication systems, thereby limiting their ability to infect plants. In plant defense mechanisms, glycosides act as a primary line of defense, often stored in an inactive form that is activated upon pathogen invasion. Their rapid response upon damage or attack ensures that plants can swiftly react to biotic stressors. Furthermore, glycosides frequently work in synergy with other secondary metabolites such as flavonoids, alkaloids, and saponins. This interaction enhances the efficacy of the plant's overall defense system, providing a more robust and layered protection against environmental stresses and pathogens.

The applications of glycosides in sustainable agriculture have garnered increasing attention due to their potential to replace synthetic agrochemicals. Their natural antimicrobial properties make them suitable for use in integrated pest management (IPM) strategies, offering eco-friendly alternatives for controlling plant diseases and reducing the reliance on harmful pesticides. By promoting the use of glycosides in agricultural systems, it is possible to foster more sustainable practices that benefit both crop yield and environmental health. However, challenges remain in the large-scale production and consistent application of glycosides in agriculture. The variability in glycoside content across plant species, as well as the difficulties in their extraction and application, are obstacles that must be addressed. Research in biotechnological advancements, such as genetic engineering and synthetic biology, holds promise for overcoming these challenges. By enhancing glycoside production pathways and optimizing their scalability, it will be possible to expand their utilization in sustainable crop protection.

References

Abdel-Massih, R. M., Debs, E., Othman, L., Attieh, J., & Cabrerizo, F. M. (2023). Glucosinolates, a natural chemical arsenal: More to tell than the myrosinase story. *Frontiers in Microbiology, 14*, 1130208.

Agati, G., Azzarello, E., Pollastri, S., & Tattini, M. (2012). Flavonoids as antioxidants in plants: Location and functional significance. *Plant Science, 196*, 67-76.

Bakkali, F., Averbeck, S., Averbeck, D., & Idaomar, M. (2008). Biological effects of essential oils: A review. *Food and Chemical Toxicology, 46*(2), 446-475.

Cheeke, P. R. (1996). Biological effects of feed and forage saponins and their impacts on animal production. Advances in Experimental Medicine and Biology, *405*, 377-385.

Copolovici, L., Väärtnõu, F., Portillo Estrada, M., & Niinemets, Ü. (2014). Oak powdery mildew (Erysiphe alphitoides)-induced volatile emissions scale with the degree of infection in *Quercus robur. Tree Physiology*, *34*(12), 1399-410.

Cushnie, T. P., & Lamb, A. J. (2011). Recent advances in understanding the antibacterial properties of flavonoids. *International Journal of Antimicrobial Agents, 38*(2), 99-107.

Drabczyk, A., Kudłacik-Kramarczyk, S., Jamroży, M,. & Krzan, M. (2024). Biomaterials in drug delivery: advancements in cancer and diverse therapies-Review. *International Journal of Molecular Sciences*, *25*(6), 3126.

Durrant, W. E., & Dong, X. (2004). Systemic acquired resistance. *Annual Review of Phytopathology, 42*, 185-209.

He, M., Zhou, X., & Wang, X. (2024). Glycosylation: mechanisms, biological functions and clinical implications. *Signal Transduction and Targeted Therapy*, *9*, 194.

Hussain, M., Debnath, B., Qasim, M., Bamisile, B. S., Islam, W., Hameed, M. S., Wang, L., & Qiu, D. (2019). Role of saponins in plant defense against specialist herbivores. *Molecules*, *24*(11), 2067.

Gupta, P., Sharma, S., & Saxena, S. (2015). Biomass yield and steviol glycoside production in callus and suspension culture of Stevia rebaudiana treated with proline and polyethylene glycol. *Applied Biochemistry and Biotechnology*, *176*, 863-874.

Karaca, M., & Erbaş, O. (2024). Solanine poisoning: Effects, risks, and management strategies. *Journal of Experimental and Basic Medical Sciences*, *5*(2), 189-193.

Libik-Konieczny, M., Capecka, E., Tuleja, M., & Konieczny, R. (2021). Synthesis and production of steviol glycosides: recent research trends and perspectives. *Applied Microbiology and Biotechnology*, *105*(10), 3883-3900.

Metsämuuronen, S., & Sirén, H. (2019). Bioactive phenolic compounds, metabolism and properties: a review on valuable chemical compounds in Scots pine and Norway spruce. *Phytochemistry Reviews*, *18*, 623-664.

Møller, B. L. (2010). Functional diversifications of cyanogenic glucosides. *Current Opinion in Plant Biology*, 13(3), 337-346.

Mordue, A. J., & Nisbet, A. J. (2000). Azadirachtin from the neem tree *Azadirachta indica*: Its action against insects. *Anais da Sociedade Entomológica do Brasil, 29*(4), 615-632.

Moses, T., Papadopoulou, K. K., & Osbourn, A. (2014). Metabolic and functional diversity of saponins, biosynthetic intermediates, and semi-synthetic derivatives. *Critical Reviews in Biochemistry and Molecular Biology*, *49*(6), 439-462.

Osbourn, A. E. (2003). Saponins in cereals. *Phytochemistry, 62*(1), 1-4.

Papadopoulou, K., Melton, R. E., Leggett, M., Daniels, M. J., & Osbourn, A. E. (1999). Compromised disease resistance in saponin-deficient plants. *Proceedings of the National Academy of Sciences, 96*(22), 12923-12928.

Priyashantha, A. K. H., Dai, D. Q., Bhat, D. J., Stephenson, S. L., Promputtha, I., Kaushik, P., Tibpromma, S., & Karunarathna, S. C. (2023). Plant-Fungi Interactions: Where It Goes? *Biology (Basel)*, *12*(6), 809.

Sharma, K., Kaur, R., Kumar, S., Saini, R. K., Sharma, S., Pawde, S. V., & Kumar, V. (2023). Saponins: A concise review on food related aspects, applications and health implications. *Food Chemistry Advances*, *2*, 100191.

Ting, H.-M., Cheah, B. H., Chen, Y.-C., Yeh, P.-M., Cheng, C.-P., Yeo, F. K. S., Vie, A. K., Rohloff, J., Winge, P., Bones, A. M., & Kissen, R. (2020). The role of a glucosinolate-derived nitrile in plant immune responses. *Frontiers in Plant Sciences*, *11*, 257.

Vlot, A. C., Dempsey, D. A., & Klessig, D. F. (2009). Salicylic acid, a multifaceted hormone to combat disease. *Annual Review of Phytopathology, 47*, 177-206.

War, A. R., Taggar, G. K., Hussain, B., Taggar, M. S., Nair, R. M., & Sharma, H. C. (2018). Plant defence against herbivory and insect adaptations. *AoB Plants*, *10*(4), ply037.

Chapter - 8

Role of Tannins in Plant Defense and Pathogen Suppression

Sheetanshu Gupta[1], Pratima Raypa[2], Vinod Chouhan[3], Susheel Kumar[4] and Zuby Gohar Ansari[5]

[1]Department of Biochemistry, G.B. Pant University of Agriculture & Technology, Pantnagar, Uttarakhand, India

[2]Department of Biochemistry, H.N.B. Garhwal University, Srinagar, Uttarakhand, India

[3]Division of Plant Pathology, Indian Council of Agricultural Research-Indian Agricultural Research Institute (ICAR-IARI), New Delhi, India

[4]Plant Molecular Virology Laboratory, CPMB Division, CSIR-National Botanical Research Institute, Rana Pratap Marg, Lucknow, Uttar Pradesh, India

[5]B.M. College of Agriculture, Jaswadi Road Khandwa, Madhya Pradesh, India

Abstract

Tannins are a diverse group of polyphenolic compounds that play a crucial role in plant defense and pathogen suppression. These compounds are synthesized via the phenylpropanoid pathway and are classified into two major groups: condensed tannins (proanthocyanidins) and hydrolyzable tannins. Tannins exhibit significant biological activities, including antioxidant and antimicrobial properties, which contribute to their effectiveness in plant defense mechanisms. Their antioxidant activity helps neutralize reactive oxygen species (ROS) generated during environmental stress, thus protecting plant tissues from oxidative damage. Tannins also act as potent antimicrobial agents, targeting a wide range of pathogens, including bacteria, fungi, and viruses, by disrupting microbial cell walls, inhibiting enzymatic activity, and interfering with microbial communication systems. Additionally, tannins play a vital role in herbivore deterrence and insect resistance, reducing herbivory by binding to proteins and decreasing the digestibility of plant tissues for herbivores. The presence of tannins in plants enhances their resistance to environmental stressors such as drought, salinity, and temperature extremes, further strengthening their overall defense mechanisms. In agriculture, tannins hold promise as natural biopesticides and eco-friendly alternatives to synthetic chemicals for disease and pest management. They can be incorporated into integrated pest management (IPM) strategies, offering sustainable solutions for crop protection and pathogen control. However, challenges remain in terms of large-scale

DOI: 10.1201/9781003717454-8

production, extraction, and application of tannins in agricultural systems. Research in biotechnological advancements, including genetic engineering and metabolic pathway optimization, offers potential solutions to enhance tannin production and utilization. As interest in sustainable agriculture grows, tannins are gaining attention for their natural efficacy in plant defense and pathogen suppression, making them valuable tools for environmentally sustainable disease management strategies.

Keywords: Tannins, plant defense, antioxidant, antimicrobial, pathogen suppression, sustainable agriculture.

Introduction

Tannins are a diverse group of polyphenolic compounds naturally present in many plant species, playing crucial roles in plant defense, growth regulation, and ecological interactions. These compounds are characterized by their ability to bind proteins and other organic molecules, contributing to their role in deterring herbivores, pathogens, and environmental stressors. Tannins can be divided into two major categories: hydrolyzable tannins and condensed tannins. Hydrolyzable tannins consist of gallic acid or ellagic acid units esterified with a core sugar molecule, while condensed tannins, also known as proanthocyanidins, are polymers of flavonoid units. Both types of tannins are synthesized through complex metabolic pathways, primarily involving the phenylpropanoid and flavonoid biosynthetic routes. These pathways are regulated by various environmental factors, such as light, temperature, and nutrient availability, leading to variability in tannin composition and concentration across different plant species and tissues (Barbehenn & Constabel, 2011). The classification and biosynthesis of tannins are central to understanding their diverse roles in plant biology, particularly their contributions to defense against biotic and abiotic stressors.

Tannins are well-known for their potent antioxidant properties, which play a significant role in plant defense mechanisms. These compounds scavenge reactive oxygen species (ROS) generated during environmental stress, such as UV radiation, drought, or pathogen attack, thereby protecting plant cells from oxidative damage. The antioxidant activity of tannins is attributed to their polyphenolic structure, which allows them to donate electrons or hydrogen atoms to neutralize ROS. In addition to their direct antioxidant action, tannins can also modulate the activity of antioxidant enzymes, such as superoxide dismutase and catalase, further enhancing the plant's ability to cope with oxidative stress (Salminen & Karonen, 2011). This protective role is particularly important in leaves and fruits, where tannins accumulate in response to environmental challenges, providing a critical defense mechanism that enhances plant resilience.

Beyond their antioxidant properties, tannins also serve as effective antimicrobial agents, protecting plants from a wide range of pathogens, including bacteria, fungi, and viruses. The antimicrobial activity of tannins is primarily based on their ability to disrupt microbial cell membranes and interfere with enzyme activity. Tannins can form complexes with microbial proteins, leading to the inhibition of essential metabolic processes such as nutrient uptake and cell wall synthesis. For instance, condensed tannins have been shown to inhibit the growth of pathogenic fungi, such as *Botrytis cinerea* and *Fusarium oxysporum*, by disrupting their cell membranes and preventing spore germination (Scalbert, 1991). Additionally, hydrolyzable tannins, like ellagitannins, exhibit strong antibacterial activity by binding to bacterial cell walls and inhibiting enzyme activity, further contributing to plant defense. The dual role of tannins as antioxidants and antimicrobial agents makes them versatile defense compounds that enhance plant survival in pathogen-rich environments.

Tannins also play a critical role in deterring herbivores and promoting insect resistance in plants. Their ability to bind to dietary proteins and digestive enzymes in herbivores' guts reduces the nutritional value of plant tissues, making them less palatable to herbivores. This anti-nutritional effect of tannins is particularly important in defending plants against herbivores such as insects, mammals, and even certain bird species. The binding of tannins to proteins in the herbivore gut forms indigestible complexes that interfere with protein digestion, leading to reduced growth and development of herbivores (War et al., 2018). For example, condensed tannins in leguminous plants, such as alfalfa and clover, have been shown to significantly deter feeding by insect pests like aphids and caterpillars by inhibiting digestive enzymes. Moreover, the deterrent effect of tannins is often enhanced by their interaction with other plant defense compounds, such as alkaloids and flavonoids, resulting in a more robust defense strategy that limits herbivore damage (Al-Khayri et al., 2023).

In addition to their role in herbivore deterrence, tannins contribute to plant resistance against environmental stressors, such as drought, salinity, and extreme temperatures. The accumulation of tannins in response to abiotic stress is thought to protect plant tissues by stabilizing proteins and membranes under harsh conditions. For instance, during drought stress, tannins help maintain cellular integrity by reducing water loss and protecting membrane lipids from oxidative damage. Furthermore, tannins have been shown to interact with plant hormones, such as abscisic acid, which regulates stress responses in plants. This interaction enhances the plant's ability to adapt to changing environmental conditions by modulating stress-related gene expression and physiological responses (Haghpanah et al., 2024). The role of tannins in enhancing plant resilience to environmental stress highlights their importance not only in biotic defense but also in supporting plant survival in challenging habitats.

The application of tannins in sustainable agriculture and pathogen control is gaining increasing attention due to their natural origin, biodegradability, and low environmental impact. Tannins are being explored as natural biopesticides and biofungicides, offering an eco-friendly alternative to synthetic agrochemicals. For example, tannin extracts from plants such as chestnut (*Castanea sativa*) and quebracho (*Schinopsis spp.*) have been used to control fungal diseases in crops like grapes, tomatoes, and wheat. These tannin-based formulations inhibit fungal growth by disrupting spore germination and hyphal growth, providing a natural solution to managing plant diseases (Chung et al., 1998). Additionally, tannins have been shown to improve soil health by enhancing microbial diversity and promoting the growth of beneficial microorganisms, such as nitrogen-fixing bacteria and mycorrhizal fungi. This contributes to the long-term sustainability of agricultural systems by reducing the need for chemical inputs and supporting soil fertility. The future of tannins in sustainable agriculture lies in their potential to be integrated into broader integrated pest management (IPM) strategies, where they can be used in combination with other natural products and practices to reduce the reliance on synthetic chemicals and promote environmental conservation.

1. Biosynthesis and Classification of Tannins in Plants

Tannins are a diverse group of polyphenolic compounds found abundantly in the plant kingdom. These compounds play a critical role in plant defense, growth regulation, and ecological interactions.

Tannins are primarily classified into two major groups: hydrolyzable tannins and condensed tannins (proanthocyanidins), each with distinct biosynthetic pathways and structural characteristics. The biosynthesis of tannins is a complex process that involves several key metabolic pathways, primarily the phenylpropanoid and flavonoid biosynthetic routes.

1.1 Biosynthesis of Tannins

Tannins are synthesized through well-established metabolic pathways, with the biosynthesis of both hydrolyzable and condensed tannins beginning from common precursor molecules derived from the shikimate and phenylpropanoid pathways. These pathways are essential for the production of various aromatic compounds, including phenolics, flavonoids, and lignins, all of which contribute to the formation of tannins. Environmental factors such as light, temperature, and nutrient availability significantly influence tannin biosynthesis, leading to variations in tannin content and composition across different plant species and tissues (Barbehenn & Constabel, 2011).

1.1.1 Biosynthesis of Hydrolyzable Tannins

Hydrolyzable tannins are composed of a central sugar core, usually glucose, esterified with gallic acid or ellagic acid units. The biosynthesis of hydrolyzable tannins begins with the formation of gallic acid through the shikimate pathway, which involves the conversion of 3-dehydroshikimate into gallic acid by the enzyme shikimate dehydrogenase (Salminen & Karonen, 2011). Once gallic acid is synthesized, it is esterified to a sugar molecule, typically glucose, to form gallotannins. Ellagitannins are formed when gallotannins undergo oxidative coupling of the galloyl groups, producing hexahydroxydiphenoyl (HHDP) units, which are characteristic of ellagitannins (Chung et al., 1998). The structural complexity of hydrolyzable tannins arises from the number of galloyl or HHDP groups esterified to the sugar core. These tannins can be hydrolyzed by acids or enzymes to release gallic acid or ellagic acid, which are responsible for their astringent properties and biological activities. Hydrolyzable tannins are commonly found in species like oak (*Quercus*), chestnut (*Castanea*), and pomegranate (*Punica granatum*), where they serve as potent defense compounds against herbivores and pathogens (Mora et al., 2022).

1.1.2 Biosynthesis of Condensed Tannins (Proanthocyanidins)

Condensed tannins, also known as proanthocyanidins, are polymers of flavan-3-ol units, such as catechin and epicatechin. The biosynthesis of condensed tannins is closely linked to the flavonoid biosynthetic pathway, which originates from phenylalanine through the phenylpropanoid pathway. The key intermediates in this pathway include p-coumaroyl-CoA, which is converted into naringenin, a precursor of flavonoids. Naringenin is further modified by enzymes such as flavonoid 3'-hydroxylase (F3'H) and dihydroflavonol reductase (DFR) to produce flavan-3-ol monomers like catechin and epicatechin (Dong & Lin, 2003). Condensed tannins are formed when these flavan-3-ol units undergo polymerization, resulting in a range of oligomeric and polymeric structures. The degree of polymerization can vary greatly, leading to a diversity of condensed tannins with different molecular weights and biological activities. These tannins are resistant to hydrolysis, unlike hydrolyzable tannins, and are commonly found in plants such as grapevines (*Vitis vinifera*), tea (*Camellia sinensis*), and legumes like alfalfa (*Medicago sativa*). The astringency of condensed tannins is attributed to their ability to bind proteins and other macromolecules, which is also the basis of their role in plant defense against herbivores.

1.2 Classification of Tannins

Tannins are broadly classified into two main categories based on their chemical structure: hydrolyzable tannins and condensed tannins. Each class of tannins has distinct structural characteristics and biological functions, contributing to their roles in plant defense and ecological interactions.

1.2.1 Hydrolyzable Tannins

Hydrolyzable tannins are esters of a sugar molecule, usually glucose, with phenolic acids such as gallic acid or ellagic acid. They are called hydrolyzable because they can be broken down into their constituent phenolic acids and sugars through hydrolysis, either by acids or enzymes. Hydrolyzable tannins are further subdivided into gallotannins and ellagitannins based on the type of phenolic acid present. These tannins consist of gallic acid units esterified to a sugar core. Gallotannins are commonly found in plants such as sumac (*Rhus* species) and oak (*Quercus* species). They are known for their strong astringent properties and their ability to form complexes with proteins, contributing to their role in plant defense (Scalbert, 1991). Ellagitannins are formed by the oxidative coupling of galloyl groups, producing hexahydroxydiphenoyl (HHDP) units. These tannins are found in plants like pomegranate and chestnut and are known for their potent antioxidant and antimicrobial properties. Ellagitannins have been extensively studied for their health benefits in humans, particularly in the prevention of chronic diseases (Salminen & Karonen, 2011).

1.2.2 Condensed Tannins (Proanthocyanidins)

Condensed tannins, or proanthocyanidins, are polymers of flavan-3-ol units such as catechin and epicatechin. These tannins are resistant to hydrolysis and are commonly found in a wide variety of plants, including grapes, tea, cocoa, and legumes. Condensed tannins are known for their ability to bind proteins and other macromolecules, making them effective deterrents against herbivores. The polymeric nature of condensed tannins allows them to form complexes with proteins in the digestive systems of herbivores, reducing the digestibility of plant tissues and thus serving as an anti-nutritional defense mechanism (Barbehenn & Constabel, 2011). Condensed tannins are also known for their strong antioxidant activity, which contributes to their role in protecting plant tissues from oxidative damage caused by environmental stressors such as UV radiation and pathogen attack. In addition to their defensive properties, condensed tannins are important contributors to the flavor and astringency of many fruits and beverages, such as wine, tea, and cocoa.

1.3 Ecological and Biological Roles of Tannins

Tannins serve a wide range of ecological and biological functions in plants, particularly in defense against herbivores, pathogens, and environmental stressors. Their ability to bind proteins, metals, and other organic molecules makes them effective at deterring herbivory and preventing microbial infections. For example, the astringent taste of tannins reduces the palatability of plant tissues to herbivores, while their ability to inhibit digestive enzymes further discourages feeding (Adamczyk et al., 2017). In addition to their defensive roles, tannins contribute to the structural integrity of plant tissues by cross-linking with cell wall proteins and polysaccharides. This structural reinforcement helps plants withstand physical damage and environmental stress, such as drought or extreme temperatures. Tannins also play a role in regulating nutrient cycling in ecosystems, as they influence the decomposition of plant litter and the availability of nutrients to soil microorganisms (Bogati & Walczak, 2022). The biosynthesis and classification of tannins underscore their importance in plant biology and their potential applications in sustainable agriculture, medicine, and food industries. As research continues to uncover the complex regulatory networks involved in tannin biosynthesis and their diverse ecological roles, the potential for harnessing tannins for various commercial applications will continue to grow.

2. Antioxidant Properties of Tannins in Plant Defense

Tannins play a vital role in the antioxidant defense mechanisms of plants, helping to protect them from the oxidative damage caused by environmental stressors such as ultraviolet (UV) radiation, pathogen invasion, and herbivore attack. As polyphenolic compounds, tannins can scavenge reactive oxygen species (ROS), which are highly reactive molecules generated during environmental stresses and normal metabolic processes.

2.1 Mechanism of Antioxidant Action: Scavenging Reactive Oxygen Species (ROS)

Reactive oxygen species (ROS), such as superoxide anions (O_2^-), hydrogen peroxide (H_2O_2), and hydroxyl radicals (OH·), are byproducts of normal cellular metabolism but are produced in excess under stress conditions. These molecules can cause significant oxidative damage to cellular components, including lipids, proteins, and nucleic acids, leading to impaired cell function and even cell death. The antioxidant activity of tannins plays a crucial role in neutralizing ROS and preventing oxidative stress in plants (Andrés et al., 2023).

2.1.1 ROS Scavenging by Hydrolyzable Tannins

Hydrolyzable tannins, particularly gallotannins, and ellagitannins, are effective in scavenging ROS due to their high content of phenolic hydroxyl groups. These hydroxyl groups donate hydrogen atoms or electrons to neutralize ROS, converting them into less harmful species such as water or oxygen. The multiple phenolic groups present in hydrolyzable tannins enable them to engage in several redox reactions simultaneously, making them highly efficient antioxidants (Chung et al., 1998). For example, ellagitannins, such as those found in pomegranate (*Punica granatum*) and chestnut (*Castanea sativa*), have been shown to exhibit strong antioxidant activity by scavenging hydrogen peroxide and superoxide anions. These tannins not only neutralize ROS but also prevent the propagation of free radicals, which can lead to further oxidative damage. The ability of hydrolyzable tannins to form complexes with metal ions, such as iron and copper, also contributes to their antioxidant properties by inhibiting the metal-catalyzed generation of ROS through Fenton reactions (Žitek Makoter et al., 2024). This mechanism is particularly important in mitigating oxidative stress in plant tissues exposed to high levels of environmental stress.

2.1.2 ROS Scavenging by Condensed Tannins (Proanthocyanidins)

Condensed tannins, also known as proanthocyanidins, are another major class of tannins with potent antioxidant properties. These compounds are polymers of flavan-3-ol units, such as catechin and epicatechin, and are widely distributed in plants like grapes (*Vitis vinifera*), tea (*Camellia sinensis*), and cocoa (*Theobroma cacao*). The antioxidant activity of condensed tannins is primarily attributed to their ability to donate electrons and hydrogen atoms, neutralizing ROS and preventing oxidative damage to plant tissues (Dixon & Sarnala 2020). The polymeric nature of condensed tannins enhances their antioxidant capacity by increasing the number of reactive sites available for ROS scavenging. For instance, studies have shown that proanthocyanidins in grape seeds can effectively neutralize hydrogen peroxide and superoxide radicals, protecting plant cells from oxidative damage caused by UV radiation and pathogen invasion (Nie et al., 2023). Additionally, the high molecular weight of condensed tannins allows them to form strong complexes with metal ions, further reducing the generation of ROS through metal-catalyzed reactions.

2.2 Role of Tannins in Mitigating Oxidative Stress Induced by Environmental Stressors

Tannins play a critical role in protecting plants from oxidative stress induced by various environmental factors, including UV radiation, drought, and pathogen attack. By scavenging ROS and regulating the activity of antioxidant enzymes, tannins help plants maintain cellular homeostasis and enhance their resilience to environmental challenges.

2.2.1 Protection Against UV Radiation

Exposure to ultraviolet (UV) radiation, particularly UV-B (280-315 nm), can lead to the generation of ROS in plant tissues, resulting in oxidative damage to DNA, proteins, and lipids. Plants have evolved various protective mechanisms to mitigate UV-induced oxidative stress, and tannins are an integral part of this defense system. Tannins accumulate in the epidermal tissues of leaves and fruits, where they act as UV filters, absorbing harmful radiation and preventing the formation of ROS (Barbehenn & Constabel, 2011). Hydrolyzable tannins, such as gallotannins and ellagitannins, are particularly effective in protecting plants from UV-induced oxidative stress. These compounds absorb UV radiation and dissipate the energy as heat, thereby preventing the generation of ROS. In addition to their direct UV-absorbing properties, tannins also enhance the activity of antioxidant enzymes, such as superoxide dismutase (SOD) and catalase, which further reduce ROS levels in plant tissues. Studies on grapevines (*Vitis vinifera*) have demonstrated that the accumulation of proanthocyanidins in grape skins provides significant protection against UV-B radiation, reducing oxidative damage and improving fruit quality (Agati et al., 2012).

2.2.2 Drought Tolerance and Antioxidant Defense

Drought stress is another environmental factor that triggers the production of ROS in plants, leading to oxidative damage and impaired cellular function. Tannins play a key role in enhancing drought tolerance by mitigating the effects of oxidative stress. During drought conditions, tannins accumulate in plant tissues, where they act as antioxidants and protect cellular structures from ROS-induced damage. Condensed tannins, in particular, have been shown to enhance drought tolerance by stabilizing cell membranes and protecting membrane lipids from peroxidation. By scavenging ROS, condensed tannins prevent the oxidation of unsaturated fatty acids in membrane lipids, maintaining membrane integrity and preventing cell death (Nie et al., 2023). The antioxidant properties of tannins also contribute to the regulation of stomatal closure, a physiological response that reduces water loss during drought conditions. By minimizing oxidative damage, tannins enable plants to maintain cellular homeostasis and survive prolonged periods of water deficit.

2.2.3 Defense Against Pathogen-Induced Oxidative Stress

Pathogen invasion is a major source of oxidative stress in plants, as it triggers the rapid production of ROS as part of the plant's immune response. While ROS plays a crucial role in signaling and activating defense pathways, excessive ROS levels can lead to oxidative damage to plant tissues. Tannins help modulate the balance between ROS production and scavenging during pathogen attack, ensuring that ROS levels remain within a range that is beneficial for defense without causing collateral damage to plant cells (Scalbert, 1991). The antioxidant properties of tannins are particularly important in defending plants against fungal pathogens, which often produce ROS to weaken plant cell walls and facilitate infection. For example, condensed tannins in grapevines have been shown to inhibit the growth of fungal pathogens such as *Botrytis cinerea* and *Plasmopara viticola* by scavenging ROS and preventing oxidative damage to plant tissues (Das et al., 2020). Additionally, tannins can enhance the activity of plant defense enzymes, such as peroxidases and polyphenol oxidases, which further contribute to the detoxification of ROS and the strengthening of plant cell walls.

2.3 Interaction of Tannins with Antioxidant Enzymes

In addition to their direct ROS-scavenging activity, tannins also interact with and modulate the activity of antioxidant enzymes, which play a critical role in maintaining redox homeostasis in plants. These enzymes include superoxide dismutase (SOD), catalase (CAT), and ascorbate peroxidase (APX), all of which are involved in the detoxification of ROS.

2.3.1 Enhancement of Superoxide Dismutase (SOD) Activity

Superoxide dismutase (SOD) is a key enzyme that catalyzes the conversion of superoxide anions (O_2^-) into hydrogen peroxide (H_2O_2), which is less reactive and can be further detoxified by catalase or peroxidases. Tannins have been shown to enhance the activity of SOD in plants exposed to oxidative stress, improving the plant's ability to detoxify superoxide radicals (Bele et al., 2010). For instance, studies on tea plants (*Camellia sinensis*) have demonstrated that the accumulation of condensed tannins in leaves is associated with increased SOD activity, which helps protect the plant from oxidative damage caused by environmental stressors such as UV radiation and pathogen attack (Agati et al., 2012). The interaction between tannins and SOD highlights the importance of tannins not only as direct antioxidants but also as modulators of enzymatic defense pathways.

2.3.2 Modulation of Catalase (CAT) and Peroxidase (POD) Activity

Catalase (CAT) and peroxidases (POD) are enzymes that play a crucial role in detoxifying hydrogen peroxide (H_2O_2), a ROS that is generated during both normal metabolic processes and stress responses. Tannins have been shown to modulate the activity of these enzymes, enhancing the plant's ability to neutralize H_2O_2 and prevent oxidative damage. Hydrolyzable tannins, such as ellagitannins, have been found to increase the activity of catalase in plants exposed to oxidative stress, reducing the accumulation of H_2O_2 and mitigating its harmful effects (Chung et al., 1998). Similarly, tannins can enhance the activity of peroxidases, which are involved in both ROS detoxification and the cross-linking of cell wall components, strengthening the plant's structural defenses against pathogens.

2.4 Ecological Implications of Tannins as Antioxidants

The antioxidant properties of tannins not only benefit individual plants but also have broader ecological implications. By protecting plants from oxidative stress, tannins contribute to the stability and resilience of ecosystems, particularly in environments where plants are exposed to high levels of abiotic and biotic stressors. For example, tannin-rich plants in forest ecosystems play a critical role in regulating nutrient cycling and soil health. The antioxidant activity of tannins influences the decomposition of plant litter, as tannins can inhibit the activity of microbial decomposers and slow the breakdown of organic matter. This regulation of decomposition rates helps maintain soil fertility and supports the long-term sustainability of ecosystems (Huang et al., 2018). Moreover, tannins contribute to plant-pollinator and plant-herbivore interactions by modulating the oxidative status of plant tissues. The presence of tannins in flowers, fruits, and leaves can influence the behavior of pollinators and herbivores, shaping ecological dynamics and promoting plant survival.

3. Tannins as Antimicrobial Agents: Mechanisms Against Pathogens

Tannins are highly effective antimicrobial agents that play a critical role in defending plants against a broad spectrum of pathogens, including bacteria, fungi, and viruses. Their antimicrobial activity stems from their ability to interact with and disrupt key components of microbial cells, including proteins, membranes, and enzymes. Tannins exhibit multiple mechanisms of action against pathogens, ranging from disrupting microbial cell walls to interfering with essential metabolic processes. 3

3.1 Disruption of Microbial Cell Membranes

One of the most prominent antimicrobial mechanisms of tannins is their ability to disrupt the integrity of microbial cell membranes. This activity is particularly important in defending plants against bacterial and fungal pathogens. Tannins, due to their polyphenolic nature, can interact with lipid bilayers in microbial cell membranes, causing structural damage and increased membrane permeability. This leads to the leakage of intracellular components such as ions, proteins, and nucleotides, ultimately resulting in cell death (Scalbert, 1991).

3.1.1 Hydrolyzable Tannins and Bacterial Membrane Disruption

Hydrolyzable tannins, such as gallotannins and ellagitannins, have been shown to exhibit strong antibacterial properties through their ability to bind to bacterial cell membranes and disrupt their structure. These tannins interact with the lipid components of bacterial membranes, destabilizing the membrane's architecture and increasing its permeability (Lobiuc et al., 2023). For instance, ellagitannins, which are abundant in pomegranate (*Punica granatum*), can bind to the cell membranes of Gram-positive bacteria like *Staphylococcus aureus* and *Bacillus subtilis*, leading to the leakage of essential ions and cellular components. This disruption of membrane integrity results in the rapid death of bacterial cells, making hydrolyzable tannins potent antibacterial agents. The antimicrobial action of hydrolyzable tannins is further enhanced by their ability to chelate metal ions, such as iron, that are essential for bacterial growth. By sequestering these metal ions, tannins inhibit microbial metabolic processes, effectively starving the pathogen of the nutrients it needs to survive (Bialonska et al., 2009).

3.1.2 Condensed Tannins and Fungal Membrane Permeability

Condensed tannins (proanthocyanidins), found in plants such as grapevines (*Vitis vinifera*), tea (*Camellia sinensis*), and cocoa (*Theobroma cacao*), are particularly effective against fungal pathogens. These tannins interact with fungal cell membranes by binding to membrane proteins and lipids, causing increased membrane permeability and loss of cellular homeostasis. This leads to the leakage of ions and metabolites, ultimately resulting in fungal cell death (Chiocchio et al., 2021). Studies have shown that condensed tannins from grape seeds can inhibit the growth of fungal pathogens such as *Botrytis cinerea*, which causes gray mold in grapevines. By disrupting the fungal cell membrane, proanthocyanidins prevent the pathogen from establishing an infection and spreading throughout the plant (Jian et al., 2024). Additionally, condensed tannins have been found to inhibit spore germination in fungal pathogens, further reducing the spread of infection.

3.2 Inhibition of Microbial Enzymes

Another important mechanism by which tannins exert their antimicrobial effects is the inhibition of microbial enzymes. Pathogens rely on a variety of enzymes to invade plant tissues, degrade cell walls, and acquire nutrients from host cells. Tannins can inhibit these enzymes, preventing pathogens from successfully infecting and colonizing plants.

3.2.1 Inhibition of Fungal Cell Wall-Degrading Enzymes

Fungal pathogens produce a range of extracellular enzymes, such as cellulases, pectinases, and glucanases, which degrade plant cell walls and facilitate infection. Tannins, particularly hydrolyzable tannins, can bind to these enzymes and inhibit their activity, preventing the breakdown of plant cell walls and reducing the ability of fungi to invade plant tissues (Scalbert, 1991). For example, gallotannins and ellagitannins have been shown to inhibit the activity of fungal cellulases and pectinases, enzymes that are crucial for the degradation of plant cell walls during infection. By inhibiting these enzymes, tannins limit the ability of fungal pathogens like *Fusarium oxysporum* and *Rhizoctonia solani* to infect plant roots and stems, providing an effective defense against root rot and wilt diseases.

3.2.2 Inhibition of Bacterial Virulence Factors

In addition to their ability to inhibit fungal enzymes, tannins can also interfere with bacterial virulence factors, such as proteases and toxins, which are essential for bacterial pathogenesis. Bacterial pathogens often produce proteases that degrade plant cell walls and disrupt host immune responses, allowing the pathogen to invade plant tissues. Tannins, particularly condensed tannins, have been found to inhibit the activity of these proteases, reducing bacterial virulence and limiting the severity of infections (Chung et al., 1998). For instance, condensed tannins from tea leaves have been shown to inhibit the activity of proteases produced by *Pseudomonas syringae*, a bacterial pathogen that causes leaf spot and blight in various crops. By inhibiting these virulence factors, tannins reduce the ability of the pathogen to degrade plant tissues and evade the plant's immune system, providing an additional layer of defense.

3.3 Formation of Complexes with Microbial Proteins

Tannins are known for their ability to form stable complexes with proteins, a property that is central to their antimicrobial activity. When tannins bind to microbial proteins, they cause protein denaturation and precipitation, rendering the proteins non-functional. This mechanism is particularly important in inhibiting microbial enzymes and structural proteins that are essential for the survival and virulence of pathogens (Lobiuc et al., 2023).

3.3.1 Protein Precipitation and Enzyme Inhibition

The protein-binding ability of tannins allows them to precipitate microbial enzymes, effectively inhibiting their catalytic activity. This is especially relevant for enzymes that are secreted by pathogens to degrade plant cell walls or evade host immune responses. Tannins can bind to the active sites of these enzymes, preventing them from interacting with their substrates and halting their activity.

For example, gallotannins from oak (*Quercus* species) have been shown to precipitate cellulases and pectinases produced by fungal pathogens, reducing their ability to degrade plant cell walls and invade host tissues (Salminen & Karonen, 2011). This protein-binding activity is not limited to enzymes; tannins can also bind to structural proteins in microbial cell walls and membranes, destabilizing these structures and leading to cell death.

3.3.2 Disruption of Microbial Adhesion

In addition to their ability to precipitate proteins, tannins can also inhibit microbial adhesion to plant surfaces, a critical step in the infection process. Pathogens, particularly bacteria, rely on adhesion proteins such as adhesins and pili to attach to plant cell walls and establish infections. Tannins can interfere with these adhesion mechanisms by binding to microbial adhesion proteins, preventing the pathogen from adhering to the plant surface (Chung et al., 1998). For instance, ellagitannins from pomegranate have been shown to inhibit the adhesion of *Xanthomonas campestris*, a bacterial pathogen that causes black rot in cruciferous crops. By preventing bacterial adhesion, tannins reduce the ability of the pathogen to colonize plant tissues and initiate infection, providing an effective defense against bacterial diseases.

3.4 Antiviral Activity of Tannins

In addition to their antibacterial and antifungal properties, tannins have been found to exhibit antiviral activity, making them valuable in protecting plants from viral pathogens. The antiviral activity of tannins is primarily based on their ability to interfere with viral replication and prevent the attachment of viruses to host cells. Tannins can inhibit viral enzymes, such as reverse transcriptase and DNA polymerase, which are essential for viral replication, thereby reducing viral load in infected tissues (Scalbert, 1991).

3.4.1 Inhibition of Viral Enzymes

Tannins, particularly hydrolyzable tannins like ellagitannins, have been shown to inhibit the activity of viral enzymes involved in nucleic acid synthesis and replication. For example, ellagitannins from pomegranate have been found to inhibit reverse transcriptase, an enzyme used by retroviruses to replicate their genetic material. By inhibiting this enzyme, tannins reduce the replication of viruses such as the tobacco mosaic virus (TMV), which affects many crop species (Banc et al., 2023). The ability of tannins to interfere with viral replication makes them valuable tools in managing viral diseases in agriculture. Although viruses are generally more difficult to control than bacterial or fungal pathogens, the antiviral properties of tannins offer a promising avenue for developing natural antiviral treatments for crops.

3.4.2 Prevention of Viral Attachment

In addition to inhibiting viral enzymes, tannins can also prevent the attachment of viruses to host cells by binding to viral surface proteins. This mechanism is particularly important in preventing the initial stages of viral infection, where the virus must attach to plant cell receptors to enter the host cell. By binding to viral proteins, tannins block the interaction between the virus and the host cell, reducing the likelihood of infection (Scalbert, 1991). For example, condensed tannins from grapevines have been shown to prevent the attachment of cucumber mosaic virus (CMV) to plant cells, reducing the spread of the virus in infected tissues. This antiviral activity is further enhanced by the ability of tannins to strengthen plant cell walls, making it more difficult for viruses to penetrate plant tissues and establish infections.

3.5 Synergistic Effects of Tannins with Other Plant Metabolites

Tannins often work in synergy with other plant secondary metabolites, such as alkaloids, flavonoids, and terpenoids, to enhance their antimicrobial activity. These synergistic interactions result in more effective defense mechanisms that target multiple aspects of pathogen physiology, including membrane integrity, enzyme activity, and replication. By combining tannins with other antimicrobial compounds, plants can deploy a more robust and comprehensive defense against pathogens (Barbehenn & Constabel, 2011).

3.5.1 Synergy Between Tannins and Flavonoids

Flavonoids, another class of polyphenolic compounds, are known for their strong antimicrobial and antioxidant properties. When tannins and flavonoids are combined, they exhibit enhanced antimicrobial activity, particularly against fungal pathogens. For instance, the combination of proanthocyanidins (condensed tannins) and quercetin (a flavonoid) has been shown to inhibit the growth of fungal pathogens such as *Fusarium oxysporum* more effectively than either compound alone (Salminen & Karonen, 2011).

This synergy is based on the complementary mechanisms of action of tannins and flavonoids. While tannins primarily disrupt microbial membranes and precipitate proteins, flavonoids interfere with microbial metabolism and oxidative stress responses. Together, these compounds provide a more comprehensive defense against pathogens, reducing the likelihood of infection and promoting plant health.

3.5.2 Interaction with Terpenoids and Alkaloids

Tannins also interact synergistically with terpenoids and alkaloids, two other classes of secondary metabolites with strong antimicrobial properties. For example, the combination of tannins with saponins (a class of terpenoids) has been shown to enhance the antifungal activity of both compounds, resulting in more effective inhibition of fungal pathogens like *Phytophthora infestans* (War et al., 2018). Similarly, tannins and alkaloids can work together to inhibit bacterial growth and reduce the virulence of pathogens. These synergistic interactions highlight the importance of tannins in the integrated defense strategies of plants. By working in concert with other secondary metabolites, tannins provide a multi-faceted defense system that targets multiple aspects of pathogen biology, ensuring the survival of the plant in hostile environments.

4. Impact of Tannins on Herbivore Deterrence and Insect Resistance

Tannins are key components of plant defense mechanisms against herbivores, including insects, mammals, and even birds. Their astringent taste, ability to form complexes with proteins, and anti-nutritional effects make them highly effective deterrents against herbivores. By influencing the palatability, digestibility, and nutritional quality of plant tissues, tannins play a significant role in reducing herbivory and promoting insect resistance in plants.

4.1 Tannins and Herbivore Digestive Processes

The primary mechanism by which tannins deter herbivores is through their interaction with dietary proteins and digestive enzymes. Tannins, particularly condensed tannins (proanthocyanidins), have a strong affinity for proteins, and this ability to bind to proteins has profound implications for herbivore digestion. By forming complexes with proteins in both plant tissues and the herbivore gut, tannins reduce the digestibility and nutritional value of the plant material, making it less appealing to herbivores (Barbehenn & Constabel, 2011).

4.1.1 Protein Precipitation and Reduced Nutritional Value

When herbivores consume tannin-rich plants, tannins bind to dietary proteins in the gut, forming insoluble complexes that are resistant to enzymatic breakdown. This binding renders the proteins indigestible, reducing the nutritional benefits of the plant material for the herbivore. For example, condensed tannins in leguminous plants such as alfalfa (*Medicago sativa*) and clover (*Trifolium spp.*) have been shown to bind to proteins in the digestive system of herbivores, including livestock and insects, significantly reducing protein absorption and growth rates in herbivores (War et al., 2018). The anti-nutritional effects of tannins are particularly detrimental to insect herbivores, whose protein requirements are high for growth and development. By limiting the availability of essential amino acids, tannins reduce the growth and reproductive success of insect pests, making them effective defense compounds against herbivory. In agricultural settings, the presence of tannins in forage crops like alfalfa helps protect these crops from insect pests such as aphids and caterpillars, reducing the need for chemical insecticides.

4.1.2 Inhibition of Digestive Enzymes

In addition to binding to dietary proteins, tannins also inhibit the activity of digestive enzymes in the herbivore gut, further impairing nutrient absorption. Digestive enzymes such as proteases (e.g., trypsin and chymotrypsin), amylases, and lipases are essential for breaking down proteins, carbohydrates, and fats in the diet. Tannins can bind to these enzymes and inhibit their catalytic activity, preventing the efficient digestion of food (Salminen & Karonen, 2011). For example, studies have shown that condensed tannins from grape seeds (*Vitis vinifera*) and tea leaves (*Camellia sinensis*) can inhibit the activity of proteases in insect herbivores, reducing protein digestion and assimilation. This inhibition of digestive enzymes leads to reduced growth and survival of insect pests, making tannins valuable in pest-resistant crop varieties. In mammals, the presence of tannins in forage crops has been linked to reduced feed efficiency and weight gain, which can have both positive and negative implications depending on the agricultural context (Barbehenn & Constabel, 2011).

4.2 Influence of Tannins on Herbivore Feeding Behavior

Tannins not only affect herbivore digestion but also influence feeding behavior through their astringency and bitterness. The astringent taste of tannins, caused by their protein-binding properties, makes plant tissues less palatable to herbivores, leading to reduced feeding rates. This deterrent effect is particularly important in deterring insect herbivores, whose feeding preferences are highly influenced by the taste and texture of plant tissues.

4.2.1 Astringency and Feeding Deterrence

The astringency of tannins is a result of their ability to bind to salivary proteins, causing a drying and puckering sensation in the mouth. This unpleasant taste discourages herbivores from consuming tannin-rich plants, reducing the likelihood of repeated feeding. For example, studies have shown that high tannin concentrations in tree species such as oaks (*Quercus* spp.) and willows (*Salix* spp.) significantly reduce the feeding activity of herbivorous insects like caterpillars and aphids (Volf, 2018). In addition to their astringency, tannins can also interfere with the gustatory receptors of insects, altering their perception of food quality. This sensory disruption reduces the attractiveness of tannin-containing plants to herbivores, leading to lower feeding rates and decreased herbivory. In many cases, insect herbivores will actively avoid tannin-rich plants in favor of more palatable, low-tannin species.

4.2.2 Deterrence of Specific Insect Pests

Tannins are particularly effective in deterring insect herbivores, which rely heavily on chemical cues to locate and assess food sources. The presence of tannins in leaves, stems, and seeds can deter a wide range of insect pests, including aphids, caterpillars, and beetles. For example, condensed tannins in the leaves of certain leguminous plants have been shown to reduce feeding by aphids (*Aphis spp.*), while hydrolyzable tannins in tree bark deter bark beetles (*Scolytinae*) (Scalbert, 1991). The deterrent effects of tannins are not limited to foliar herbivores; they also extend to seed-feeding insects. In species like chestnut (*Castanea sativa*) and oak, tannins in seeds and acorns provide a chemical defense against seed predators, reducing the likelihood of seed consumption and improving seed survival rates. The bitter taste of tannins in seeds discourages herbivores from consuming them, while the anti-nutritional effects further reduce their attractiveness as food sources.

4.3 Role of Tannins in Induced Resistance

In addition to their constitutive presence in plant tissues, tannins can also be synthesized in response to herbivore attack, a phenomenon known as induced resistance. When plants are damaged by herbivores, they often upregulate the production of tannins and other secondary metabolites as part of their defense response. This increase in tannin concentration makes the plant less palatable to herbivores and reduces the extent of further damage.

4.3.1 Herbivore-Induced Tannin Synthesis

Herbivore-induced tannin synthesis is a well-documented defense mechanism in many plant species.

For example, studies on willows (*Salix* spp.) have shown that herbivory by caterpillars induces the production of condensed tannins in leaves, which deters further feeding and reduces caterpillar growth rates (War et al., 2018). Similarly, oak trees (*Quercus* spp.) increase the production of hydrolyzable tannins in response to herbivory by leaf-eating insects, providing a chemical defense that limits damage. The induction of tannins in response to herbivory is often mediated by signaling molecules such as jasmonic acid and salicylic acid, which regulate the expression of genes involved in tannin biosynthesis. This rapid upregulation of tannin production allows plants to respond quickly to herbivore attack, enhancing their resistance to further damage.

4.3.2 Long-Term Effects of Induced Tannins on Herbivore Populations

The presence of tannins in plants not only deters individual herbivores but can also have long-term effects on herbivore populations. By reducing the palatability and nutritional quality of plant tissues, tannins can decrease the growth, reproduction, and survival of herbivore populations over time. This is particularly important in managing insect pests in agricultural systems, where the use of tannin-rich crop varieties can reduce pest populations without the need for chemical insecticides (Barbehenn & Constabel, 2011). For example, tannin-rich forage crops like alfalfa and clover have been shown to reduce populations of insect pests such as aphids and caterpillars, improving crop yield and reducing the need for chemical interventions. In forest ecosystems, the presence of tannins in tree leaves and bark can limit the population growth of defoliating insects, contributing to the long-term health and stability of the forest.

4.4 Tannins and Insect Resistance in Agricultural Systems

The application of tannins in agricultural systems is gaining increasing attention as a natural and environmentally friendly strategy for pest management. Tannin-rich plant varieties are being explored for their potential to enhance crop resistance to insect pests, reduce the need for synthetic insecticides, and promote sustainable farming practices.

4.4.1 Use of Tannin-Rich Varieties in Crop Protection

Several crop species, including forage crops like alfalfa and clover, have been bred to contain higher levels of condensed tannins to improve resistance to insect pests. These tannin-rich varieties are particularly effective in deterring insect herbivores and reducing pest damage in the field. For example, tannin-rich alfalfa cultivars have been shown to reduce feeding by aphids and caterpillars, improving forage quality and yield (War et al., 2018). In addition to forage crops, tannin-rich fruit and vegetable crops are being explored for their potential to enhance pest resistance. Grapevines, for example, produce high levels of proanthocyanidins in their leaves and skins, which provide natural protection against insect pests like leafhoppers and mites. The use of tannin-rich grape cultivars in vineyards reduces the need for chemical pesticides, contributing to more sustainable viticulture practices.

4.4.2 Potential for Tannins in Integrated Pest Management (IPM)

Tannins have significant potential for integration into broader pest management strategies, particularly in the context of integrated pest management (IPM). IPM focuses on using a combination of biological, cultural, and chemical control methods to manage pests in a sustainable and environmentally friendly manner. Tannins, as natural deterrents of herbivores and insect pests, can be incorporated into IPM strategies to reduce the reliance on synthetic insecticides and promote ecological balance in agricultural systems. For example, tannin-rich cover crops, such as legumes and grasses, can be planted alongside cash crops to deter insect pests and improve soil health. The presence of tannins in these cover crops reduces pest pressure on the main crop, while also contributing to the overall sustainability of the farming system.

4.5 Ecological Implications of Tannins in Herbivore Resistance

The impact of tannins on herbivore resistance extends beyond individual plants to broader ecological interactions. By deterring herbivores and reducing herbivory, tannins influence plant community dynamics, herbivore populations, and predator-prey interactions in ecosystems.

4.5.1 Tannins and Plant Community Structure

In ecosystems where tannin-rich plants are prevalent, herbivore populations may shift their feeding preferences toward less defended plant species, altering the composition of plant communities. This selective feeding pressure can lead to changes in plant abundance and diversity, with tannin-rich species gaining a competitive advantage in herbivore-dominated ecosystems (Salminen & Karonen, 2011).

4.5.2 Tannins and Predator-Prey Dynamics

Tannins also play a role in shaping predator-prey dynamics by influencing the quality of herbivores as prey for predators. Herbivores that consume tannin-rich plants may experience reduced growth and reproductive success, making them less attractive or less available as prey for predators. This can have cascading effects on food webs, influencing the population dynamics of both herbivores and their predators.

5. Tannins in Enhancing Plant Resistance to Environmental Stressors

Tannins play a crucial role in enhancing plant resistance to various environmental stressors, including drought, salinity, extreme temperatures, and oxidative stress. These secondary metabolites contribute to plant resilience by protecting cellular structures, modulating stress-related signaling pathways, and enhancing the plant's ability to cope with abiotic challenges. The antioxidant, osmoprotective, and protective properties of tannins allow plants to survive and thrive under adverse environmental conditions.

5.1 Tannins and Drought Tolerance

Drought is one of the most significant environmental stressors affecting plant growth and productivity. During periods of water scarcity, plants experience oxidative stress due to the overproduction of reactive oxygen species (ROS), which can damage cellular components such as membranes, proteins, and DNA. Tannins, particularly condensed tannins (proanthocyanidins), play a vital role in protecting plants from the damaging effects of drought by acting as antioxidants, stabilizing cellular membranes, and reducing water loss through enhanced water retention mechanisms (Barbehenn & Constabel, 2011).

5.1.1 Antioxidant Role of Tannins in Drought Conditions

Under drought stress, the accumulation of ROS in plant cells leads to oxidative damage, which compromises cell function and survival. Tannins, with their strong antioxidant properties, help mitigate this oxidative damage by scavenging ROS and preventing the peroxidation of membrane lipids. Condensed tannins, in particular, are known for their ability to neutralize hydrogen peroxide and superoxide radicals, which are commonly produced during drought stress (Chung et al., 1998). By reducing oxidative damage, tannins maintain membrane integrity and enhance the plant's ability to tolerate prolonged periods of water scarcity. Additionally, tannins have been shown to enhance the activity of antioxidant enzymes such as superoxide dismutase (SOD), catalase (CAT), and peroxidases (POD), which are involved in detoxifying ROS. By modulating the expression and activity of these enzymes, tannins further strengthen the plant's defense against oxidative stress during drought conditions. For example, studies on grapevines (*Vitis vinifera*) have demonstrated that the presence of condensed tannins in leaves helps protect against drought-induced oxidative damage, improving plant survival and fruit quality (Agati et al., 2012).

5.1.2 Osmoprotective Effects of Tannins

In addition to their antioxidant role, tannins also contribute to drought tolerance by acting as osmoprotectants. Osmoprotectants are compounds that help maintain cellular water balance and protect cells from osmotic stress caused by dehydration. Tannins can bind to proteins and other cellular macromolecules, stabilizing their structure and preventing denaturation under low water conditions (Salminen & Karonen, 2011). By protecting cellular proteins and membranes from dehydration-induced damage, tannins enhance the plant's ability to retain water and maintain cellular homeostasis during drought. This osmoprotective function is particularly important in protecting sensitive tissues such as leaves and roots, which are often the first to experience dehydration under water-limited conditions. The accumulation of tannins in these tissues helps plants withstand prolonged drought periods by reducing water loss and preserving cellular integrity.

5.2 Role of Tannins in Salinity Tolerance

Salinity is another major abiotic stressor that affects plant growth, particularly in arid and semi-arid regions where salt accumulation in soils is common. High salt concentrations in the soil lead to osmotic stress, ion toxicity, and oxidative damage in plants. Tannins contribute to salinity tolerance by mitigating the harmful effects of salt stress through their antioxidant activity and their ability to regulate ion balance in plant cells (Scalbert, 1991).

5.2.1 Tannins as Antioxidants in Saline Conditions

Similar to drought stress, salinity stress leads to the overproduction of ROS in plant cells, causing oxidative damage to membranes, proteins, and nucleic acids. Tannins help mitigate this oxidative damage by scavenging ROS and protecting cellular structures from salt-induced oxidative stress. The antioxidant properties of tannins are particularly important in protecting membrane lipids from peroxidation, which is a common consequence of salinity stress (Chung et al., 1998). Studies have shown that plants with higher levels of tannins are more resistant to salt-induced oxidative damage. For example, the accumulation of condensed tannins in halophytic plants (salt-tolerant species) has been linked to increased salinity tolerance, as these tannins protect the plant from oxidative damage and improve ion homeostasis (War et al., 2018). By enhancing the plant's antioxidant defense system, tannins enable plants to survive in saline environments and reduce the detrimental effects of salt accumulation in plant tissues.

5.2.2 Regulation of Ion Balance by Tannins

Salinity stress disrupts ion homeostasis in plants by causing the excessive accumulation of sodium (Na^+) and chloride (Cl^-) ions in cells, which leads to ion toxicity and impaired cellular function. Tannins have been shown to play a role in regulating ion balance in plant cells, particularly by modulating the uptake and distribution of Na^+ and Cl^- ions. By reducing the accumulation of toxic ions in plant tissues, tannins help maintain cellular homeostasis and protect plants from the harmful effects of salinity stress (Salminen & Karonen, 2011). For instance, in salt-tolerant plants like mangroves and certain legumes, tannins have been found to interact with ion transporters in the root and shoot systems, reducing the uptake of Na^+ and promoting the exclusion of toxic ions from sensitive tissues. This ion-regulatory function of tannins helps plants maintain nutrient balance and reduces the physiological damage caused by salt accumulation, enhancing their ability to thrive in saline environments.

5.3 Tannins and Temperature Extremes

Plants are frequently exposed to temperature extremes, including both high heat and freezing temperatures, which can cause cellular damage and disrupt normal metabolic processes. Tannins contribute to plant resistance to temperature extremes by protecting cells from thermal damage, stabilizing cellular membranes, and reducing the production of ROS under heat and cold stress conditions.

5.3.1 Protection Against Heat Stress

High temperatures can lead to the denaturation of proteins, disruption of membrane integrity, and increased production of ROS, all of which contribute to heat stress in plants. Tannins, particularly condensed tannins, play a protective role in mitigating the effects of heat stress by stabilizing cellular membranes and proteins, reducing oxidative damage, and enhancing the activity of heat shock proteins (HSPs) (Scalbert, 1991). Heat shock proteins are molecular chaperones that help prevent protein denaturation and assist in the refolding of damaged proteins during heat stress. Tannins have been shown to enhance the expression and activity of HSPs, improving the plant's ability to cope with high temperatures. Additionally, the antioxidant properties of tannins help neutralize ROS generated during heat stress, preventing oxidative damage to cellular structures.

5.3.2 Cold Tolerance and Membrane Stability

Low temperatures, particularly freezing conditions, can cause membrane damage, ice formation, and osmotic stress in plants. Tannins contribute to cold tolerance by protecting cellular membranes from freezing-induced damage and stabilizing membrane lipids. Condensed tannins, in particular, have been found to prevent the peroxidation of membrane lipids during cold stress, maintaining membrane integrity and preventing the leakage of cellular contents (Barbehenn & Constabel, 2011). In cold-tolerant plant species, tannins accumulate in leaves and stems during periods of low temperature, providing a protective barrier against freezing damage. The ability of tannins to bind to and stabilize membrane proteins and lipids helps plants survive freezing temperatures by preventing the formation of ice crystals and reducing osmotic stress.

5.4 Tannins and Oxidative Stress

Oxidative stress is a common consequence of various environmental stressors, including drought, salinity, temperature extremes, and pathogen attack. Tannins play a crucial role in mitigating oxidative stress by acting as potent antioxidants that neutralize ROS and protect cellular components from oxidative damage. Their ability to scavenge free radicals and modulate the activity of antioxidant enzymes makes tannins essential for maintaining redox homeostasis in plants under stress conditions.

5.4.1 Scavenging of Reactive Oxygen Species (ROS)

As polyphenolic compounds, tannins have multiple hydroxyl groups that can donate electrons or hydrogen atoms to neutralize ROS, such as superoxide anions, hydrogen peroxide, and hydroxyl radicals. This ROS-scavenging activity prevents the accumulation of oxidative species in plant cells and reduces the risk of oxidative damage to membranes, proteins, and nucleic acids (Chung et al., 1998). By neutralizing ROS, tannins help protect plant cells from the detrimental effects of oxidative stress and improve plant survival under challenging environmental conditions.

5.4.2 Modulation of Antioxidant Enzymes

In addition to their direct antioxidant activity, tannins also enhance the plant's enzymatic antioxidant defense system by modulating the activity of key enzymes involved in ROS detoxification. These enzymes include superoxide dismutase (SOD), catalase (CAT), and ascorbate peroxidase (APX), all of which play critical roles in detoxifying ROS and maintaining redox balance in plant cells. Tannins have been shown to increase the activity of these enzymes, further strengthening the plant's ability to cope with oxidative stress (Salminen & Karonen, 2011). For example, studies on tea plants (*Camellia sinensis*) have demonstrated that the accumulation of condensed tannins in leaves is associated with increased SOD and CAT activity, which helps protect the plant from oxidative damage caused by environmental stressors such as drought and UV radiation. This enhancement of antioxidant enzyme activity by tannins contributes to the overall resilience of plants in stressful environments.

5.5 Ecological and Agricultural Implications of Tannins in Stress Resistance

The role of tannins in enhancing plant resistance to environmental stressors has important ecological and agricultural implications. By improving plant resilience to drought, salinity, temperature extremes, and oxidative stress, tannins contribute to the survival and productivity of plants in challenging environments. This is particularly important in natural ecosystems, where tannin-rich plants often dominate in stress-prone habitats such as arid regions, salt marshes, and alpine environments.

In agricultural systems, tannins can be harnessed to improve crop resistance to abiotic stress, reduce the need for chemical inputs, and promote sustainable farming practices. The development of tannin-rich crop varieties with enhanced drought tolerance, salinity resistance, and temperature resilience has the potential to improve agricultural productivity in regions affected by climate change and environmental degradation. By incorporating tannin-rich plants into crop rotations, agroforestry systems, and integrated pest management strategies, farmers can enhance the sustainability of their farming systems while reducing the reliance on synthetic chemicals. The use of natural plant defenses, such as tannins, represents a promising approach to promoting resilience in agricultural systems and supporting the long-term health of ecosystems.

6. Applications of Tannins in Sustainable Agriculture and Pathogen Control

Tannins have gained significant attention in sustainable agriculture due to their multifunctional properties, particularly their role in enhancing plant resistance to pests and diseases while reducing the need for synthetic chemical inputs. Their natural origin, biodegradability, and wide range of biological activities make them valuable tools in promoting ecological balance and improving crop productivity. Tannins serve as effective agents in pathogen control, pest deterrence, soil health management, and improving plant resilience against environmental stressors.

6.1 Tannins as Natural Biopesticides

One of the most promising applications of tannins in sustainable agriculture is their use as natural biopesticides. Unlike synthetic pesticides, tannins are biodegradable, environmentally friendly, and pose minimal risk to non-target organisms such as beneficial insects, pollinators, and soil microorganisms. Tannins exhibit strong insecticidal, antifungal, and antibacterial properties, making them effective in controlling a wide range of agricultural pests and pathogens.

6.1.1 Insecticidal Properties of Tannins

Tannins, particularly condensed tannins (proanthocyanidins), have been shown to deter insect herbivores by affecting their feeding behavior, digestion, and growth. Tannins reduce the palatability of plant tissues by binding to dietary proteins and digestive enzymes in the insect gut, resulting in reduced nutrient absorption and growth. This anti-nutritional effect is especially effective against insects such as aphids, caterpillars, and beetles (Barbehenn & Constabel, 2011). For example, studies have shown that tannin-rich grapevines (*Vitis vinifera*) exhibit natural resistance to insect pests such as leafhoppers and mites, reducing the need for synthetic insecticides. In forage crops like alfalfa (*Medicago sativa*) and clover (*Trifolium spp.*), the presence of condensed tannins deters aphid feeding and limits insect damage, leading to improved crop quality and yield (War et al., 2018). The use of tannins as biopesticides in crop protection offers a sustainable alternative to chemical pesticides, reducing the environmental impact of agricultural practices.

6.1.2 Antifungal and Antibacterial Properties

Tannins also exhibit potent antifungal and antibacterial activity, making them valuable in managing plant diseases caused by fungal and bacterial pathogens. The antimicrobial mechanisms of tannins involve disrupting microbial cell membranes, inhibiting enzyme activity, and interfering with pathogen adhesion to plant surfaces (Scalbert, 1991). By targeting key components of microbial cells, tannins provide an effective defense against a wide range of plant pathogens.

For instance, tannins from chestnut (*Castanea sativa*) and quebracho (*Schinopsis spp.*) have been used to control fungal diseases in crops such as grapes, tomatoes, and wheat. These tannin-based formulations inhibit fungal spore germination and mycelial growth, reducing the spread of diseases like powdery mildew and blight. The ability of tannins to form complexes with microbial proteins and enzymes further enhances their antifungal activity, providing a natural solution to managing fungal infections in crops (Chung et al., 1998). In addition to their antifungal properties, tannins are effective against bacterial pathogens that cause diseases such as bacterial leaf spot, blight, and crown gall. Hydrolyzable tannins, such as ellagitannins, exhibit strong antibacterial activity by binding to bacterial cell walls and inhibiting the activity of bacterial enzymes. This action prevents the colonization and spread of bacterial pathogens, reducing the severity of bacterial infections in crops (Czerkas et al., 2024).

6.2 Tannins in Soil Health and Fertility

Beyond their role in pest and disease management, tannins contribute to soil health and fertility by enhancing soil structure, promoting beneficial microbial activity, and improving nutrient cycling. Tannins interact with soil organic matter and minerals, influencing the chemical and physical properties of soils, which is essential for maintaining soil fertility and crop productivity.

6.2.1 Enhancement of Soil Structure

Tannins, particularly condensed tannins, have been shown to improve soil structure by promoting the formation of soil aggregates. Soil aggregation is crucial for maintaining soil porosity, water infiltration, and root penetration, all of which are essential for healthy plant growth. Tannins enhance soil aggregation by interacting with soil particles and organic matter, forming stable complexes that improve the structural integrity of soils (Barbehenn & Constabel, 2011). Improved soil structure also contributes to better water retention and drainage, reducing the risk of soil erosion and nutrient leaching. In agricultural systems, the application of tannin-rich plant residues or tannin-based soil amendments can enhance soil health and reduce the need for synthetic soil conditioners, promoting sustainable farming practices.

6.2.2 Promotion of Beneficial Microbial Activity

Tannins play a significant role in shaping soil microbial communities by promoting the growth of beneficial microorganisms, such as nitrogen-fixing bacteria and mycorrhizal fungi. These microorganisms are essential for nutrient cycling and plant growth, as they facilitate the uptake of essential nutrients like nitrogen and phosphorus. Tannins, by interacting with soil organic matter, create an environment that supports the proliferation of these beneficial microbes, improving soil fertility and plant productivity (Chung et al., 1998). For example, studies have shown that tannin-rich plant species, such as legumes, enhance the activity of nitrogen-fixing bacteria in the rhizosphere, increasing nitrogen availability to plants. Similarly, tannins have been found to stimulate the colonization of plant roots by mycorrhizal fungi, which improve phosphorus uptake and enhance plant resilience to environmental stressors such as drought and salinity (War et al., 2018). The role of tannins in promoting beneficial microbial activity highlights their importance in sustainable agriculture, where the focus is on reducing the reliance on chemical fertilizers and promoting natural nutrient cycling processes.

6.3 Tannins in Integrated Pest Management (IPM)

Tannins have great potential for integration into Integrated Pest Management (IPM) strategies, which focus on combining biological, cultural, and chemical control methods to manage pests in a sustainable and environmentally friendly manner. The use of tannins as natural biopesticides, soil health enhancers, and plant resilience boosters aligns with the goals of IPM, providing farmers with effective tools to reduce pest pressure without relying on synthetic chemicals.

6.3.1 Tannins as a Component of IPM Programs

In IPM programs, tannins can be used in combination with other natural products, such as essential oils, microbial biocontrol agents, and plant-based extracts, to develop comprehensive pest management strategies. For example, tannin-rich plant residues can be incorporated into crop rotations or used as cover crops to deter insect pests and improve soil health. The use of tannin-based biopesticides in combination with microbial-based products can enhance the efficacy of pest control strategies, reducing the need for synthetic insecticides and fungicides (Barbehenn & Constabel, 2011). In vineyards, for example, the use of tannin-rich grape varieties and tannin-based foliar sprays has been integrated into IPM strategies to manage pests such as leafhoppers and fungal diseases like powdery mildew. By reducing pest pressure and improving plant resilience, tannins contribute to more sustainable and productive agricultural systems.

6.3.2 Potential for Tannins in Organic Farming

Tannins are particularly valuable in organic farming systems, where the use of synthetic pesticides and fertilizers is restricted. Their natural origin and biodegradability make them ideal candidates for use in organic agriculture, where the focus is on maintaining ecological balance and promoting soil health. Tannins can be used as natural biopesticides, soil amendments, and plant growth enhancers in organic farming systems, providing farmers with effective tools to manage pests and improve crop productivity without compromising environmental sustainability (Chung et al., 1998). In organic farming, tannins can be applied through various methods, including the use of tannin-rich cover crops, composts, and foliar sprays. These applications help control pests, enhance soil fertility, and improve plant resilience to environmental stressors, contributing to the long-term sustainability of organic farming systems.

6.4 Tannins and Climate-Resilient Agriculture

The role of tannins in enhancing plant resilience to environmental stressors, such as drought, salinity, and temperature extremes, makes them valuable in climate-resilient agriculture. As climate change continues to impact agricultural productivity, the use of natural compounds like tannins to improve crop resilience and reduce environmental degradation is becoming increasingly important.

6.4.1 Enhancing Plant Resilience to Climate Change

Tannins, with their antioxidant, osmoprotective, and antimicrobial properties, contribute to plant resilience in the face of climate change-induced stressors. By protecting plants from oxidative damage, improving water retention, and enhancing nutrient uptake, tannins help plants withstand periods of drought, salinity, and heat stress (Salminen & Karonen, 2011). The use of tannin-rich crop varieties and tannin-based soil amendments can improve the adaptability of agricultural systems to changing environmental conditions, reducing the impact of climate change on food security.

6.4.2 Reducing the Carbon Footprint of Agriculture

In addition to their role in improving plant resilience, tannins can also contribute to reducing the carbon footprint of agriculture. By promoting soil health and reducing the need for synthetic fertilizers and pesticides, tannins help lower the greenhouse gas emissions associated with conventional farming practices. The use of tannin-based biopesticides and soil conditioners can reduce the reliance on energy-intensive chemical inputs, contributing to more sustainable and environmentally friendly agricultural systems.

6.5 Future Directions in Tannin Research and Application

While tannins have shown great promise in sustainable agriculture, further research is needed to fully understand their mechanisms of action and optimize their use in crop protection and soil management. Future research should focus on identifying tannin-rich plant species with enhanced pest resistance, developing tannin-based formulations for biopesticides and soil amendments, and exploring the synergistic effects of tannins with other natural compounds in integrated pest management strategies (Barbehenn & Constabel, 2011).

Conclusion

Tannins, as polyphenolic compounds, play an essential role in plant defense and pathogen suppression, owing to their diverse biosynthesis and classification. Produced through the phenylpropanoid pathway, tannins are categorized into condensed and hydrolyzable forms, each with unique structural features that contribute to their wide range of biological functions. Their role as antioxidants is crucial in protecting plants from oxidative damage by scavenging reactive oxygen species (ROS), especially under environmental stress conditions like drought, salinity, and temperature fluctuations. By neutralizing ROS, tannins help maintain cellular integrity and support the overall health and resilience of plants. Beyond their antioxidant properties, tannins are powerful antimicrobial agents that disrupt the growth and function of a broad spectrum of pathogens, including bacteria, fungi, and viruses. Their antimicrobial mechanisms include disruption of microbial cell walls, inhibition of key enzymatic activities, and interference with microbial communication pathways. This makes tannins highly effective in controlling plant pathogens, contributing significantly to plant survival and pathogen resistance. Tannins also play a crucial role in herbivore deterrence by reducing the digestibility of plant tissues through protein-binding activities, thus preventing herbivory and limiting damage from insects and other herbivores. Their ability to enhance insect resistance strengthens plants' natural defense systems.

Tannins are also pivotal in enhancing plant resistance to various environmental stressors. Their presence helps plants adapt to challenging conditions by reinforcing structural barriers and contributing to biochemical defense responses. This multifaceted role makes tannins an integral part of the plant's strategy to cope with both biotic and abiotic stress factors. In agriculture, tannins offer promising applications as natural biopesticides and components of eco-friendly pathogen control strategies. They can be incorporated into integrated pest management (IPM) systems to reduce reliance on synthetic chemicals and promote sustainable agricultural practices. However, the large-scale application of tannins in agriculture is met with challenges, including difficulties in consistent extraction, production, and application across diverse agricultural settings. Advances in biotechnology, including genetic engineering and metabolic optimization, present potential solutions for increasing tannin availability and ensuring their effective use in agriculture. As research continues to explore the full potential of tannins, these compounds are poised to play a significant role in developing natural, sustainable approaches to disease management and crop protection. Their antioxidant, antimicrobial, and insect-repellent properties, combined with their capacity to enhance plant stress tolerance, underscore their importance in future agricultural innovations aimed at sustainability and environmental stewardship.

References

Adamczyk, B., Simon, J., Kitunen, V., Adamczyk, S., & Smolander, A. (2017). Tannins and their complex interaction with different organic nitrogen compounds and enzymes: Old paradigms versus recent advances. *ChemistryOpen*, *6*(5), 610-614.

Agati, G., Azzarello, E., Pollastri, S., & Tattini, M. (2012). Flavonoids as antioxidants in plants: Location and functional significance. *Plant Science, 196*, 67-76.

Al-Khayri, J. M., Rashmi, R., Toppo, V., Chole, P. B., Banadka, A., Sudheer, W. N., Nagella, P., Shehata, W. F., Al-Mssallem, M. Q., Alessa, F. M., Almaghasla, M. I., & Rezk, A. A. (2023). Plant secondary metabolites: The weapons for biotic stress management. *Metabolites*, *13*(6), 716.

Andrés, C. M. C., Pérez de la Lastra, J. M., Andrés Juan, C., Plou, F. J., & Pérez-Lebeña, E. (2023). Superoxide anion chemistry-its role at the core of the innate immunity. *International Journal of Molecular Sciences*, *24*(3), 1841.

Banc, R., Rusu, M. E., Filip, L., & Popa, D. -S. (2023). The impact of ellagitannins and their metabolites through gut microbiome on the gut health and brain wellness within the gut–brain axis. *Foods*, *12*(2), 270.

Barbehenn, R. V., & Constabel, C. P. (2011). Tannins in plant-herbivore interactions. *Phytochemistry, 72*(13), 1551-1565.

Bele, A. A., Jadhav, V. M., & Kadam, V. J. (2010). Potential of tannnins: a review. *Asian Journal of Plant Sciences*, *9*(4), 209-214.

Bialonska, D., Kasimsetty, S. G., Schrader, K. K., & Ferreira, D. (2009). The effect of pomegranate (Punica granatum L.) byproducts and ellagitannins on the growth of human gut bacteria. *Journal of Agricultural Food Chemistry*, *57*(18), 8344-8349.

Bogati, K., & Walczak, M. (2022). The impact of drought stress on soil microbial community, enzyme activities and plants. *Agronomy*, *12*(1), 189.

Chiocchio, I., Mandrone, M., Tomasi, P., Marincich, L., & Poli, F. (2021). Plant secondary metabolites: An opportunity for circular economy. *Molecules*, *26*(2), 495.

Chung, K. T., Wong, T. Y., Wei, C. I., Huang, Y. W., & Lin, Y. (1998). Tannins and human health: A review. *Critical Reviews in Food Science and Nutrition, 38*(6), 421-464.

Czerkas, K., Olchowik-Grabarek, E., Łomanowska, M., Abdulladjanova, N., & Sękowski, S. (2024). Antibacterial activity of plant polyphenols belonging to the tannins against streptococcus mutans-potential against dental caries. *Molecules*, *29*(4), 879.

Das A. K., Islam, M. N., Faruk, M. O., Ashaduzzaman, M., & Dungani, R. (2020). Review on tannins: Extraction processes, applications and possibilities. *South African Journal of Botany*, *135*, 58-70.

Dixon, R. A., & Sarnala, S. (2020). Proanthocyanidin biosynthesis—a matter of protection. *Plant Physiology*, *184*(2), 579-591.

Dong, N. Q., & Lin, H. X. (2021). Contribution of phenylpropanoid metabolism to plant development and plant-environment interactions. *Journal of Integrative Plant Biology*, *63*, 180-209.

Haghpanah, M., Hashemipetroudi, S., Arzani, A., & Araniti, F. (2024). Drought tolerance in plants: Physiological and molecular responses. *Plants (Basel)*, *13*(21), 2962.

Huang, Q., Liu, X., Zhao, G., Hu, T., & Wang, Y. (2018). Potential and challenges of tannins as an alternative to in-feed antibiotics for farm animal production. *Animal Nutrition*, *4*(2), 137-150.

Jian, Y., Gong, D., Wang, Z., Liu, L., He, J., Han, X., & Tsuda, K. (2024). How plants manage pathogen infection. *EMBO Reports*, *25*(1), 31-44.

Lobiuc, A., Pavăl, N. -E., Mangalagiu, I. I., Gheorghiță, R., Teliban, G. -C., Amăriucăi-Mantu, D., & Stoleru, V. (2023). Future antimicrobials: Natural and functionalized phenolics. *Molecules*, *28*(3), 1114.

Mora, J., Pott, D. M., Osorio, S., & Vallarino, J. G. (2022). Regulation of plant tannin synthesis in crop species. *Frontiers in Genetics*, *13*, 870976.

Nie, F., Liu, L., Cui, J., Zhao, Y., Zhang, D., Zhou, D., Wu, J., Li, B., Wang, T., Li, M., & Yan, M. (2023). Oligomeric proanthocyanidins: An updated review of their natural sources, synthesis, and potentials. *Antioxidants*, *12*(5), 1004.

Salminen, J. P., & Karonen, M. (2011). Chemical ecology of tannins: Recent developments in tannin chemistry reveal new structures and structure-activity patterns. *Chemistry European Journal*, *17*, 2806-2816.

Scalbert, A. (1991). Antimicrobial properties of tannins. *Phytochemistry*, *30*(12), 3875-3883.

Volf, M. (2018). Differential response of herbivores to plant defence. In: Merillon, JM., Ramawat, K. (eds) Co-evolution of secondary metabolites. *Reference Series in Phytochemistry*. Springer, Cham.

War, A. R., Taggar, G. K., Hussain, B., Taggar, M. S., Nair, R. M., & Sharma, H. C. (2018). Plant defence against herbivory and insect adaptations. *AoB Plants*, *10*(4), ply037.

Žitek Makoter, T., Knez Marevci, M., & Knez, Ž. (2024). Ellagitannin content in extracts of the chestnut wood aesculus. *Molecules*, *29*(17), 4015.

Chapter - 9

Phytoalexins: Inducible Antimicrobials in Plant-Pathogen Interactions

Praneet Chauhan[1], Reshma Jadhav[2], Susheel Kumar[3], Vinod Chouhan[4] and Sheetanshu Gupta[5]

[1]Department of Plant Pathology, Dr. Khem Singh Gill Akal College of Agriculture, Eternal University, Baru Sahib, Sirmour, Himachal Pradesh, India
[2]Department of Biochemistry, Mahatma Phule Krishi Vidyapeeth, Rahuri, Ahmednagar, Maharashtra, India
[3]Plant Molecular Virology Laboratory, CPMB Division, CSIR-National Botanical Research Institute, Rana Pratap Marg, Lucknow, Uttar Pradesh, India
[4]Division of Plant Pathology, Indian Council of Agricultural Research-Indian Agricultural Research Institute (ICAR-IARI), New Delhi, India
[5]Department of Biochemistry, G.B. Pant University of Agriculture & Technology, Pantnagar, Uttarakhand, India

Abstract

Phytoalexins are a diverse group of inducible antimicrobial compounds produced by plants in response to pathogen attacks. These secondary metabolites are synthesized through complex biochemical pathways, including the phenylpropanoid, terpenoid, and alkaloid pathways, depending on the plant species and the specific pathogen encountered. The induction of phytoalexins is a critical component of the plant's innate immune system, triggered by pathogen-associated molecular patterns (PAMPs) or specific pathogen effectors. Once produced, phytoalexins exhibit potent antimicrobial properties that inhibit the growth and spread of a wide range of pathogens, including bacteria, fungi, and viruses. Their mechanisms of action include disrupting pathogen cell membranes, inhibiting key enzymatic processes, and interfering with pathogen signaling systems, effectively suppressing infection and promoting plant survival. In addition to their direct antimicrobial activity, phytoalexins play an important role in plant defense signaling pathways, often working in concert with other plant defense mechanisms such as systemic acquired resistance (SAR) and the hypersensitive response (HR). These compounds frequently interact synergistically with other plant secondary metabolites, enhancing the overall effectiveness of the plant's defense system. In sustainable agriculture, phytoalexins hold great potential as natural tools for disease management, offering an eco-friendly alternative to synthetic pesticides. By understanding

DOI: 10.1201/9781003717454-9

and harnessing the biosynthesis and activity of phytoalexins, researchers and agricultural practitioners can develop innovative strategies for improving crop resilience and controlling pathogens in a more environmentally sustainable way. However, challenges such as variability in phytoalexin production among different plant species and the complexity of their induction require further research to fully realize their potential.

Keywords: Phytoalexins, plant defense, inducible antimicrobials, pathogen interactions, sustainable agriculture, disease management.

Introduction

Phytoalexins are a diverse group of antimicrobial secondary metabolites produced by plants in response to pathogen attack and other environmental stressors. They play a crucial role in the plant's innate immune system, forming part of the rapid defense response to infections caused by bacteria, fungi, viruses, and other pathogens. Phytoalexins are generally synthesized only upon the detection of pathogen-associated molecular patterns (PAMPs) or damage-associated molecular patterns (DAMPs), making them part of an inducible defense mechanism. The biosynthesis of phytoalexins varies significantly among plant species, with different types of phytoalexins being characteristic of specific plant families. This variation reflects the complexity and specificity of plant-pathogen interactions. Phytoalexins can be broadly classified into several chemical groups, including terpenoids, flavonoids, stilbenes, and alkaloids, each of which has distinct biosynthetic pathways. These compounds are typically synthesized via pathways involving enzymes from the phenylpropanoid, mevalonate, and isoprenoid biosynthesis routes (Ahuja et al., 2012). The structural diversity of phytoalexins across plant species highlights the evolutionary adaptation of plants to their unique pathogenic environments.

During a pathogen attack, the induction of phytoalexin biosynthesis is part of a highly coordinated and complex signaling cascade that begins with the recognition of pathogen-associated molecules by pattern recognition receptors (PRRs) located on the plant cell surface. This recognition activates a series of intracellular signaling pathways, including the production of reactive oxygen species (ROS), activation of mitogen-activated protein kinases (MAPKs), and the release of phytohormones such as salicylic acid, jasmonic acid, and ethylene. These signaling molecules play pivotal roles in regulating the expression of genes involved in phytoalexin biosynthesis (Dodds et al., 2024). Upon activation of these pathways, specific transcription factors are upregulated, leading to the expression of key enzymes required for the production of phytoalexins. The rapid and localized accumulation of phytoalexins at the site of infection is crucial for restricting pathogen growth and spread. This inducible defense mechanism provides an efficient and targeted response, minimizing resource expenditure under non-stress conditions.

The antimicrobial properties of phytoalexins are highly effective in defending plants against a wide range of pathogens. These compounds exhibit diverse modes of action, including disrupting pathogen cell membranes, inhibiting enzyme activity, and interfering with the pathogen's metabolic processes. For instance, in grapevines (*Vitis vinifera*), the phytoalexin resveratrol has been shown to inhibit the growth of fungal pathogens such as *Botrytis cinerea* by disrupting cell wall integrity and preventing spore germination (Jeandet et al., 2010). Similarly, the soybean phytoalexin glyceollin is effective against a variety of bacterial and fungal pathogens, acting by disrupting membrane potential and leading to cell death. The broad-spectrum antimicrobial activity of phytoalexins makes them a critical component of the plant's defense arsenal, providing protection against a wide array of biotic threats.

Beyond their direct antimicrobial functions, phytoalexins also play a significant role in modulating plant defense signaling pathways. They are involved in cross-communication with other defense-related molecules, such as phytohormones, and contribute to the amplification of defense signals. For example, the accumulation of phytoalexins has been linked to the enhanced production of salicylic acid, a key hormone involved in systemic acquired resistance (SAR). This interaction not only enhances local defense but also triggers a systemic response, preparing the plant for potential future attacks (Li et al., 2015). Furthermore, phytoalexins interact with signaling molecules involved in jasmonic acid- and ethylene-mediated defense pathways, which are essential for resistance against necrotrophic pathogens and herbivores. This intricate network of signaling pathways ensures that plants can mount an appropriate and effective defense response tailored to the specific nature of the attacking pathogen.

The effectiveness of phytoalexins is further enhanced by their synergistic interactions with other plant defense compounds, such as phenolics, flavonoids, and terpenoids. These interactions amplify the overall antimicrobial activity and increase the efficacy of the plant's defense system. For example, the co-accumulation of phytoalexins and flavonoids has been observed in several plant species, leading to stronger inhibition of pathogen growth compared to either compound alone (Shomali et al., 2022). Additionally, the combination of phytoalexins with oxidative bursts generated during pathogen attack can create a hostile environment for pathogens, further inhibiting their growth and spread. The synergistic effects of phytoalexins with other secondary metabolites underline the importance of multi-layered defense mechanisms in plants, which provide robust protection against a wide range of threats.

Given their potent antimicrobial properties and role in defense signaling, phytoalexins are increasingly being explored for their applications in sustainable agriculture. The use of phytoalexin-inducing agents, such as elicitors, has been shown to enhance plant resistance to pathogens in a variety of crops, reducing the need for chemical pesticides (Walters et al., 2013). Elicitors, such as chitosan, salicylic acid analogs, and plant extracts, can stimulate the production of phytoalexins in crops, leading to enhanced disease resistance without the negative environmental impacts associated with synthetic chemicals. Furthermore, breeding programs aimed at enhancing phytoalexin production in crops offer promising avenues for developing disease-resistant varieties. The integration of phytoalexin-based strategies into integrated pest management (IPM) systems holds great potential for reducing the reliance on chemical inputs and promoting more sustainable agricultural practices.

1. Biosynthesis and Types of Phytoalexins in Plants

Phytoalexins are low molecular weight antimicrobial compounds that are synthesized de novo by plants in response to pathogen attack, environmental stress, or physical damage. These compounds serve as a crucial part of the plant's innate immune system, functioning as toxins to invading pathogens or herbivores. The biosynthesis of phytoalexins involves complex metabolic pathways that are activated upon recognition of pathogen-associated molecular patterns (PAMPs) or damage-associated molecular patterns (DAMPs). Phytoalexins are highly diverse in their chemical structures, which reflect the evolutionary adaptation of different plant species to specific pathogens. They can be classified into several chemical categories, including terpenoids, flavonoids, stilbenes, and alkaloids, each with distinct biosynthetic origins.

1.1 Biosynthesis of Phytoalexins

The biosynthesis of phytoalexins typically begins with the activation of signaling pathways that are triggered by pathogen detection. This recognition occurs through pattern recognition receptors (PRRs) on the plant cell surface, which detect PAMPs or DAMPs released by pathogens or damaged cells. Upon detection, a cascade of defense-related signaling pathways is activated, leading to the induction of genes responsible for the biosynthesis of phytoalexins (Ahuja et al., 2012). Phytoalexin biosynthesis generally follows well-established metabolic routes, such as the phenylpropanoid, isoprenoid, and flavonoid pathways, depending on the chemical nature of the phytoalexin. The key precursors for these pathways are derived from primary metabolic intermediates such as acetyl-CoA, mevalonic acid, and phenylalanine. These intermediates undergo enzymatic transformations to produce phytoalexins with complex structures tailored to combat specific pathogens.

1.1.1 Terpenoid Phytoalexins

Terpenoid phytoalexins are among the most common types produced by plants. These compounds are synthesized via the mevalonate or methylerythritol phosphate (MEP) pathway, which leads to the production of isoprene units (C5) that serve as the building blocks for terpenoids. Terpenoids can vary widely in their structure and include monoterpenes, diterpenes, and sesquiterpenes, all of which play significant roles in plant defense (Ninkuu et al., 2021). For instance, the phytoalexin capsidiol, a sesquiterpene produced by *Nicotiana* species in response to fungal pathogens like *Phytophthora*, is synthesized via the mevalonate pathway. Capsidiol disrupts fungal cell membranes, inhibiting pathogen growth. Similarly, rice produces momilactones, diterpenoid phytoalexins that accumulate in response to fungal pathogens, providing an effective chemical barrier to infection (Ahuja et al., 2012).

1.1.2 Flavonoid Phytoalexins

Flavonoids, another class of phytoalexins, are derived from the phenylpropanoid pathway. These compounds are synthesized from phenylalanine, which is converted to p-coumaroyl-CoA by the enzyme phenylalanine ammonia-lyase (PAL). P-coumaroyl-CoA then enters the flavonoid biosynthetic pathway, where it undergoes enzymatic modifications to produce a wide range of flavonoid phytoalexins (Dixon & Paiva, 1995). For example, in legumes such as soybeans (*Glycine max*), the phytoalexin glyceollin is produced in response to fungal pathogens like *Phytophthora sojae*. Glyceollin biosynthesis involves the flavonoid pathway, where the intermediate daidzein is converted into glyceollin by the action of cytochrome P450 enzymes. Glyceollin disrupts fungal cell walls and inhibits spore germination, providing a rapid defense response to pathogen invasion.

1.1.3 Stilbene Phytoalexins

Stilbenes, such as resveratrol, are another important group of phytoalexins. These compounds are synthesized via the phenylpropanoid pathway, similar to flavonoids. Resveratrol, produced by grapevines (*Vitis vinifera*) in response to fungal pathogens like *Botrytis cinerea*, is a well-known stilbene phytoalexin. It is synthesized from p-coumaroyl-CoA through the action of stilbene synthase (Jeandet et al.,

2010). Resveratrol functions by disrupting the cell wall integrity of fungal pathogens and inhibiting their growth, thereby protecting grapevines from infections.

1.1.4 Alkaloid Phytoalexins

Alkaloid phytoalexins are nitrogen-containing compounds that are often derived from amino acid precursors. These compounds are synthesized via complex pathways involving decarboxylation and amination reactions. Alkaloid phytoalexins exhibit potent antimicrobial properties due to their ability to interfere with pathogen metabolism and disrupt cellular functions. One example is the phytoalexin camalexin, an indole alkaloid produced by *Arabidopsis thaliana* in response to bacterial and fungal pathogens. Camalexin is synthesized from tryptophan via the action of cytochrome P450 enzymes. It has been shown to inhibit fungal spore germination and disrupt pathogen growth, providing an effective defense against a wide range of pathogens (Much et al., 2019).

1.2 Types of Phytoalexins in Plants

The diversity of phytoalexins is a reflection of the wide variety of plant species and their specific pathogen challenges. Based on their chemical structure, phytoalexins can be classified into several major groups: terpenoids, flavonoids, stilbenes, alkaloids, and others. Each group of phytoalexins plays a distinct role in plant defense, often targeting specific types of pathogens.

1.2.1 Terpenoid Phytoalexins

Terpenoid phytoalexins, such as capsidiol and momilactones, are primarily produced by solanaceous plants and cereals. These compounds are highly effective against fungal pathogens due to their ability to disrupt membrane integrity and inhibit fungal growth. Terpenoid phytoalexins are typically produced in large quantities during pathogen attack, providing an immediate chemical barrier against infection (Ahuja et al., 2012).

1.2.2 Flavonoid Phytoalexins

Flavonoid phytoalexins, such as glyceollin in soybeans and pisatin in peas (*Pisum sativum*), are common in leguminous plants. These compounds are highly active against fungal pathogens and play a critical role in protecting legumes from diseases like root rot and leaf blight. Flavonoid phytoalexins often act by inhibiting fungal spore germination and disrupting fungal cell walls (Dixon & Paiva, 1995).

1.2.3 Stilbene Phytoalexins

Stilbene phytoalexins, such as resveratrol, are produced by plants in the Vitaceae family, particularly grapevines. Resveratrol is one of the most studied stilbenes due to its potent antifungal properties and its role in protecting grapevines from diseases like downy mildew and powdery mildew. Stilbenes are also known for their antioxidant properties, contributing to plant resilience against environmental stressors (Jeandet et al., 2013).

1.2.4 Alkaloid Phytoalexins

Alkaloid phytoalexins, such as camalexin in *Arabidopsis* and phaseollin in beans (*Phaseolus vulgaris*), are produced in response to both bacterial and fungal pathogens. These nitrogen-containing compounds often interfere with pathogen metabolism and disrupt cellular processes, providing a broad-spectrum defense against a variety of pathogens (Mucha et al., 2019).

1.3 Evolutionary Significance and Adaptation

The diversity in the types and biosynthetic pathways of phytoalexins is a result of evolutionary adaptation to different environmental challenges and pathogen pressures. Plants have developed highly specialized mechanisms to produce phytoalexins that are tailored to the specific pathogens they encounter. For example, legumes have evolved to produce flavonoid phytoalexins as a defense against root-infecting fungi, while solanaceous plants rely on terpenoid phytoalexins to combat foliar pathogens. This chemical diversity ensures that plants can respond effectively to a wide range of pathogens, enhancing their survival and fitness in diverse ecological niches (Ahuja et al., 2012). The study of phytoalexins has significant implications for agriculture, as understanding the biosynthesis and regulation of these compounds can lead to the development of disease-resistant crop varieties. By manipulating the pathways involved in phytoalexin production, it may be possible to enhance the natural defense mechanisms of crops, reducing the reliance on chemical pesticides and promoting sustainable agricultural practices.

2. Mechanisms of Phytoalexin Induction During Pathogen Attack

The induction of phytoalexins in plants is a highly sophisticated process that is rapidly activated in response to pathogen attack. Phytoalexins, as antimicrobial secondary metabolites, are not stored in plant tissues but are synthesized de novo when plants detect invading pathogens. This inducible defense mechanism is triggered by the recognition of pathogen-associated molecular patterns (PAMPs), effectors, or damage-associated molecular patterns (DAMPs), which activate a cascade of signaling pathways that culminate in the production of phytoalexins. These signaling pathways involve a complex network of molecular signals, including reactive oxygen species (ROS), phytohormones such as salicylic acid (SA), jasmonic acid (JA), and ethylene (ET), as well as the activation of mitogen-activated protein kinases (MAPKs) and transcription factors.

2.1 Recognition of Pathogens and Activation of Defense Signaling

The first step in the induction of phytoalexins is the recognition of invading pathogens by the plant's immune system. Plants have evolved pattern recognition receptors (PRRs) located on the cell surface, which detect conserved molecular signatures associated with pathogens, known as PAMPs. These include molecules such as flagellin from bacterial flagella and chitin from fungal cell walls (Boller & Felix, 2009). Upon recognition of PAMPs, PRRs activate PAMP-triggered immunity (PTI), which involves the production of ROS and the activation of MAPK cascades, leading to the transcriptional reprogramming necessary for phytoalexin production. In addition to PTI, plants also employ effector-triggered immunity (ETI), a more specific immune response triggered by the recognition of pathogen effector proteins by plant resistance (R) proteins. ETI is often associated with a stronger and more localized defense response, including the hypersensitive response (HR), where localized cell death occurs at the site of infection to limit pathogen spread. Both PTI and ETI can activate the biosynthesis of phytoalexins, though ETI often leads to a more robust phytoalexin accumulation (Naveed et al., 2020).

2.1.1 Role of Pattern Recognition Receptors (PRRs)

The role of PRRs in phytoalexin induction is central to the plant's ability to detect a wide range of pathogens. PRRs, such as receptor-like kinases (RLKs) and receptor-like proteins (RLPs), bind to PAMPs and initiate downstream signaling cascades. For instance, the flagellin-sensing 2 (FLS2) receptor in *Arabidopsis thaliana* recognizes bacterial flagellin and triggers the MAPK cascade, which in turn activates the biosynthesis of phytoalexins like camalexin (Macho & Zipfel, 2014). The activation of PRRs leads to the rapid production of ROS, which act as both direct antimicrobial agents and signaling molecules that further amplify the defense response. The ROS burst also contributes to the strengthening of cell walls by promoting the cross-linking of cell wall proteins, thereby forming a physical barrier against pathogens. This early defense response is critical for the activation of the biosynthetic pathways involved in phytoalexin production.

2.2 Signal Transduction Pathways in Phytoalexin Induction

Once a pathogen has been detected, the plant's defense signaling network is activated, leading to the induction of phytoalexin biosynthesis. Key components of this signaling network include MAPK cascades, calcium signaling, and the production of phytohormones such as SA, JA, and ET. These signals converge to activate transcription factors that regulate the expression of genes involved in phytoalexin biosynthesis.

2.2.1 MAPK Cascades and Calcium Signaling

MAPK cascades play a pivotal role in the transduction of pathogen-induced signals. Upon activation by PRRs or R proteins, MAPKs phosphorylate downstream transcription factors, which then bind to the promoters of defense-related genes, including those involved in phytoalexin biosynthesis. For example, in *Arabidopsis*, the MPK3 and MPK6 MAPKs are activated in response to pathogen attack, leading to the production of the phytoalexin camalexin (Galletti et al., 2011). This MAPK-mediated regulation of phytoalexin biosynthesis ensures a rapid and coordinated defense response. Calcium (Ca^{2+}) signaling is another critical component of the defense response. Pathogen recognition triggers an influx of Ca^{2+} ions into the cytosol, which activates calcium-dependent protein kinases (CDPKs). CDPKs, in turn, regulate various defense responses, including the activation of phytoalexin biosynthesis. Studies have shown that Ca^{2+} signaling is involved in the production of phytoalexins like sakuranetin in rice, highlighting the importance of this pathway in plant defense (Katsumata et al., 2018).

2.2.2 Phytohormonal Regulation: Salicylic Acid, Jasmonic Acid, and Ethylene

Phytohormones, particularly salicylic acid (SA), jasmonic acid (JA), and ethylene (ET), play central roles in regulating the biosynthesis of phytoalexins. These hormones act as signaling molecules that modulate the expression of defense-related genes in response to pathogen attacks. The specific role of each hormone varies depending on the type of pathogen encountered. SA is primarily involved in defense against biotrophic pathogens, which derive nutrients from living host cells. SA signaling leads to the induction of systemic acquired resistance (SAR), a long-lasting defense response that involves the accumulation of phytoalexins such as camalexin in *Arabidopsis* and scopoletin in tobacco (*Nicotiana tabacum*) (Dempsey & Klessig, 2012). SA also plays a key role in the activation of the transcription factor WRKY33, which regulates camalexin biosynthesis genes. JA and ET are more commonly associated with defense against necrotrophic pathogens and herbivores. JA and ET signaling pathways often work in synergy to activate the production of phytoalexins such as phytoalexin-deficient 2 (PAD2) in *Arabidopsis* and momilactones in rice (*Oryza sativa*) (Bari & Jones, 2009). The JA and ET pathways activate transcription factors such as ERF1, which regulate the expression of genes involved in secondary metabolite biosynthesis, including phytoalexins.

2.3 Transcriptional Regulation of Phytoalexin Biosynthesis

The biosynthesis of phytoalexins is tightly regulated at the transcriptional level by a network of transcription factors that respond to pathogen-induced signaling pathways. These transcription factors bind to the promoters of phytoalexin biosynthetic genes and modulate their expression in response to pathogen attack.

2.3.1 MYB and WRKY Transcription Factors

MYB and WRKY transcription factors are key regulators of phytoalexin biosynthesis in many plant species. WRKY transcription factors, such as WRKY33 in *Arabidopsis*, are activated in response to pathogen attack and play a crucial role in the induction of genes involved in camalexin biosynthesis (Birkenbihl et al., 2012). WRKY33 binds to the promoters of genes encoding enzymes such as cytochrome P450s and glutathione S-transferases, which are essential for camalexin production. Similarly, MYB transcription factors regulate the biosynthesis of flavonoid phytoalexins in legumes. For instance, the MYB transcription factor GmMYB29 in soybeans (*Glycine max*) is involved in the regulation of glyceollin biosynthesis in response to fungal infection (Zhou et al., 2022). MYB transcription factors activate the expression of key enzymes in the flavonoid biosynthetic pathway, including chalcone synthase and isoflavone reductase, leading to the production of glyceollins.

2.3.2 Role of Basic Helix-Loop-Helix (bHLH) Transcription Factors

Basic helix-loop-helix (bHLH) transcription factors are also involved in the regulation of phytoalexin biosynthesis. In rice, the bHLH transcription factor OsTGAP1 is required for the biosynthesis of diterpenoid phytoalexins such as momilactones and phytocassanes. OsTGAP1 regulates the expression of genes encoding diterpene cyclases, which are key enzymes in the biosynthesis of these phytoalexins (Okada et al., 2009). The coordinated action of MYB, WRKY, and bHLH transcription factors ensures that phytoalexin biosynthesis is rapidly and specifically induced in response to pathogen attack. This tight regulation allows plants to deploy phytoalexins precisely where and when they are needed, minimizing resource expenditure and maximizing defense efficiency.

2.4 Role of Reactive Oxygen Species (ROS) in Phytoalexin Induction

ROS, such as hydrogen peroxide (H_2O_2) and superoxide (O_2^-), play dual roles in plant defense, functioning as both signaling molecules and direct antimicrobial agents. The accumulation of ROS at the site of pathogen infection is one of the earliest responses to pathogen recognition and is crucial for the induction of phytoalexin biosynthesis.

2.4.1 ROS-Mediated Activation of Defense Genes

ROS act as secondary messengers that activate defense-related genes, including those involved in phytoalexin biosynthesis. In *Arabidopsis*, the production of ROS in response to pathogen attack leads to the activation of the transcription factor WRKY33, which regulates the expression of camalexin biosynthetic genes (Birkenbihl et al., 2012). Similarly, ROS production in rice is associated with the activation of genes involved in the biosynthesis of momilactones, highlighting the conserved role of ROS in phytoalexin induction across plant species.

2.4.2 ROS and Phytoalexin Accumulation

In addition to their signaling role, ROS directly contribute to the accumulation of phytoalexins by triggering oxidative stress responses in plants. Oxidative stress induced by pathogen attack leads to the activation of antioxidant enzymes and secondary metabolite biosynthesis pathways, including those involved in phytoalexin production. For example, the accumulation of sakuranetin, a flavonoid phytoalexin in rice, is associated with ROS production in response to fungal infection (Katsumata et al., 2018).

3. Antimicrobial Properties of Phytoalexins Against Plant Pathogens

Phytoalexins are crucial antimicrobial secondary metabolites that plants synthesize in response to pathogen invasion. These compounds exhibit potent antimicrobial activities against a wide range of plant pathogens, including bacteria, fungi, and viruses, by disrupting vital cellular processes within the pathogens. The diverse chemical structures of phytoalexins allow them to target multiple aspects of pathogen physiology, including cell wall integrity, membrane stability, enzyme function, and gene expression. By interfering with these critical processes, phytoalexins inhibit pathogen growth, spread, and infection, providing a rapid and effective defense mechanism for plants. The antimicrobial properties of phytoalexins are highly specific, often targeting pathogens that pose significant threats to the survival of the plant species that produce them.

3.1 Disruption of Pathogen Cell Walls and Membranes

One of the most common mechanisms by which phytoalexins exert their antimicrobial effects is through the disruption of pathogen cell walls and membranes. This disruption impairs the structural integrity of pathogens, leading to cell lysis and death. Phytoalexins such as terpenoids, flavonoids, and stilbenes are known for their ability to interact with the lipid bilayers of microbial membranes, increasing membrane permeability and causing the leakage of essential cellular contents.

3.1.1 Terpenoid Phytoalexins and Membrane Disruption

Terpenoid phytoalexins, such as capsidiol and momilactones, are well-documented for their membrane-disruptive properties. Capsidiol, produced by tobacco (*Nicotiana tabacum*) in response to infection by fungal pathogens such as *Phytophthora* species, integrates into fungal cell membranes, disrupting their lipid bilayer structure. This disruption increases membrane permeability, allowing ions and other small molecules to leak out of the fungal cells, ultimately leading to cell death (Kuc, 1995). Similarly, momilactones produced by rice (*Oryza sativa*) disrupt fungal cell walls, preventing the establishment and growth of pathogens like *Magnaporthe oryzae*, the causative agent of rice blast disease (Valletta et al., 2023). In addition to their fungicidal effects, terpenoid phytoalexins also exhibit activity against bacterial pathogens by targeting bacterial cell membranes. Studies have shown that terpenoids can increase the fluidity of bacterial membranes, leading to the loss of membrane integrity and the release of intracellular contents. This mechanism is particularly effective against Gram-positive bacteria, which have a thicker peptidoglycan layer that is susceptible to terpenoid-induced disruption (Boncan et al., 2020).

3.1.2 Flavonoid Phytoalexins and Cell Wall Integrity

Flavonoid phytoalexins, such as glyceollin produced by soybeans (*Glycine max*) in response to fungal pathogens, also play a crucial role in disrupting pathogen cell walls. Glyceollin has been shown to interfere with the synthesis and assembly of fungal cell walls, inhibiting the growth of pathogens such as *Phytophthora sojae* and *Rhizoctonia solani* (Deice Raasch-Fernandes et al., 2019). By preventing the proper formation of the cell wall, glyceollin weakens the structural integrity of the pathogen, making it more susceptible to environmental stresses and host defenses. In addition to disrupting cell walls, flavonoid phytoalexins can also inhibit enzymes involved in cell wall biosynthesis. For example, glyceollin has been found to inhibit the activity of chitin synthase, an enzyme responsible for the polymerization of chitin, a key component of fungal cell walls. This inhibition prevents the proper assembly of the fungal cell wall, leading to cell death (Lenardon et al., 2010).

3.2 Inhibition of Pathogen Metabolic Pathways

Phytoalexins also target critical metabolic pathways within pathogens, disrupting their ability to grow, reproduce, and infect host plants. By inhibiting key enzymes and metabolic processes, phytoalexins interfere with the synthesis of essential biomolecules, such as nucleic acids, proteins, and lipids, ultimately leading to pathogen death.

3.2.1 Alkaloid Phytoalexins and Enzyme Inhibition

Alkaloid phytoalexins, such as camalexin produced by *Arabidopsis thaliana*, are particularly effective at inhibiting key enzymes involved in pathogen metabolism. Camalexin has been shown to inhibit enzymes responsible for detoxifying reactive oxygen species (ROS) in pathogens, such as superoxide dismutase and catalase. By inhibiting these enzymes, camalexin increases the accumulation of ROS within the pathogen, causing oxidative damage to cellular components and leading to cell death (Khare et al., 2017). Additionally, alkaloid phytoalexins can interfere with the synthesis of nucleic acids and proteins in pathogens. For example, phaseollin, an alkaloid phytoalexin produced by beans (*Phaseolus vulgaris*), inhibits RNA polymerase, preventing the transcription of essential genes in fungal pathogens (Westrick et al., 2021). This inhibition of RNA synthesis disrupts the ability of the pathogen to produce vital proteins, leading to reduced growth and virulence.

3.2.2 Stilbene Phytoalexins and Metabolic Interference

Stilbene phytoalexins, such as resveratrol produced by grapevines (*Vitis vinifera*), exhibit potent antimicrobial activity by interfering with the metabolic processes of pathogens. Resveratrol inhibits key enzymes involved in the synthesis of ergosterol, a vital component of fungal cell membranes. By disrupting ergosterol synthesis, resveratrol compromises the integrity of fungal cell membranes, leading to increased permeability and cell death (Jeandet et al., 2013). This mechanism is particularly effective against fungal pathogens such as *Botrytis cinerea*, which causes gray mold in grapevines. In addition to its effects on ergosterol synthesis, resveratrol has been shown to inhibit bacterial enzymes involved in DNA replication and repair. By targeting these enzymes, resveratrol disrupts bacterial cell division and reduces the ability of bacterial pathogens to infect host plants (Montero et al., 2003).

3.3 Induction of Reactive Oxygen Species (ROS) and Oxidative Stress in Pathogens

Phytoalexins often induce the accumulation of ROS within pathogens, leading to oxidative stress and cell death. ROS, such as hydrogen peroxide (H_2O_2) and superoxide radicals (O_2^-), are highly reactive molecules that can damage cellular components, including lipids, proteins, and nucleic acids. Phytoalexins can either directly generate ROS within pathogens or inhibit pathogen enzymes responsible for detoxifying ROS, leading to an accumulation of these toxic molecules.

3.3.1 ROS Generation by Phytoalexins

Several phytoalexins have been shown to directly generate ROS within pathogens, contributing to their antimicrobial activity. For example, the phytoalexin sakuranetin, produced by rice (*Oryza sativa*) in response to fungal infection, induces the production of ROS in fungal pathogens such as *Magnaporthe oryzae*. The accumulation of ROS leads to oxidative damage to fungal cell membranes and proteins, ultimately resulting in cell death (Hasegawa et al., 2014). In addition to their direct generation of ROS, phytoalexins can enhance the plant's oxidative burst, rapid production of ROS at the site of pathogen infection. This oxidative burst not only damages the pathogen but also strengthens the plant's cell walls, making it more difficult for pathogens to penetrate and establish an infection.

3.3.2 Inhibition of Pathogen Antioxidant Defenses

Many pathogens possess antioxidant enzymes, such as superoxide dismutase, catalase, and peroxidase, which help detoxify ROS and protect the pathogen from oxidative stress. Phytoalexins can inhibit these antioxidant defenses, leading to an accumulation of ROS and increased susceptibility to oxidative damage. For instance, the flavonoid phytoalexin glyceollin has been shown to inhibit the activity of superoxide dismutase in fungal pathogens, preventing the detoxification of superoxide radicals. This inhibition leads to the accumulation of ROS within the pathogen, causing oxidative damage to cellular components and promoting cell death (Montero et al., 2003). Similarly, resveratrol has been found to inhibit catalase activity in bacterial pathogens, reducing their ability to detoxify hydrogen peroxide and increasing their susceptibility to oxidative stress (Jeandet et al., 2013).

3.4 Synergistic Effects of Phytoalexins with Other Plant Defense Compounds

Phytoalexins often work in concert with other plant defense compounds, such as phenolics, terpenoids, and pathogenesis-related proteins, to enhance their antimicrobial efficacy. These synergistic interactions allow plants to mount a multi-faceted defense response, targeting multiple aspects of pathogen physiology and increasing the likelihood of successfully repelling an infection.

3.4.1 Synergy with Phenolic Compounds

Phenolic compounds, such as tannins and flavonoids, are known to work synergistically with phytoalexins to enhance their antimicrobial activity. For example, the combination of resveratrol with other phenolic compounds, such as quercetin, has been shown to increase the inhibition of fungal growth in grapevines compared to resveratrol alone (Shomali et al., 2022). This synergistic effect is likely due to the complementary mechanisms of action of these compounds, with resveratrol disrupting fungal cell membranes and phenolics interfering with fungal metabolism.

3.4.2 Interaction with Pathogenesis-Related Proteins

Pathogenesis-related (PR) proteins, such as chitinases and glucanases, are often co-expressed with phytoalexins during pathogen attack. These proteins degrade the cell walls of fungal and bacterial pathogens, making them more susceptible to the antimicrobial effects of phytoalexins. The combined action of phytoalexins and PR proteins creates a hostile environment for pathogens, reducing their ability to infect the plant. For instance, in tobacco (*Nicotiana tabacum*), the co-expression of the PR protein chitinase and the phytoalexin capsidiol has been shown to enhance resistance to fungal pathogens, such as *Phytophthora infestans*. The chitinase degrades the fungal cell wall, allowing capsidiol to penetrate the pathogen more effectively and disrupt its membrane (Han & Schneiter, 2024).

3.5 Evolutionary Adaptations of Pathogens to Phytoalexins

While phytoalexins provide a robust defense mechanism for plants, many pathogens have evolved strategies to detoxify or evade the effects of these compounds. Pathogens can produce enzymes that degrade phytoalexins or modify their metabolic pathways to resist the toxic effects of these compounds.

3.5.1 Phytoalexin-Degrading Enzymes

Some pathogens produce enzymes that can degrade phytoalexins, rendering them ineffective. For example, the fungal pathogen *Botrytis cinerea* produces laccases, enzymes that degrade stilbene phytoalexins such as resveratrol, allowing the fungus to overcome the plant's chemical defenses (Westrick et al., 2021).

3.5.2 Metabolic Adaptation to Phytoalexins

Other pathogens have evolved metabolic adaptations that allow them to tolerate high levels of phytoalexins. For instance, some bacterial pathogens can modify their cell membranes to reduce the permeability of phytoalexins, limiting their ability to disrupt membrane integrity. These evolutionary adaptations highlight the ongoing arms race between plants and pathogens, as plants evolve new phytoalexins and pathogens develop countermeasures to overcome them.

4. Role of Phytoalexins in Plant Defense Signaling Pathways

Phytoalexins play a critical role not only as antimicrobial agents but also as integral components of the plant defense signaling pathways. These secondary metabolites are synthesized in response to pathogen detection and contribute to a broad and robust defense mechanism. Their production is tightly regulated by an intricate network of signaling molecules, including reactive oxygen species (ROS), calcium ions, and phytohormones such as salicylic acid (SA), jasmonic acid (JA), and ethylene (ET). These signaling molecules orchestrate a series of molecular events that lead to the expression of genes involved in phytoalexin biosynthesis.

4.1 Activation of Phytoalexin Biosynthesis Through Salicylic Acid (SA) Pathway

The salicylic acid (SA) signaling pathway is primarily associated with defense responses against biotrophic pathogens, which feed on living host tissues. SA plays a crucial role in systemic acquired resistance (SAR), a form of long-lasting, broad-spectrum immunity that is activated following localized pathogen infection. Phytoalexins are key components of the SA-mediated defense response, as they are rapidly synthesized at the site of infection to inhibit pathogen growth and spread.

4.1.1 SA-Dependent Phytoalexin Accumulation

SA signaling leads to the induction of several defense-related genes, including those responsible for the biosynthesis of phytoalexins. For example, in *Arabidopsis thaliana*, the production of the phytoalexin camalexin is strongly induced by SA signaling. The activation of the transcription factor WRKY33, a critical regulator of the SA pathway, is necessary for the expression of camalexin biosynthetic genes such as *CYP71A13* and *PAD3* (Li et al., 2015). WRKY33 binds to the promoters of these genes, enhancing their transcription and leading to the accumulation of camalexin, which acts to inhibit pathogen proliferation at the site of infection. Additionally, the SA-mediated induction of systemic acquired resistance involves the spread of defense signals to uninfected tissues, preparing them for potential pathogen attack. In this systemic response, phytoalexins contribute to establishing an antimicrobial environment throughout the plant, bolstering its overall defense capability.

4.1.2 Cross-Talk with Other Signaling Molecules

SA signaling often operates in conjunction with other plant signaling molecules, including ROS and nitric oxide (NO), to enhance phytoalexin production. ROS, in particular, act as secondary messengers that amplify SA signaling, leading to a more robust defense response. The interplay between SA and ROS ensures rapid and localized production of phytoalexins, such as scopoletin in tobacco (*Nicotiana tabacum*) and glyceollin in soybeans (*Glycine max*), which are essential for inhibiting bacterial and fungal pathogens (Esaki et al., 1998). This interaction highlights the importance of coordinated signaling in regulating phytoalexin production and optimizing plant defense mechanisms.

4.2 Jasmonic Acid (JA) and Ethylene (ET) in Phytoalexin Signaling

Jasmonic acid (JA) and ethylene (ET) are phytohormones that play central roles in defense against necrotrophic pathogens and herbivores. Unlike biotrophic pathogens, which require living host cells, necrotrophic pathogens kill host cells and feed on the dead tissue. The JA and ET signaling pathways are typically activated in response to necrotrophic pathogens, mechanical damage, and herbivore attack, and they regulate the expression of defense genes, including those involved in phytoalexin biosynthesis.

4.2.1 JA and ET Synergy in Phytoalexin Production

The interaction between JA and ET signaling pathways enhances the production of phytoalexins in response to pathogen attacks. In *Arabidopsis*, the phytoalexin camalexin is not only regulated by SA but is also induced by JA and ET in response to necrotrophic pathogens such as *Botrytis cinerea*. The transcription factors ERF1 and MYC2, which are activated by JA and ET signaling, regulate camalexin biosynthesis by modulating the expression of genes involved in its production (Lorenzo et al., 2003). The coordination between JA and ET ensures that camalexin accumulates rapidly at the site of infection, effectively limiting pathogen spread. Similarly, rice (*Oryza sativa*) responds to fungal infection by producing diterpenoid phytoalexins, such as momilactones and phytocassanes, which are regulated by JA signaling. The expression of genes involved in diterpenoid biosynthesis, including *OsCPS2* and *OsKSL4*, is upregulated by JA and leads to the accumulation of these phytoalexins, providing effective protection against rice blast disease caused by *Magnaporthe oryzae* (Anh et al., 2023). This example demonstrates the importance of JA in regulating phytoalexin biosynthesis in response to fungal pathogens.

4.2.2 Role of JA and ET in Herbivore-Induced Phytoalexins

JA and ET are also critical for inducing phytoalexins in response to herbivory. When plants are attacked by herbivores, such as caterpillars or aphids, JA and ET signaling pathways are activated, leading to the production of phytoalexins that deter further herbivore feeding. For instance, in maize (*Zea mays*), the production of zealexins and kauralexins, two types of terpenoid phytoalexins, is induced by JA in response to herbivore attack. These phytoalexins accumulate in the damaged tissues, acting as toxins to herbivores and preventing further feeding (Schmelz et al., 2014). The induction of phytoalexins by JA and ET demonstrates their importance not only in pathogen defense but also in plant resistance to herbivores.

4.3 Reactive Oxygen Species (ROS) and Phytoalexin Induction

Reactive oxygen species (ROS), such as hydrogen peroxide (H_2O_2) and superoxide radicals (O_2^-), play dual roles in plant defense, functioning as both antimicrobial agents and signaling molecules. The production of ROS, also known as the oxidative burst, is one of the earliest responses to pathogen attack and is crucial for the induction of phytoalexins.

4.3.1 ROS as Signaling Molecules for Phytoalexin Biosynthesis

ROS serve as key signaling molecules that activate downstream defense responses, including the biosynthesis of phytoalexins. In response to pathogen recognition, ROS are rapidly generated at the site of infection by NADPH oxidases and other oxidant-producing enzymes. These ROS act as secondary messengers, triggering the activation of transcription factors that regulate phytoalexin biosynthetic genes (Walters et al., 2013). For example, in rice, the accumulation of ROS in response to fungal infection leads to the induction of sakuranetin, a flavonoid phytoalexin. ROS acts upstream of the transcription factor OsWRKY45, which binds to the promoters of sakuranetin biosynthetic genes and enhances their expression. This ROS-mediated signaling ensures a rapid accumulation of sakuranetin, which inhibits fungal growth and restricts infection (Ahuja et al., 2012).

4.3.2 ROS-Mediated Cross-Talk with Phytohormones

ROS also interacts with phytohormones such as SA, JA, and ET to regulate the production of phytoalexins. In the case of biotrophic pathogens, ROS amplifies SA signaling, leading to an increased production of phytoalexins such as camalexin in *Arabidopsis* and scopoletin in tobacco. In contrast, in response to necrotrophic pathogens, ROS works in concert with JA and ET to induce phytoalexin biosynthesis, enhancing the plant's ability to defend against a wide range of pathogens (Walters et al., 2013). The cross-talk between ROS and phytohormones highlights the complexity of plant defense signaling networks, ensuring that phytoalexins are produced in a manner that is tailored to the specific pathogen or environmental stress encountered by the plant.

4.4 Calcium Signaling in Phytoalexin Induction

Calcium ions (Ca^{2+}) are essential secondary messengers in plant defense signaling. The influx of Ca^{2+} into the cytosol is one of the earliest events following pathogen recognition and plays a pivotal role in the regulation of phytoalexin biosynthesis.

4.4.1 Calcium-Dependent Protein Kinases (CDPKs)

Calcium signaling is mediated by calcium-dependent protein kinases (CDPKs), which are activated by the binding of Ca^{2+} ions. CDPKs, in turn, phosphorylate downstream targets, including transcription factors that regulate the expression of phytoalexin biosynthetic genes. For instance, in rice, CDPKs are involved in the induction of diterpenoid phytoalexins in response to pathogen infection (Katsumata et al., 2018). The activation of CDPKs leads to the phosphorylation of transcription factors that control the expression of genes such as *OsCPS2* and *OsKSL7*, which are essential for the biosynthesis of phytoalexins like momilactones and phytocassanes.

4.4.2 Calcium Oscillations and Phytoalexin Production

In addition to CDPK activation, the frequency and amplitude of Ca^{2+} oscillations play a crucial role in fine-tuning phytoalexin biosynthesis. Studies have shown that specific patterns of Ca^{2+} oscillations are associated with the induction of phytoalexins in response to different types of pathogens (Han & Schneiter, 2024). For example, high-frequency Ca^{2+} oscillations are linked to the rapid production of phytoalexins in response to fungal infection, while low-frequency oscillations are associated with herbivore-induced phytoalexin production.

4.5 Integration of Phytoalexins in Systemic Acquired Resistance (SAR)

Phytoalexins also play a role in systemic acquired resistance (SAR), a form of long-lasting, broad-spectrum immunity that is induced following localized infection. SAR is characterized by the accumulation of defense-related compounds, including phytoalexins, in tissues distant from the initial infection site, providing protection against subsequent pathogen attacks.

4.5.1 Long-Distance Signaling and Phytoalexin Accumulation

The induction of SAR involves the transmission of long-distance signals, such as SA, JA, and azelaic acid, from the site of infection to uninfected tissues. These signals activate the expression of defense-related genes, including those involved in phytoalexin biosynthesis, throughout the plant (Dempsey & Klessig, 2012). As a result, phytoalexins such as camalexin in *Arabidopsis* and scopoletin in tobacco are synthesized in both infected and uninfected tissues, providing enhanced resistance to subsequent pathogen attacks.

4.5.2 Role of Phytoalexins in Enhancing SAR

The accumulation of phytoalexins in uninfected tissues contributes to the establishment of an antimicrobial environment that inhibits pathogen colonization. For example, camalexin has been shown to accumulate in both infected and systemic tissues during SAR, providing protection against a wide range of bacterial and fungal pathogens (Li et al., 2015). The role of phytoalexins in SAR highlights their importance not only as localized defense compounds but also as integral components of the plant's systemic immune response.

5. Synergistic Interactions Between Phytoalexins and Other Plant Defense Compounds

In the intricate web of plant defense mechanisms, phytoalexins play a crucial role in direct antimicrobial activity. However, their effectiveness is significantly enhanced through synergistic interactions with other plant defense compounds, such as pathogenesis-related (PR) proteins, phenolics, flavonoids, and terpenoids. These interactions form an integrated defense network that amplifies the plant's capacity to resist pathogen invasion. This synergistic action allows plants to mount a multi-layered defense system, optimizing the utilization of their biochemical resources while minimizing energy expenditure.

5.1 Interaction with Pathogenesis-Related (PR) Proteins

Pathogenesis-related (PR) proteins are a group of plant defense proteins that are induced in response to pathogen attack. These proteins, including chitinases, glucanases, and defensins, exhibit direct antimicrobial activity by degrading pathogen cell walls or inhibiting pathogen growth. Phytoalexins often work in synergy with PR proteins, creating a dual defense mechanism that enhances the plant's resistance to pathogens.

For instance, in tobacco (*Nicotiana tabacum*), the co-expression of PR proteins such as chitinase and the phytoalexin scopoletin has been shown to significantly enhance the plant's defense against fungal pathogens like *Phytophthora infestans*. The PR protein chitinase degrades fungal cell walls, exposing the pathogen to the direct antimicrobial effects of scopoletin, which further inhibits fungal growth by disrupting membrane integrity (Boncan et al., 2020). This combination of enzymatic degradation by PR proteins and chemical inhibition by phytoalexins creates a more hostile environment for pathogens, reducing their ability to infect and proliferate. In another example, in *Arabidopsis thaliana*, the phytoalexin camalexin works synergistically with PR proteins during a pathogen attack. While PR proteins such as β-1,3-glucanase degrade the polysaccharide components of pathogen cell walls, camalexin acts as a secondary layer of defense by targeting and disrupting pathogen membranes (Khare et al., 2017). The combined action of these two defense components enhances the plant's ability to quickly and effectively neutralize invading pathogens.

5.2 Synergy with Phenolics and Flavonoids

Phenolic compounds, including flavonoids, are another important class of plant defense molecules. These compounds exhibit strong antioxidant properties, protecting plants from oxidative stress and providing direct antimicrobial activity. The co-accumulation of phenolics and phytoalexins during pathogen attack has been shown to significantly enhance the antimicrobial efficacy of both types of compounds. For example, the combination of the stilbene phytoalexin resveratrol and the flavonoid quercetin in grapevines (*Vitis vinifera*) has been found to exhibit a stronger inhibitory effect on fungal pathogens than either compound alone (Shomali et al., 2020). Resveratrol disrupts fungal cell membranes, while quercetin interferes with fungal metabolism and enzyme activity. This synergistic action between resveratrol and quercetin results in a more comprehensive inhibition of pathogen growth, providing enhanced protection against diseases like powdery mildew and downy mildew. In soybeans (*Glycine max*), the phytoalexin glyceollin also interacts with phenolic compounds during pathogen defense. Glyceollin's antifungal properties are complemented by the antioxidant activity of phenolics such as coumestrol, which helps reduce oxidative stress in infected tissues. This interaction not only enhances the direct antimicrobial activity of glyceollin but also protects plant cells from pathogen-induced oxidative damage, allowing the plant to maintain cellular integrity while mounting a robust defense response (Esaki et al., 1998).

5.3 Synergistic Enhancement of Defense Signaling Pathways

Phytoalexins also play a role in enhancing the effectiveness of plant defense signaling pathways, particularly those mediated by phytohormones such as salicylic acid (SA), jasmonic acid (JA), and ethylene (ET). The accumulation of phytoalexins during pathogen attack can amplify these signaling pathways, leading to a stronger and more coordinated defense response. For instance, in *Arabidopsis*, the production of the phytoalexin camalexin is tightly regulated by the SA signaling pathway. The accumulation of camalexin during pathogen attack not only inhibits pathogen growth but also enhances the expression of SA-responsive genes involved in systemic acquired resistance (SAR) (Li et al., 2015). This amplification of SA signaling by camalexin leads to the establishment of a stronger systemic defense response, providing long-lasting protection against future pathogen attacks. Similarly, phytoalexins such as sakuranetin in rice (*Oryza sativa*) work in synergy with the JA and ET signaling pathways during pathogen defense. Sakuranetin, a flavonoid phytoalexin, enhances the expression of JA- and ET-responsive genes involved in the production of antimicrobial peptides and secondary metabolites (Hasegawa et al., 2014). This interaction between phytoalexins and defense signaling pathways ensures that the plant mounts a rapid and effective response to both biotic and abiotic stressors.

5.4 Interaction with Structural Defense Compounds

Phytoalexins also contribute to the reinforcement of plant structural defenses, particularly by interacting with lignin and callose, two compounds involved in fortifying plant cell walls. The accumulation of phytoalexins at the site of pathogen infection can enhance the deposition of lignin and callose, creating a physical barrier that prevents pathogen penetration.

For example, in rice, the accumulation of diterpenoid phytoalexins such as momilactones has been shown to promote the deposition of lignin in infected tissues (Anh et al., 2023). Lignin fortifies the plant cell wall, making it more resistant to enzymatic degradation by pathogens. In addition, phytoalexins can enhance the synthesis of callose, a polysaccharide that is deposited at the site of pathogen infection to block pathogen entry and movement within the plant. The interaction between phytoalexins and structural defense compounds highlights the multifaceted role of phytoalexins in plant defense. By both directly inhibiting pathogen growth and reinforcing physical barriers, phytoalexins contribute to a comprehensive defense strategy that enhances plant resilience against a wide range of pathogens.

6. Applications of Phytoalexins in Sustainable Disease Management

The growing interest in sustainable agriculture has led to a renewed focus on natural plant defense mechanisms, including the use of phytoalexins in managing plant diseases. Phytoalexins, with their potent antimicrobial properties and ability to be synthesized by plants in response to pathogen attack, offer an environmentally friendly alternative to synthetic chemical pesticides. In sustainable agriculture, phytoalexins can be harnessed to reduce the reliance on chemical inputs, improve crop resilience, and promote ecological balance. This section explores the various applications of phytoalexins in sustainable disease management, including their use in elicitor treatments, breeding programs, and integrated pest management (IPM) strategies.

6.1 Elicitation of Phytoalexin Production for Disease Resistance

One of the most promising applications of phytoalexins in sustainable agriculture is their use in elicitor treatments. Elicitors are natural or synthetic compounds that stimulate the production of phytoalexins and other defense-related compounds in plants, enhancing their resistance to pathogens without the need for chemical pesticides. For instance, chitosan, a natural polysaccharide derived from chitin, is a well-known elicitor that induces phytoalexin production in a variety of crops. In grapevines (*Vitis vinifera*), chitosan treatment has been shown to stimulate the accumulation of resveratrol, a stilbene phytoalexin that inhibits fungal pathogens such as *Botrytis cinerea* (Abourehab et al., 2017). Similarly, in rice, chitosan induces the production of diterpenoid phytoalexins, such as momilactones, which enhance resistance to rice blast disease. The use of elicitors to stimulate phytoalexin production offers a sustainable approach to disease management by reducing the need for chemical fungicides and promoting natural plant defenses. Other elicitors, such as salicylic acid analogs and plant extracts, have also been used to enhance phytoalexin production in crops like soybeans (*Glycine max*) and tomatoes (*Solanum lycopersicum*). These treatments lead to the accumulation of glyceollin and tomatine, respectively, which protect against fungal and bacterial pathogens (Walters et al., 2013). The ability of elicitors to induce phytoalexin production across different plant species highlights their potential for broad application in sustainable agriculture.

6.2 Breeding for Enhanced Phytoalexin Production

Breeding programs aimed at enhancing phytoalexin production in crops represent another important application of phytoalexins in sustainable disease management. By selecting crop varieties that naturally produce higher levels of phytoalexins, plant breeders can develop disease-resistant cultivars that require fewer chemical inputs. For example, in grapevines, breeding programs have focused on selecting varieties that produce higher levels of stilbene phytoalexins, such as resveratrol and pterostilbene. These compounds are highly effective against fungal pathogens like powdery mildew and downy mildew, reducing the need for fungicide applications (De Filippis et al., 2019). Similarly, in rice, breeding efforts have focused on enhancing the production of momilactones and phytocassanes, diterpenoid phytoalexins that provide resistance to rice blast and bacterial blight (Kato-Noguchi, 2023). By incorporating these phytoalexin-producing traits into commercial crop varieties, breeders can contribute to the development of more resilient and sustainable agricultural systems.

6.3 Phytoalexins in Integrated Pest Management (IPM) Strategies

Phytoalexins also have great potential for integration into integrated pest management (IPM) strategies, which aim to combine biological, cultural, and chemical control methods to manage pests and diseases in a sustainable and environmentally friendly manner. The use of phytoalexins in IPM can reduce the reliance on synthetic pesticides and promote ecological balance by enhancing natural plant defenses. For example, in vineyards, the use of phytoalexin-inducing elicitors, such as methyl jasmonate, has been incorporated into IPM strategies to manage fungal diseases like powdery mildew. Methyl jasmonate induces the production of resveratrol and other phytoalexins in grapevines, providing effective protection against fungal pathogens while reducing the need for chemical fungicides (Shomali et al., 2022). The integration of elicitor treatments with other IPM practices, such as crop rotation and biological control, enhances the sustainability of disease management in vineyards. In rice production, the combination of phytoalexin-inducing treatments with biological control agents, such as *Trichoderma* species, has been shown to enhance resistance to fungal and bacterial pathogens. *Trichoderma* produces enzymes that degrade pathogen cell walls, while phytoalexins such as sakuranetin and momilactones provide chemical defense, creating a synergistic effect that improves disease resistance (Valletta et al., 2023). The use of phytoalexins in IPM strategies offers a promising approach to managing plant diseases in a sustainable and environmentally friendly manner.

6.4 Potential for Organic Farming Systems

Phytoalexins hold particular promise for use in organic farming systems, where the use of synthetic pesticides is restricted. Their natural origin, biodegradability, and low toxicity make them ideal candidates for incorporation into organic disease management practices. In organic grape production, for instance, the application of phytoalexin-inducing elicitors has been used to manage fungal diseases without the need for chemical fungicides. Chitosan and other natural elicitors have been shown to enhance the production of phytoalexins like resveratrol, providing effective protection against pathogens while adhering to organic farming standards (Abourehab et al., 2022). Similarly, in organic tomato production, the use of phytoalexin-inducing treatments has been shown to reduce the incidence of bacterial and fungal diseases, improving crop yields and reducing losses (Walters et al., 2013). The integration of phytoalexins into organic farming systems offers a sustainable alternative to chemical inputs, promoting plant health and resilience while maintaining ecological balance.

6.5 Future Prospects and Challenges in Phytoalexin-Based Disease Management

While the potential applications of phytoalexins in sustainable disease management are promising, several challenges remain. One of the main challenges is the variability in phytoalexin production among different plant species and cultivars. Some crops naturally produce low levels of phytoalexins, limiting their effectiveness in disease management. Breeding programs aimed at enhancing phytoalexin production, as well as the development of more effective elicitors, will be essential for overcoming this challenge. Another challenge is the potential for pathogens to evolve resistance to phytoalexins. Just as pathogens have evolved resistance to synthetic pesticides, they may develop mechanisms to detoxify or evade the effects of phytoalexins. Ongoing research into the mechanisms of phytoalexin resistance in pathogens will be important for developing strategies to mitigate this risk and ensure the long-term effectiveness of phytoalexin-based disease management.

Conclusion

Phytoalexins play a pivotal role in plant defense by acting as inducible antimicrobial compounds that are synthesized in response to pathogen attack. Their biosynthesis varies among plant species and is primarily derived from pathways such as the phenylpropanoid, terpenoid, and alkaloid pathways, giving rise to a wide range of phytoalexin types. These compounds are critical to a plant's innate immune response, with their production being triggered by specific pathogen-associated molecular patterns (PAMPs) or pathogen effectors. The ability of plants to rapidly synthesize phytoalexins upon pathogen detection is essential for limiting pathogen spread and damage. The antimicrobial properties of phytoalexins make them highly effective against a broad spectrum of pathogens, including bacteria, fungi, and viruses. Their modes of action, such as disrupting pathogen cell membranes, inhibiting essential enzymatic functions, and interfering with microbial signaling pathways, highlight their effectiveness in preventing and suppressing infections. Beyond their direct antimicrobial action, phytoalexins play a crucial role in plant defense signaling pathways, reinforcing other defense mechanisms such as systemic acquired resistance (SAR) and the hypersensitive response (HR). Their involvement in these pathways underscores their importance not just in direct pathogen defense, but also in coordinating the plant's broader immune responses.

Phytoalexins frequently act synergistically with other plant defense compounds, such as flavonoids, saponins, and phenolics, enhancing the overall effectiveness of the plant's defense system. These interactions create a robust network of chemical defenses that improve the plant's ability to resist a wide range of biotic stressors. Such synergy between different defense compounds maximizes the plant's resilience and strengthens its natural defense mechanisms, making phytoalexins an indispensable component of the plant's response to pathogens. In the context of sustainable agriculture, phytoalexins offer significant potential as natural alternatives to synthetic pesticides. Their ability to be naturally produced in plants and their efficacy against pathogens provide a promising foundation for developing eco-friendly disease management strategies. By harnessing phytoalexins, integrated pest management (IPM) programs can reduce reliance on harmful chemicals, promoting healthier ecosystems and more sustainable agricultural practices. However, challenges such as variability in phytoalexin production and the complexities of their induction require further research and biotechnological advancements to optimize their use in agriculture.

References

Abourehab, M. A. S., Pramanik, S., Abdelgawad, M. A., Abualsoud, B. M., Kadi, A., Ansari, M. J., & Deepak, A. (2022). Recent advances of chitosan formulations in biomedical applications. *International Journal of Molecular Sciences, 23*(18), 10975.

Ahuja, I., Kissen, R., & Bones, A. M. (2012). Phytoalexins in defense against pathogens. *Trends in Plant Science, 17*(2), 73-90.

Ahuja, I., Kissen, R., & Bones, A. M. (2012). Phytoalexins in defense against pathogens. *Trends in Plant Science, 17*(2), 73-90.

Anh, L. H., Khanh, T. D., & Xuan, T. D. (2023). Biological roles of momilactones: achievements, challenges, and promising approaches to exploit their beneficial properties. *Frontiers in Natural Products, 2*, 1245869.

Bari, R., & Jones, J. D. (2009). Role of plant hormones in plant defense responses. *Plant Molecular Biology, 69*(4), 473-488.

Birkenbihl, R. P., Diezel, C., & Somssich, I. E. (2012). Arabidopsis WRKY33 is a key transcriptional regulator of hormonal and metabolic responses toward *Botrytis cinerea* infection. *Plant Physiology, 159*(1), 266-285.

Boncan, D. A. T., Tsang, S. S. K., Li, C., Lee, I. H. T., Lam, H. M., Chan, T. F., & Hui, J. H. L. (2020). Terpenes and terpenoids in plants: interactions with environment and insects. *International Journal of Molecular Sciences, 21*(19), 7382.

De Filippis, B., Ammazzalorso, A., Amoroso, R., & Giampietro, L. (2019). Stilbene derivatives as new perspective in antifungal medicinal chemistry. *Drug Development and Research*, *80*, 285-293.

Deice Raasch-Fernandes, L., Bonaldo, S. M., de Jesus Rodrigues, D., Magela Vieira-Junior, G., Regina Freitas Schwan-Estrada, K., Rocco da Silva, C., Gabriela Araújo Verçosa, A., Lopes de Oliveira, D., & Wender Debiasi, B. (2019). Induction of phytoalexins and proteins related to pathogenesis in plants treated with extracts of cutaneous secretions of southern *Amazonian Bufonidae* amphibians. *PLoS One*, *14*(1), e0211020.

Dempsey, D. A., & Klessig, D. F. (2012). SOS-too many signals for systemic acquired resistance? *Trends in Plant Science*, *17*(9), 538-545.

Dixon, R. A., & Paiva, N. L. (1995). Stress-induced phenylpropanoid metabolism. *Plant Cell*, *7*(7), 1085-1097.

Dodds, P. N., Chen, J., & Outram, M. A. (2024). Pathogen perception and signaling in plant immunity, *The Plant Cell*, *36*(5), 1465-1481.

Esaki, H., Onozaki, H., Morimitsu, Y., & Kawakishi, S. (1998). Potent antioxidative isoflavones isolated from soybeans fermented with *Aspergillus saitoi*. *Biosciences Biotechnology and Biochemistry*, *62*, 740-746.

Galletti, R., Ferrari, S., & De Lorenzo, G. (2011). Arabidopsis MPK3 and MPK6 play different roles in basal and oligogalacturonide- or flagellin-induced resistance against *Botrytis cinerea*. *Plant Physiology*, *157*(2), 804-814.

Han, Z., & Schneiter, R. (2024). Dual functionality of pathogenesis-related proteins: defensive role in plants versus immunosuppressive role in pathogens. Frontiers in Plant Sciences, *15*, 1368467.

Hasegawa, M., Mitsuhara, I., Seo, S., Okada, K., Yamane, H., Iwai, T., & Ohashi, Y. (2014). Analysis on blast fungus-Responsive characters of a flavonoid phytoalexin sakuranetin; Accumulation in infected rice leaves, antifungal activity and detoxification by fungus. *Molecules*, *19*(8), 11404-11418.

Jeandet, P., Clément, C., Courot, E., & Cordelier, S. (2013). Modulation of phytoalexin biosynthesis in engineered plants for disease resistance. *International Journal of Molecular Sciences*, *14*(7), 14136-1470.

Jeandet, P., Delaunois, B., Conreux, A., Donnez, D., Nuzzo, V., Cordelier, S., Clément, C., & Courot, E. (2010). Biosynthesis, metabolism, molecular engineering and biological functions of stilbene phytoalexins in plants. *BioFactors*, *36*, 331-341.

Kato-Noguchi, H. (2023). Defensive molecules momilactones A and B: Function, biosynthesis, induction and occurrence. *Toxins*, *15*(4), 241.

Katsumata, S., Toshima, H., & Hasegawa, M. (2018). Xylosylated detoxification of the rice flavonoid phytoalexin sakuranetin by the rice sheath blight fungus *Rhizoctonia solani*. *Molecules*, *23*(2), 276.

Khare, D., Choi, H., Huh, S.U., Bassin, B., Kim, J., Martinoia, E., Sohn, K. H., Paek, K., & Lee, Y. (2017). Arabidopsis ABCG34 contributes to defense against necrotrophic pathogens by mediating the secretion of camalexin. *Proceedings of National Academy of Sciences U.S.A.*, *114*(28), E5712-E5720.

Kuc, J. (1995). Phytoalexins, stress metabolism, and disease resistance in plants. *Annual Review of Phytopathology, 33*(1), 275-297.

Lenardon, M. D., Munro, C. A., & Gow, N. A. (2010). Chitin synthesis and fungal pathogenesis. *Current Opinion in Microbiology*, *13*(4), 416-23.

Li, R., Tee, CS., Jiang, & Y. L. et al. (2015). A terpenoid phytoalexin plays a role in basal defense of *Nicotiana benthamiana* against Potato virus X. *Scientific Reports*, *5*, 9682.

Lorenzo, O., Piqueras, R., & Solano, R. (2003). Ethylene response factor1 integrates signals from ethylene and jasmonic acid pathways in plant defense. *Plant Cell, 15*(1), 165-178.

Macho, A. P., & Zipfel, C. (2014). Plant PRRs and the activation of innate immune signaling. *Molecular Cell, 54*(2), 263-272.

Montero, C., Cristescu, S. M., Jiménez, J. B., Orea, J. M., te Lintel Hekkert, S., Harren, F. J., & González Ureña, A. (2003). trans-Resveratrol and grape disease resistance. A dynamical study by high-resolution laser-based techniques. *Plant Physiology*, *131*(1), 129-38.

Mucha, S., Heinzlmeir, S., Kriechbaumer, V., Strickland, B., Kirchhelle, C., Choudhary, M., Kowalski, N., Eichmann, R., Hückelhoven, R., Grill, E., Kuster, B., & Glawischnig, E. (2019). The formation of a camalexin biosynthetic metabolon. *Plant Cell*, *31*(11), 2697-2710.

Naveed, Z. A., Wei, X., Chen, J., Mubeen, H., & Ali, G. S. (2020) The PTI to ETI continuum in phytophthora-plant interactions. *Frontiers in Plant Sciences*, *11*, 593905.

Ninkuu, V., Zhang, L., Yan, J., Fu, Z., Yang, T., & Zeng, H. (2021). Biochemistry of terpenes and recent advances in plant protection. *International Journal of Molecular Sciences*, *22*(11), 5710.

Okada, A., Shimizu, T., Okada, K., & Kuzuyama, T. (2009). OsTGAP1 functions as a positive regulator of diterpenoid phytoalexins production in rice. *Journal of Biological Chemistry*, *284*(39), 26510-26518.

Schmelz, E.A., Huffaker, A., Sims, J.W., Christensen, S.A., Lu, X., Okada, K., & Peters, R.J. (2014), Biosynthesis, elicitation and roles of monocot terpenoid phytoalexins. *Plant Journal*, *79*, 659-678.

Shomali, A., Das, S., Arif, N., Sarraf, M., Zahra, N., Yadav, V., Aliniaeifard, S., Chauhan, D. K., & Hasanuzzaman, M. (2022). Diverse physiological roles of flavonoids in plant environmental stress responses and tolerance. *Plants*, *11*(22), 3158.

Valletta, A., Iozia, L. M., Fattorini, L., & Leonelli, F. (2023). Rice phytoalexins: Half a century of amazing discoveries; Part I: Distribution, biosynthesis, chemical synthesis, and biological activities. *Plants*, *12*(2), 260.

Walters, D. R., Ratsep, J., & Havis, N. D. (2013). Controlling crop diseases using induced resistance: Challenges for the future. *Journal of Experimental Botany, 64*(5), 1263-1280.

Westrick, N. M., Smith, D. L., & Kabbage, M. (2021). Disarming the host: Detoxification of plant defense compounds during fungal necrotrophy. Frontiers in Plant Sciences, *12*, 651716.

Zhou, J., Mu, Q., Wang, X., Zhang, J., Yu, H., Huang, T., He, Y., Dai, S., & Meng, X. (2022). Multilayered synergistic regulation of phytoalexin biosynthesis by ethylene, jasmonate, and MAPK signaling pathways in *Arabidopsis*. *Plant Cell*, *34*(8), 3066-3087.

Chapter - 10

Lignins and Plant Structural Defenses Against Pathogens

Geethanjali P A[1], Praneet Chauhan[2], Susheel Kumar[3] and Pratima Raypa[4]

[1]Department of Microbiology, Field Marshal K M Cariappa College, Madikeri, Kodagu, Karnataka, India
[2]Department of Plant Pathology, Dr. Khem Singh Gill Akal College of Agriculture, Eternal University, Baru Sahib, Sirmour, Himachal Pradesh, India
[3]Plant Molecular Virology Laboratory, CPMB Division, CSIR-National Botanical Research Institute, Rana Pratap Marg, Lucknow, Uttar Pradesh, India
[4]Department of Biochemistry, H.N.B. Garhwal University, Srinagar, Uttarakhand, India

Abstract

Lignins are complex phenolic polymers that play a critical role in strengthening plant cell walls and forming a structural barrier against various pathogens. Their biosynthesis occurs through the phenylpropanoid pathway, resulting in the production of three primary monolignol precursors: p-coumaryl, coniferyl, and sinapyl alcohols, which polymerize to form lignin in plant tissues. These compounds contribute to the rigidity and impermeability of the cell walls, thereby preventing pathogen entry and the spread of infection. Lignin accumulation is particularly important as an inducible defense mechanism when plants are under biotic stress, such as during a fungal, bacterial, or viral attack. Pathogens attempting to invade plant tissues face significant challenges due to lignin's ability to physically block penetration and disrupt enzymatic activity aimed at cell wall degradation. Lignin deposition is tightly regulated by complex signaling networks involving plant hormones like jasmonic acid, salicylic acid, and ethylene, which mediate the plant's defense response. Moreover, the process of lignification is not only a passive physical defense but also an active biochemical defense, where the plant strategically increases lignin levels in response to early pathogen recognition. Despite lignin's effectiveness as a defense strategy, some pathogens have evolved mechanisms to degrade or bypass this barrier, utilizing lignin-degrading enzymes like laccases and peroxidases to break down lignin polymers and facilitate tissue invasion. This co-evolutionary arms race between plants and pathogens continues to influence the development of more resistant crop varieties. In recent years, advances in genetic engineering have allowed scientists to manipulate lignin biosynthesis pathways to enhance

DOI: 10.1201/9781003717454-10

disease resistance in economically important crops. By altering the expression of key enzymes involved in lignin production, plants can be made more resistant to pathogens without compromising their growth and development. However, the modification of lignin pathways must be done carefully to avoid negative effects, such as reduced digestibility in forage crops or excessive lignin accumulation leading to decreased plant flexibility. Ongoing research in this area holds promise for sustainable agricultural practices by improving crop resilience to pathogens while minimizing the use of chemical pesticides.

Keywords: Lignins, Cell Wall, Pathogen Resistance, Inducible Defense, Lignification, Genetic Engineering.

Introduction

Lignins are complex phenolic polymers that play a vital role in the structural integrity and defense mechanisms of plants. These macromolecules are primarily localized within the cell walls of vascular plants, where they provide rigidity and hydrophobicity. Lignins are synthesized through the polymerization of monolignols, which are hydroxycinnamyl alcohol derivatives such as p-coumaryl, coniferyl, and sinapyl alcohol. The polymerization occurs via oxidative coupling, predominantly catalyzed by peroxidases and laccases. The chemical structure of lignins is highly heterogeneous, consisting of various linkages between the monolignol units, including β-O-4, β-β, and β-5 linkages. This variability in structure allows lignins to form a network that is highly resistant to degradation, making it a key component in the plant's defense against both abiotic and biotic stresses (Boerjan et al., 2003).

The biosynthesis of lignins occurs via the phenylpropanoid pathway, which is central to the production of various secondary metabolites in plants. The pathway begins with the deamination of phenylalanine by phenylalanine ammonia-lyase (PAL), followed by a series of enzymatic reactions that lead to the formation of monolignols. Cinnamate-4-hydroxylase (C4H) and 4-coumarate ligase (4CL) are key enzymes that regulate the early steps of lignin biosynthesis. The final polymerization of lignins involves peroxidase- and laccase-mediated radical coupling, a process that results in the formation of a complex three-dimensional polymer (Bonawitz & Chapple, 2010). This biosynthetic pathway is tightly regulated by both genetic and environmental factors, ensuring that lignins are produced in response to specific developmental cues or in response to pathogen attack.

One of the primary roles of lignins in plants is the strengthening of the cell wall. By embedding within the cellulose and hemicellulose matrix of the secondary cell walls, lignins enhance the mechanical strength of the plant's tissues, making them more resistant to mechanical stress and pathogen invasion. The deposition of lignins in the cell walls of xylem vessels also contributes to water transport efficiency by preventing the collapse of these conduits under the negative pressure generated during transpiration (Barros et al., 2015). Furthermore, the hydrophobic nature of lignins provides a barrier to water loss, which is particularly important in environments where water availability is limited. This dual role in both mechanical reinforcement and water retention highlights the importance of lignins in plant survival. Lignification is not only a constitutive feature of plant development but also serves as an inducible defense mechanism. Upon pathogen invasion, plants can increase the deposition of lignins at the site of infection, a process known as lignification, which reinforces the cell wall and limits pathogen spread (Miedes et al., 2014). This localized lignin deposition is often accompanied by the accumulation of other defense-related compounds, such as phenolic acids and flavonoids, which further strengthen the cell wall and inhibit pathogen growth. The inducibility of lignification is regulated by signaling pathways involving hormones such as jasmonic acid, salicylic acid, and ethylene, which activate transcription factors that upregulate lignin biosynthesis genes. For example, the WRKY and MYB families of transcription factors have been implicated in the regulation of lignin biosynthesis in response to biotic stress (Schluttenhofer & Yuan, 2015).

The interaction between lignins and pathogens is multifaceted, involving both direct and indirect mechanisms of inhibition. One of the primary ways in which lignins inhibit pathogens is by physically restricting their access to plant tissues. The dense, cross-linked structure of lignins forms a barrier that is difficult for most pathogens to penetrate. In addition, lignins can directly inhibit the activity of pathogen-derived enzymes, such as cellulases and pectinases, which are necessary for the degradation of plant cell walls (Lee et al., 2019). The oxidative environment created by lignin biosynthesis also generates reactive oxygen species (ROS), which can damage pathogen cells. Furthermore, lignins and their precursors can serve as substrates for the generation of antimicrobial compounds, such as lignans and phenolic acids, which further inhibit pathogen growth and development (Choi et al., 2023).

Despite the effectiveness of lignins in inhibiting pathogen invasion, some pathogens have evolved mechanisms to overcome lignin-based defenses. One such strategy is the production of lignin-degrading enzymes, such as laccases and peroxidases, which allow pathogens to break down lignins and access plant tissues (de Gonzalo et al., 2016). Fungal pathogens, in particular, are known to produce a wide array of ligninolytic enzymes that facilitate their invasion of lignified tissues. In response, plants may alter the composition and structure of their lignins to make them more resistant to degradation. For example, an increase in the proportion of guaiacyl lignins, which are more cross-linked and resistant to enzymatic degradation, has been observed in plants under pathogen attack (Berthet et al., 2011).

Genetic engineering of lignin biosynthesis pathways has emerged as a promising strategy for enhancing disease resistance in plants. By manipulating key enzymes involved in lignin biosynthesis, researchers have been able to increase lignin content and modify its composition in ways that enhance the plant's defense capabilities. For instance, overexpression of enzymes such as PAL and C4H has been shown to increase lignin deposition and improve resistance to fungal pathogens (Zhang et al., 2023). Similarly, downregulation of genes involved in lignin degradation, such as laccases, has been used to enhance lignin stability and reduce susceptibility to pathogen attack (Zhang et al., 2019). However, genetic manipulation of lignin pathways must be approached with caution, as excessive lignin accumulation can negatively impact plant growth and development. Balancing the trade-offs between enhanced disease resistance and plant fitness is a key challenge in the genetic engineering of lignins.

1. Chemical Structure of Lignins

Lignins are a fundamental component of plant secondary cell walls, providing rigidity and structural support that allow plants to grow tall and withstand environmental pressures. These biopolymers are composed of three primary monolignols-p-coumaryl alcohol, coniferyl alcohol, and sinapyl alcohol-which polymerize to form p-hydroxyphenyl (H), guaiacyl (G), and syringyl (S) units, respectively. The diversity of lignins is primarily driven by the varying ratios of these monolignols across different plant species, tissues, and environmental conditions.

1.1 Structural Complexity and Linkages

The structure of lignins is characterized by the diverse range of bonds formed during polymerization, the most common being the β-O-4 (aryl ether) linkage, which constitutes approximately 50% of all linkages. These bonds form long, flexible polymer chains that give lignins their ability to enhance the mechanical strength of the plant's cell walls. In addition to β-O-4 linkages, other types of bonds, including β-5 (phenylcoumaran), β-β (resinol), and β-1 (spirodienone), introduce branching and create a complex, three-dimensional structure that further enhances lignins' recalcitrance to degradation (Vanholme et al., 2010). The variability in lignin composition among plant species has significant functional implications. For instance, **gymnosperms** primarily produce G-type lignins, which are highly cross-linked and contribute to the rigidity of wood. **Angiosperms**, on the other hand, tend to have a higher proportion of S-units, which results in more linear, less cross-linked lignins. This difference in lignin composition is thought to play a role in the more flexible and diverse growth habits of angiosperms compared to gymnosperms (Chen et al., 2009).

1.2 Functional Implications of Lignin Structure

The structural heterogeneity of lignins allows them to perform a range of essential functions in plants, including mechanical support, defense, and water transport. The dense, cross-linked lignin matrix within the secondary cell walls of vascular tissues provides the rigidity necessary for plants to support their weight and maintain an upright posture. This mechanical strength is particularly crucial in tall plants, such as trees, where lignins are responsible for maintaining structural integrity under the force of gravity and environmental stressors like wind. Furthermore, lignins play a vital role in protecting plants from pathogens and herbivores. The irregular and highly cross-linked nature of lignin makes it resistant to enzymatic degradation by microbes, creating a physical barrier that prevents pathogens from penetrating the cell wall and accessing plant nutrients. Additionally, lignins contribute to the impermeability of plant tissues, reducing the likelihood of water loss and preventing the invasion of waterborne pathogens (Barros et al., 2015). The hydrophobic properties of lignins also play a critical role in the efficient transport of water through the xylem. By reinforcing the xylem vessels and preventing their collapse under negative pressure during transpiration, lignins ensure the continuous flow of water from the roots to the leaves. This function is essential for maintaining water balance and nutrient transport in vascular plants (Wang et al., 2018).

2. Biosynthesis of Lignins

Lignin biosynthesis is a multi-step process that involves the conversion of simple phenylpropanoid precursors into complex, cross-linked polymers. The biosynthesis of lignins begins with the phenylpropanoid pathway, a metabolic route that produces the aromatic precursors needed for monolignol formation. This pathway is initiated by the enzyme phenylalanine ammonia-lyase (PAL), which catalyzes the deamination of phenylalanine to produce cinnamic acid. Cinnamic acid then undergoes hydroxylation and methylation reactions to generate hydroxycinnamic acids, such as p-coumaric acid, ferulic acid, and sinapic acid, which are precursors to the three main lignin monomers (Zhang et al., 2023).

2.1 Enzymatic Steps in Lignin Biosynthesis

Following the production of hydroxycinnamic acids, the 4-coumarate ligase (4CL) enzyme activates these acids by converting them into their corresponding CoA esters. This activation step is essential for the subsequent reactions that lead to monolignol formation. Cinnamate-4-hydroxylase (C4H) and caffeoyl-CoA O-methyltransferase (CCoAOMT) are key enzymes that further modify the hydroxycinnamic acids, eventually leading to the formation of the monolignols p-coumaryl alcohol, coniferyl alcohol, and sinapyl alcohol (Bonawitz & Chapple, 2010). The final polymerization of monolignols into lignins occurs through radical-mediated oxidative coupling reactions. This process is catalyzed by laccases and peroxidases, which oxidize the monolignols to generate free radicals. These radicals then undergo random coupling, leading to the formation of a heterogeneous polymer with a wide variety of bond types. The non-enzymatic nature of the coupling reactions results in a highly branched and irregular lignin structure, which is crucial for its role in reinforcing cell walls and resisting degradation (Miedes et al., 2014).

2.2 Regulation of Lignin Biosynthesis

The regulation of lignin biosynthesis is controlled by a network of transcription factors and signaling pathways that ensure lignin production is coordinated with the plant's developmental needs and environmental conditions. **MYB** and **WRKY** transcription factors are key regulators of lignin biosynthetic genes, including **PAL**, **C4H**, and **4CL**. These transcription factors are activated in response to developmental signals, such as cell differentiation, or environmental stressors, such as pathogen attack and mechanical injury (Wang et al., 2018). In response to pathogen invasion, plants can rapidly upregulate lignin biosynthesis to reinforce the cell walls at the site of infection. This localized lignification serves as a barrier that limits the spread of the pathogen and prevents further tissue damage. In many cases, this defense response is accompanied by the accumulation of other phenolic compounds, such as **flavonoids** and **phenolic acids**, which further enhance the cell wall's resistance to degradation (Schluttenhofer & Yuan, 2015).

3. Role of Lignins in Plant Cell Wall Strengthening

Lignins play a crucial role in the reinforcement and strengthening of plant cell walls, serving as a vital component of plant structural integrity and defense mechanisms. These phenolic polymers are deposited mainly in the secondary cell walls of vascular tissues, where they interact with other cell wall components like cellulose and hemicellulose to create a rigid and durable matrix. The presence of lignins allows plants to grow upright, resist environmental stresses, and combat pathogen invasion.

3.1 Interaction of Lignins with Cellulose and Hemicellulose

The plant cell wall is a complex and dynamic structure, composed of a network of polysaccharides, proteins, and phenolic compounds. Cellulose, a crystalline polymer made up of β-1,4-linked glucose units, forms the backbone of the cell wall structure, while hemicellulose, a more amorphous polysaccharide, links cellulose microfibrils to each other. Lignins interact with both cellulose and hemicellulose, embedding themselves within this polysaccharide matrix and creating a highly cross-linked and rigid structure. The deposition of lignins within the cell wall matrix contributes to the overall stiffness and tensile strength of the plant. Lignins act as a cementing agent, binding cellulose and hemicellulose fibers together and preventing the separation of these components under mechanical stress. The cross-linking of lignins with hemicellulose in particular forms a highly recalcitrant matrix that is resistant to microbial degradation and physical disruption (Xie et al., 20180).

3.2 Mechanical Reinforcement and Plant Height

One of the primary roles of lignins in plant cell wall strengthening is their contribution to mechanical support. Plants rely on lignins to maintain structural rigidity, particularly in stems and woody tissues, which must bear the plant's weight and support vertical growth. This mechanical reinforcement is especially important in tall plants, such as trees, where the strength of lignified cell walls allows for increased plant height and greater access to sunlight for photosynthesis. The rigid and hydrophobic properties of lignins also prevent cell wall collapse, especially in xylem vessels, which are responsible for transporting water and nutrients. During transpiration, xylem vessels experience negative pressure, and the presence of lignins ensures that these vessels remain open and functional despite the tension created by water movement. This structural fortification enables efficient long-distance water transport, which is essential for plant growth and survival (Xiao et al., 2021).

3.3 Contribution to Compression Strength

Lignins significantly enhance the compression strength of plant cell walls, a property that is essential for woody plants. Compression strength refers to the ability of a material to withstand forces that would reduce its size. Lignins, with their complex, irregular structure, provide this resistance by filling in the spaces between cellulose and hemicellulose fibers, thus forming a more compact cell wall matrix. This increased cell wall density enhances the plant's ability to resist both internal turgor pressure and external mechanical forces, such as wind, gravity, or herbivory. In trees and other woody plants, lignins are primarily responsible for the development of durable wood tissues, which not only support the plant but also protect it from external damage. This role is especially evident in the thickening of the secondary cell walls in xylem tissues, where lignin deposition provides the necessary strength for these cells to function as the plant's vascular system and structural backbone (Gibson, 2012).

3.4 Lignification and Cell Wall Modifications in Response to Environmental Stress

Plants can dynamically modify the lignin content of their cell walls in response to environmental challenges. This process, known as **lignification**, is particularly important in plants facing biotic or abiotic stresses, such as pathogen invasion, drought, or mechanical injury. In such cases, lignins are deposited more rapidly in cell walls, especially in areas that are exposed to stress, providing an immediate increase in mechanical strength and resistance. For example, during pathogen attacks, lignin deposition at the site of infection can create a physical barrier that prevents the spread of the pathogen by reinforcing the cell walls of surrounding tissues. This defense mechanism is often accompanied by an increase in the production of reactive oxygen species (ROS), which can further strengthen cell walls and inhibit pathogen penetration (Cesarino, 2019). Lignins also interact with other secondary metabolites, such as flavonoids and phenolic acids, to create a chemically hostile environment for invading organisms.

3.5 Influence of Lignin Composition on Cell Wall Properties

The composition of lignins—specifically the ratio of the three main lignin subunits, p-hydroxyphenyl (H), guaiacyl (G), and syringyl (S)—plays a critical role in determining the mechanical properties of plant cell walls. Different plant species and tissues exhibit varying ratios of these subunits, which influence the overall rigidity and flexibility of the lignin polymer.

- Guaiacyl lignins (G-units): Found predominantly in gymnosperms, G-units are highly cross-linked and contribute to the dense, rigid nature of wood. This makes G-type lignins particularly important in the development of hard, strong tissues.
- Syringyl lignins (S-units): Found primarily in angiosperms, S-units form more linear polymers with fewer cross-links, which results in a more flexible and less dense structure. This flexibility allows for greater elasticity in tissues that require both strength and pliability.
- p-Hydroxyphenyl lignins (H-units): These lignins are typically found in herbaceous plants and grasses, where they provide a balance between rigidity and flexibility.

The precise ratio of these subunits is critical for the mechanical properties of plant cell walls. For example, a higher proportion of S-units tends to make lignins more amenable to degradation, whereas a higher proportion of G-units results in more recalcitrant, rigid lignins. This variability allows plants to optimize their lignin composition based on their specific environmental needs and growth strategies (Barros et al., 2015).

3.6 Implications for Industrial Applications

The role of lignins in plant cell wall strengthening has significant implications for various industrial applications, particularly in the fields of bioenergy, paper production, and biomaterials. Lignins' ability to form strong, recalcitrant polymers makes them a valuable resource for producing durable materials, such as carbon fibers, adhesives, and bioplastics. However, their resistance to degradation also presents challenges in industries that rely on the breakdown of plant biomass, such as biofuel production. Lignins' strong cross-linked structure makes it difficult to access cellulose and hemicellulose, which are needed for fermentation into biofuels. As a result, research is ongoing to develop methods for altering lignin composition and reducing its recalcitrance, making plant biomass more amenable to enzymatic breakdown. Genetic engineering approaches, such as modifying the expression of lignin biosynthesis genes or altering the ratio of S- to G-units, have shown promise in creating plants with more easily degradable lignins without compromising their structural integrity (Li et al., 2016).

4. Lignification as an Inducible Defense Mechanism

Lignification, the process by which lignins are deposited into plant cell walls, is not solely a developmental feature; it is also a highly inducible defense mechanism. Plants have evolved sophisticated defense strategies to combat biotic stresses such as pathogen attack, insect herbivory, and environmental injuries. One of the most effective responses is the rapid deposition of lignins in and around the cell walls at the site of infection or injury, reinforcing the cell walls and creating a physical barrier that limits pathogen invasion and spread. This dynamic process of **induced lignification** plays a vital role in the plant's ability to resist diseases and other environmental challenges.

4.1 Lignin Deposition in Response to Pathogen Attack

Lignification is a key component of the plant's defense system, particularly in response to pathogenic attacks, such as those from fungi, bacteria, and viruses. When a pathogen breaches the plant's surface, a cascade of defense mechanisms is triggered, including the localized synthesis and deposition of lignins in the affected tissues. This rapid lignification reinforces the structural integrity of the cell wall, making it more resistant to enzymatic degradation by pathogens. Pathogen recognition triggers the plant's innate immune responses through the activation of pattern recognition receptors (PRRs) that detect pathogen-associated molecular patterns (PAMPs). This interaction activates a signaling cascade involving mitogen-activated protein kinases (MAPKs) and the production of defense-related hormones such as salicylic acid (SA), jasmonic acid (JA), and ethylene. These hormones act as signals to upregulate the expression of lignin biosynthesis genes, including those encoding for enzymes such as phenylalanine ammonia-lyase (PAL), cinnamate-4-hydroxylase (C4H), and 4-coumarate ligase (4CL) (Miedes et al., 2014).

4.2 Reinforcement of the Cell Wall as a Physical Barrier

The primary function of lignification during pathogen attack is to fortify the plant cell wall. As lignins are deposited within the matrix of the cell wall, they increase the overall mechanical strength and impermeability of the wall, preventing pathogens from penetrating deeper into plant tissues. The cross-linking of lignins with polysaccharides, particularly hemicelluloses, further strengthens this barrier, creating a dense, recalcitrant structure that limits the movement of pathogens through intercellular spaces (Lee et al., 2019). The mechanical reinforcement provided by lignification is crucial in defending against necrotrophic pathogens, which kill plant cells to extract nutrients. By lignifying cell walls around the site of infection, plants can limit cell death and prevent the spread of necrosis, effectively containing the pathogen within a localized region. In this way, lignification serves as both a preventive and a containment strategy.

4.3 Induction of Lignification Through Signaling Pathways

Lignification as a defense mechanism is regulated by several signaling pathways, with salicylic acid (SA) and jasmonic acid (JA) being the most prominent players. Salicylic acid is primarily associated with resistance to biotrophic pathogens, which feed on living cells, while jasmonic acid is more closely linked to defense against necrotrophic pathogens and herbivorous insects. These signaling molecules activate specific transcription factors that bind to the promoters of lignin biosynthesis genes, thereby increasing the production of lignins at the site of infection or injury (Chen et al., 2009). Research has shown that WRKY and MYB transcription factors play pivotal roles in the regulation of lignin biosynthesis during stress conditions. These transcription factors are activated in response to pathogen attack or mechanical wounding and subsequently enhance the expression of genes encoding lignin biosynthetic enzymes. For example, WRKY33, a well-characterized transcription factor in Arabidopsis, is known to activate PAL, C4H, and 4CL in response to pathogen infection (Schluttenhofer & Yuan, 2015). This coordinated upregulation of lignin biosynthesis allows for the rapid fortification of the cell wall and the formation of a physical barrier against pathogen penetration.

4.4 Reactive Oxygen Species and Cross-Linking with Lignins

In addition to mechanical reinforcement, lignification is often associated with the production of reactive oxygen species (ROS), which further enhance the plant's defense capacity. ROS, such as hydrogen peroxide (H_2O_2), are generated as part of the oxidative burst that occurs during pathogen recognition. These reactive molecules not only serve as signaling intermediates but also directly participate in the cross-linking of lignins with other cell wall components. Peroxidases, which are enzymes that catalyze the polymerization of lignins, utilize H_2O_2 as a cofactor to facilitate the oxidative coupling of monolignols into complex lignin polymers. The simultaneous production of ROS and the upregulation of lignin biosynthesis creates a highly oxidative environment at the site of infection, leading to the rapid polymerization and deposition of lignins (Sattler & Funnell-Harris, 2013). The oxidative cross-linking of lignins with cell wall polysaccharides adds an additional layer of strength to the reinforced cell wall, making it even more resistant to pathogen attack.

4.5 Role of Lignification in Wound Healing and Mechanical Stress

Apart from its role in pathogen defense, lignification is also induced in response to mechanical stress or injury. When plants are wounded, whether by herbivores, environmental factors, or human activity, they initiate a lignification response to repair and fortify damaged tissues. This is particularly important in protecting vulnerable areas from secondary infections by opportunistic pathogens. The lignification process in wound healing is similar to that in pathogen defense, involving the upregulation of lignin biosynthetic genes and the deposition of lignins in the damaged cell walls. In addition to wound healing, plants exposed to mechanical stress, such as bending or compression, may also exhibit increased lignification in their stems or trunks. This reinforces the structural integrity of the plant, allowing it to withstand physical forces that could otherwise cause damage or reduce its stability. For example, trees subjected to high wind pressures often develop thicker, more lignified cell walls in their trunks to resist the mechanical stress (Ma, 2024).

4.6 Pathogen-Specific Lignification Responses

Different pathogens elicit varying degrees of lignification based on their mode of attack and interaction with the plant. For example, biotrophic pathogens, which feed on living plant cells, often trigger a moderate lignification response, as the plant seeks to reinforce cell walls without killing the affected tissues. In contrast, necrotrophic pathogens, which kill plant cells, tend to induce a much stronger lignification response, aimed at isolating and containing the dead cells to prevent further pathogen spread (Cesarino, 2019). Fungal pathogens, such as *Verticillium dahliae* and *Botrytis cinerea*, are particularly known for inducing strong lignification responses in their host plants. Infected plants often show significant lignin accumulation in their vascular tissues as a means of blocking the spread of the fungal hyphae. Similarly, bacterial pathogens can trigger localized lignification to confine the bacterial colony and limit nutrient access (de Gonzalo et al., 2016).

4.7 Implications for Crop Improvement and Disease Resistance

The mechanisms behind lignification as an inducible defense open avenues for crop improvement strategies aimed at enhancing disease resistance. Genetic engineering approaches that upregulate lignin biosynthesis in response to pathogen attacks could lead to the development of crops with increased resistance to a broad range of pathogens. However, it is important to balance lignification with overall plant fitness, as excessive lignin accumulation can interfere with plant growth and biomass production. By manipulating key transcription factors or signaling pathways involved in lignification, such as MYB or WRKY, researchers can fine-tune the lignification response to optimize both disease resistance and plant development. Furthermore, identifying crop varieties with natural tendencies for inducible lignification could provide valuable genetic resources for breeding programs focused on improving stress tolerance and yield (Zhang et al., 2023).

5. Lignin-Pathogen Interactions: Mechanisms of Inhibition

Lignins serve as one of the most effective biochemical and physical barriers that plants use to protect themselves from pathogen invasion. Their structural complexity, dense polymer network, and ability to cross-link with other cell wall components create an impenetrable barrier that most pathogens struggle to overcome. However, the interaction between lignins and pathogens is multifaceted, involving both direct inhibition of pathogen entry and growth, and indirect effects through plant signaling and immune responses. These interactions provide insight into how plants defend themselves from a variety of pathogens, including bacteria, fungi, and viruses.

5.1 Physical Barrier Against Pathogen Invasion

The primary mechanism by which lignins inhibit pathogens is by reinforcing the plant cell wall, thus creating a physical barrier that limits pathogen access to the internal tissues. This physical defense is particularly effective against pathogens that rely on mechanical penetration, such as fungi and some bacteria. The highly cross-linked, rigid structure of lignins makes the cell wall recalcitrant to degradation by enzymatic or mechanical means. Fungal pathogens, for example, often produce cell wall-degrading enzymes (CWDEs) such as cellulases, pectinases, and hemicellulases to break down plant cell walls and access the nutrients inside. However, the presence of lignins significantly reduces the effectiveness of these enzymes. Lignins not only increase the rigidity of the cell wall, making it difficult for fungal hyphae to penetrate, but they also interfere with the action of CWDEs by blocking the enzymes from reaching their target polysaccharides (Lee et al., 2019).

5.2 Chemical Defense and Cross-Linking with Pathogen-Derived Enzymes

Lignins also contribute to chemical defenses against pathogens. During pathogen invasion, plants produce reactive oxygen species (ROS) such as hydrogen peroxide (H_2O_2), which are involved in the lignin polymerization process. ROS are toxic to pathogens and play a crucial role in inhibiting their growth by damaging their cellular components. The generation of ROS during lignification helps to create a hostile environment for the pathogen, making it more difficult for the pathogen to survive and proliferate within the plant tissues (Miedes et al., 2014). Additionally, lignins can trap pathogen-derived enzymes, such as lytic enzymes that attempt to degrade the plant cell wall. By cross-linking with these enzymes, lignins inhibit their activity, preventing them from breaking down the cell wall matrix. This cross-linking is facilitated by peroxidases and laccases, which are key enzymes involved in the oxidative coupling of lignins and their interaction with other molecules in the cell wall (Ma, 2024). This mechanism not only enhances the physical integrity of the cell wall but also directly impedes the pathogen's enzymatic arsenal.

5.3 Reactive Oxygen Species and Lignin Synthesis in Defense

A critical aspect of lignin-pathogen interactions is the role of reactive oxygen species (ROS) in both pathogen inhibition and lignin biosynthesis. Upon pathogen attack, plants generate a burst of ROS in the affected tissues. ROS not only act as signaling molecules that trigger defense responses but also facilitate the rapid polymerization of lignins, which reinforces the cell wall at the site of infection. In this context, laccases and peroxidases are activated to catalyze the polymerization of monolignols into lignin polymers. These oxidative enzymes use ROS as cofactors to drive the lignin biosynthesis process, which strengthens the cell wall and enhances the plant's resistance to pathogen penetration (Cesarino, 2019). ROS also have a direct antimicrobial effect, as they can damage the cell membranes, proteins, and DNA of invading pathogens, further inhibiting their growth and spread.

5.4 Lignin Deposition as a Localized Defense Response

Lignification is often induced as a **localized defense response** to pathogen attack, a process that can be observed in many plants facing biotic stress. This targeted lignin deposition, known as **lignification** or **lignin fortification**, typically occurs around the site of infection, where lignins are rapidly synthesized and deposited into the cell walls. This localized defense response restricts the pathogen to a limited area, preventing it from spreading to other parts of the plant. For example, during infection by fungal pathogens like *Verticillium dahliae* or *Fusarium oxysporum*, plants deposit lignins in the xylem tissues, where the fungi attempt to colonize and block water transport. By reinforcing the xylem vessels with lignins, the plant not only prevents the pathogen from moving through the vascular system but also maintains the integrity of water transport, which is essential for the plant's survival under stress conditions (Vanholme et al., 2010). This ability to compartmentalize the pathogen through lignin deposition is crucial for minimizing damage and preserving plant function.

5.5 Interaction with Pathogen-Secreted Effectors

Some pathogens secrete specific effector molecules to suppress plant defenses and facilitate infection. These effectors can target various components of the plant immune system, including the lignin biosynthesis pathway. For instance, some fungal pathogens produce ligninolytic enzymes, such as laccases and peroxidases, which are capable of degrading lignins. These enzymes allow the pathogen to break down the lignin barrier and access the underlying plant tissues. However, plants have evolved counter-defense mechanisms to neutralize the effects of these pathogen-secreted enzymes. One such mechanism is the modification of lignin composition to make it more resistant to degradation. For example, some plants increase the proportion of guaiacyl (G) lignins, which are more cross-linked and resistant to enzymatic breakdown, in response to pathogen attack (Berthet et al., 2011). This dynamic modification of lignin composition serves as an adaptive defense mechanism that enhances the plant's ability to withstand pathogen assaults.

5.6 Systemic Acquired Resistance and Lignification

Lignification also plays a role in systemic acquired resistance (SAR), a long-lasting defense response that protects the plant from subsequent attacks by a broad spectrum of pathogens. SAR is typically triggered by a localized infection, and the signal is transmitted throughout the plant, leading to the upregulation of defense-related genes in distal tissues. Lignin biosynthesis genes are among those activated during SAR, resulting in an overall increase in lignin content in both the infected and uninfected tissues. This systemic lignification response fortifies the plant's defenses, making it more resistant to future pathogen invasions. The upregulation of lignin biosynthesis during SAR is mediated by signaling molecules such as salicylic acid (SA) and jasmonic acid (JA), which are critical for the coordination of defense responses across different parts of the plant (Sattler & Funnell-Harris, 2013). By enhancing lignin deposition in both local and systemic tissues, plants create a more robust barrier that inhibits pathogen spread and improves overall resistance.

5.7 Pathogen Evasion Strategies

While lignification is an effective defense mechanism, some pathogens have evolved strategies to overcome lignin-based defenses. Certain fungal and bacterial pathogens produce lignin-degrading enzymes that allow them to breach the lignin barrier and penetrate plant tissues. For example, **white-rot fungi** are known to produce a variety of ligninolytic enzymes, including **laccases** and **peroxidases**, which degrade lignins and enable the fungi to access the nutrients within the lignified tissues (de Gonzalo et al., 2016). Other pathogens may avoid triggering lignification by secreting effectors that suppress plant defense responses. These effectors can interfere with the signaling pathways that activate lignin biosynthesis, preventing the plant from reinforcing its cell walls in time to block the pathogen. In response to these evasion strategies, plants may modify their lignin composition or upregulate other defense mechanisms to counteract the pathogen's efforts.

6. Lignin Modifications and Pathogen Evasion

Lignins are central to plant defense due to their structural complexity and role in strengthening cell walls, which act as physical and chemical barriers against pathogen invasion. However, pathogens have developed diverse strategies to circumvent lignin-based defenses, leading to an ongoing evolutionary arms race between plants and their attackers. In this context, lignin modifications by both plants and pathogens play a crucial role. Plants may adjust lignin composition and structure to reinforce their defenses, while pathogens deploy enzymes and effectors to degrade or bypass lignin barriers.

6.1 Lignin Composition and Structural Modifications by Plants

Plants can modify the composition and structure of lignins in response to pathogen attack or environmental stress. These modifications typically involve changes in the ratio of the three main lignin subunits—p-hydroxyphenyl (H), guaiacyl (G), and syringyl (S) units—which affect the rigidity and recalcitrance of the lignin polymer. G-lignins, for example, are more highly cross-linked and resistant to enzymatic degradation, making them particularly effective against pathogens that secrete cell wall-degrading enzymes. By increasing the proportion of G-units in response to pathogen attack, plants can create a more robust barrier that pathogens find difficult to penetrate (Chen et al., 2009). In some cases, plants also increase the deposition of p-hydroxyphenyl (H)-type lignins, which are common in herbaceous plants and provide an additional layer of structural reinforcement. These lignins are often associated with regions of the plant that are under direct attack, such as the vascular tissues in the xylem, which pathogens often target to disrupt water and nutrient transport. By modifying lignin composition in these critical areas, plants can effectively limit the spread of pathogens and maintain vascular function (Miedes et al., 2014).

6.2 Alteration of Lignin Biosynthesis Pathways in Response to Stress

Plants can dynamically regulate lignin biosynthesis pathways in response to biotic and abiotic stresses. Several key enzymes involved in lignin biosynthesis, including phenylalanine ammonia-lyase (PAL), cinnamate-4-hydroxylase (C4H), and 4-coumarate ligase (4CL), are upregulated during pathogen infection or physical damage. These enzymes facilitate the conversion of phenylpropanoid precursors into monolignols, which are the building blocks of lignins. One important aspect of this regulation is the activation of specific transcription factors, such as MYB and WRKY proteins, which bind to the promoters of lignin biosynthesis genes. These transcription factors are often triggered by hormonal signaling pathways involving salicylic acid (SA) and jasmonic acid (JA), which are central to the plant's defense responses. Under the influence of these signaling molecules, the plant can fine-tune the lignin biosynthesis pathway to produce lignins that are more resistant to pathogen-derived enzymes or that are more effective at reinforcing weakened cell walls (Cesarino, 2019).

6.3 Pathogen Strategies for Lignin Degradation

Despite the effectiveness of lignin modifications by plants, many pathogens have evolved sophisticated mechanisms to degrade lignins or bypass the lignified cell walls. Fungal pathogens, in particular, are known to produce a range of ligninolytic enzymes, including laccases, peroxidases, and lignin peroxidases, which break down lignins and allow the pathogen to penetrate the plant cell wall. These enzymes are especially prevalent in necrotrophic fungi such as Botrytis cinerea and Sclerotinia sclerotiorum, which kill plant cells and feed on the dead tissue (de Gonzalo et al., 2016). White-rot fungi, such as those in the genus Phanerochaete, are particularly adept at lignin degradation. These fungi produce an array of oxidative enzymes that can depolymerize the complex lignin structure, allowing the fungi to access the cellulose and hemicellulose beneath the lignin layer. By breaking down lignin, these pathogens are able to colonize lignified tissues, such as the xylem, which are otherwise highly resistant to microbial invasion (Ma, 2024).

6.4 Effector-Mediated Suppression of Lignin Biosynthesis

In addition to producing lignin-degrading enzymes, some pathogens use **effectors** to suppress the plant's lignin biosynthesis pathways. These effectors are small proteins or molecules secreted by the pathogen that manipulate the plant's cellular machinery to weaken its defenses. For example, certain bacterial pathogens, such as **Pseudomonas syringae**, secrete effectors that interfere with the plant's hormonal signaling pathways, particularly those involving salicylic acid (SA) and jasmonic acid (JA). By disrupting these pathways, the pathogen prevents the upregulation of lignin biosynthesis genes, thereby making the plant more susceptible to infection (Sattler & Funnell-Harris, 2013). Some fungal pathogens, like **Ustilago maydis**, also produce effectors that target lignin biosynthesis directly. These effectors may bind to key transcription factors, such as MYB proteins, and prevent them from activating lignin biosynthesis genes. As a result, the plant's ability to produce lignins in response to pathogen attack is compromised, allowing the pathogen to establish an infection more easily (Xie et al., 2018).

6.5 Modulation of Lignin Composition for Pathogen Resistance

To counteract pathogen strategies, plants can modulate the composition and deposition of lignins in a way that confers increased resistance. For instance, some plants increase the proportion of **S-lignins (syringyl units)** in their cell walls during pathogen attack. S-lignins form more linear polymers that are easier to depolymerize but are also less cross-linked, which may allow for quicker modification and reinforcement of the cell wall in response to dynamic pathogen activity (Cesarino, 2019). This adaptive lignification process not only strengthens the plant's physical barriers but also makes it more difficult for pathogens to degrade the lignins using their enzymes. By adjusting the ratio of S-lignins to G-lignins, the plant can optimize its defense based on the specific pathogen it is encountering. For example, G-lignins may be more effective against fungi that produce laccases, whereas S-lignins might offer better protection against pathogens that target other cell wall components (Barros et al., 2015).

6.6 Genetic Engineering of Lignin Pathways for Enhanced Resistance

Given the critical role of lignin modifications in plant defense, genetic engineering has emerged as a promising tool for enhancing lignin-based pathogen resistance in crops. By overexpressing key enzymes in the lignin biosynthesis pathway, such as PAL, C4H, and 4CL, researchers have been able to increase lignin content and alter lignin composition in transgenic plants. This strategy has been shown to improve resistance to a wide range of pathogens, including fungal, bacterial, and viral invaders (Zhang et al., 2023). In some cases, researchers have also engineered plants to produce lignins with altered monomeric compositions, such as increased G-lignins, to make the cell walls more resistant to enzymatic degradation by pathogens. Another approach involves the downregulation of lignin-degrading enzymes in the plant, preventing pathogens from accessing the enzymes they need to break down lignins. These genetic modifications have the potential to enhance crop resilience and reduce the need for chemical pesticides, making them an attractive solution for sustainable agriculture (Xiao et al., 2021).

6.7 Implications for Crop Protection

The ability to manipulate lignin composition and biosynthesis in plants offers significant opportunities for improving crop protection. As pathogens continue to evolve strategies to degrade or bypass lignin barriers, researchers must stay ahead by developing new approaches to enhance lignification in crops. Through selective breeding and genetic engineering, it is possible to create plant varieties with enhanced lignin-based defenses that can resist a wider array of pathogens. Moreover, understanding the interaction between lignins and pathogen effectors will provide further insights into how pathogens evade plant defenses. By targeting the molecular mechanisms that pathogens use to suppress lignin biosynthesis, researchers can develop new strategies for reinforcing plant defenses and mitigating the impact of plant diseases on global food security (Sattler & Funnell-Harris, 2013).

7. Genetic Engineering of Lignin Pathways for Enhanced Disease Resistance

Lignins play an essential role in plant defense, contributing to the structural integrity of cell walls and acting as a physical and chemical barrier against pathogen invasion. However, the intrinsic complexity of lignin biosynthesis has made it a challenge to manipulate for improving plant resistance. With advances in genetic engineering, researchers have begun to explore how modifications to the lignin biosynthesis pathways can enhance disease resistance in crops.

7.1 The Lignin Biosynthesis Pathway

The lignin biosynthesis pathway is a multi-step metabolic process involving several enzymes that convert phenylpropanoids into monolignols, the building blocks of lignin. The pathway begins with the deamination of phenylalanine by phenylalanine ammonia-lyase (PAL), followed by a series of hydroxylation and methylation reactions catalyzed by enzymes such as cinnamate-4-hydroxylase (C4H), 4-coumarate ligase (4CL), and caffeic acid O-methyltransferase (COMT). These enzymes produce p-coumaryl alcohol, coniferyl alcohol, and sinapyl alcohol, which are then polymerized by oxidative enzymes like laccases and peroxidases to form the lignin polymer (Bonawitz & Chapple, 2010). Given the complexity of this pathway, genetic engineering efforts often target key enzymes involved in the early and late stages of lignin biosynthesis to modify lignin composition and structure.

These modifications can either increase lignin content, alter its composition, or redirect metabolic flux towards other defense-related compounds, thus enhancing the plant's resistance to pathogens.

7.2 Engineering Lignin Composition for Disease Resistance

One of the primary goals of genetic engineering in lignin pathways is to enhance disease resistance by altering the composition of lignins in the cell wall. As mentioned earlier, different types of lignins (e.g., G-lignins, S-lignins, and H-lignins) confer varying degrees of rigidity and resistance to enzymatic degradation. By manipulating the ratios of these lignin subunits, researchers can create plants that are more resistant to pathogens. For example, increasing the proportion of G-lignins (guaiacyl units) in the cell wall has been shown to improve resistance to pathogens that produce lignin-degrading enzymes such as laccases and peroxidases. G-lignins are highly cross-linked and form more rigid polymers, making them more resistant to enzymatic attack by fungi and bacteria. Overexpression of enzymes such as PAL, C4H, and 4CL has been demonstrated to increase G-lignin content, thereby enhancing the mechanical strength of the cell wall and limiting pathogen access (Barros et al., 2015). On the other hand, increasing the levels of S-lignins (syringyl units) can provide a more flexible lignin structure, which may be beneficial for tissues that need to be responsive to environmental changes or pathogen attacks. Some plants have been engineered to overexpress F5H (ferulate 5-hydroxylase), an enzyme that leads to an increased production of S-lignins. These S-lignins form linear polymers that, although less cross-linked than G-lignins, can still strengthen the cell wall and make it more difficult for pathogens to penetrate (Ralph et al., 2019).

7.3 Reducing Lignin Content to Enhance Biomass Degradation

While increasing lignin content can improve pathogen resistance, it can also have unintended consequences for plant growth and biomass utilization. High lignin levels can reduce the digestibility of plant biomass, posing challenges for industries such as biofuel production, where the breakdown of cellulose and hemicellulose is essential for generating fermentable sugars. Therefore, another approach to engineering lignins for disease resistance involves reducing lignin content without compromising the plant's ability to resist pathogens. For example, researchers have used RNA interference (RNAi) and CRISPR-Cas9 gene editing technologies to downregulate lignin biosynthesis genes such as COMT and CAD (cinnamyl alcohol dehydrogenase). This reduction in lignin content has been shown to improve the digestibility of plant biomass for bioenergy production while maintaining or even enhancing pathogen resistance (Zhang et al., 2023). In some cases, the reduction in lignin content is compensated for by the increased deposition of other phenolic compounds, such as lignans and flavonoids, which can also contribute to the plant's defense against pathogens.

7.4 Redirecting Metabolic Flux for Enhanced Defense

Another promising strategy in lignin engineering is the redirection of metabolic flux within the phenylpropanoid pathway to produce defense-related compounds other than lignins. By downregulating certain steps in lignin biosynthesis, researchers can redirect the metabolic precursors toward the synthesis of other compounds, such as phytoalexins, flavonoids, or suberin, which have antimicrobial properties and can enhance disease resistance. For instance, downregulating COMT in some plants has been shown to reduce lignin content while simultaneously increasing the accumulation of ferulic acid and other phenolic acids. These compounds can cross-link with cell wall polysaccharides and proteins, creating a chemically resistant matrix that limits pathogen penetration. Moreover, the accumulation of these phenolics can inhibit the growth of fungal and bacterial pathogens directly by disrupting their cellular functions (Miedes et al., 2014).

7.5 Engineering Transcription Factors to Regulate Lignin Biosynthesis

In addition to targeting individual enzymes in the lignin biosynthesis pathway, researchers have also focused on engineering transcription factors that regulate the expression of these enzymes. Transcription factors such as MYB, NAC, and WRKY proteins are key regulators of lignin biosynthesis and are activated in response to developmental cues and environmental stresses, including pathogen attacks. By manipulating these transcription factors, it is possible to achieve a more coordinated regulation of lignin biosynthesis across different tissues and stages of plant development. For example, overexpression of MYB58, a transcription factor that positively regulates lignin biosynthesis, has been shown to increase lignin content and improve pathogen resistance in several plant species. Conversely, silencing MYB58 or related transcription factors can reduce lignin deposition and make the plant more susceptible to infection, highlighting the importance of precise control over lignin biosynthesis for effective disease resistance (Schluttenhofer & Yuan, 2015).

7.6 Case Studies: Genetic Engineering of Lignin Pathways in Crops

Several case studies illustrate the potential of genetic engineering to enhance disease resistance through lignin modification. One notable example is the genetic engineering of poplar trees to overexpress PAL, C4H, and 4CL, resulting in higher lignin content and increased resistance to fungal pathogens such as Melampsora medusae, which causes leaf rust. These transgenic poplars showed reduced fungal colonization and improved survival rates under disease pressure (Zhang et al., 2023). Another case study involves the engineering of sorghum for improved lignin composition. By downregulating COMT, researchers were able to produce sorghum plants with altered lignin composition that not only improved their resistance to Colletotrichum sublineolum, the causal agent of anthracnose, but also enhanced their biomass digestibility for biofuel production (Sattler & Funnell-Harris, 2013). These examples demonstrate the versatility of lignin engineering in enhancing disease resistance while balancing other agricultural traits.

7.7 Balancing Trade-offs: Growth, Yield, and Disease Resistance

While genetic engineering of lignin pathways offers exciting potential for improving disease resistance, it also poses significant challenges in balancing trade-offs between plant growth, yield, and resistance. Lignins play a crucial role in maintaining plant structure and water transport, and excessive lignification can lead to stunted growth, reduced flexibility, and lower yields. Therefore, a key focus of current research is to develop strategies that optimize lignin modifications without negatively impacting other aspects of plant physiology. One approach to achieving this balance is the spatial and temporal regulation of lignin biosynthesis. By using tissue-specific promoters or inducible systems, researchers can target lignin modifications to specific tissues (e.g., stems, leaves, or roots) or during particular stages of development (e.g., after pathogen attack), thereby minimizing the impact on overall growth and productivity (Cesarino, 2019).

Conclusion

Lignins, as integral components of plant cell walls, play a multifaceted role in plant defense, offering both structural support and protective functions against a wide range of pathogens. The chemical structure of lignins is highly complex, comprising cross-linked phenolic polymers derived from three primary monolignols: p-coumaryl alcohol, coniferyl alcohol, and sinapyl alcohol. These monolignols are synthesized through the phenylpropanoid pathway, which begins with phenylalanine, and are polymerized within the cell wall to form lignins. The degree of polymerization and the ratio of different monolignol units determine the specific properties of the lignin produced, with guaiacyl-rich lignins providing stronger mechanical strength and syringyl-rich lignins offering more flexibility. The variability in the chemical structure of lignins across different plant species, and even within tissues of the same plant, is a testament to their adaptability in providing both mechanical and defensive functions.

In the context of modern agriculture, lignins have garnered significant attention for their potential in enhancing disease resistance in crops. Genetic engineering of lignin biosynthesis pathways offers promising avenues for improving plant resilience to pathogens. By altering the expression of key genes involved in lignin production, such as those encoding PAL, CAD, or other lignin biosynthetic enzymes, researchers can develop crop varieties with enhanced lignin content or modified lignin composition. These modifications can result in stronger cell walls that are more resistant to pathogen attack, particularly in staple crops that are prone to fungal, bacterial, or viral infections. For example, overexpressing genes involved in monolignol biosynthesis can lead to the accumulation of more robust lignins, enhancing the plant's ability to withstand pathogenic invasion. However, the genetic modification of lignin biosynthesis is not without its challenges. One of the primary concerns is the potential trade-offs between increased lignin content and plant growth or development. Excessive lignin accumulation can lead to reduced cell wall flexibility, which may impair processes such as stem elongation, vascular transport, or seed dispersal. Moreover, in forage crops, elevated lignin content can negatively affect digestibility, reducing their nutritional value for livestock.

Nonetheless, the potential benefits of genetically enhancing lignin biosynthesis for disease resistance are substantial, particularly in the face of increasing global food security concerns. By creating crops that are less susceptible to pathogens, farmers can reduce their reliance on chemical pesticides, which not only lowers production costs but also mitigates the environmental impact of pesticide use. Furthermore, genetically engineered lignin pathways can provide long-term, sustainable solutions to the persistent challenge of pathogen resistance in crops. This is especially important in light of the growing threat posed by climate change, which is expected to exacerbate the spread and severity of plant diseases. As temperatures rise and weather patterns become more unpredictable, the ability of crops to withstand pathogen pressure will become even more critical. In this regard, lignin-based defense mechanisms, particularly those that can be enhanced through genetic engineering, offer a promising path forward for the development of resilient, disease-resistant agricultural systems.

References

Barros, J., Serk, H., Granlund, I., & Pesquet, E. (2015). The cell biology of lignification in higher plants. *Annals of Botany, 115*(7), 1053–1074.

Berthet, S., Demont-Caulet, N., Pollet, B., Bidzinski, P., Cézard, L., Le Bris, P. & Jouanin, L. (2011). Disruption of LACCASE4 and 17 results in tissue-specific alterations to lignification of Arabidopsis thaliana stems. *The Plant Cell, 23*(3), 1124-1137.

Boerjan, W., Ralph, J., & Baucher, M. (2003). Lignin biosynthesis. *Annual Review of Plant Biology, 54*(1), 519-546.

Bonawitz, N. D., & Chapple, C. (2010). The genetics of lignin biosynthesis: connecting genotype to phenotype. *Annual Review of Genetics, 44*, 337-363.

Cesarino, I. (2019). Structural features and regulation of lignin deposited upon biotic and abiotic stresses. *Current Opinion in Biotechnology, 56*, 209-214.

Chen, F., Liu, C., Tschaplinski, T., & Zhao, N. (2009). Genomics of secondary metabolism in populus: Interactions with biotic and abiotic environments. *Critical Reviews in Plant Sciences, 28*, 375-392.

Choi, S. J., Lee, Z., Kim, S., Jeong, E., & Shim, J. S. (2023). Modulation of lignin biosynthesis for drought tolerance in plants. *Frontiers in Plant Sciences, 14*, 1116426.

de Gonzalo, G., Colpa, D. I., Habib, M. H., & Fraaije, M. W. (2016). Bacterial enzymes involved in lignin degradation. *Journal of Biotechnology, 236*, 110-119.

Gibson, L. J. (2012). The hierarchical structure and mechanics of plant materials. *Journal of the Royal Society Interface, 9*(76), 2749-2766.

Lee, M. H., Jeon, H. S., Kim, S. H., Chung, J. H., Roppolo, D., Lee, H. J., Cho, H. J., Tobimatsu, Y., Ralph, J., & Park, O.K. (2019). Lignin-based barrier restricts pathogens to the infection site and confers resistance in plants. *EMBO Journal, 38*(23), e101948.

Li, M., Pu, Y., & Ragauskas, A. J. (2016). Current understanding of the correlation between lignin structure, degradation, and cell wall recalcitrance. *Frontiers in Chemistry, 4*, 45.

Ma, Q. H. (2024). Lignin biosynthesis and its diversified roles in disease resistance. *Genes (Basel), 15*(3), 295.

Miedes, E., Vanholme, R., Boerjan, W., & Molina, A. (2014). The role of the secondary cell wall in plant resistance to pathogens. *Frontiers in Plant Science, 5*, 358.

Ralph, J., Lapierre, C., & Boerjan, W. (2019). Lignin structure and its engineering. *Current Opinion in Biotechnology, 56*, 240-249.

Sattler, S. E., & Funnell-Harris, D. L. (2013). Modifying lignin to improve bioenergy feedstocks: strengthening the barrier against pathogens. *Frontiers in Plant Science, 4*, 70.

Schluttenhofer, C., & Yuan, L. (2015). Regulation of specialized metabolism by WRKY transcription factors. *Plant Physiology, 167*(2), 295-306.

Vanholme, R., Demedts, B., Morreel, K., Ralph, J., & Boerjan, W. (2010). Lignin biosynthesis and structure. *Plant Physiology, 153*(3), 895-905.

Wang, X., Yao, B., & Su, X. (2018). Linking enzymatic oxidative degradation of lignin to organics detoxification. *International Journal of Molecular Sciences*, *19*(11), 3373.

Xiao, R., Zhang, C., Guo, X., Li, H., & Lu, H. (2021). MYB Transcription factors and its regulation in secondary cell wall formation and lignin biosynthesis during xylem development. *International Journal of Molecular Sciences*, *22*(7), 3560.

Xie, M., Zhang, J., Tschaplinski, T. J., Tuskan, G. A., Chen, J. -G., & Muchero, W. (2018). Regulation of lignin biosynthesis and its role in growth-defense tradeoffs. *Frontiers in Plant Sciences*, *9*, 1427.

Zhang, J., Guo, M., Wu, H., Hou, L., Li, S., Chen, G., Liu, L., Liu, Z., Lu, L., Kabir, N., & Yang, Z. (2023). GhPAS1, a bHLH transcription factor in upland cotton (*Gossypium hirsutum*), positively regulates Verticillium dahlia resistance. *Industrial Crops and Products*, *192*, 116077.

Zhang, Y., Wu, L., Wang, X., Chen, B., Zhao, J., Cui, J., Li, Z., Yang, J., Wu, L., Wu, J., Zhang, G., & Ma, Z. (2019). The cotton laccase gene GhLAC15 enhances Verticillium wilt resistance via an increase in defence-induced lignification and lignin components in the cell walls of plants. *Molecular Plant Pathology*, *20*(3), 309-322.

Chapter - 11

Coumarins and their Role in Suppressing Plant Pathogens

Vinod Chouhan[1], Praneet Chauhan[2], Susheel Kumar[3] and Sheetanshu Gupta[4]

[1]Division of Plant Pathology, Indian Council of Agricultural Research-Indian Agricultural Research Institute (ICAR-IARI), New Delhi, India

[2]Department of Plant Pathology, Dr. Khem Singh Gill Akal College of Agriculture, Eternal University, Baru Sahib, Sirmour, Himachal Pradesh, India

[3]Plant Molecular Virology Laboratory, CPMB Division, CSIR-National Botanical Research Institute, Rana Pratap Marg, Lucknow, Uttar Pradesh, India

[4]Department of Biochemistry, G.B. Pant University of Agriculture & Technology, Pantnagar, Uttarakhand, India

Abstract

Coumarins are a diverse class of secondary metabolites with significant roles in plant defense against various pathogens. These naturally occurring phenolic compounds are synthesized through the shikimate pathway and are commonly found in many plant families, including Apiaceae and Rutaceae. Coumarins exhibit strong antimicrobial properties, which make them effective in suppressing a wide range of pathogens, including fungi, bacteria, and viruses. Their mode of action involves several mechanisms, such as disrupting cell membrane integrity, interfering with microbial enzyme activity, and generating reactive oxygen species that can damage pathogen cells. Additionally, coumarins can inhibit quorum sensing, thereby preventing microbial communication and biofilm formation. Plants can enhance the synthesis and accumulation of coumarins in response to pathogen attack, highlighting the inducible nature of this defense mechanism. Moreover, certain coumarins have been shown to work synergistically with other plant defense compounds, providing a multifaceted approach to pathogen suppression. The biosynthesis of coumarins is tightly regulated by environmental factors, including light and stress signals, as well as by hormonal pathways like jasmonic acid and salicylic acid.

Keywords: coumarins, antimicrobial, pathogen suppression, inducible defense, quorum sensing inhibition, genetic modification.

DOI: 10.1201/9781003717454-11

Introduction

Coumarins are a versatile class of naturally occurring secondary metabolites that play a vital role in plant defense systems, particularly against microbial pathogens. Structurally, coumarins are derived from the phenylpropanoid pathway, a crucial metabolic route in plants. These compounds are characterized by a benzopyrone nucleus, and their structural diversity is primarily due to variations in hydroxylation, methoxylation, and glycosylation patterns (Bourgaud et al., 2006). The biosynthesis of coumarins begins with the amino acid phenylalanine, which undergoes a series of enzymatic transformations. The core structure is synthesized via the shikimate pathway, leading to the production of umbelliferone, esculetin, and scopoletin—key precursors to many naturally occurring coumarins (Dini & Grumetto, 2022). Environmental factors, such as light exposure and pathogen infection, can significantly influence the biosynthesis of these compounds, highlighting the adaptive capacity of plants to modulate coumarin production in response to external stressors (Kai et al., 2006).

The antimicrobial properties of coumarins are well-documented, making them potent agents in the suppression of a wide range of plant pathogens, including fungi, bacteria, and viruses (Thakur et al., 2015). These compounds exhibit their antimicrobial effects through multiple mechanisms. They can disrupt the integrity of microbial cell membranes, leading to cell lysis and death, or inhibit crucial microbial enzymes necessary for pathogen survival and replication (Kovács et al., 2008). Additionally, coumarins can interfere with microbial metabolic pathways, resulting in the generation of reactive oxygen species (ROS), which further damages pathogen cells (Cuffaro et al., 2024). The broad-spectrum efficacy of coumarins against a variety of pathogens makes them indispensable components of the plant's chemical defense arsenal.

Coumarins are not merely constitutive defenses but are also involved in inducible defense mechanisms, where their synthesis is upregulated in response to pathogen attack (Dixon & Paiva, 1995). This inducible nature is governed by complex signaling pathways involving phytohormones such as jasmonic acid and salicylic acid, which are critical regulators of plant immunity (Farmer et al., 2003). Upon pathogen recognition, plants initiate a cascade of biochemical reactions, leading to the enhanced biosynthesis of coumarins, which accumulate at the infection site and act to limit pathogen spread (Dini & Grumetto, 2022). This rapid and localized response is essential for minimizing the impact of infections and ensuring plant survival.

Coumarins exert their pathogen-inhibiting effects through several key mechanisms. In addition to disrupting cell membranes and inhibiting enzymes, coumarins have been shown to interfere with microbial communication systems, specifically quorum sensing. By inhibiting quorum sensing, coumarins prevent microbial populations from coordinating their activities, such as biofilm formation or the secretion of virulence factors, thereby reducing the pathogenicity of infections (Thaku et al., 2015). This multi-pronged mode of action enhances the overall effectiveness of coumarins as defense compounds. Another significant aspect of coumarin's function in plant defense is their ability to work synergistically with other secondary metabolites. Studies have shown that coumarins can enhance the activity of compounds like flavonoids and phenolic acids, creating a more robust antimicrobial defense (Harborne, 1999).

1. Chemical Structure and Biosynthesis of Coumarins

1.1 Chemical Structure of Coumarins

Coumarins, a group of benzopyrones, are phenolic compounds recognized for their vast structural diversity and bioactivity. The fundamental structure of coumarins consists of a fused benzene and α-pyrone ring system, classified as 1,2-benzopyrone. This core structure provides the basis for various structural modifications, such as hydroxylation, glycosylation, methylation, and prenylation, resulting in different coumarin derivatives with unique properties. The simplest coumarin molecule is represented by 2H-1-benzopyran-2-one, where the carbonyl group of the pyrone ring is essential for the compound's aromatic and biochemical properties (Bourgaud et al., 2006).

Coumarins can be broadly divided into two major classes based on their structural features: simple coumarins and complex coumarins. Simple coumarins, such as umbelliferone and esculetin, possess a basic benzopyrone ring with few modifications, whereas complex coumarins include additional functional groups or multiple coumarin units linked together, forming compounds like furanocoumarins, pyranocoumarins, and bicoumarins (Kai et al., 2006).

1. **Simple Coumarins**: Simple coumarins are the most basic form of coumarins, with few modifications to the benzopyrone ring. Examples include umbelliferone (7-hydroxycoumarin), esculetin (6,7-dihydroxycoumarin), and scopoletin (6-methoxy-7-hydroxycoumarin). These compounds are widespread in plant tissues and serve various ecological roles, particularly in defense against pathogens (Akwu et al., 2023).
2. **Furanocoumarins**: Furanocoumarins contain an additional furan ring fused to the benzopyrone nucleus, which significantly alters their chemical properties and biological activity. Furanocoumarins, such as psoralen and bergapten, are known for their ability to crosslink with DNA, making them effective in defending plants against herbivores and pathogens. However, they also possess phototoxic properties, which can be harmful under excessive light exposure (Agrawal & Zhang, 2021).
3. **Pyranocoumarins**: Pyranocoumarins have a pyran ring fused to the benzopyrone structure. These compounds are less common than furanocoumarins but exhibit significant biological activity, particularly in their interactions with microbial pathogens. Xanthyletin and seselin are examples of pyranocoumarins, often found in plants within the Rutaceae family (Shan et al., 2023).
4. **Glycosylated Coumarins**: In many plant species, coumarins exist in glycosylated forms, where sugar moieties are attached to the hydroxyl groups of the coumarin backbone. Glycosylation enhances the solubility and transport of coumarins within plant tissues. Compounds like esculin (the glucoside of esculetin) are commonly found in plant defense systems, where they act as precursors to more active free coumarins when hydrolyzed by β-glucosidase in response to pathogen attack (Stringlis et al., 2019).

1.2 Pathways Involved in Coumarin Biosynthesis

The biosynthesis of coumarins occurs primarily through the phenylpropanoid pathway, a highly conserved metabolic route responsible for the production of numerous phenolic compounds, including flavonoids, lignins, and stilbenes (Dixon & Paiva, 1995). The precursor for coumarin biosynthesis is the amino acid phenylalanine, which is deaminated by the enzyme phenylalanine ammonia-lyase (PAL), converting it into trans-cinnamic acid. This compound then undergoes a series of hydroxylation, methylation, and lactonization reactions to form the coumarin skeleton (Bourgaud et al., 2006).

1.2.1 Initial Steps: Formation of Cinnamic Acid

The first step in coumarin biosynthesis is the conversion of phenylalanine to trans-cinnamic acid, catalyzed by the enzyme PAL. This step is common to all phenylpropanoids and represents a critical control point in the production of secondary metabolites (Korkina, 2007). Cinnamic acid serves as a key intermediate in the synthesis of various coumarin derivatives.

1.2.2 Hydroxylation and Methylation

Cinnamic acid is hydroxylated at the C-4 position by cinnamate-4-hydroxylase (C4H), a cytochrome P450 monooxygenase, to form p-coumaric acid. This step is critical as it introduces the necessary hydroxyl groups required for further coumarin biosynthesis. Additional hydroxylation reactions may occur at various positions on the aromatic ring, catalyzed by enzymes like 4-coumarate-3-hydroxylase (C3H) and flavonoid 3',5'-hydroxylase (F3'5'H), resulting in hydroxylated derivatives like umbelliferone and esculetin (Dini & Grumetto, 2022). In some cases, hydroxylated intermediates undergo methylation by O-methyltransferases, leading to the formation of methylated coumarins, such as scopoletin (6-methoxy-7-hydroxycoumarin) (Kai et al., 2006). Methylation is important in regulating the solubility and bioactivity of coumarins, influencing their interactions with pathogens.

1.2.3 Lactonization: The Formation of Coumarins

The final step in coumarin biosynthesis is lactonization, where the hydroxylated phenylpropanoid intermediates undergo intramolecular cyclization to form the characteristic benzopyrone ring structure. This reaction is facilitated by specific enzymes, though the exact mechanisms remain poorly understood (Bourgaud et al., 2006). The resulting coumarins can then undergo further modification through glycosylation, hydroxylation, or methylation to produce a wide array of structurally diverse compounds.

1.3 Role of Light and Environmental Factors in Coumarin Biosynthesis

Coumarin biosynthesis is not a static process; it is highly influenced by environmental stimuli, particularly light exposure. UV light, in particular, plays a significant role in regulating the production of coumarins in plants. UV-B radiation induces the expression of genes involved in the phenylpropanoid pathway, leading to an increased accumulation of coumarins like scopoletin and esculetin in the epidermal tissues (Kai et al., 2006). This response is thought to protect plants from UV-induced oxidative stress by quenching reactive oxygen species (ROS) and shielding cellular components from damage. Apart from light, other environmental factors such as nutrient availability, temperature, and pathogen attack can also modulate coumarin biosynthesis. For instance, nutrient stress, particularly low iron availability, has been shown to trigger the production of coumarins in Arabidopsis thaliana as part of the plant's strategy to improve iron uptake (Fourcroy et al., 2014). This phenomenon illustrates the multifaceted role of coumarins, not only in defense but also in plant nutrition.

1.4 Regulation of Coumarin Biosynthesis by Plant Hormones

The production of coumarins in plants is tightly regulated by phytohormones, including jasmonic acid (JA), salicylic acid (SA), and ethylene, which are key players in plant defense signaling. These hormones act as signals that activate the expression of genes involved in coumarin biosynthesis in response to biotic and abiotic stresses.

1.4.1 Jasmonic Acid Pathway

Jasmonic acid is a key hormone involved in plant responses to wounding and pathogen attack. The JA signaling pathway has been shown to upregulate the biosynthesis of coumarins, particularly in response to insect herbivory and fungal infections. Studies have demonstrated that exogenous application of JA can induce the accumulation of scopoletin and other coumarins in various plant species, enhancing their resistance to pathogens (Farmer et al., 2003).

1.4.2 Salicylic Acid Pathway

Salicylic acid is another important hormone in plant defense, particularly against biotrophic pathogens. The SA signaling pathway is closely linked to the production of coumarins, with SA playing a crucial role in the induction of defense-related genes. Scopoletin and esculetin have been found to accumulate in high concentrations in plants undergoing pathogen attack, particularly in response to SA-mediated signaling (Dixon & Paiva, 1995).

1.4.3 Ethylene Signaling

Ethylene is a gaseous hormone that regulates various aspects of plant growth and stress responses. Ethylene signaling has been implicated in the regulation of coumarin biosynthesis, particularly during pathogen-induced stress. The interplay between ethylene and other hormones, such as JA and SA, further fine-tunes the production of coumarins, ensuring that the plant mounts an effective defense response (Thakur et al., 2015).

1.5 Ecological Functions of Coumarins in Plant Defense

Coumarins are not only structural compounds but also serve essential ecological functions in protecting plants from a variety of biotic stressors, particularly microbial pathogens and herbivores. Their antimicrobial and allelopathic properties make them key players in the plant's defense mechanisms. Coumarins exhibit potent antimicrobial activity against a wide range of pathogens, including bacteria, fungi, and viruses. Their mode of action involves disrupting microbial cell membranes, inhibiting enzyme activity, and interfering with microbial metabolic processes. For example, scopoletin has been shown to possess antifungal properties, inhibiting the growth of plant pathogens such as *Fusarium* and *Alternaria* species (Cuffaro et al., 2024). Additionally, coumarins can act as signaling molecules, triggering the production of defense-related enzymes and compounds in response to microbial invasion (Dini & Grumetto, 2022). Beyond pathogen suppression, coumarins also play a role in allelopathy, where they inhibit the growth of competing plant species by releasing phytotoxic compounds into the soil. This allelopathic activity helps plants secure their territory and reduce competition for resources. Compounds like umbelliferone and esculin have been reported to exhibit strong allelopathic effects, influencing the germination and growth of neighboring plants (Shan et al., 2023).

1.6 Future Perspectives: Genetic Engineering of Coumarin Pathways

With growing interest in sustainable agriculture and crop protection, the genetic manipulation of coumarin biosynthesis offers a promising avenue for enhancing plant resistance to pathogens. By overexpressing key enzymes involved in coumarin production or modifying regulatory pathways, researchers aim to develop crops with increased levels of these bioactive compounds, providing enhanced protection against microbial diseases. One approach to increasing coumarin content in plants is the overexpression of key biosynthetic genes, such as PAL, C4H, and 4-coumarate CoA ligase (4CL). By upregulating these enzymes, plants can increase the flux through the phenylpropanoid pathway, leading to elevated coumarin levels. Another strategy is the suppression of genes involved in coumarin catabolism, preventing the breakdown of these compounds and maintaining higher concentrations in plant tissues (Kai et al., 2006). The genetic engineering of coumarin pathways has the potential to improve crop resistance to pathogens without relying on chemical pesticides. However, there are challenges associated with this approach, including potential trade-offs between increased coumarin production and plant growth or development. Additionally, the ecological impacts of genetically modified plants producing high levels of coumarins on non-target organisms must be carefully assessed (Bourgaud et al., 2006).

2. Antimicrobial Properties of Coumarins

Coumarins are well known for their broad-spectrum antimicrobial activities, making them potent natural defense compounds against a wide range of microbial pathogens, including bacteria, fungi, viruses, and even certain parasites. The antimicrobial properties of coumarins are derived from their ability to disrupt various cellular processes in pathogens, leading to inhibition of growth, cellular dysfunction, or cell death. These properties have made coumarins essential components of plant defense mechanisms and have attracted attention in agriculture, medicine, and pharmaceutical sciences due to their potential applications as natural antimicrobial agents (Thakur et al., 2015).

2.1 Mechanisms of Antimicrobial Action

The antimicrobial activity of coumarins can be attributed to multiple mechanisms. These mechanisms vary depending on the pathogen type, and coumarins often exhibit a combination of direct antimicrobial effects and interference with microbial virulence factors. The most commonly studied mechanisms of action include disruption of microbial cell membranes, interference with nucleic acid and protein synthesis, inhibition of microbial enzymes, and modulation of quorum sensing, which is essential for microbial communication and virulence.

2.1.1 Disruption of Cell Membrane Integrity

One of the primary modes of action of coumarins against microbial pathogens is their ability to disrupt cell membrane integrity. The lipid bilayers of microbial membranes are particularly vulnerable to coumarins because these compounds can interact with membrane lipids, causing structural disorganization. This interaction leads to increased membrane permeability, loss of cellular contents, and ultimately cell lysis (Cowan, 1999). Studies have demonstrated that coumarins like scopoletin and umbelliferone have the capacity to disrupt both fungal and bacterial cell membranes, causing leakage of essential intracellular ions and metabolites, leading to the death of the pathogen (Cuffaro et al., 2024). This membrane-disrupting action is especially important in controlling fungal pathogens, which have rigid cell walls but are still susceptible to perturbations in their membrane integrity.

2.1.2 Inhibition of Enzymatic Activity

Coumarins are known to inhibit various microbial enzymes that are critical for pathogen survival, metabolism, and pathogenicity. For example, coumarins can inhibit ATPases, which are essential for energy production in microorganisms, thus impeding their ability to carry out vital cellular processes (Kovács et al., 2008). Additionally, certain coumarins act as inhibitors of microbial proteases, which are enzymes that help pathogens degrade host proteins during infection. By blocking these enzymes, coumarins prevent pathogens from effectively invading host tissues and suppress their virulence. For instance, scopoletin has been shown to inhibit extracellular proteases secreted by fungal pathogens such as *Aspergillus* and *Fusarium*, reducing their ability to break down host cell walls (Cuffaro et al., 2024).

In bacterial pathogens, coumarins can inhibit DNA gyrase and topoisomerase IV, enzymes that are essential for DNA replication and transcription. This inhibition results in the suppression of bacterial proliferation, effectively controlling bacterial infections. Umbelliferone and esculetin have been reported to show such inhibitory activities against bacterial pathogens like *Escherichia coli* and *Staphylococcus aureus*, reducing their ability to multiply and spread within plant tissues (Basile et al., 2009). The ability of coumarins to target these vital enzymes highlights their effectiveness as antimicrobial agents against both bacterial and fungal pathogens.

2.1.3 Generation of Reactive Oxygen Species (ROS)

Another critical mechanism through which coumarins exhibit antimicrobial properties is the generation of reactive oxygen species (ROS). When plants produce coumarins in response to pathogen attack, these compounds can induce oxidative stress in microbial cells by increasing the levels of ROS, such as hydrogen peroxide and superoxide radicals. These reactive molecules cause oxidative damage to cellular components, including proteins, lipids, and DNA, leading to cell death (Cowan, 1999). ROS-mediated antimicrobial activity is particularly effective against fungal pathogens, as the oxidative stress can disrupt fungal spores and hyphae, inhibiting their ability to grow and spread. Scopoletin and its derivatives are known to induce oxidative stress in various fungal pathogens, making them powerful tools in controlling fungal infections in plants (Thakur et al., 2015).

2.1.4 Inhibition of Quorum Sensing

Quorum sensing is a bacterial communication system that regulates the expression of virulence factors in response to population density. Pathogens use quorum sensing to coordinate activities such as biofilm formation, toxin production, and motility, all of which are crucial for infection and survival within the host (Rasmussen & Givskov, 2006). Coumarins have been shown to interfere with quorum sensing pathways, thereby disrupting the ability of bacterial pathogens to regulate their virulence. For example, scopoletin has been demonstrated to inhibit quorum sensing in *Pseudomonas aeruginosa*, a plant and human pathogen, leading to reduced biofilm formation and decreased production of virulence factors like elastase and pyocyanin (Zhang et al., 2018). By blocking quorum sensing, coumarins effectively reduce the pathogenicity of bacterial infections without necessarily killing the bacteria outright, which can help reduce the likelihood of resistance development.

2.1.5 Inhibition of Biofilm Formation

Biofilms are complex microbial communities that adhere to surfaces and are embedded in a self-produced matrix of extracellular polymeric substances. Biofilms protect pathogens from environmental stressors, including antimicrobial agents, making infections more persistent and difficult to treat. Coumarins have been shown to inhibit biofilm formation, reducing the ability of pathogens to establish long-lasting infections (Zhang et al., 2018). This activity is particularly important for controlling bacterial pathogens like *Pseudomonas aeruginosa* and *Staphylococcus aureus*, which form biofilms as part of their infection strategy. By preventing biofilm formation, coumarins enhance the effectiveness of the plant's immune response and other antimicrobial agents.

2.2 Antifungal Activity of Coumarins

Fungal pathogens represent a significant threat to plant health, causing diseases such as rusts, blights, mildews, and rots. Coumarins have been extensively studied for their antifungal properties, which are attributed to their ability to disrupt fungal cell walls, interfere with spore germination, and inhibit the growth of fungal hyphae. Fungi rely on the integrity of their cell walls and membranes to maintain homeostasis, and coumarins target these structures to inhibit fungal growth (Thakur et al., 2015).

2.2.1 Disruption of Fungal Cell Walls

Coumarins, particularly those with hydroxyl groups like esculetin and scopoletin, have been shown to interfere with the synthesis of fungal cell walls, which are composed of chitin and glucans. By disrupting the biosynthesis of these structural polysaccharides, coumarins weaken the fungal cell wall, rendering it more susceptible to environmental stresses and immune responses. This mechanism is particularly effective against pathogenic fungi such as *Fusarium*, *Aspergillus*, and *Botrytis*, which rely on their cell walls for protection and virulence (Cuffaro et al., 2024).

2.2.2 Inhibition of Spore Germination

Fungal spores are highly resistant structures that play a crucial role in the spread of fungal infections. Coumarins have been shown to inhibit the germination of fungal spores, preventing the initiation of infection. Studies have demonstrated that scopoletin, esculetin, and umbelliferone inhibit the germination of *Fusarium oxysporum* and *Alternaria alternata* spores, two common plant pathogens responsible for wilts and blights, respectively (Dini & Grumetto, 2022). By preventing spore germination, coumarins limit the ability of fungal pathogens to establish infections in plant tissues.

2.2.3 Suppression of Hyphal Growth

Once a fungal spore germinates, it develops into a hypha, which is the vegetative growth form of most fungi. Hyphal growth is essential for the spread of fungal infections within plant tissues. Coumarins can inhibit the elongation and branching of fungal hyphae, reducing the spread of the infection. For example, studies have shown that scopoletin inhibits hyphal growth in *Fusarium solani*, a pathogen responsible for root rot in various crops (Basile et al., 2009). The inhibition of hyphal growth effectively contains the infection and allows the plant's immune system to neutralize the remaining fungal cells.

2.3 Antibacterial Activity of Coumarins

In addition to their antifungal properties, coumarins exhibit potent antibacterial activity against both Gram-positive and Gram-negative bacteria. Their ability to inhibit bacterial growth and pathogenicity through multiple mechanisms makes them versatile agents in plant defense. The antibacterial activity of coumarins is particularly valuable in controlling bacterial pathogens responsible for common plant diseases such as bacterial blight, wilt, and soft rot (Rasmussen & Givskov, 2006).

2.3.1 Activity Against Gram-Positive Bacteria

Gram-positive bacteria, such as *Staphylococcus aureus* and *Bacillus subtilis*, are characterized by thick peptidoglycan cell walls that provide structural integrity and resistance to environmental stressors. Coumarins have been shown to disrupt the cell wall synthesis of Gram-positive bacteria, leading to cell lysis and death. Esculetin and umbelliferone have been reported to inhibit the growth of *Bacillus cereus* and *Listeria monocytogenes*, two pathogens responsible for foodborne illnesses and plant infections (Kovács et al., 2008).

2.3.2 Activity Against Gram-Negative Bacteria

Gram-negative bacteria, such as *Escherichia coli* and *Pseudomonas aeruginosa*, are known for their more complex cell walls, which consist of an outer membrane that provides an additional layer of protection against antimicrobial agents. Despite this added protection, coumarins have been shown to exhibit significant activity against Gram-negative bacteria by targeting multiple aspects of their physiology. One of the key mechanisms by which coumarins exert their effects is through the disruption of the bacterial membrane integrity, leading to the loss of cellular contents and eventual cell death. For instance, umbelliferone and scopoletin have demonstrated inhibitory effects against *E. coli*, preventing bacterial proliferation (Basile et al., 2009). In addition to membrane disruption, coumarins also inhibit critical enzymes required for bacterial replication and metabolism. By inhibiting DNA gyrase and topoisomerase IV, coumarins prevent bacterial cells from effectively replicating their DNA, thereby halting cell division. This mechanism has been particularly effective against *Pseudomonas aeruginosa*, a common plant and human pathogen that forms biofilms and exhibits resistance to many traditional antibiotics (Alt et al., 2011). Coumarins' ability to bypass these resistance mechanisms and interfere with bacterial cell physiology makes them valuable tools in combating Gram-negative bacterial infections.

2.4 Antiviral Activity of Coumarins

While the majority of research on coumarins has focused on their antibacterial and antifungal properties, recent studies have also explored their potential antiviral activities. Viruses, unlike bacteria and fungi, do not have cellular structures that can be directly targeted; instead, they rely on the host's cellular machinery for replication. Coumarins, however, have been shown to interfere with viral replication cycles, making them effective antiviral agents in certain contexts.

One of the mechanisms by which coumarins exhibit antiviral activity is through the inhibition of viral entry into host cells. Certain coumarins, such as esculetin and its derivatives, have been reported to block the attachment and entry of viruses into plant cells by binding to viral proteins essential for this process. For example, studies have shown that coumarins can inhibit the entry of tobacco mosaic virus (TMV) and cucumber mosaic virus (CMV), two major plant viruses responsible for significant agricultural losses (Dini & Grumetto, 2022). By preventing viral attachment, coumarins disrupt the infection process at an early stage, reducing the severity of viral diseases. Additionally, coumarins may also interfere with viral replication once the virus has entered the host cell. Scopoletin, in particular, has been shown to inhibit the replication of plant viruses by modulating the expression of host defense genes and inducing the production of reactive oxygen species (ROS), which can damage viral RNA and proteins (Basile et al., 2009). This dual-action mechanism highlights the potential of coumarins as antiviral agents that can both prevent viral entry and disrupt replication.

2.5 Activity Against Parasitic Organisms

Beyond their efficacy against bacteria, fungi, and viruses, coumarins have also shown activity against parasitic organisms that pose a threat to plant health. Parasitic nematodes, for example, are responsible for severe damage to crops, reducing yield and quality. Coumarins, particularly those with potent bioactivity such as scopoletin and umbelliferone, have been reported to exhibit nematicidal activity by disrupting the normal functioning of nematode physiology (Degroote et al., 2024). This activity is crucial in agricultural systems, where nematode infestations can lead to significant economic losses. Coumarins exert their nematicidal effects by interfering with the nervous systems of parasitic nematodes, leading to paralysis and death. Moreover, coumarins have been shown to disrupt the life cycle of nematodes, inhibiting their ability to reproduce and establish in plant roots. This anti-nematode activity highlights the versatility of coumarins in protecting plants from a broad spectrum of pathogens, including parasitic organisms (Wuyts et al., 2006).

2.6 Synergistic Effects of Coumarins with Other Plant Defense Compounds

One of the unique aspects of coumarins is their ability to work synergistically with other plant secondary metabolites to enhance the overall antimicrobial defense response. Plants produce a wide array of secondary metabolites, including phenolic acids, flavonoids, alkaloids, and terpenoids, each contributing to the plant's defense against pathogens. Coumarins, when combined with these other compounds, can enhance their antimicrobial efficacy, leading to a more robust defense system. For instance, studies have shown that coumarins like scopoletin can enhance the activity of phenolic acids such as caffeic acid and ferulic acid, amplifying their antimicrobial properties against bacterial and fungal pathogens (Harborne, 1999). This synergistic interaction occurs through multiple mechanisms, including the enhancement of membrane-disrupting activity and the potentiation of ROS generation, which causes oxidative damage to pathogen cells. Additionally, coumarins can interact with flavonoids, another group of phenolic compounds, to inhibit the growth of pathogens more effectively than either compound alone (Harborne, 1999).

These synergistic effects are not limited to the chemical interactions between secondary metabolites. Coumarins also enhance the activity of defense-related proteins, such as pathogenesis-related (PR) proteins, which play a crucial role in the plant immune response. For example, scopoletin has been reported to increase the expression of PR proteins in response to pathogen infection, leading to a more coordinated and effective defense response (Lee et al., 2011). This ability to act synergistically with other defense compounds and proteins underscores the importance of coumarins as integral components of plant immunity.

2.7 Potential for Use in Agriculture and Medicine

The antimicrobial properties of coumarins have significant implications for both agriculture and medicine. In agriculture, the use of coumarins as natural antimicrobial agents offers a sustainable alternative to synthetic pesticides and fungicides, which are often associated with environmental pollution and the development of resistance in pathogens. By enhancing the natural production of coumarins in crops through genetic engineering or plant breeding, farmers can develop crops that are more resistant to diseases without the need for harmful chemicals (Thakur et al., 2015). Additionally, coumarins can be used as biopesticides in integrated pest management (IPM) systems, reducing the reliance on synthetic chemicals and promoting ecological balance. In medicine, coumarins have shown promise as potential therapeutic agents for treating infections caused by drug-resistant pathogens. Their ability to inhibit microbial growth through multiple mechanisms makes them attractive candidates for drug development. Furthermore, the anti-quorum sensing activity of coumarins offers a novel approach to controlling bacterial infections without exerting selective pressure for resistance, a growing concern in modern healthcare (Rasmussen & Givskov, 2006). Coumarins may also serve as lead compounds for the development of new antimicrobial drugs, particularly in the fight against multidrug-resistant bacterial and fungal pathogens.

2.8 Limitations and Future Directions

While the antimicrobial properties of coumarins are well established, there are still several challenges and limitations to their use in agriculture and medicine. One of the main challenges is the potential for phytotoxicity, where high concentrations of coumarins may have detrimental effects on the plants themselves or non-target organisms in the environment (Akwu et al., 2023). Further research is needed to optimize the concentrations and formulations of coumarins used in agricultural applications to minimize their negative impacts while maximizing their antimicrobial efficacy. In terms of human medicine, the bioavailability and toxicity of coumarins must be carefully evaluated. While coumarins are generally considered safe at low concentrations, certain coumarins, such as furanocoumarins, can cause phototoxic reactions and may pose health risks if consumed in large quantities or applied topically without proper precautions (Agrawal & Zhang, 2021). Therefore, the development of coumarin-based antimicrobial drugs will require thorough pharmacological studies to ensure their safety and efficacy. Future research should focus on identifying new coumarin derivatives with enhanced antimicrobial properties and reduced toxicity. Advances in genetic engineering and synthetic biology offer exciting opportunities to modify the biosynthetic pathways of coumarins in plants, allowing for the production of novel coumarin analogs with improved antimicrobial activity. Additionally, research into the mechanisms of resistance to coumarins in microbial pathogens could provide insights into how to overcome or prevent the development of resistance, ensuring the long-term effectiveness of coumarin-based antimicrobial strategies.

3. Inducible Defense Mechanisms Involving Coumarins

Coumarins are not only constitutively produced in plants but also form part of inducible defense mechanisms activated in response to biotic stress, particularly pathogen attacks. When plants are exposed to pathogens such as bacteria, fungi, or viruses, or when they face environmental challenges such as mechanical wounding or herbivory, they trigger a cascade of biochemical processes that lead to the increased synthesis and accumulation of defense compounds like coumarins. This inducible nature allows plants to adapt their defense responses based on the intensity and type of threat, optimizing their energy expenditure and improving survival (Dini & Grumetto, 2022). Inducible defense mechanisms involving coumarins are regulated by complex signaling networks, primarily mediated by plant hormones such as jasmonic acid (JA), salicylic acid (SA), and ethylene. These hormones play a crucial role in signaling the need for enhanced production of coumarins, leading to a rapid and localized increase in their concentration at the site of pathogen invasion. The inducible synthesis of coumarins, such as scopoletin, esculetin, and umbelliferone, is an essential part of the plant's defensive arsenal, allowing for both local and systemic protection (Kai et al., 2006).

3.1 Hormonal Regulation of Coumarin Production

The biosynthesis of coumarins as part of an inducible defense response is intricately regulated by phytohormones that orchestrate the plant's overall immune response. Among the most critical signaling molecules involved in this process are jasmonic acid, salicylic acid, and ethylene. These hormones interact in complex networks to determine the timing, intensity, and location of coumarin production, allowing plants to mount an appropriate defense against diverse pathogens.

3.1.1 Jasmonic Acid and Coumarin Biosynthesis

Jasmonic acid (JA) is one of the primary plant hormones involved in the regulation of defense responses, particularly against necrotrophic pathogens and insect herbivores. JA is synthesized in response to mechanical damage, herbivory, and pathogen attacks, and it plays a key role in the activation of defense-related genes, including those involved in coumarin biosynthesis (Farmer et al., 2003). Studies have demonstrated that jasmonic acid can upregulate the expression of phenylalanine ammonia-lyase (PAL), a key enzyme in the phenylpropanoid pathway, which is responsible for the production of coumarin precursors (Dixon & Paiva, 1995). Upon pathogen detection, JA signaling activates a series of transcription factors that induce the production of coumarins like scopoletin and esculetin. These coumarins are then transported to the site of infection, where they accumulate in large quantities, enhancing the plant's ability to suppress pathogen growth and spread. For example, scopoletin has been shown to accumulate rapidly in response to JA treatment in tobacco plants, leading to increased resistance against fungal pathogens such as *Fusarium* and *Alternaria* (Zhang et al., 2018).

3.1.2 Salicylic Acid and Systemic Acquired Resistance

Salicylic acid (SA) is another key hormone in plant defense, particularly effective against biotrophic pathogens that feed on living host cells. SA is closely associated with the activation of systemic acquired resistance (SAR), a defense mechanism that provides long-lasting protection against a broad spectrum of pathogens (Durrant & Dong, 2004). The role of coumarins in SA-mediated defense is well established, as they are often synthesized in response to SA signaling and contribute to the establishment of both local and systemic resistance.

SA triggers the expression of genes involved in the biosynthesis of coumarins, leading to their accumulation in infected tissues. In particular, scopoletin and esculetin have been found to increase significantly during SA-mediated defense responses, where they function as antimicrobial agents that inhibit pathogen growth and spread (Kai et al., 2006). Additionally, SA-mediated pathways can interact with JA pathways in a complex crosstalk mechanism that ensures a balanced and efficient response to pathogen attacks.

3.1.3 Ethylene and Stress-Induced Coumarin Production

Ethylene is a gaseous hormone involved in regulating a wide range of physiological processes in plants, including defense responses. Ethylene is often produced in response to wounding, pathogen attack, and other environmental stresses, and it plays a critical role in modulating the production of secondary metabolites such as coumarins (Thakur et al., 2015). Ethylene signaling is particularly important for the activation of defense genes in response to necrotrophic pathogens, which require rapid and localized defense responses to limit damage. Ethylene interacts with both JA and SA signaling pathways to fine-tune the plant's defense response, ensuring that coumarins are produced at optimal levels. In many cases, ethylene can enhance the production of coumarins by synergistically activating key enzymes in the phenylpropanoid pathway, such as PAL and cinnamate-4-hydroxylase (C4H). For instance, ethylene-induced scopoletin production has been observed in several plant species, contributing to enhanced resistance against both bacterial and fungal pathogens (Farmer et al., 2003).

3.2 Localization of Coumarins in Response to Pathogen Attack

One of the critical aspects of inducible defense mechanisms involving coumarins is their ability to localize precisely at the site of pathogen invasion. When a pathogen breaches the plant's outer defenses, the plant responds by transporting coumarins to the site of infection, where they accumulate in high concentrations. This localized accumulation is essential for limiting the spread of pathogens, as coumarins act as both physical and chemical barriers that prevent further colonization (Dini & Grumetto, 2022).

3.2.1 Role of Scopoletin in Localized Defense

Scopoletin is one of the most well-studied coumarins in terms of its role in localized defense responses. Upon pathogen detection, scopoletin is rapidly synthesized and transported to the infection site, where it accumulates in high concentrations. This accumulation occurs primarily in the cell walls and intercellular spaces, where scopoletin functions as an antimicrobial agent that inhibits the growth of invading pathogens (Cuffaro et al., 2024). In tobacco plants, for example, scopoletin has been shown to accumulate in response to infection by *Phytophthora nicotianae*, a fungal pathogen responsible for black shank disease (Dini & Grumetto, 2022). In addition to its antimicrobial properties, scopoletin also plays a role in reinforcing the structural integrity of plant cell walls. By cross-linking with cell wall components, scopoletin helps to strengthen the physical barrier against pathogen penetration, further enhancing the plant's resistance to infection. This dual action—both antimicrobial and structural—makes scopoletin an essential component of the plant's inducible defense arsenal.

3.2.2 Esculetin and Systemic Defense Responses

While scopoletin is primarily involved in localized defense, esculetin plays a broader role in systemic defense responses. Esculetin, a hydroxylated coumarin, is produced in response to pathogen attack and can be transported to distant tissues through the plant's vascular system. This systemic transport allows esculetin to provide protection not only at the site of infection but also in other parts of the plant that may be vulnerable to secondary infections (Dixon & Paiva, 1995).

Esculetin has been shown to inhibit the growth of various pathogens, including bacteria, fungi, and viruses, by disrupting their cellular functions and metabolic pathways. In Arabidopsis thaliana, for example, esculetin accumulates in response to infection by *Botrytis cinerea*, a fungal pathogen that causes gray mold disease (Kai et al., 2006). The systemic accumulation of esculetin enhances the plant's ability to resist infection, providing a long-lasting defense response that is essential for plant survival.

3.3 Coumarins as Phytoalexins

Coumarins can also function as phytoalexins, a class of antimicrobial compounds that are synthesized de novo in response to pathogen attack. Phytoalexins are produced only when plants are exposed to biotic stress, and they play a crucial role in slowing down pathogen progression and allowing the plant's immune system to mount a more robust defense. Coumarins such as scopoletin, umbelliferone, and esculetin are often classified as phytoalexins due to their inducible nature and potent antimicrobial properties (Harborne, 1999).

3.3.1 Rapid Synthesis of Phytoalexins

One of the hallmarks of phytoalexins is their rapid synthesis in response to pathogen attack. Unlike constitutively produced defense compounds, phytoalexins are synthesized only when needed, allowing plants to conserve energy and resources under non-stress conditions. The synthesis of coumarins as phytoalexins is triggered by pathogen-associated molecular patterns (PAMPs), which are recognized by pattern recognition receptors (PRRs) on the surface of plant cells. Upon recognition of these molecular patterns, plants activate a series of defense-related signaling pathways, leading to the production of coumarins (Cowan, 1999).

3.3.2 Antimicrobial Activity of Coumarin Phytoalexins

As phytoalexins, coumarins exhibit potent antimicrobial activity against a wide range of pathogens, including bacteria, fungi, and viruses. For example, scopoletin has been shown to inhibit the growth of *Fusarium oxysporum*, a fungal pathogen responsible for vascular wilt diseases in many crops (Dini & Grumetto, 2022). Similarly, umbelliferone has been reported to suppress the growth of *Phytophthora infestans*, the causative agent of late blight in potatoes (Kai et al., 2006). These antimicrobial properties make coumarins essential components of the plant's inducible defense system, allowing plants to rapidly respond to pathogen attacks and minimize damage.

3.4 Synergistic Effects of Coumarins with Other Inducible Defenses

Coumarins often work synergistically with other inducible defense mechanisms, enhancing the overall effectiveness of the plant's immune response. For example, coumarins can enhance the activity of reactive oxygen species (ROS), which are produced as part of the plant's oxidative burst during pathogen attack. ROS play a critical role in limiting pathogen spread by damaging microbial cells and signaling the activation of further defense responses. Coumarins, such as scopoletin and esculetin, have been shown to potentiate ROS production, amplifying their antimicrobial effects (Cuffaro et al., 2024). In addition to working with ROS, coumarins can also enhance the activity of defense-related proteins, such as pathogenesis-related (PR) proteins. These proteins are produced in response to pathogen infection and contribute to the plant's immune system by breaking down pathogen cell walls and inhibiting their growth. By acting in concert with PR proteins, coumarins provide a multi-layered defense strategy that significantly improves the plant's ability to resist infections.

4. Coumarins and Pathogen Inhibition: Mechanisms of Action

Coumarins play a significant role in plant defense by inhibiting the growth and spread of various pathogens, including bacteria, fungi, and viruses. These secondary metabolites exhibit a broad spectrum of antimicrobial activities due to their unique chemical structures, which allow them to interact with multiple cellular targets in pathogenic organisms. The mechanisms through which coumarins inhibit pathogens are diverse and include membrane disruption, interference with enzyme function, inhibition of nucleic acid synthesis, disruption of microbial communication, and enhancement of the plant's immune response. These multifaceted modes of action make coumarins potent and versatile defense compounds that can suppress pathogen growth through a combination of direct and indirect effects (Thakur et al., 2015).

4.1 Membrane Disruption and Integrity Loss

One of the most prominent mechanisms by which coumarins inhibit pathogens is through the disruption of cell membrane integrity. The lipid bilayer of microbial cell membranes is essential for maintaining cellular homeostasis and protecting the cell from external stressors. Coumarins, particularly hydroxylated derivatives such as esculetin and scopoletin, are known to interact with membrane lipids, destabilizing the membrane structure and leading to increased permeability. This disruption allows ions and small molecules to leak out of the cell, resulting in the loss of vital cellular contents and the eventual death of the pathogen (Basile et al., 2009).

Membrane disruption is especially critical in controlling fungal pathogens, which have robust cell walls but still rely on membrane integrity for survival. Studies have shown that scopoletin can compromise the cell membranes of fungal species such as *Aspergillus niger* and *Candida albicans*, leading to cytoplasmic leakage and inhibition of fungal growth (Cuffaro et al., 2024). This membrane-targeting action makes coumarins particularly effective against fungi, which are often difficult to control using conventional fungicides. Additionally, coumarins have been reported to inhibit the formation of biofilms, which are protective layers formed by bacterial communities. Biofilms provide bacteria with enhanced resistance to environmental stressors and antimicrobial agents. By disrupting bacterial membrane integrity and biofilm formation, coumarins such as umbelliferone and scopoletin reduce the ability of bacterial pathogens to establish persistent infections in plant tissues (Zhang et al., 2018).

4.2 Inhibition of Microbial Enzymes

Coumarins are also effective in inhibiting the activity of key microbial enzymes that are essential for pathogen survival, growth, and virulence. Enzymes play crucial roles in various metabolic pathways within pathogens, including those responsible for DNA replication, protein synthesis, and cell wall degradation. By targeting these enzymes, coumarins can disrupt the normal functioning of pathogens, reducing their ability to cause infection (Cowan, 1999).

One of the key enzymes targeted by coumarins is DNA gyrase, an essential enzyme involved in the supercoiling and replication of bacterial DNA. Inhibition of DNA gyrase prevents bacterial cells from properly replicating their DNA, leading to cell cycle arrest and eventual death. Umbelliferone, a common coumarin derivative, has been shown to inhibit DNA gyrase in several bacterial species, including *Escherichia coli* and *Staphylococcus aureus*, resulting in reduced bacterial proliferation and virulence (Basile et al., 2009). Additionally, coumarins can inhibit microbial proteases, which are enzymes that degrade host proteins during infection. Proteases allow pathogens to break down host tissues and evade immune defenses, making them critical for pathogenicity. By inhibiting these proteases, coumarins such as scopoletin and esculetin interfere with the ability of pathogens to invade and colonize plant tissues. This inhibition has been particularly well-documented in fungal pathogens, where coumarins block the activity of extracellular proteases, reducing the ability of fungi to degrade plant cell walls and spread throughout the host (Dini & Grumetto, 2022).

4.3 Inhibition of Nucleic Acid Synthesis

In addition to targeting enzymes, coumarins can directly interfere with nucleic acid synthesis in pathogens. This mechanism is particularly effective against viruses, which rely on the host's cellular machinery to replicate their genetic material. Coumarins such as scopoletin and esculetin have been shown to inhibit the replication of viral RNA and DNA by disrupting the activity of viral polymerases and other enzymes involved in nucleic acid synthesis (Kovács et al., 2008).

By inhibiting viral replication, coumarins prevent the spread of viral infections within plant tissues. For example, scopoletin has been reported to inhibit the replication of tobacco mosaic virus (TMV), a plant virus that causes significant damage to tobacco and other crops. By interfering with the synthesis of viral RNA, scopoletin reduces the production of viral particles, limiting the severity of the infection and enhancing the plant's ability to recover (Dini & Grumetto, 2022).

This mode of action is not limited to viruses. Coumarins can also disrupt the nucleic acid synthesis of bacteria and fungi, further inhibiting their ability to multiply and cause infections. In bacterial pathogens, coumarins have been shown to interfere with both DNA replication and transcription, preventing the production of essential proteins required for cell survival (Basile et al., 2009).

4.4 Disruption of Quorum Sensing and Microbial Communication

Quorum sensing is a microbial communication system that allows bacteria to coordinate their behavior in response to population density. This system regulates the expression of virulence factors, biofilm formation, and motility, all of which are essential for establishing infections and evading host defenses. Coumarins have been shown to interfere with quorum sensing pathways, disrupting microbial communication and reducing the pathogenicity of bacterial infections (Rasmussen & Givskov, 2006).

Scopoletin, in particular, has been demonstrated to inhibit quorum sensing in *Pseudomonas aeruginosa*, a bacterial pathogen that infects both plants and humans. By blocking quorum sensing, scopoletin prevents *P. aeruginosa* from producing key virulence factors such as elastase and pyocyanin, which are necessary for damaging host tissues and evading immune responses. This disruption of microbial communication reduces the ability of *P. aeruginosa* to form biofilms and establish persistent infections, making it more vulnerable to both plant immune responses and antimicrobial agents (Zhang et al., 2018).

The ability of coumarins to target quorum sensing offers a promising strategy for controlling bacterial infections without exerting selective pressure for resistance. Since coumarins disrupt communication rather than directly killing bacteria, they may reduce the likelihood of resistance development, making them valuable tools in sustainable agriculture and antimicrobial therapy (Rasmussen & Givskov, 2006).

4.5 Enhancement of Plant Immune Responses

In addition to their direct antimicrobial activities, coumarins also enhance the plant's immune responses by triggering the production of reactive oxygen species (ROS) and activating defense-related genes. Reactive oxygen species, such as hydrogen peroxide and superoxide radicals, play a critical role in the plant's oxidative burst, a defense mechanism that damages microbial cells and signals the activation of further immune responses (Cowan, 1999). Coumarins such as scopoletin and esculetin have been shown to enhance ROS production in response to pathogen attack, amplifying the oxidative burst and increasing the plant's ability to inhibit pathogen growth. By potentiating ROS generation, coumarins create an inhospitable environment for pathogens, causing oxidative damage to microbial cell membranes, proteins, and nucleic acids (Basile et al., 2009).

Furthermore, coumarins can activate the expression of pathogenesis-related (PR) proteins, which are produced in response to pathogen infection and contribute to the plant's immune defense. PR proteins, such as chitinases and glucanases, degrade the cell walls of invading pathogens, weakening their structure and making them more susceptible to other defense mechanisms. By enhancing the production of PR proteins, coumarins reinforce the plant's ability to resist infections and limit pathogen spread (Dini & Grumetto, 2022).

4.6 Synergistic Effects with Other Plant Defense Compounds

Coumarins often work synergistically with other plant secondary metabolites, enhancing their antimicrobial activity and providing a more comprehensive defense against pathogens. For example, coumarins can interact with phenolic acids and flavonoids, amplifying their ability to disrupt microbial membranes and inhibit pathogen growth. This synergistic action is particularly important in controlling fungal infections, where coumarins and flavonoids work together to inhibit hyphal growth and spore germination (Harborne, 1999). Additionally, coumarins can enhance the activity of defense-related enzymes, such as phenylalanine ammonia-lyase (PAL) and peroxidases, which are involved in the synthesis of other antimicrobial compounds. By boosting the activity of these enzymes, coumarins increase the overall production of defense compounds, further strengthening the plant's immune response (Thakur et al., 2015).

5. Synergistic Effects of Coumarins with Other Plant Defense Compounds

Coumarins, as vital secondary metabolites in plants, not only function independently as antimicrobial agents but also demonstrate significant synergistic interactions with other plant defense compounds. These synergistic effects are crucial for optimizing the plant's immune response and enhancing its resistance to a wide range of pathogens, including bacteria, fungi, and viruses. By working in conjunction with other secondary metabolites, proteins, and defense-related enzymes, coumarins contribute to a multifaceted defense strategy that significantly improves plant resilience against pathogens. This synergism amplifies the antimicrobial efficacy of plant defense compounds, reduces energy expenditure for the plant, and enhances the overall efficiency of the plant's immune system (Harborne, 1999).

5.1 Coumarins and Phenolic Acids

Phenolic acids, such as ferulic acid, caffeic acid, and chlorogenic acid, are well-known defense compounds in plants, offering antimicrobial, antioxidant, and allelopathic properties. When combined with coumarins, these phenolic acids can produce a more potent defense response, particularly in the suppression of pathogen growth. Both phenolic acids and coumarins are derived from the phenylpropanoid pathway, which allows for their co-accumulation in plant tissues during pathogen attacks. This shared biosynthetic origin also facilitates their ability to work together in a complementary manner, enhancing each other's effectiveness (Harborne, 1999).

5.1.1 Amplification of Antimicrobial Activity

Coumarins, when combined with phenolic acids, can synergistically amplify their antimicrobial activity. Phenolic acids disrupt microbial cell walls, weaken pathogen defenses, and increase membrane permeability. Coumarins complement this action by further destabilizing the microbial membrane and inhibiting essential enzymatic processes within the pathogen (Cowan, 1999). For example, scopoletin and ferulic acid together have been shown to exhibit stronger antifungal activity against *Botrytis cinerea* and *Fusarium oxysporum* than when applied separately. This combined effect occurs because phenolic acids create initial damage to the fungal cell wall, allowing coumarins to penetrate more effectively and inhibit intracellular processes such as enzyme activity and nucleic acid synthesis (Dini & Grumetto, 2022).

5.1.2 ROS Generation and Oxidative Stress Enhancement

Coumarins are known to enhance the production of reactive oxygen species (ROS) in response to pathogen attack, which contributes to the oxidative burst, a critical part of the plant's defense. ROS are highly reactive molecules that cause oxidative damage to microbial cell structures, leading to cell death. Phenolic acids also contribute to ROS production, and their combination with coumarins can lead to a stronger oxidative burst, overwhelming the pathogen's ability to neutralize these reactive molecules (Basile et al., 2009). Studies have shown that when scopoletin and caffeic acid are combined, ROS levels increase significantly compared to when either compound is applied alone, leading to enhanced control of fungal and bacterial infections (Cuffaro et al., 2024).

5.2 Synergy Between Coumarins and Flavonoids

Flavonoids are another class of phenolic compounds that play an essential role in plant defense. They have broad-spectrum antimicrobial properties and serve as antioxidants, UV protectants, and signaling molecules in plant-pathogen interactions. Coumarins and flavonoids often act in concert, enhancing each other's effectiveness in suppressing pathogen growth and reinforcing plant defenses.

5.2.1 Complementary Antimicrobial Action

Coumarins and flavonoids have complementary modes of action that, when combined, result in more effective pathogen inhibition. While coumarins primarily disrupt microbial cell membranes and interfere with enzyme activity, flavonoids inhibit microbial nucleic acid synthesis and scavenger free radicals. This dual-action approach leads to a more comprehensive attack on the pathogen's vital systems, making it more difficult for the pathogen to adapt or develop resistance. For instance, the combination of quercetin (a flavonoid) and umbelliferone has been shown to inhibit the growth of *E. coli* and *S. aureus* more effectively than either compound alone, highlighting the benefits of their synergistic interaction (Harborne, 1999).

5.2.2 Enhancement of Plant Immune Signaling

Flavonoids are also involved in the regulation of plant immune responses by modulating signaling pathways such as those mediated by jasmonic acid (JA), salicylic acid (SA), and ethylene. Coumarins can enhance the production and activity of flavonoids, creating a feedback loop that amplifies plant immune signaling. This synergistic relationship strengthens the plant's systemic acquired resistance (SAR) and primes the plant for a quicker and more robust response to subsequent pathogen attacks (Dini & Grumetto, 2022). For example, the combination of scopoletin and kaempferol (a flavonoid) has been found to increase the expression of pathogenesis-related (PR) proteins, which play a key role in the plant's immune defense against pathogens (Farmer et al., 2003).

5.3 Coumarins and Alkaloids

Alkaloids are nitrogen-containing compounds that exhibit potent antimicrobial activity and serve as toxic deterrents against herbivores and pathogens. When combined with coumarins, alkaloids can contribute to a more effective and multi-pronged defense strategy, particularly against bacterial and fungal pathogens.

5.3.1 Enhanced Antibacterial Activity

Alkaloids such as berberine and nicotine are known for their strong antibacterial properties, particularly against Gram-positive bacteria. When coupled with coumarins like scopoletin, these alkaloids exhibit enhanced antibacterial activity by targeting multiple aspects of bacterial physiology. Alkaloids disrupt bacterial cell wall synthesis and interfere with DNA replication, while coumarins inhibit bacterial enzymes and increase membrane permeability. This synergistic interaction not only kills bacteria more efficiently but also reduces the likelihood of resistance development, as the bacteria are attacked on multiple fronts (Kovács et al., 2008).

5.3.2 Inhibition of Fungal Growth

Alkaloids are also effective antifungal agents, and their combination with coumarins results in enhanced inhibition of fungal growth. For example, the combination of scopoletin and berberine has been shown to inhibit the growth of *Candida albicans* more effectively than either compound alone. This synergistic effect is due to the complementary mechanisms of action: alkaloids inhibit fungal DNA synthesis, while coumarins disrupt fungal cell membranes and inhibit enzyme activity. Together, they provide a more comprehensive antifungal defense (Cuffaro et al., 2024).

5.4 Synergistic Interaction with Defense-Related Proteins

Coumarins also interact with defense-related proteins in plants, such as pathogenesis-related (PR) proteins, chitinases, and glucanases. These proteins play a crucial role in breaking down the cell walls of invading pathogens, weakening their structure and making them more vulnerable to antimicrobial compounds. Coumarins enhance the production and activity of these proteins, further improving the plant's ability to resist pathogen attacks (Farmer et al., 2003).

5.4.1 Activation of PR Proteins

PR proteins are a group of proteins that are induced in response to pathogen infection and are involved in degrading microbial cell walls. Coumarins can stimulate the production of PR proteins, enhancing their ability to break down the structural components of pathogens. For instance, the combination of scopoletin and PR proteins such as chitinase has been shown to degrade the cell walls of fungal pathogens like *Fusarium* and *Botrytis*, leading to a significant reduction in fungal colonization and spread (Dixon & Paiva, 1995). This synergistic effect allows the plant to mount a more effective defense response by combining chemical and enzymatic strategies to eliminate the pathogen.

5.4.2 Enhancement of Enzyme Activity

Coumarins also interact with enzymes involved in the synthesis of other antimicrobial compounds, such as phenylalanine ammonia-lyase (PAL) and peroxidases. These enzymes are essential for the production of phenolic compounds, lignins, and other secondary metabolites that reinforce the plant's structural defenses. By enhancing the activity of these enzymes, coumarins contribute to the production of a broader range of antimicrobial compounds, providing the plant with a more comprehensive defense system (Dini & Grumetto, 2022).

5.5 Role of Synergistic Interactions in Sustainable Agriculture

The synergistic effects of coumarins with other plant defense compounds have important implications for sustainable agriculture. By leveraging these synergistic interactions, farmers can develop crop varieties that are more resistant to pathogens without relying on synthetic pesticides. The use of natural defense compounds like coumarins, in combination with phenolic acids, flavonoids, alkaloids, and defense-related proteins, can reduce the need for chemical inputs and promote more environmentally friendly agricultural practices (Thakur et al., 2015). Furthermore, understanding these synergistic interactions can help in the development of biopesticides that utilize natural plant compounds to enhance crop protection. By combining coumarins with other plant-derived antimicrobial agents, it is possible to create formulations that are highly effective against a broad spectrum of pathogens while minimizing the risk of resistance development and environmental harm.

6. Genetic Engineering of Coumarin Pathways for Enhanced Pathogen Resistance

Coumarins, as naturally occurring secondary metabolites, have long been recognized for their significant role in plant defense against a wide range of pathogens, including bacteria, fungi, and viruses. Given their potent antimicrobial properties, there has been increasing interest in manipulating coumarin biosynthetic pathways through genetic engineering to enhance plant resistance to diseases. Genetic engineering of coumarin biosynthesis involves modifying specific genes or regulatory elements to increase coumarin production, improve the effectiveness of these compounds, or introduce coumarin pathways into crops that may not naturally produce high levels of these compounds. This approach is promising in the context of sustainable agriculture, where the development of pathogen-resistant crops is essential for reducing the dependence on chemical pesticides and enhancing crop yields under biotic stress (Kai et al., 2006).

6.1 Key Enzymes in the Coumarin Biosynthesis Pathway

The genetic engineering of coumarin biosynthesis centers around several key enzymes involved in the phenylpropanoid pathway, which is responsible for the production of coumarins and other phenolic compounds. Understanding the roles of these enzymes is crucial for identifying the genetic targets for modification.

6.1.1 Phenylalanine Ammonia-Lyase (PAL)

Phenylalanine ammonia-lyase (PAL) is the first enzyme in the phenylpropanoid pathway, catalyzing the deamination of phenylalanine to produce cinnamic acid. This step is critical for the synthesis of coumarins, as cinnamic acid serves as the precursor for many coumarin derivatives (Dixon & Paiva, 1995). Genetic manipulation of PAL expression has been shown to increase the flux through the phenylpropanoid pathway, leading to enhanced production of coumarins and other defense-related compounds. Overexpression of PAL in various plant species has resulted in increased accumulation of coumarins such as scopoletin and esculetin, which are known to provide enhanced resistance to fungal and bacterial pathogens (Bourgaud et al., 2006).

6.1.2 Cinnamate-4-Hydroxylase (C4H)

Cinnamate-4-hydroxylase (C4H) is another critical enzyme in coumarin biosynthesis, responsible for the hydroxylation of cinnamic acid to p-coumaric acid. This step is essential for the formation of key coumarin precursors such as umbelliferone, which can be further modified to produce a wide array of coumarins (Kai et al., 2006). Genetic engineering of C4H has been explored as a means of increasing the production of hydroxylated coumarins, which are particularly effective against pathogens due to their enhanced solubility and bioavailability. Overexpression of C4H in model plants like *Arabidopsis thaliana* has led to increased resistance against fungal pathogens such as *Botrytis cinerea* and *Fusarium oxysporum* (Dini & Grumetto, 2022).

6.1.3 4-Coumarate-CoA Ligase (4CL)

4-Coumarate-CoA ligase (4CL) catalyzes the conversion of p-coumaric acid into its CoA ester, which is a crucial intermediate in the biosynthesis of both lignins and coumarins. Modifying the activity of 4CL can influence the balance between lignin production and the synthesis of soluble phenolic compounds like coumarins. Genetic manipulation of 4CL expression can be used to channel more metabolic intermediates toward coumarin biosynthesis, thereby enhancing the plant's ability to produce antimicrobial coumarins in response to pathogen attacks (Kai et al., 2006). This approach has been applied in crop species such as rice and wheat, where increased coumarin production has led to improved resistance to fungal diseases like rice blast and wheat rust (Thakur et al., 2015).

6.2 Manipulation of Regulatory Genes for Coumarin Pathways

In addition to modifying the expression of biosynthetic enzymes, genetic engineering can also target regulatory genes that control the overall flux through the coumarin biosynthesis pathway. These regulatory genes include transcription factors and other regulatory elements that modulate the expression of multiple genes involved in coumarin production.

6.2.1 Transcription Factors

Transcription factors are proteins that bind to specific DNA sequences and regulate the transcription of genes. Several transcription factors have been identified that regulate the expression of key enzymes in the phenylpropanoid pathway, including PAL, C4H, and 4CL. Genetic engineering efforts have focused on manipulating these transcription factors to enhance coumarin production in response to pathogen attacks. For example, MYB transcription factors have been shown to regulate phenylpropanoid metabolism and coumarin biosynthesis in plants. Overexpression of MYB transcription factors in *Arabidopsis* has led to increased scopoletin and umbelliferone production, resulting in enhanced resistance to bacterial and fungal pathogens (Bourgaud et al., 2006). Additionally, WRKY transcription factors, which are involved in regulating plant defense responses, have been linked to coumarin biosynthesis. Engineering plants to overexpress WRKY transcription factors can upregulate the expression of coumarin biosynthetic genes, leading to increased production of antimicrobial coumarins and improved pathogen resistance (Thakur,et al., 2015).

6.2.2 Signal Transduction Pathways

Coumarin biosynthesis is also regulated by various signal transduction pathways that are activated in response to environmental stimuli, such as pathogen attacks or abiotic stress. These signaling pathways, which involve plant hormones such as jasmonic acid (JA), salicylic acid (SA), and ethylene, play a crucial role in modulating the expression of coumarin biosynthetic genes. Genetic engineering can be used to manipulate these signaling pathways to enhance coumarin production. For example, overexpression of genes involved in the JA signaling pathway has been shown to increase the production of scopoletin and other coumarins in response to wounding and pathogen infection. This enhanced coumarin production correlates with increased resistance to necrotrophic pathogens, which rely on killing host cells to feed on the dead tissue (Farmer et al., 2003). Similarly, manipulating the SA signaling pathway can enhance systemic acquired resistance (SAR) and lead to increased production of coumarins such as esculetin, providing broad-spectrum resistance against biotrophic pathogens (Durrant & Dong, 2004).

6.3 Genetic Engineering for Disease-Resistant Crop Varieties

The ultimate goal of genetic engineering of coumarin pathways is to develop crop varieties with enhanced resistance to diseases. By increasing coumarin production, plants can effectively suppress pathogen growth and limit the damage caused by infections. Several strategies have been explored to achieve this goal, including the overexpression of biosynthetic genes, the manipulation of regulatory elements, and the introduction of coumarin biosynthetic pathways into crop species that naturally produce low levels of these compounds.

6.3.1 Overexpression of Coumarin Biosynthetic Genes

One of the most direct approaches to enhancing coumarin production in crops is the overexpression of key coumarin biosynthetic genes. For example, overexpressing PAL, C4H, or 4CL in crops such as rice, wheat, and maize has been shown to increase the accumulation of coumarins like scopoletin and umbelliferone, resulting in improved resistance to a wide range of pathogens, including *Magnaporthe oryzae* (rice blast) and *Puccinia graminis* (wheat stem rust) (Kai et al., 2006). This strategy has been particularly effective in crop species where coumarin production is naturally low, as it provides an additional layer of defense that was previously underutilized. By boosting coumarin levels, genetically engineered crops are better equipped to fend off infections and reduce yield losses caused by disease outbreaks.

6.3.2 Introduction of Coumarin Pathways into Non-Coumarin-Producing Crops

In some cases, genetic engineering can be used to introduce coumarin biosynthetic pathways into crop species that do not naturally produce high levels of these compounds. This approach involves the transfer of genes encoding key coumarin biosynthetic enzymes from plants that produce high levels of coumarins (such as *Petroselinum crispum*, parsley) into crop species that lack robust coumarin pathways (Thakur, et al., 2015). By introducing these pathways, researchers can create transgenic plants that produce coumarins as part of their defense strategy. For example, researchers have successfully introduced the coumarin biosynthesis pathway into *Solanum lycopersicum* (tomato) plants, which naturally produce low levels of coumarins. The transgenic tomato plants exhibited increased production of scopoletin and umbelliferone, leading to improved resistance to fungal pathogens such as *Phytophthora infestans*, the causative agent of late blight (Dini & Grumetto, 2022). This approach opens up new possibilities for enhancing disease resistance in a wide range of crops that do not typically rely on coumarins for defense.

6.4 Challenges and Future Directions

While the genetic engineering of coumarin pathways holds great promise for enhancing pathogen resistance in crops, several challenges remain. One of the primary challenges is the potential for unintended effects on plant growth and development. Coumarins, while beneficial for pathogen resistance, can also affect plant physiology in ways that may impact growth, reproduction, and nutrient uptake. Therefore, it is essential to carefully balance the production of coumarins to avoid negative trade-offs (Bourgaud et al., 2006). Additionally, the ecological impacts of genetically engineered crops that produce high levels of coumarins need to be considered. Coumarins can have allelopathic effects on neighboring plants and may influence soil microbial communities. Therefore, field trials and ecological assessments are necessary to evaluate the long-term sustainability of these genetically engineered crops. Future research should focus on fine-tuning the regulation of coumarin biosynthesis to ensure that plants produce these compounds in response to pathogen attacks without compromising other physiological processes. Advances in synthetic biology and gene editing technologies, such as CRISPR-Cas9, offer new opportunities for precisely manipulating coumarin pathways and achieving optimal levels of pathogen resistance in crops (Thakur et al., 2015).

Conclusion

Coumarins, a diverse group of secondary metabolites, play an integral role in plant defense systems, particularly through their antimicrobial properties and involvement in inducible defense mechanisms. Their chemical structure and biosynthesis contribute to their versatility and biological activity. Coumarins are synthesized via the phenylpropanoid pathway, allowing plants to produce a range of derivatives that can act as potent inhibitors against various pathogens. This pathway is highly regulated and can be modified under stress conditions, showcasing the adaptability of coumarins as part of a plant's defense arsenal. The antimicrobial properties of coumarins are well-documented, as they have demonstrated activity against bacteria, fungi, and viruses. Their mechanisms of action include disrupting microbial cell membranes, inhibiting enzyme activity, and generating oxidative stress within the pathogens. Coumarins also show selective toxicity, targeting pathogenic organisms without significantly harming the plant itself. Furthermore, coumarins can act synergistically with other defense-related metabolites, enhancing their efficacy and broadening their antimicrobial spectrum. In terms of inducible defense mechanisms, coumarins are produced in response to biotic stress, particularly during pathogen attack. Upon recognition of pathogen-associated molecular patterns (PAMPs), plants upregulate the biosynthesis of coumarins, which are then deployed at the infection site to limit pathogen growth. This localized defense response is often accompanied by the generation of reactive oxygen species (ROS), which further enhances the antimicrobial potential of coumarins. Coumarins inhibit pathogens through multiple mechanisms. They can chelate essential metal ions, disrupt cellular homeostasis, and inhibit critical enzymatic processes, such as DNA replication and protein synthesis. Their ability to generate oxidative stress and interfere with pathogen signaling pathways further underscores their role in pathogen inhibition. Some pathogens have developed evasion strategies, but coumarins remain effective by constantly modulating their biosynthesis and chemical structure. The synergistic effects of coumarins with other plant defense compounds, such as phenolics, flavonoids, and terpenoids, contribute to a multifaceted defense strategy. When combined with these compounds, coumarins enhance the overall defensive capacity of plants, creating a hostile environment for invading pathogens. This synergy can also lead to a reduction in the amount of each compound required for effective defense, making the response more energy-efficient for the plant. Genetic engineering offers significant promise in enhancing coumarin-based defense mechanisms. By overexpressing key enzymes in the coumarin biosynthesis pathway or manipulating transcription factors that regulate their production, researchers can develop plants with increased resistance to pathogens. Advances in CRISPR-Cas9 technology and other gene-editing tools make it possible to fine-tune the coumarin pathways for both increased pathogen resistance and minimal impact on plant growth and productivity.

References

Agrawal, A. A. & Zhang, X. (2021). The evolution of coevolution in the study of species interactions. *Evolution*, 75, 1594-1606.

Akwu, N. A., Lekhooa, M., Deqiang, D., & Aremu, A. O. (2023). Antidepressant effects of coumarins and their derivatives: A critical analysis of research advances. *European Journal of Pharmacology, 956*, 175958

Alt, S., Mitchenall, L. A., Maxwell, A., & Heide, L. (2011) Inhibition of DNA gyrase and DNA topoisomerase IV of Staphylococcus aureus and Escherichia coli by aminocoumarin antibiotics. *Journal of Antimicrobial Chemotherapy*, *66*(9), 2061-2069.

Basile, A., Sorbo, S., Spadaro, V., Bruno, M., Maggio, A., Faraone, N., & Rosselli, S. (2009). Antibacterial and antifungal activity of coumarins from the roots of *Ferulago campestris* (Apiaceae). *Molecules, 14*(3), 939-952.

Bourgaud, F., Hehn, A., Larbat, R., Doerper, S., Gontier, E., Kellner, S., & Matern, U. (2006). Biosynthesis of coumarins in plants: A major pathway still to be unravelled for cytochrome P450 enzymes. *Phytochemistry Reviews*, *5*(2-3), 293-308.

Cowan, M. M. (1999). Plant products as antimicrobial agents. *Clinical Microbiology Reviews, 12*(4), 564-582.

Cuffaro, D., Bertolini, A., Silva, A. M., Rodrigues, F., Gabbia, D., De Martin, S., Saba, A., Bertini, S., Digiacomo, M., & Macchia, M. (2024). Comparative analysis on polyphenolic composition of different olive mill wastewater and related extra virgin olive oil extracts and evaluation of nutraceutical properties by cell-based studies. *Foods, 13*(20), 3312.

Degroote, E., Schoorens, C., Pockelé, S., Stojilković, B., Demeestere, K., Mangelinckx, S., & Kyndt, T. (2024). A combination of plant-based compounds and extracts acts nematicidal and induces resistance against Meloidogyne incognita in tomato. *Frontiers in Plant Sciences*, *15*, 1411825.

Dini, I., & Grumetto, L. (2022). Recent advances in natural polyphenol research. *Molecules*, *27*(24), 8777.

Dixon, R. A., & Paiva, N. L. (1995). Stress-induced phenylpropanoid metabolism. *The Plant Cell*, *7*(7), 1085-1097.

Durrant, W. E., & Dong, X. (2004). Systemic acquired resistance. *Annual Review of Phytopathology*, *42*, 185-209.

Farmer, E. E., Almeras, E., & Krishnamurthy, V. (2003). Jasmonates and related oxylipins in plant responses to pathogenesis and herbivory. *Current Opinion in Plant Biology*, *6*(4), 372-378.

Fourcroy, P., Sisó-Terraza, P., Sudre, D., Savirón, M., Reyt, G., Gaymard, F., & Briat, J. F. (2014). Involvement of the ABCG37 transporter in secretion of scopoletin and derivatives by Arabidopsis roots in response to iron deficiency. *New Phytologist*, *201*(1), 155-167.

Harborne, J. B. (1999). The comparative biochemistry of phytoalexin induction in plants. *Biochemical Systematics and Ecology*, *27*(4), 335-367.

Kai, K., Shimizu, B., Mizutani, M., Watanabe, K., & Sakata, K. (2006). Accumulation of coumarins in Arabidopsis thaliana. *Phytochemistry*, *67*(4), 379-386.

Korkina, L. (2007). Phenylpropanoids as naturally occurring antioxidants: from plant defense to human health. *Cellular and Molecular Biology*, *53*(1), 15-25.

Kovács, A., Vasas, A., & Forgo, P. (2008). Natural phenanthrenes and their biological activity. *Phytochemistry*, *69*(5), 1084-1110.

Lee, D. H., Choi, H. W., & Hwang. B. K. (2011) The pepper E3 ubiquitin ligase RING1 gene, CaRING1, is required for cell death and the salicylic acid-dependent defense response. *Plant Physiology*, *156*(4), 2011-25.

Rasmussen, T. B., & Givskov, M. (2006). Quorum sensing inhibitors: A bargain of effects. *Microbiology*, *152*(4), 896-904.

Shan, Z., Zhou, S., Shah, A., Arafat, Y., Arif Hussain Rizvi, S., & Shao, H. (2023). Plant allelopathy in response to biotic and abiotic factors. *Agronomy*, 13(9), 2358.

Stringlis, I. A., de Jonge, R., & Pieterse, C. M. J. (2019). The Age of coumarins in plant-microbe interactions. *Plant Cell Physiology*, *60*(7), 1405-1419.

Thakur, A., Singla, R., & Jaitak, V. (2015). Coumarins as anticancer agents: A review on synthetic strategies, mechanism of action and SAR studies. *European Journal of Medicinal Chemistry*, *101*, 476-495.

Wuyts, N., Swennen, R., & De Waele, D. (2006). Effects of plant phenylpropanoid pathway products and selected terpenoids and alkaloids on the behaviour of the plant-parasitic nematodes *Radopholus similis*, *Pratylenchus penetrans* and *Meloidogyne incognita*. *Nematology*, *8*, 89-101.

Zhang, Y., Sass, A., van Acker, H., Wille, J., Verhasselt, B., van Nieuwerburgh, F., Kaever, V., Crabbé, A., & Coenye, T. (2018). Coumarin reduces virulence and biofilm formation in pseudomonas aeruginosa by affecting quorum sensing, type III secretion and C-di-GMP levels. Frontiers in Microbiology, *9*, 1952.

Chapter - 12

Defense Signaling Pathways Involving Secondary Metabolites

Susheel Kumar[1], Zuby Gohar Ansari[2], Reshma Jadhav[3], Praneet Chauhan[4] and Sheetanshu Gupta[5]

[1]Plant Molecular Virology Laboratory, CPMB Division, CSIR-National Botanical Research Institute, Rana Pratap Marg, Lucknow, Uttar Pradesh, India

[2]B.M. College of Agriculture, Jaswadi Road Khandwa, Madhya Pradesh, India

[3]Department of Biochemistry, Mahatma Phule Krishi Vidyapeeth, Rahuri, Ahmednagar, Maharashtra, India

[4]Department of Plant Pathology, Dr. Khem Singh Gill Akal College of Agriculture, Eternal University, Baru Sahib, Sirmour, Himachal Pradesh, India

[5]Department of Biochemistry, G.B. Pant University of Agriculture & Technology, Pantnagar, Uttarakhand, India

Abstract

Plant secondary metabolites, including flavonoids, terpenoids, alkaloids, and phenolics, play a crucial role in defense mechanisms by acting as antimicrobial agents, deterrents against herbivores, and signaling molecules. These compounds are tightly regulated by various defense signaling pathways that respond to biotic and abiotic stress. Among the most critical defense-related signaling pathways in plants are those mediated by jasmonic acid (JA), salicylic acid (SA), and ethylene (ET), which are activated upon pathogen recognition and orchestrate the production of secondary metabolites. These pathways enable plants to mount rapid, localized responses to pathogen invasion, as well as systemic responses that fortify tissues far from the infection site. Secondary metabolites are often induced upon the detection of pathogen-associated molecular patterns (PAMPs) through pattern recognition receptors (PRRs) on plant cells. This recognition triggers complex signal transduction pathways that activate transcription factors and defense-related genes, leading to the production and accumulation of secondary metabolites at the site of infection. Furthermore, secondary metabolites are not only defensive compounds but also play an integral role in signaling, coordinating various aspects of the plant's immune response. Jasmonic acid and its derivatives (jasmonates) regulate the production of terpenoids and alkaloids in response to herbivory and necrotrophic pathogens, while salicylic acid is essential for systemic acquired resistance (SAR) against biotrophic pathogens. Ethylene

DOI: 10.1201/9781003717454-12

works synergistically with both JA and SA to fine-tune the plant's defense strategy. Other signaling molecules, such as abscisic acid (ABA), auxins, and reactive oxygen species (ROS), also play supporting roles in modulating defense responses. In recent years, advances in genetic engineering have allowed scientists to manipulate these signaling pathways to enhance the production of secondary metabolites in crops, providing an avenue for developing more robust, disease-resistant varieties.

Keywords: Defense Signaling, Secondary Metabolites, Jasmonic Acid, Salicylic Acid, Ethylene, Systemic Acquired Resistance.

Introduction

Plants have evolved a sophisticated array of defense mechanisms to combat both biotic and abiotic stresses, relying heavily on the production and regulation of secondary metabolites. These metabolites, which include alkaloids, flavonoids, terpenoids, phenolics, and coumarins, play a central role in defending against pathogens, herbivores, and environmental challenges. Unlike primary metabolites, which are essential for growth and development, secondary metabolites serve specialized functions that are largely defensive. They not only act as direct inhibitors of pathogens but also serve as signaling molecules, initiating and amplifying the plant's immune responses. The orchestration of secondary metabolite production is controlled by defense signaling pathways mediated by key plant hormones such as jasmonic acid (JA), salicylic acid (SA), and ethylene (ET). These hormones are crucial in tailoring the plant's defense responses to specific threats, ensuring that resources are allocated efficiently. Jasmonic acid, for instance, is primarily associated with the plant's response to necrotrophic pathogens and herbivory, inducing the synthesis of metabolites such as glucosinolates, alkaloids, and terpenoids that directly deter herbivores and pathogens (Wasternack & Strnad, 2018). Conversely, salicylic acid is the key regulator of systemic acquired resistance (SAR), a long-lasting, broad-spectrum defense response activated by biotrophic pathogens. SA promotes the production of pathogenesis-related (PR) proteins and antimicrobial compounds such as phenolics and flavonoids, which strengthen the plant's cell walls and inhibit pathogen growth (Vlot et al., 2009; Klessig et al., 2018).

Ethylene plays a synergistic role in defense, particularly in coordination with jasmonic acid, to amplify responses against necrotrophic pathogens and herbivores. Ethylene enhances the production of defensive compounds like lignins and terpenoids, reinforcing the structural integrity of cell walls and generating a hostile environment for invading pathogens (Broekaert et al., 2006). The role of ethylene is not limited to pathogen defense; it also modulates responses to abiotic stresses, including mechanical damage and drought, ensuring a more integrated defense strategy (Khan et al., 2015).

A key feature of plant defense is the crosstalk between these hormonal pathways, which allows for the fine-tuning of immune responses. The antagonistic relationship between salicylic acid (SA) and jasmonic acid (JA) is a well-studied example of this crosstalk. Biotrophic pathogens, which depend on living tissue, often induce SA-dependent responses, which suppress JA signaling. In contrast, necrotrophic pathogens and herbivores trigger JA and ethylene (ET) responses, which inhibit SA pathways, allowing the plant to prioritize its defense resources depending on the nature of the threat (Pieterse et al., 2012). This dynamic balance ensures that plants can mount effective defenses without wasting resources.

In addition to serving as direct defense agents, secondary metabolites themselves influence these hormonal pathways. For example, flavonoids, which are well-known for their antimicrobial properties, also function as modulators of plant hormone signaling. Phenolics and terpenoids are similarly involved, contributing to the activation of signaling pathways and enhancing the plant's overall immune response (Bari & Jones, 2009). This intricate interplay between secondary metabolites and hormonal signaling amplifies the defense mechanisms, ensuring rapid and efficient responses to biotic stressors.

Advances in genetic engineering have made it possible to manipulate these signaling pathways to enhance the production of defense-related secondary metabolites. For instance, overexpression of transcription factors such as WRKY or MYB, which are key regulators of salicylic acid and jasmonic acid signaling, has been shown to significantly boost the production of secondary metabolites such as flavonoids and terpenoids, thereby improving pathogen resistance in genetically modified crops (Li et al., 2019). Similarly, CRISPR-Cas9 gene-editing technology has enabled targeted manipulation of key enzymes in the jasmonic acid and salicylic acid pathways, resulting in enhanced resistance to necrotrophic pathogens through increased production of defense-related compounds (Howe, 2004). This ability to fine-tune the plant's immune system through genetic modification offers promising avenues for developing more resilient and disease-resistant crops.

The integration of defense signaling pathways with secondary metabolite production illustrates the complexity and efficiency of plant immune systems. By tightly regulating the production of specialized metabolites and leveraging hormonal cross-talk, plants can deploy tailored defenses against a variety of threats. The potential for manipulating these pathways to enhance crop resistance is immense, providing a sustainable alternative to chemical pesticides and contributing to food security in the face of growing environmental challenges (Pieterse et al., 2012). As research continues to unravel the complexities of these pathways, the role of secondary metabolites in plant defense will undoubtedly remain a focal point in efforts to improve agricultural resilience.

1. Overview of Defense Signaling Pathways in Plants

Plants, being sessile organisms, have developed intricate defense mechanisms to survive and thrive in an environment where they are constantly exposed to pathogens, herbivores, and abiotic stressors. One of the key aspects of their defense is the signaling pathways that regulate their immune responses. These pathways orchestrate the activation of defensive measures, including the production of secondary metabolites, which serve as antimicrobial agents, deterrents to herbivores, and modulators of stress responses. The primary defense signaling pathways in plants are mediated by hormonal signals such as salicylic acid (SA), jasmonic acid (JA), and ethylene (ET). These hormones not only regulate defense gene expression but also modulate the synthesis and accumulation of secondary metabolites that enhance the plant's ability to resist a wide range of biotic challenges (Bari & Jones, 2009).

1.1 Pattern-Triggered Immunity (PTI) and Effector-Triggered Immunity (ETI)

The plant immune system is structured into two major layers: Pattern-Triggered Immunity (PTI) and Effector-Triggered Immunity (ETI). PTI is the plant's first line of defense and is activated when pattern recognition receptors (PRRs) on the plant cell surface detect conserved pathogen-associated molecular patterns (PAMPs). These PAMPs are molecular signatures common to many classes of microbes, such as bacterial flagellin or fungal chitin. Once PRRs recognize these PAMPs, a signaling cascade is initiated, leading to the activation of defense responses such as the production of reactive oxygen species (ROS), cell wall fortification, and the expression of defense-related genes (Tsuda & Katagiri, 2010). In contrast, Effector-Triggered Immunity (ETI) is a more specialized immune response, activated when intracellular immune receptors known as Nucleotide-binding Leucine-Rich Repeat (NLR) proteins recognize specific pathogen effectors. Effectors are molecules secreted by pathogens that attempt to suppress PTI and promote infection. ETI is often stronger and more robust than PTI, leading to a hypersensitive response (HR), which includes localized cell death at the infection site to prevent the pathogen from spreading (Jones & Dangl, 2006). Both PTI and ETI are critical in regulating the downstream production of secondary metabolites and other defensive compounds that directly inhibit pathogen growth or strengthen the plant's physical defenses.

1.2 Salicylic Acid, Jasmonic Acid, and Ethylene in Defense Signaling

The salicylic acid (SA), jasmonic acid (JA), and ethylene (ET) signaling pathways are central to the regulation of plant defense responses. Each of these hormones plays a unique role in modulating the plant's immune system, often acting in response to specific types of pathogens or environmental stresses. Salicylic acid is primarily associated with defense against biotrophic pathogens, which feed on living plant tissue. SA promotes Systemic Acquired Resistance (SAR), a long-lasting, broad-spectrum immune response that protects uninfected parts of the plant from future attacks. SAR is characterized by the accumulation of pathogenesis-related (PR) proteins and the synthesis of antimicrobial secondary metabolites such as phenolics, flavonoids, and coumarins, which help inhibit pathogen growth and strengthen the cell wall (Vlot et al., 2009). In contrast, jasmonic acid and ethylene are more closely associated with defense against necrotrophic pathogens (which kill plant cells) and herbivorous insects. Jasmonic acid (JA) plays a pivotal role in regulating the production of secondary metabolites such as terpenoids, glucosinolates, and alkaloids, which act as chemical deterrents against herbivores and pathogens (Howe, 2004). Ethylene often works synergistically with JA, amplifying its effects and promoting defenses that include lignin deposition, which strengthens the cell wall against pathogen penetration (Broekaert et al., 2006).

1.3 Integration of Defense Signaling Pathways

The integration and coordination between the SA, JA, and ET pathways are critical for the plant to mount an effective immune response. The cross-talk between these pathways allows plants to fine-tune their defenses according to the specific type of threat they face. For example, SA and JA signaling are often antagonistic, with SA pathways being activated in response to biotrophic pathogens, while JA and ET pathways dominate in response to necrotrophs and herbivory. This antagonism ensures that the plant does not waste resources by activating both defense responses simultaneously, which would be metabolically costly and inefficient (Pieterse et al., 2012). Furthermore, the cross-talk is influenced by secondary metabolites themselves, which can modulate the intensity and duration of hormonal signaling. For instance, flavonoids are known to interact with auxin transport, influencing the balance between SA and JA responses. This interaction between signaling pathways and secondary metabolites ensures that the plant's defense responses are not only robust but also highly adaptable to a wide range of biotic stresses (Klessig et al., 2016).

1.4 Systemic Signaling and the Role of Secondary Metabolites

One of the remarkable features of plant defense is its ability to trigger systemic responses. Localized pathogen infection or herbivore damage leads to the production of signaling molecules such as salicylic acid, which are transported through the plant's vascular system to activate defense responses in distal tissues. This systemic signaling is key to the establishment of Systemic Acquired Resistance (SAR), where the entire plant is primed for enhanced defense following a localized attack. Secondary metabolites, particularly phenolics and terpenoids, are crucial components of this systemic defense, acting both as signaling molecules and as antimicrobial agents in tissues far from the original site of infection (Vlot et al., 2009). The ability of plants to activate defense responses systemically is also seen in Induced Systemic Resistance (ISR), which is primarily regulated by jasmonic acid and ethylene. ISR is typically triggered by beneficial microbes in the soil, such as rhizobacteria, which prime the plant for enhanced defense against a broad spectrum of pathogens. The production of secondary metabolites such as alkaloids and glucosinolates is a key feature of ISR, providing long-lasting protection against both pathogens and herbivores (Bari & Jones, 2009).

2. Role of Jasmonic Acid in Secondary Metabolite Production

Jasmonic acid (JA) is a critical plant hormone that regulates a wide array of physiological and developmental processes, but its most prominent role is in the regulation of defense responses against necrotrophic pathogens and herbivorous insects. JA is central to the biosynthesis and accumulation of secondary metabolites, which are crucial for plant defense and survival under stress. These metabolites include a diverse range of compounds such as alkaloids, glucosinolates, phenolics, flavonoids, and terpenoids, all of which play significant roles in protecting the plant from biotic stressors by either deterring herbivores, inhibiting pathogen growth, or modulating the plant's immune system. The regulation of these compounds by JA involves a complex network of signaling pathways and transcriptional regulation, making jasmonic acid one of the most critical components in the plant's secondary metabolism.

2.1 Jasmonic Acid Biosynthesis and Signaling

The biosynthesis of jasmonic acid begins in the chloroplast, where linolenic acid, a fatty acid derived from membrane lipids, is converted into 12-oxo-phytodienoic acid (OPDA) via the activity of lipoxygenase (LOX), allene oxide synthase (AOS), and allene oxide cyclase (AOC). OPDA is then transported to the peroxisome, where it undergoes reduction by OPDA reductase and subsequent rounds of β-oxidation to form jasmonic acid (JA). JA can be further conjugated with amino acids such as isoleucine to form jasmonoyl-isoleucine (JA-Ile), which is the biologically active form of the hormone (Wasternack & Strnad, 2018). Once synthesized, JA-Ile binds to its receptor, COI1 (CORONATINE INSENSITIVE 1), which forms part of the SCFCOI1 ubiquitin ligase complex. This binding triggers the degradation of JAZ (Jasmonate ZIM-domain) proteins, which are repressors of jasmonic acid-responsive genes. The degradation of JAZ proteins frees transcription factors such as MYC2, MYC3, and MYC4, which then activate the expression of jasmonic acid-responsive genes involved in the production of secondary metabolites (Thines et al., 2007). This mechanism ensures a rapid and targeted response to environmental cues, allowing plants to adjust their metabolism according to the specific biotic stress they face.

2.2 Role of Jasmonic Acid in Alkaloid Production

Alkaloids are nitrogen-containing secondary metabolites that serve as potent deterrents to herbivores and pathogens due to their toxicity. These compounds include well-known plant defense chemicals such as nicotine, morphine, quinine, and caffeine. The biosynthesis of alkaloids is tightly regulated by jasmonic acid, especially in response to herbivory. For instance, in Nicotiana tabacum (tobacco), herbivore damage triggers the JA pathway, leading to the upregulation of genes responsible for nicotine biosynthesis. Nicotine is synthesized in the roots of the plant and transported to the leaves, where it acts as a neurotoxin to deter herbivores, particularly insects (Baldwin, 1999). In Catharanthus roseus, the production of terpenoid indole alkaloids (TIAs), which have pharmaceutical significance, is also regulated by jasmonic acid. JA signaling activates transcription factors that upregulate the expression of genes involved in the TIA biosynthesis pathway, such as strictosidine synthase (STR), a key enzyme in the biosynthesis of TIAs (El-Sayed & Verpoorte, 2007). This regulation underscores the role of jasmonic acid as a master regulator of alkaloid biosynthesis, ensuring that these toxic metabolites are produced in response to biotic stress.

2.3 Jasmonic Acid and Glucosinolate Biosynthesis

Glucosinolates are sulfur-containing secondary metabolites predominantly found in plants of the Brassicaceae family, including Arabidopsis thaliana and Brassica crops. These compounds are known for their role in deterring herbivores and protecting plants from pathogens due to their ability to break down into toxic compounds, such as isothiocyanates, upon tissue damage. The production of glucosinolates is tightly regulated by jasmonic acid signaling, particularly in response to insect attacks. In Arabidopsis, jasmonic acid stimulates the expression of MYB transcription factors, specifically MYB28, MYB29, and MYB76, which are key regulators of glucosinolate biosynthesis. These transcription factors activate the expression of glucosinolate biosynthetic genes, leading to the accumulation of both aliphatic and indole glucosinolates (Sønderby et al., 2010). This JA-regulated glucosinolate production provides the plant with an effective defense against a wide range of herbivores, including caterpillars and aphids, by releasing volatile toxic compounds when the plant tissue is damaged.

2.4 Jasmonic Acid and Phenolic Compound Biosynthesis

Phenolic compounds are another important group of secondary metabolites involved in plant defense. These include flavonoids, lignins, and tannins, which provide structural support to the plant cell wall and act as deterrents to herbivores and pathogens. Lignin, for example, strengthens the cell wall, making it more difficult for pathogens to penetrate and establish infections. The biosynthesis of phenolic compounds is also influenced by jasmonic acid, particularly under conditions of biotic stress. In response to herbivory, jasmonic acid induces the accumulation of flavonoids and tannins in various plant species. For instance, in Medicago truncatula, JA signaling has been shown to upregulate the expression of genes involved in flavonoid biosynthesis, leading to the accumulation of flavonoids in response to insect herbivory (Liu et al., 2021). These compounds not only deter herbivores by making plant tissues less palatable but also have antimicrobial properties that help the plant resist pathogen infections. Lignin biosynthesis is another process regulated by JA signaling. The deposition of lignin in response to pathogen attack is a critical defense mechanism, as it reinforces the plant cell wall and prevents pathogen entry. JA triggers the expression of genes involved in the phenylpropanoid pathway, leading to the synthesis of lignin monomers, which are then polymerized and incorporated into the cell wall (Miedes et al., 2014). This process is particularly important during pathogen infection, where rapid lignification can restrict the spread of the pathogen within the plant.

2.5 Jasmonic Acid in Terpenoid Biosynthesis

Terpenoids are one of the largest classes of secondary metabolites, with diverse roles in plant defense. These compounds include monoterpenes, sesquiterpenes, and diterpenes, which can act as toxins, feeding deterrents, or signal molecules that attract predators or parasitoids of herbivores. Volatile terpenoids are particularly important in indirect plant defense, where they attract natural enemies of herbivores, such as parasitic wasps, to herbivore-damaged plants. Jasmonic acid plays a key role in regulating the biosynthesis of terpenoids. In tomato (*Solanum lycopersicum*), JA signaling induces the expression of genes involved in the production of sesquiterpenes and monoterpenes, which are emitted from the damaged tissues to repel herbivores and attract predatory insects (Tholl, 2006). This indirect defense mechanism highlights the ecological importance of jasmonic acid-regulated terpenoid production in maintaining the balance of plant-herbivore interactions. In addition to volatile terpenoids, JA also regulates the production of non-volatile diterpenoids such as gibberellins and phytoalexins. These compounds play essential roles in strengthening the plant's immune system by inhibiting pathogen growth and signaling other defense pathways (Schwachtje & Baldwin, 2008).

2.6 Jasmonic Acid and Crosstalk with Other Hormones

The role of jasmonic acid in regulating secondary metabolite production does not occur in isolation. JA interacts with other hormonal signaling pathways, such as those involving salicylic acid (SA), ethylene (ET), and abscisic acid (ABA), to modulate the plant's defense response. For example, salicylic acid typically antagonizes JA signaling, prioritizing the plant's defense against biotrophic pathogens, whereas JA and ethylene often act synergistically to promote defense against necrotrophic pathogens and herbivores (Pieterse et al., 2012). This hormonal crosstalk ensures that the plant can fine-tune its secondary metabolite production in response to different types of biotic stress. For instance, in response to herbivory, JA induces the production of defense-related terpenoids and alkaloids, while the concurrent suppression of SA-mediated responses prevents unnecessary energy expenditure on defenses against biotrophic pathogens (Thines et al., 2007). Understanding these intricate hormonal interactions is crucial for developing strategies to enhance secondary metabolite production in crops through genetic engineering.

2.7 Genetic Manipulation of Jasmonic Acid Pathways for Enhanced Metabolite Production

The importance of jasmonic acid in regulating secondary metabolite production has spurred interest in genetically manipulating JA signaling pathways to enhance plant defense and improve resistance to pests and pathogens. Through the overexpression of JA biosynthetic genes, such as LOX, AOS, and OPR, researchers have been able to increase the levels of JA and JA-regulated secondary metabolites in several crop species, leading to enhanced resistance to herbivores and diseases (Li et al., 2016). CRISPR-Cas9 technology has also been used to target JA signaling components, such as JAZ repressors, to promote the constitutive activation of JA-responsive genes. This genetic manipulation has led to increased production of secondary metabolites such as nicotine in tobacco and glucosinolates in Arabidopsis, improving the plant's overall defense capacity (Howe, 2004).

3. Salicylic Acid and Systemic Acquired Resistance (SAR)

Salicylic acid (SA) is a crucial phytohormone that regulates a wide array of plant defense responses, particularly in the context of Systemic Acquired Resistance (SAR). SAR is a form of long-lasting, broad-spectrum immune response that is activated when a plant encounters a pathogen. The accumulation of SA in both local and systemic tissues plays a pivotal role in orchestrating this resistance, which is effective against a wide range of pathogens, particularly biotrophic and hemibiotrophic pathogens, which require living host tissue to survive. SA promotes the production of pathogenesis-related (PR) proteins, enhances the synthesis of antimicrobial secondary metabolites, and primes the plant for enhanced defense responses against subsequent infections (Vlot et al., 2009; Klessig et al., 2016).

3.1 Biosynthesis and Accumulation of Salicylic Acid

The biosynthesis of salicylic acid in plants primarily occurs through two pathways: the isochorismate pathway and the phenylalanine ammonia-lyase (PAL) pathway. The isochorismate pathway, which occurs in plastids, is the major route for SA production in many plants during pathogen attack. Isochorismate synthase (ICS) catalyzes the conversion of chorismate into isochorismate, which is then converted into SA by isochorismate pyruvate lyase (IPL). In parallel, the PAL pathway, which is part of the phenylpropanoid pathway, also contributes to SA production. In this pathway, phenylalanine ammonia-lyase catalyzes the conversion of phenylalanine to trans-cinnamic acid, which is then metabolized into SA (Lefevere et al., 2020). Upon pathogen detection, SA levels increase rapidly at the infection site and in distant tissues. This local and systemic accumulation of SA triggers the activation of SAR, a phenomenon that protects the plant from further pathogen attack, even in tissues far from the initial site of infection. SA acts as a signaling molecule, coordinating the induction of defense-related genes and the production of antimicrobial compounds, thus preparing the plant for enhanced immunity.

3.2 Role of Salicylic Acid in Systemic Acquired Resistance (SAR)

Systemic Acquired Resistance (SAR) is one of the most effective defense mechanisms that plants use to protect themselves from a broad range of pathogens. Once SAR is triggered by a local infection, typically involving biotrophic pathogens, SA mediates the spread of the defense signal throughout the plant, resulting in the activation of defense responses in both the initially infected tissue and uninfected distal tissues. SAR is characterized by the expression of pathogenesis-related (PR) proteins, which have antimicrobial properties and help suppress pathogen growth (Vlot et al., 2009). The establishment of SAR involves not only the accumulation of SA but also the generation of mobile signals that propagate the immune response throughout the plant. Several molecules, including methyl salicylate (MeSA), glycerol-3-phosphate (G3P), and azelaic acid, have been identified as mobile signals that contribute to the systemic induction of SAR (Klessig et al., 2016). Among these, methyl salicylate, the methylated form of SA, plays a central role. Produced at the site of infection, methyl salicylate is transported to distant tissues, where it is converted back to SA by salicylic acid-binding protein 2 (SABP2), thus activating the SAR response in those tissues (Park et al., 2007)

3.3 Pathogenesis-Related (PR) Proteins and Antimicrobial Compounds

One of the hallmarks of SAR is the upregulation of pathogenesis-related (PR) proteins. These proteins, which include PR-1, PR-2 (β-1,3-glucanases), and PR-5 (thaumatin-like proteins), are produced in response to pathogen infection and have antimicrobial activity. PR proteins can degrade the cell walls of invading pathogens or inhibit their growth by disrupting key cellular processes. The accumulation of these proteins in both infected and uninfected tissues as part of the SAR response is a key mechanism by which plants restrict pathogen spread (Vlot et al., 2009). In addition to PR proteins, SA induces the production of secondary metabolites, such as flavonoids, coumarins, and phenolics, which play direct roles in pathogen inhibition. These compounds contribute to the plant's defense by reinforcing the cell wall, producing reactive oxygen species (ROS), and exhibiting antimicrobial properties. For example, coumarins, which are phenolic compounds, are known to inhibit fungal and bacterial growth, while flavonoids can scavenge free radicals and contribute to cell wall fortification (Mandal et al., 2010). This multifunctional role of secondary metabolites highlights the importance of SA in mediating not only protein-based but also chemical defenses during SAR.

3.4 Cross-Talk Between Salicylic Acid and Other Hormones in Defense Signaling

The role of salicylic acid in plant defense does not function in isolation. It interacts with other hormonal signaling pathways, most notably jasmonic acid (JA) and ethylene (ET). While SA is predominantly involved in defense against biotrophic pathogens, JA and ET are typically activated in response to necrotrophic pathogens and herbivores. The relationship between SA and JA is often antagonistic, meaning that the activation of SA-dependent defenses can suppress JA-dependent responses, and vice versa (Pieterse et al., 2012). This antagonistic relationship allows the plant to fine-tune its defense strategy based on the type of pathogen it encounters. For instance, when the plant faces a biotrophic pathogen, it prioritizes SA-mediated defenses, while suppressing JA signaling to avoid unnecessary energy expenditure on defenses against necrotrophs (Glazebrook, 2005). However, there are situations where synergistic interactions between SA, JA, and ET occur. For example, in some cases, SA and JA work together to mediate resistance against certain pathogens, such as Pseudomonas syringae, highlighting the complexity of the hormonal cross-talk involved in plant defense (Thaler et al., 2012).

3.5 Priming and the Role of Salicylic Acid in Memory of Defense

An important aspect of Systemic Acquired Resistance (SAR) is the concept of priming, where the plant's defense system is "pre-activated" to respond more rapidly and robustly to future pathogen attacks. Once SAR is triggered, plants enter a primed state, in which defense responses are enhanced without requiring additional SA accumulation. This phenomenon is particularly significant because it allows the plant to conserve energy while maintaining a heightened state of readiness for future attacks. Priming is associated with the deposition of chromatin marks on defense-related genes, facilitating faster transcriptional activation of these genes upon subsequent pathogen exposure (Conrath, 2011). In addition to PR proteins and antimicrobial compounds, SA-dependent priming involves the accumulation of callose, a polysaccharide that reinforces the cell wall, further restricting pathogen penetration. The priming effect of SA ensures that plants can maintain a durable and efficient defense system, enabling them to survive in pathogen-rich environments.

3.6 Genetic Engineering and Manipulation of Salicylic Acid Pathways

The importance of salicylic acid in plant defense has made it a key target for genetic engineering aimed at improving crop resistance to pathogens. Overexpression of genes involved in SA biosynthesis, such as

ICS1 or PAL, has been shown to enhance SA levels and activate SAR, leading to increased resistance to both biotrophic and hemibiotrophic pathogens in several crop species. For example, overexpression of ICS1 in Arabidopsis and rice has been demonstrated to confer resistance to pathogens such as *Pseudomonas syringae* and *Magnaporthe oryzae* (Shine et al., 2016). Recent advances in CRISPR-Cas9 technology have also allowed for the precise manipulation of SA biosynthesis and signaling pathways. By targeting negative regulators of SA signaling, researchers can enhance SA-dependent defense responses while minimizing the impact on plant growth and development. This genetic manipulation of SA pathways holds great promise for developing crop varieties with enhanced disease resistance, offering a sustainable alternative to chemical pesticides.

4. Ethylene and Its Synergistic Role in Plant Defense

Ethylene (ET) is a gaseous plant hormone that is well known for its role in regulating fruit ripening, senescence, and other developmental processes. However, ethylene also plays a crucial role in plant defense responses, especially when it comes to defense against necrotrophic pathogens and herbivores. Ethylene works both independently and in synergy with other key defense hormones such as jasmonic acid (JA) and salicylic acid (SA) to orchestrate an integrated and adaptive defense response. This hormonal cross-talk allows the plant to fine-tune its defenses depending on the nature of the stressor, ensuring that the correct metabolic pathways are activated at the right time. In addition to its signaling functions, ethylene is also involved in inducing secondary metabolites such as terpenoids, phenolics, and alkaloids that enhance the plant's ability to deter herbivores, inhibit pathogen growth, and reinforce structural defenses.

4.1 Ethylene Biosynthesis and Perception

Ethylene biosynthesis begins with the conversion of methionine to S-adenosylmethionine (SAM), which is then converted into 1-aminocyclopropane-1-carboxylic acid (ACC) by the enzyme ACC synthase (ACS). ACC is subsequently converted into ethylene by ACC oxidase (ACO) (Wang et al., 2002). Once produced, ethylene is perceived by ethylene receptors (ETR1, ETR2, ERS1, ERS2, and EIN4), which are located in the endoplasmic reticulum. These receptors work in conjunction with downstream signaling components such as CTR1 and EIN2, leading to the activation of EIN3/EIL transcription factors that regulate ethylene-responsive genes (Binder, 2020). In response to biotic stress, such as pathogen infection or herbivore attack, ethylene biosynthesis is upregulated, leading to the activation of ethylene-responsive genes involved in defense, secondary metabolism, and stress responses.

4.2 Ethylene and Jasmonic Acid Synergy in Defense Against Necrotrophic Pathogens

One of the most critical roles of ethylene in plant defense is its synergistic interaction with jasmonic acid (JA). Together, these hormones regulate defenses against necrotrophic pathogens, which kill plant cells to extract nutrients. Necrotrophic pathogens, such as Botrytis cinerea and Alternaria brassicicola, require plants to undergo cell death for successful colonization, making the JA-ET pathway essential for mounting an effective defense response. The synergy between ethylene and jasmonic acid is crucial in coordinating the plant's defense against these pathogens. Ethylene enhances JA-induced responses by promoting the expression of JA-responsive transcription factors, such as ERF1 (Ethylene Response Factor 1), which integrate signals from both pathways to activate defense genes (Lorenzo et al., 2003). ERF1, in particular, regulates the expression of pathogenesis-related (PR) genes, as well as genes involved in the synthesis of secondary metabolites like phenolics and alkaloids, which provide additional layers of defense by directly inhibiting pathogen growth and reinforcing the plant cell wall. In Arabidopsis, studies have shown that plants with impaired ethylene signaling are more susceptible to necrotrophic pathogens, highlighting the critical role of ethylene in regulating JA-dependent defenses. For instance, ET-insensitive mutants, such as etr1-1 and ein2, show increased susceptibility to Botrytis cinerea, confirming that proper ethylene signaling is essential for resistance to necrotrophic pathogens (Thomma et al., 2001). This evidence underscores the importance of ET-JA synergy in promoting the production of defense-related secondary metabolites and other immune responses that are crucial for necrotrophic pathogen resistance.

4.3 Ethylene and Crosstalk with Salicylic Acid (SA) in Biotrophic Defense

The interaction between ethylene and salicylic acid (SA) is more complex than its synergy with JA, as these two pathways often exhibit antagonistic interactions depending on the pathogen and the context. While salicylic acid is primarily associated with defense against biotrophic pathogens (which require living host tissue), ethylene is typically involved in necrotrophic defenses. However, in some cases, ethylene can synergize with SA to enhance defenses, especially when dealing with pathogens that exhibit both biotrophic and necrotrophic phases during their infection cycle, such as *Pseudomonas syringae* and *Xanthomonas campestris* (Broekaert et al., 2006). Ethylene also plays a role in modulating the hypersensitive response (HR), a defense mechanism triggered by biotrophic pathogens. HR involves localized cell death at the infection site to limit pathogen spread. While SA is the primary hormone driving HR, ethylene can enhance this response by promoting the expression of defense-related genes involved in ROS production and cell wall reinforcement (Mur et al., 2006). This highlights the dual role of ethylene as both a modulator of cell death and a promoter of defenses against a broader range of pathogens. In some instances, ethylene can inhibit SA-dependent responses, as is the case with certain biotrophic pathogens, where too much ethylene signaling may suppress SA-driven defenses, leaving the plant more vulnerable to biotrophs (Glazebrook, 2005). The delicate balance between SA and ET signaling ensures that plants mount appropriate responses based on the type of pathogen encountered.

4.4 Ethylene-Induced Production of Secondary Metabolites

Ethylene plays a critical role in inducing the production of secondary metabolites, which are essential for plant defense against a variety of biotic stresses. These metabolites, which include phenolics, terpenoids, and alkaloids, are produced in response to pathogen attack or herbivory and serve to reinforce the cell wall, inhibit pathogen growth, or deter herbivores. For example, phenolic compounds such as lignins and flavonoids are important components of the cell wall that strengthen its structure and make it more resistant to pathogen penetration. Ethylene signaling enhances the production of these compounds by upregulating the phenylpropanoid pathway, which is responsible for the biosynthesis of phenolic acids and lignins (Miedes et al., 2014). Lignification of cell walls is particularly important during pathogen attack, as it creates a physical barrier that limits pathogen invasion and spread. Similarly, terpenoids and alkaloids are secondary metabolites that serve as direct deterrents to herbivores and pathogens. Ethylene has been shown to promote the production of sesquiterpenes and monoterpenes in plants like tomato and Arabidopsis, which are known to repel herbivorous insects and inhibit fungal growth (Tholl, 2006). These compounds can also act as volatile organic compounds (VOCs), attracting natural enemies of herbivores, such as parasitoid wasps, to the site of herbivory. Ethylene's role in regulating these secondary metabolites underscores its importance in both direct and indirect plant defenses, making it a versatile component of the plant immune system.

4.5 Ethylene in Abiotic Stress Responses and Defense Priming

In addition to its role in biotic stress, ethylene also contributes to the plant's response to abiotic stress, such as drought, flooding, and mechanical damage. Ethylene helps regulate the expression of stress-related genes that enhance plant resilience under adverse environmental conditions. For example, ethylene can prime plants for enhanced defense by inducing the production of secondary metabolites and reinforcing structural defenses before a pathogen or herbivore attack occurs (Van Loon et al., 2006). This concept of defense priming involves the activation of ethylene-responsive genes that prepare the plant for future stress, allowing it to respond more rapidly and effectively to subsequent attacks. Ethylene-mediated priming is an important strategy that allows plants to optimize their energy use while maintaining a heightened state of defense readiness, ensuring that they are better equipped to survive in stressful environments (Conrath, 2011).

4.6 Genetic Engineering of Ethylene Pathways for Enhanced Disease Resistance

Given its crucial role in plant defense, the genetic manipulation of ethylene signaling pathways has become a promising strategy for improving disease resistance in crops. By modulating the expression of key components of the ethylene signaling pathway, such as ACS and ACO, or by overexpressing ethylene receptors, researchers can enhance the plant's ability to produce secondary metabolites and mount robust defense responses against pathogens. For example, overexpression of ERF transcription factors, which integrate signals from both JA and ET pathways, has been shown to enhance resistance to Botrytis cinerea in tomato and Arabidopsis by promoting the production of secondary metabolites such as phenolics and terpenoids (Berrocal-Lobo et al., 2002). Similarly, genetic engineering approaches that increase ethylene biosynthesis in response to pathogen attack have been used to develop crops with improved resistance to necrotrophic pathogens (Li et al., 2016).

5. Cross-Talk Between Hormonal Pathways and Secondary Metabolites

In plant defense, the concept of hormonal cross-talk plays a pivotal role in enabling plants to fine-tune their responses to various biotic and abiotic stresses. Hormones such as salicylic acid (SA), jasmonic acid (JA), ethylene (ET), and others, such as abscisic acid (ABA), auxins, and gibberellins (GA), are involved in regulating plant immunity. These signaling molecules do not act in isolation but rather interact in complex networks, where cross-talk ensures that the plant can mount appropriate and energy-efficient defense responses. Central to this cross-talk is the regulation of secondary metabolites, which are crucial for plant defense against pathogens and herbivores. Secondary metabolites, including alkaloids, phenolics, terpenoids, and flavonoids, are often synthesized or modulated in response to hormonal signaling, acting as both direct defense compounds and signaling intermediates.

5.1 Hormonal Regulation of Secondary Metabolite Production

The synthesis of secondary metabolites is tightly regulated by hormonal signals, primarily driven by jasmonic acid (JA), salicylic acid (SA), and ethylene (ET), which modulate their production in response to biotic stress. JA is particularly involved in activating the synthesis of terpenoids, alkaloids, and glucosinolates, while SA is associated with the production of phenolics and flavonoids that are important in plant defense against biotrophic pathogens (Pieterse et al., 2012). Ethylene, on the other hand, synergizes with JA to regulate the production of certain terpenoids and phenolic compounds, especially in response to necrotrophic pathogens and herbivory (Broekaert et al., 2006). This hormonal regulation is essential for ensuring that the plant synthesizes the right secondary metabolites at the right time. For example, when a plant is attacked by a necrotrophic pathogen, the activation of the JA-ET signaling pathway leads to the accumulation of phenolic acids and terpenes, compounds known to reinforce cell walls and exhibit antimicrobial properties. In contrast, SA signaling induces the production of flavonoids and coumarins, which are particularly effective in defending against biotrophic pathogens (Glazebrook, 2005).

5.2 Antagonistic and Synergistic Hormonal Interactions

One of the most intriguing aspects of plant defense is the antagonistic and synergistic interactions between hormone signaling pathways. The classic example of this is the antagonism between salicylic acid (SA) and jasmonic acid (JA) pathways. SA is primarily involved in defenses against biotrophic pathogens, while JA and ET are central to the defense against necrotrophic pathogens and herbivores. This antagonism ensures that when a plant is attacked by a biotrophic pathogen, SA-dependent defenses are prioritized, while JA-dependent defenses are suppressed (Glazebrook, 2005). Conversely, when a necrotrophic pathogen invades, JA and ET pathways dominate, inhibiting SA signaling to avoid conflicting defense responses (Pieterse et al., 2012). For example, in Arabidopsis, Pseudomonas syringae infection activates the SA pathway, leading to the accumulation of SA-regulated secondary metabolites such as phenolics and flavonoids, which reinforce the plant's defenses against biotrophic pathogens. In contrast, during infection by Botrytis cinerea, a necrotrophic pathogen, the JA-ET pathway is activated, leading to the production of terpenoids and phenolic compounds that target cell wall fortification and microbial inhibition (Broekaert et al., 2006). This antagonistic interaction allows the plant to optimize its defense strategy according to the type of pathogen it encounters. In addition to antagonism, there are instances of synergistic interactions between hormone pathways. For example, JA and ET often work together to regulate the production of defense-related secondary metabolites in response to herbivory. This synergistic interaction enhances the production of alkaloids and glucosinolates, which act as toxic deterrents to herbivores (Howe, 2004). ERF (Ethylene Response Factor) transcription factors integrate JA and ET signals, leading to the activation of genes involved in secondary metabolite biosynthesis, such as phenylpropanoids and terpenoids (Lorenzo et al., 2003).

5.3 Secondary Metabolites as Modulators of Hormonal Cross-Talk

Interestingly, secondary metabolites not only serve as defense compounds but also function as modulators of hormonal cross-talk. For instance, flavonoids have been shown to interact with hormone signaling pathways, particularly by modulating auxin transport and influencing the balance between JA and SA signaling. Flavonoids can either inhibit or promote the transport of auxins, thereby affecting the plant's ability to respond to environmental stressors (Bari & Jones, 2009). This modulation of auxin transport by flavonoids is crucial in controlling plant growth and development, particularly under biotic stress, where the plant must balance growth with defense activation. Moreover, phenolic compounds, including lignins and tannins, are influenced by both JA and SA signaling and act as key players in cross-talk between these pathways. Lignins reinforce the cell wall, providing structural support and defense against pathogen invasion, while tannins have antimicrobial properties that inhibit pathogen growth. These phenolic compounds can feed back into the hormonal signaling network, modulating the strength and duration of defense responses (Miedes et al., 2014). The interaction between secondary metabolites and hormonal pathways highlights the complexity of plant defense responses. Secondary metabolites can amplify or suppress specific hormone signals, ensuring that the plant's immune system is finely tuned to the type of stress it encounters.

5.4 Cross-Talk in Abiotic Stress Responses and the Role of Secondary Metabolites

In addition to their roles in biotic stress, hormonal cross-talk, and secondary metabolites are also crucial in responses to abiotic stress, such as drought, salinity, and temperature extremes. Under abiotic stress, hormones like abscisic acid (ABA) become dominant, regulating processes like stomatal closure and osmoprotection. However, ABA does not work in isolation; it interacts with JA, SA, and ET to coordinate the plant's overall stress response (Pieterse et al., 2012). For example, under drought stress, ABA and JA signaling pathways often interact to enhance the production of secondary metabolites, such as terpenoids and phenolic compounds, which help to mitigate oxidative damage and reinforce the plant's defenses (Kim et al., 2021). This cross-talk between abiotic and biotic stress responses ensures that the plant maintains resilience in the face of multiple stresses. Secondary metabolites produced during abiotic stress, such as ROS-scavenging flavonoids and tannins, not only protect the plant from abiotic damage but also prime the plant for future biotic attacks.

5.5 Genetic Engineering of Hormonal Cross-Talk to Enhance Secondary Metabolite Production

Given the importance of hormonal cross-talk in regulating secondary metabolite production, there is significant interest in using genetic engineering to manipulate these pathways and enhance plant defenses. By targeting key components of hormonal signaling pathways, such as ERF transcription factors or JA biosynthetic enzymes, researchers can increase the synthesis of secondary metabolites that are crucial for plant defense (Li et al., 2016). For example, overexpression of ERF1, a key regulator of the JA-ET pathway, has been shown to enhance the production of phenolics and terpenoids in Arabidopsis, leading to increased resistance to necrotrophic pathogens like Botrytis cinerea. Similarly, CRISPR-Cas9 technology has been used to disrupt negative regulators of SA signaling, resulting in increased production of flavonoids and coumarins that improve the plant's defense against biotrophic pathogens (Howe, 2004). The potential for manipulating hormonal cross-talk to enhance secondary metabolite production offers exciting possibilities for improving crop resilience and reducing the reliance on chemical pesticides. By engineering plants to optimize their hormonal interactions, it is possible to develop crops that are better equipped to withstand biotic and abiotic stresses, leading to more sustainable agricultural practices.

6. Genetic Manipulation of Defense Pathways for Enhanced Metabolite Production

Genetic manipulation of defense pathways has become a powerful tool for enhancing the production of secondary metabolites, which are essential for plant defense against various biotic and abiotic stressors. By altering key components of the signaling networks, such as salicylic acid (SA), jasmonic acid (JA), ethylene (ET), and abscisic acid (ABA) pathways, scientists can enhance the synthesis of critical defense-related metabolites, improving the plant's resilience against pathogens, pests, and environmental challenges. This approach not only offers new opportunities for improving crop resistance but also contributes to the sustainable use of resources by reducing the need for chemical pesticides and fertilizers. Secondary metabolites, including phenolics, alkaloids, terpenoids, and flavonoids, serve as both direct antimicrobial agents and modulators of plant signaling. Genetic engineering offers the possibility to either upregulate or downregulate specific defense pathways, allowing for targeted increases in metabolite production. This section explores key advances in the genetic manipulation of defense pathways aimed at enhancing secondary metabolite production, focusing on modifications in hormone signaling, transcription factors, and metabolic engineering.

6.1 Overexpression of Transcription Factors for Enhanced Metabolite Production

One of the most direct approaches to enhancing secondary metabolite production is through the overexpression of transcription factors (TFs) involved in defense signaling. Transcription factors such as WRKY, MYB, ERF (Ethylene Response Factor), and NAC proteins regulate large sets of defense-related genes, including those involved in secondary metabolite biosynthesis. By overexpressing these transcription factors, researchers can amplify the plant's defense response, leading to increased production of key secondary metabolites. For example, overexpression of WRKY transcription factors has been shown to enhance phenolic compound production, particularly flavonoids and lignins, which reinforce plant cell walls and provide structural barriers against pathogen invasion. In Arabidopsis, overexpression of WRKY18 and WRKY53 led to increased production of lignins and phenolics, which strengthened the plant's resistance to necrotrophic pathogens like Botrytis cinerea (Birkenbihl et al., 2012). Similarly, overexpression of MYB transcription factors, such as MYB75 in Arabidopsis, enhances the biosynthesis of anthocyanins and flavonoids, which act as potent antimicrobial agents and antioxidant compounds. The ERF1 transcription factor, which integrates ethylene (ET) and jasmonic acid (JA) signaling, has also been successfully overexpressed to enhance the production of defense-related secondary metabolites. In tomato, overexpression of ERF1 resulted in increased resistance to Botrytis cinerea, due to the elevated production of terpenoids and phenolic compounds that act as antifungal agents (Berrocal-Lobo et al., 2002).

6.2 Manipulation of Hormonal Pathways for Enhanced Defense

Hormonal signaling pathways, particularly those involving SA, JA, ET, and ABA, are central to the regulation of secondary metabolite production. Genetic manipulation of these pathways allows for the fine-tuning of defense responses and the targeted increase of metabolites that play a critical role in protecting plants from pathogens and herbivores. Jasmonic acid (JA) is particularly important in regulating the biosynthesis of alkaloids, terpenoids, and glucosinolates. In Nicotiana tabacum (tobacco), the JA pathway controls the production of nicotine, an alkaloid that serves as a potent defense against herbivores. By overexpressing genes involved in the JA biosynthesis pathway, such as LOX (lipoxygenase) and AOS (allene oxide synthase), researchers have successfully increased nicotine levels in transgenic plants, leading to enhanced resistance to insect herbivores (Baldwin, 1999). Similarly, manipulation of the salicylic acid (SA) pathway has been shown to enhance the production of phenolics and flavonoids, which are crucial for defense against biotrophic pathogens. Overexpression of ICS1 (isochorismate synthase), a key enzyme in SA biosynthesis, leads to increased accumulation of SA and upregulation of pathogenesis-related (PR) proteins, which are involved in SAR (Systemic Acquired Resistance). This increased SA signaling boosts the plant's ability to produce antimicrobial secondary metabolites such as coumarins and flavonoids, enhancing resistance to pathogens like Pseudomonas syringae and Magnaporthe oryzae (Shine et al., 2019). In addition to SA and JA, manipulation of ethylene (ET) signaling has also been used to enhance metabolite production. Ethylene is known to work synergistically with JA to regulate the production of secondary metabolites such as terpenoids and phenolic acids. In tomato, overexpression of the ACO (1-aminocyclopropane-1-carboxylate oxidase) gene, which catalyzes the final step in ethylene biosynthesis, resulted in increased production of sesquiterpenes, a class of terpenoids that act as direct herbivore deterrents (Lorenzo et al., 2003).

6.3 Metabolic Engineering for Targeted Secondary Metabolite Production

Metabolic engineering involves the direct manipulation of biosynthetic pathways to enhance the production of specific secondary metabolites. This approach often involves modifying the expression of key enzymes in the biosynthetic pathways of alkaloids, flavonoids, terpenoids, and phenolics, enabling plants to produce higher levels of these defense compounds. Advances in CRISPR-Cas9 and RNA interference (RNAi) technologies have made it possible to precisely target genes involved in secondary metabolite biosynthesis, leading to enhanced production of defense compounds.For example, CRISPR-Cas9 technology has been used to knock out JAZ repressor proteins in the JA signaling pathway, leading to constitutive activation of JA-responsive genes involved in secondary metabolite production. This genetic modification results in increased accumulation of alkaloids and terpenoids in plants such as Nicotiana tabacum and Arabidopsis thaliana, improving their resistance to both herbivores and pathogens (Das et al., 2024). In Catharanthus roseus, a medicinal plant known for its production of terpenoid indole alkaloids (TIAs), metabolic engineering has been used to enhance the production of vinblastine and vincristine, compounds with significant pharmaceutical applications. By overexpressing genes involved in TIA biosynthesis, such as strictosidine synthase (STR) and geraniol 10-hydroxylase (G10H), researchers have increased the accumulation of these alkaloids, which also serve as defense compounds against herbivores and pathogens (El-Sayed & Verpoorte, 2007). Similarly, in tomato, metabolic engineering of the phenylpropanoid pathway has been used to increase the production of flavonoids and phenolic acids, which act as antioxidants and antimicrobial agents. Overexpression of CHS (chalcone synthase), a key enzyme in flavonoid biosynthesis, resulted in increased accumulation of anthocyanins, improving the plant's resistance to Botrytis cinerea and other necrotrophic pathogens (Schwachtje & Baldwin, 2008).

6.4 Enhancing Systemic Acquired Resistance (SAR) and Induced Systemic Resistance (ISR)

Systemic Acquired Resistance (SAR) and Induced Systemic Resistance (ISR) are two important defense mechanisms that involve the production of secondary metabolites in response to pathogen attack or beneficial microbial interactions. Genetic manipulation of the signaling pathways involved in SAR and ISR has been used to enhance the production of secondary metabolites and improve plant immunity. SAR, which is mediated by salicylic acid, involves the accumulation of pathogenesis-related (PR) proteins and antimicrobial compounds like phenolics, flavonoids, and coumarins. By overexpressing ICS1 or PAL (phenylalanine ammonia-lyase), key enzymes in the SA biosynthetic pathway, researchers have enhanced the plant's ability to produce these compounds, leading to increased resistance to biotrophic pathogens (Shine et al., 2019). Induced Systemic Resistance (ISR), which is regulated by jasmonic acid (JA) and ethylene (ET), can also be enhanced through genetic engineering. ISR is activated by beneficial microbes such as Pseudomonas fluorescens and leads to the accumulation of secondary metabolites such as alkaloids, terpenoids, and glucosinolates. Overexpression of MYC2, a transcription factor involved in JA signaling, has been shown to enhance ISR and increase the production of secondary metabolites in Arabidopsis and tomato, resulting in improved resistance to herbivores and necrotrophic pathogens (Pieterse et al., 2012).

6.5 Future Prospects and Challenges in Genetic Engineering of Metabolite Pathways

While genetic manipulation of defense pathways holds great promise for enhancing metabolite production, several challenges remain. The complexity of hormonal cross-talk and the potential trade-offs between growth and defense make it difficult to achieve the desired balance between resistance and yield. Additionally, overproduction of secondary metabolites can sometimes result in detrimental effects on plant growth and development, as these compounds are often energetically expensive for the plant to produce. Nevertheless, advances in synthetic biology and precision gene editing technologies, such as CRISPR-Cas9, offer new possibilities for fine-tuning metabolite pathways in a more targeted and controlled manner. Future research will likely focus on identifying new regulatory elements, optimizing metabolic flux, and minimizing unintended consequences to create crops that are both resilient and productive.

Conclusion

The defense signaling pathways in plants are highly intricate and form the cornerstone of the plant's ability to respond to biotic and abiotic stressors. Among these pathways, jasmonic acid (JA), salicylic acid (SA), and ethylene (ET) play pivotal roles in regulating the production of secondary metabolites that are essential for plant defense. These metabolites, including alkaloids, phenolics, flavonoids, and terpenoids, not only act as antimicrobial agents but also serve as deterrents to herbivores and contribute to the overall strengthening of plant structures. The interconnections and unique contributions of these signaling pathways are key to optimizing plant defense mechanisms and enhancing agricultural productivity through the development of disease-resistant crops. The jasmonic acid signaling pathway is particularly important in the plant's response to necrotrophic pathogens and herbivorous insects. Jasmonates, derivatives of JA, act as powerful regulators of secondary metabolite production, particularly terpenoids and alkaloids. These compounds are critical for defending against herbivores by disrupting their digestive processes or deterring feeding through toxic effects. Additionally, jasmonates stimulate the production of compounds that act as signaling molecules themselves, helping to initiate systemic responses throughout the plant. When plants are wounded or attacked by necrotrophic pathogens, JA signaling quickly activates genes that regulate the biosynthesis of these secondary metabolites, ensuring that the plant mounts a rapid and localized defense response. This ability to mobilize resources efficiently is a testament to the specificity and robustness of the jasmonic acid pathway in modulating plant defense.

In contrast, the salicylic acid signaling pathway is primarily associated with defense against biotrophic pathogens, which require living host tissue to survive. The hallmark of the salicylic acid response is the activation of systemic acquired resistance (SAR), a long-lasting, broad-spectrum defense mechanism that protects the entire plant from further infection. Salicylic acid is involved in the synthesis of phenolic compounds, such as flavonoids, which possess strong antimicrobial properties. In addition, SA activates pathogenesis-related (PR) proteins that degrade pathogen cell walls and hinder their ability to spread. This dual action—direct antimicrobial activity and reinforcement of the plant's immune defenses—makes the salicylic acid pathway essential for long-term plant health. Furthermore, SAR not only protects the infected tissue but also primes distant tissues, enabling the plant to respond more quickly to future pathogen invasions. This systemic nature of SA-mediated responses provides plants with a strategic advantage in managing biotrophic infections across different tissues.

Ethylene, a gaseous hormone, is another important regulator of plant defense, often working synergistically with both jasmonic acid and salicylic acid to coordinate a more comprehensive defense response. Ethylene enhances the plant's response to necrotrophic pathogens, often amplifying the effects of JA, particularly in response to mechanical damage or herbivore attacks. In many cases, ethylene acts as a modulator, balancing the actions of JA and SA depending on the nature of the pathogen. For instance, ethylene can enhance JA-induced defenses while mitigating the effects of SA when biotrophic pathogens are not the primary threat. Ethylene's role is particularly crucial in promoting the production of defense-related proteins and enzymes that fortify plant tissues, such as lignin and other structural polymers, which make it harder for pathogens to penetrate. This collaborative action between ethylene and other hormonal pathways highlights the sophisticated nature of plant defense signaling.

The crosstalk between these hormonal pathways is one of the most remarkable aspects of plant defense. Jasmonic acid, salicylic acid, and ethylene do not function in isolation but interact in complex ways that allow plants to fine-tune their responses to specific pathogens. This crosstalk ensures that the plant can allocate resources effectively, directing energy toward the most appropriate defense mechanism depending on the type of stress. For instance, while jasmonic acid and ethylene typically act together to defend against necrotrophs, they can also inhibit salicylic acid responses when biotrophic pathogens are not a threat, thereby conserving resources. This antagonistic and synergistic interaction allows plants to avoid deploying unnecessary defenses, which would otherwise deplete energy reserves. Such regulatory mechanisms underscore the efficiency of plant immune systems and demonstrate how secondary metabolites are intricately controlled by hormonal signaling.

Recent advances in genetic engineering have provided new opportunities to manipulate these defense pathways, thereby enhancing the production of secondary metabolites in crop species. By targeting specific genes involved in the jasmonic acid, salicylic acid, and ethylene pathways, scientists can increase the accumulation of key antimicrobial compounds. For example, overexpressing genes that regulate the biosynthesis of jasmonates can lead to higher production of alkaloids and terpenoids, making plants more resistant to herbivores and necrotrophic pathogens. Similarly, modifying the salicylic acid pathway can enhance the production of phenolics and PR proteins, improving a plant's ability to combat biotrophic pathogens and sustain systemic acquired resistance. The manipulation of ethylene signaling can further fine-tune the plant's response, ensuring that secondary metabolite production is increased only when needed, thereby minimizing the risk of stunted growth or reduced yields.

However, the genetic engineering of defense pathways comes with certain challenges. One of the primary concerns is the potential for unintended side effects, such as reduced plant growth, altered flowering times, or changes in metabolic processes that could affect crop yield and quality. Therefore, while enhancing secondary metabolite production through genetic engineering holds great promise, it must be approached with caution. Future research should focus on balancing defense and growth by ensuring that the enhanced production of secondary metabolites does not come at the expense of the plant's overall health or productivity. With advances in gene-editing technologies, such as CRISPR-Cas9, it is becoming increasingly feasible to make precise modifications to defense pathways, offering hope for more resilient crops that can withstand the increasing pressures of pathogens without the need for chemical pesticides.

References

Baldwin, I. T. (1999). Inducible Nicotine production in native nicotiana as an example of adaptive phenotypic plasticity. *Journal of Chemical Ecology, 25*, 3-30.

Bari, R., & Jones, J. D. G. (2009). Role of plant hormones in plant defence responses. *Plant Molecular Biology, 69*(4), 473-488.

Berrocal-Lobo, M., Molina, A., & Solano, R. (2002). Constitutive expression of ETHYLENE-RESPONSE-FACTOR1 in *Arabidopsis* confers resistance to several necrotrophic fungi. *The Plant Journal, 29*(1), 23-32.

Binder, B. M. (2020). Ethylene signaling in plants. *Journal of Biological Chemistry, 295*(22), 7710-7725.

Birkenbihl, R. P., Diezel, C., & Somssich, I. E. (2012). Arabidopsis WRKY33 is a key transcriptional regulator of hormonal and metabolic responses toward Botrytis cinerea infection. *Plant Physiology, 159*(1), 266-285.

Broekaert, W. F., Delauré, S. L., De Bolle, M. F., & Cammue, B. P. (2006). The role of ethylene in host-pathogen interactions. *Annual Review of Phytopathology, 44*, 393-416.

Conrath, U. (2011). Molecular aspects of defence priming. *Trends in Plant Science, 16*(10), 524-531.

Das, S., Kwon, M., Kim, J. Y. (2024). Enhancement of specialized metabolites using CRISPR/Cas gene editing technology in medicinal plants. *Frontiers in Plant Sciences, 15*, 1279738.

El-Sayed, M., & Verpoorte, R. (2007). Catharanthus terpenoid indole alkaloids: biosynthesis and regulation. *Phytochemistry Reviews, 6*(2), 277-305.

Glazebrook, J. (2005). Contrasting mechanisms of defense against biotrophic and necrotrophic pathogens. *Annual Review of Phytopathology, 43*(1), 205-227.

Howe, G. A. (2004). Jasmonates as signals in the wound response. *Journal of Plant Growth Regulation, 23*, 223-237.

Jones, J. D. G., & Dangl, J. L. (2006). The plant immune system. *Nature, 444*(7117), 323-329.

Khan, M. I. R., Fatma, M., Per, T. S., Anjum, N. A., & Khan, N. A. (2015). Salicylic acid-induced abiotic stress tolerance and underlying mechanisms in plants. *Frontiers in Plant Sciences, 6*, 462.

Kim, H., Seomun, S., Yoon, Y., & Jang, G. (2021). Jasmonic acid in plant abiotic stress tolerance and interaction with abscisic acid. *Agronomy, 11*(9), 1886.

Klessig, D. F., Tian, M., & Choi, H. W. (2016). Multiple targets of salicylic acid and its derivatives in plants and animals. *Frontiers in Immunology, 7*, 206.

Lefevere, H., Bauters, L., & Gheysen, G. (2020). Salicylic acid biosynthesis in plants. *Frontiers in Plant Sciences*, *11*, 338.

Li, M., Pu, Y., & Ragauskas, A. J. (2016). Current understanding of the correlation between lignin structure, degradation, and cell wall recalcitrance. *Frontiers in Chemistry, 4*, 45.

Liu, Y., Patra, B., & Singh, S.K. et al. (2021). Terpenoid indole alkaloid biosynthesis in *Catharanthus roseus*: effects and prospects of environmental factors in metabolic engineering. *Biotechnology Letters, 43*, 2085-2103.

Lorenzo, O., Piqueras, R., Sánchez-Serrano, J. J., & Solano, R. (2003). Ethylene response factor1 integrates signals from ethylene and jasmonate pathways in plant defense. *The Plant Cell, 15*(1), 165-178.

Mandal, S. M., Chakraborty, D., & Dey, S. (2010). Phenolic acids act as signaling molecules in plant-microbe symbioses. *Plant Signaling & Behavior, 5*(4), 359-368.

Miedes, E., Vanholme, R., Boerjan, W., & Molina, A. (2014). The role of the secondary cell wall in plant resistance to pathogens. *Frontiers in Plant Science, 5*, 358.

Mur, L. A. J., Kenton, P., Atzorn, R., Miersch, O., Wasternack, C., & Durner, J. (2006). The outcomes of concentration-specific interactions between salicylate and jasmonate signaling include synergy, antagonism, and oxidative stress leading to cell death. *Plant Physiology, 140*(1), 249-262.

Park, S. -W., Kaimoyo, E., Kumar, D., Mosher, S., & Klessig, D. F. (2007). Methyl salicylate is a critical mobile signal for plant systemic acquired resistance. *Science, 318*,113-116.

Pieterse, C. M., Van der Does, D., Zamioudis, C., Leon-Reyes, A., & Van Wees, S. C. (2012). Hormonal modulation of plant immunity. *Annual Review of Cell and Developmental Biology, 28*, 489-521.

Schwachtje, J., & Baldwin, I. T. (2008). Why does herbivore attack reconfigure primary metabolism? *Plant Physiology, 146*(3), 845-851.

Shine, M. B., Xiao, X., Kachroo, P., & Kachroo, A. (2019). Signaling mechanisms underlying systemic acquired resistance to microbial pathogens. *Plant Science, 279*, 81-86.

Sønderby, I. E., Geu-Flores, F., & Halkier, B. A. (2010). Biosynthesis of glucosinolates–gene discovery and beyond. *Trends in Plant Science, 15*(5), 283-290.

Thaler, J. S., Humphrey, P. T., & Whiteman, N. K. (2012). Evolution of jasmonate and salicylate signal crosstalk. *Trends in Plant Science, 17*(5), 260-270.

Thines, B., Katsir, L., Melotto, M., Niu, Y., Mandaokar, A., Liu, G., & Browse, J. (2007). JAZ repressor proteins are targets of the SCFCOI1 complex during jasmonate signalling. *Nature, 448*(7154), 661-665.

Tholl, D. (2006). Terpene synthases and the regulation, diversity and biological roles of terpene metabolism. *Current Opinion in Plant Biology, 9*(3), 297-304.

Thomma, B. P., Penninckx, I. A., Cammue, B. P., & Broekaert, W. F. (2001). The complexity of disease signaling in *Arabidopsis*. *Current Opinion in Immunology, 13*(1), 63-68.

Tsuda, K., & Katagiri, F. (2010). Comparing signaling mechanisms engaged in pattern-triggered and effector-triggered immunity. *Current Opinion in Plant Biology, 13*(4), 459-465.

Van Loon, L. C., Geraats, B. P. J., & Linthorst, H. J. M. (2006). Ethylene as a modulator of disease resistance in plants. *Trends in Plant Science, 11*(4), 184-191.

Vlot, A. C., Dempsey, D. M., & Klessig, D. F. (2009). Salicylic acid, a multifaceted hormone to combat disease. *Annual Review of Phytopathology, 47*(1), 177-206.

Wang, K. L., Li, H., & Ecker, J. R. (2002). Ethylene biosynthesis and signaling networks. *The Plant Cell, 14*(S1), S131-S151.

Wasternack, C., & Strnad, M. (2018). Jasmonates: News on occurrence, biosynthesis, metabolism, and action of an ancient group of signaling compounds. *International Journal of Molecular Sciences, 19*(9), 2539.

Chapter - 13

Environmental and Genetic Regulation of Secondary Metabolite Production

Zuby Gohar Ansari[1], Reshma Jadhav[2], Madhumita Bhattacharjee[3] and Vinod Chouhan[4]

[1]B.M. College of Agriculture, Jaswadi Road Khandwa, Madhya Pradesh, India
[2]Department of Biochemistry, Mahatma Phule Krishi Vidyapeeth, Rahuri, Ahmednagar, Maharashtra, India
[3]Department of Botany, Postgraduate Government College, Chandigarh, India
[4]Division of Plant Pathology, Indian Council of Agricultural Research-Indian Agricultural Research Institute (ICAR-IARI), New Delhi, India

Abstract

Secondary metabolite production in plants is governed by a complex interplay of environmental factors and genetic controls. These metabolites, including flavonoids, alkaloids, terpenoids, and phenolics, are crucial for plant defense, growth, and adaptation to stress. Environmental factors such as light, temperature, nutrient availability, and pathogen pressure significantly influence the biosynthesis of secondary metabolites. For example, exposure to UV radiation can enhance flavonoid production as a protective response, while nutrient deficiencies can trigger the accumulation of compounds like phenolics that assist in nutrient uptake and stress tolerance. On the genetic level, secondary metabolite production is regulated by specific biosynthetic pathways controlled by key enzymes and regulatory genes. Transcription factors, such as MYB, bHLH, and WRKY, play an essential role in modulating the expression of these pathways in response to environmental stimuli. Moreover, advancements in genetic engineering have opened new possibilities for enhancing the production of secondary metabolites by manipulating these pathways. Techniques such as gene editing, overexpression, and silencing have been used to increase the accumulation of targeted metabolites, providing new avenues for crop improvement and stress resistance. However, genetic manipulation must be carefully balanced to avoid adverse effects on plant growth and development. As the importance of secondary metabolites in agriculture and pharmaceuticals continues to grow, understanding the environmental and genetic regulation of their production is critical for the development of sustainable agricultural practices and the creation of high-value crops with enhanced resilience to biotic and abiotic stress.

DOI: 10.1201/9781003717454-13

Keywords: secondary metabolites, environmental regulation, genetic engineering, transcription factors, biosynthetic pathways, stress adaptation.

Introduction

Secondary metabolites are vital compounds produced by plants that serve diverse ecological functions, ranging from defense against pathogens and herbivores to signaling and adaptation to environmental stress. These compounds, including flavonoids, alkaloids, terpenoids, and phenolics, are not directly involved in the primary metabolic processes necessary for plant growth and development, but they play essential roles in enhancing a plant's ability to survive and thrive in fluctuating environmental conditions. The production of secondary metabolites is influenced by a complex interplay of environmental and genetic factors, with both external conditions and internal regulatory mechanisms determining the quantity and quality of these compounds in plants (Zhan et al., 2022).

Environmental factors such as light, temperature, water availability, nutrient supply, and biotic stressors (such as pathogens and herbivores) significantly impact the biosynthesis of secondary metabolites. These external cues often act as triggers that activate or suppress specific biosynthetic pathways, leading to the production of compounds that help plants adapt to changing conditions. For instance, plants exposed to high levels of ultraviolet (UV) radiation tend to produce more flavonoids, which function as UV protectants by absorbing harmful radiation and reducing oxidative damage (Di Ferdinando et al., 2012). Similarly, temperature fluctuations can alter the production of essential oils and terpenoids, as plants adjust their metabolic processes to maintain homeostasis under stress (Pant et al., 2021).

Light is one of the most crucial environmental factors influencing secondary metabolite production. Light quality, intensity, and duration all play a role in modulating plant metabolism. In particular, light serves as a signal that triggers the activation of key enzymes involved in the biosynthesis of secondary metabolites. For example, the production of anthocyanins, a type of flavonoid, is strongly regulated by light exposure, with higher light intensities promoting their synthesis to protect plants from photodamage (Rouxinol et al., 2023). The type of light—whether it is blue, red, or UV—can also differentially regulate the production of various secondary metabolites. UV light, in particular, has been shown to stimulate the accumulation of phenolic compounds that act as antioxidants and photoprotectants. This is a critical adaptation for plants growing in high-altitude or high-exposure environments where UV radiation levels are elevated (Zhang et al., 2021).

Temperature is another key environmental factor that influences secondary metabolite production. Both high and low temperatures can induce stress responses in plants, leading to changes in metabolic pathways. High temperatures, for instance, have been shown to increase the production of terpenoids and essential oils in aromatic plants such as thyme and oregano (Dobhal et al., 2024). These compounds serve as defense mechanisms against heat-induced stress and oxidative damage. Conversely, low temperatures can enhance the production of certain flavonoids and phenolic acids, which provide protection against cold-induced oxidative stress (Das et al., 2024). In some species, cold stress also stimulates the accumulation of alkaloids, which can act as antifreeze molecules or protective agents against frostbite (Qian et al., 2024).

Water availability, or the lack thereof, significantly affects the biosynthesis of secondary metabolites.

Drought stress, for example, often leads to an increase in the production of phenolic compounds and flavonoids, which help plants cope with water scarcity by mitigating oxidative stress and reducing water loss through stomatal regulation. These compounds can also enhance the structural integrity of plant cells, helping them survive prolonged periods of drought. On the other hand, excessive water availability, such as during flooding, can reduce the production of certain secondary metabolites as plants shift their energy towards survival processes like anaerobic respiration and water transport. Therefore, maintaining a balance between water supply and stress signaling is crucial for the optimal production of secondary metabolites (Ramakrishna & Ravishankar, 2011).

Nutrient availability is another crucial environmental factor that influences the production of secondary metabolites in plants. Both macronutrients, such as nitrogen, phosphorus, and potassium, and micronutrients, like iron and magnesium, play key roles in regulating the biosynthetic pathways responsible for secondary metabolite production. Nitrogen deficiency, for instance, has been shown to increase the accumulation of certain phenolic compounds, as plants divert resources from growth to defense under nutrient-limited conditions (Gill et al., 2020). Similarly, phosphorus deficiency can lead to an increase in flavonoid production, as these compounds help protect plants from oxidative stress and facilitate nutrient uptake under adverse conditions (Song et al., 2023). Conversely, high nutrient availability generally promotes primary metabolism and growth, often leading to a reduction in secondary metabolite production as plants prioritize energy towards vegetative development.

Genetic regulation of secondary metabolite production is equally important as environmental influences. The biosynthesis of these compounds is controlled by a network of genes encoding enzymes that catalyze specific steps in metabolic pathways. These genes are regulated by transcription factors, such as MYB, bHLH, and WRKY, which modulate the expression of biosynthetic enzymes in response to environmental cues. For example, MYB transcription factors have been shown to regulate flavonoid biosynthesis by activating key genes involved in the phenylpropanoid pathway (Stracke et al., 2007). Similarly, WRKY transcription factors play a critical role in regulating the production of alkaloids and terpenoids by responding to pathogen attacks and herbivore feeding (Javed & Gao, 2010). These transcription factors are crucial for ensuring that secondary metabolites are produced in appropriate amounts and at the right time to maximize their protective functions.

In recent years, advances in genetic engineering have opened new possibilities for enhancing the production of secondary metabolites by manipulating these biosynthetic pathways. Techniques such as gene editing, overexpression, and gene silencing have been used to increase the accumulation of specific secondary metabolites, improving plant resistance to pests, diseases, and environmental stressors. For example, overexpression of key regulatory genes in the phenylpropanoid pathway has been shown to increase flavonoid production, leading to enhanced resistance to UV radiation and herbivores (Zhang et al., 2021). Similarly, gene editing techniques like CRISPR-Cas9 have been employed to knock out negative regulators of secondary metabolite pathways, resulting in increased accumulation of valuable compounds such as alkaloids and terpenoids (Das et al., 2024). However, genetic manipulation must be approached carefully to avoid potential trade-offs, such as reduced growth or reproductive fitness, which can occur if metabolic resources are overly diverted towards secondary metabolite production.

Overall, the production of secondary metabolites in plants is a highly dynamic process regulated by both environmental factors and genetic mechanisms. Environmental cues such as light, temperature, and nutrient availability serve as critical triggers for the activation of biosynthetic pathways, while genetic factors, including transcription factors and biosynthetic enzymes, control the specific production of these compounds. As climate change continues to alter environmental conditions, understanding how these factors interact to regulate secondary metabolite production will be crucial for developing crops with enhanced stress tolerance and disease resistance. Moreover, advancements in genetic engineering offer exciting opportunities to optimize secondary metabolite production for agricultural and pharmaceutical applications, providing new solutions to the challenges posed by biotic and abiotic stress.

1. Impact of Environmental Factors on Secondary Metabolite Production

The production of secondary metabolites in plants is highly influenced by environmental factors, which can alter both the quantity and the quality of these compounds. Secondary metabolites, such as alkaloids, phenolics, terpenoids, and flavonoids, play essential roles in plant defense against pathogens, herbivores, and environmental stresses. Environmental conditions, including light, temperature, water availability, soil nutrients, and biotic interactions, all influence the biosynthesis pathways of these metabolites. Understanding the impact of environmental factors on secondary metabolite production is crucial for optimizing agricultural practices and improving crop resistance to stress.

1.1 Light

Light is one of the most critical environmental factors influencing the production of secondary metabolites. Light regulates plant development and photosynthesis, which provides the carbon skeletons needed for the biosynthesis of secondary metabolites. Additionally, light quality, intensity, and duration directly impact the phenylpropanoid pathway, which is responsible for the synthesis of flavonoids, anthocyanins, and phenolic acids. These compounds are often produced in response to UV radiation to protect plants from oxidative damage caused by UV exposure (Zoratti et al., 2014). For instance, flavonoid production is strongly stimulated by UV-B radiation, which induces the expression of genes involved in flavonoid biosynthesis, such as CHS (chalcone synthase) and FLS (flavonol synthase) (Bieza & Lois, 2001). Plants grown in low-light conditions tend to produce lower levels of these compounds, while plants exposed to high-light environments or supplemental UV light have higher concentrations of flavonoids and anthocyanins, which not only protect them from UV damage but also serve as antimicrobial and antioxidant compounds (Koyama et al., 2012).

1.2 Temperature

Temperature is another significant environmental factor that affects secondary metabolite production. Both high and low temperatures can influence the synthesis of secondary metabolites by altering metabolic activity and enzyme function. For example, terpenoid biosynthesis, which is crucial for defense against herbivores, is often enhanced under higher temperatures. Studies have shown that monoterpene and sesquiterpene production increases at elevated temperatures due to the increased activity of terpene synthase enzymes (Sharkey et al., 2008). Conversely, low temperatures can induce the production of phenolic compounds and alkaloids. Cold stress typically triggers the accumulation of anthocyanins and flavonoids, which protect plant cells from cold-induced oxidative stress (Dixon

& Paiva, 1995). In species like grapevine and strawberry, anthocyanin accumulation increases in response to cooler temperatures, leading to deeper pigmentation and enhanced antioxidant activity (Mancinelli, 1985).

1.3 Water Availability

Water availability significantly influences the biosynthesis of secondary metabolites, particularly under conditions of drought stress. Drought stress leads to an increase in the production of reactive oxygen species (ROS), which can cause cellular damage. To mitigate this damage, plants upregulate the production of secondary metabolites such as phenolics, flavonoids, and terpenoids, which act as antioxidants to neutralize ROS (Al-Gabbiesh et al., 2015). Drought conditions have been shown to stimulate the accumulation of phenolic compounds and alkaloids in several species. For example, in Nicotiana tabacum, drought stress increases the production of nicotine, an alkaloid that plays a defensive role against herbivores (Selmar, 2008). Similarly, terpenoid production in species like mint and basil is elevated under drought stress, resulting in higher concentrations of essential oils, which deter herbivores and pathogens (Boncan et al., 2020).

1.4 Soil Nutrients

The availability of soil nutrients plays a fundamental role in determining the levels of secondary metabolites produced by plants. Nutrient stress, particularly deficiencies in nitrogen (N), phosphorus (P), and potassium (K), can influence secondary metabolite production. Nitrogen deficiency, for example, often leads to an increase in the production of carbon-based secondary metabolites, such as phenolics and terpenoids, due to a shift in carbon allocation from growth to defense (Bryant et al., 1983). In nitrogen-limited environments, plants tend to produce higher levels of tannins and flavonoids, which are carbon-rich compounds that do not require nitrogen for their synthesis (Hodges & Nozzolillo, 1995). On the other hand, phosphorus deficiency has been shown to increase the production of alkaloids and glucosinolates, enhancing the plant's resistance to herbivores and pathogens (Penuelas & Estiarte, 1998). However, an excess of nutrients, particularly nitrogen, can reduce the production of secondary metabolites. High nitrogen availability often favors primary metabolism and growth, reducing the allocation of resources to secondary metabolite biosynthesis (Koricheva et al., 1998). This reduction can make plants more vulnerable to herbivores and pathogens, as they produce fewer defensive compounds.

1.5 Biotic Interactions

Biotic interactions, such as those with pathogens, herbivores, and beneficial microbes, also play a crucial role in regulating secondary metabolite production. Plants often increase the production of specific secondary metabolites in response to pathogen attack or herbivory. For instance, the biosynthesis of alkaloids, terpenoids, and phenolics is upregulated when plants are exposed to herbivores or microbial pathogens (Dixon, 2001). In the case of herbivore attacks, jasmonic acid (JA) and ethylene (ET) signaling pathways are activated, leading to the production of alkaloids and terpenoids, which deter herbivores and attract natural enemies of the herbivores, such as parasitic wasps (Howe & Jander, 2008). Additionally, pathogen infection triggers the salicylic acid (SA) pathway, leading to the accumulation of phenolic compounds and flavonoids, which have antimicrobial properties that inhibit pathogen growth (Pieterse et al., 2012). In addition to pathogen and herbivore interactions, beneficial microbes, such as mycorrhizal fungi and rhizobacteria, can enhance the production of secondary metabolites through the activation of induced systemic resistance (ISR). This interaction leads to increased production of defense-related secondary metabolites such as alkaloids and glucosinolates (Nishad et al., 2020).

1.6 Effects of Climate Change on Secondary Metabolite Production

With the increasing impact of climate change, shifts in environmental factors such as temperature, light, and precipitation patterns are expected to significantly alter secondary metabolite production. Rising global temperatures and increased occurrences of drought are likely to enhance the production of terpenoids and phenolics, as plants adapt to higher stress levels. Conversely, changes in light availability due to altered cloud cover or ozone depletion could influence the production of UV-protective compounds such as flavonoids (Goufo et al., 2014). In addition, the increased concentration of atmospheric CO2 is predicted to affect secondary metabolite production by altering photosynthetic rates and carbon allocation. Elevated CO2 levels often result in increased production of carbon-based secondary metabolites, such as phenolics, flavonoids, and terpenoids, as plants use excess carbon for defense-related processes (Penuelas & Estiarte, 1998).

2. Role of Light and Temperature in Metabolite Biosynthesis

Light and temperature are two critical environmental factors that significantly influence the biosynthesis of secondary metabolites in plants. Secondary metabolites, including phenolics, alkaloids, terpenoids, and flavonoids, play essential roles in plant defense mechanisms, development, and interactions with their environment. These compounds are not only involved in direct defense against herbivores and pathogens but also help plants adapt to abiotic stresses such as intense light, UV radiation, and temperature fluctuations. The impact of light and temperature on metabolite biosynthesis is well-documented, with these factors modulating the expression of key genes in metabolic pathways and the enzymatic activity required for metabolite production.

2.1 The Role of Light in Metabolite Biosynthesis

Light is one of the most influential environmental factors in regulating secondary metabolite production. It not only provides energy for photosynthesis but also acts as a signal that influences plant growth, development, and the biosynthesis of numerous metabolites. Different aspects of light, including intensity, quality (wavelength), and duration (photoperiod), directly impact the accumulation of various secondary metabolites, particularly flavonoids, anthocyanins, phenolic acids, and terpenoids.

2.1.1 Light Intensity and Metabolite Production

Increased light intensity is generally associated with higher production of secondary metabolites. For example, phenolic compounds, which include flavonoids and lignins, are often produced in greater quantities under high-light conditions. These compounds act as UV protectants and antioxidants, helping to mitigate damage caused by excessive light and UV radiation (Zoratti et al., 2014). In species like grapevine and tomato, increased light exposure has been linked to higher levels of anthocyanins and flavonols, which contribute to the plants' antioxidant capacity and improve their defense against oxidative stress (Koyama et al., 2012). Studies have shown that high light intensity promotes the expression of genes involved in the phenylpropanoid pathway, such as phenylalanine ammonia-lyase (PAL), chalcone synthase (CHS), and flavonol synthase (FLS). These genes are crucial for the biosynthesis of flavonoids and phenolics, which not only protect against UV radiation but also have roles in pathogen defense (Bieza & Lois, 2001).

2.1.2 Light Quality and Specific Wavelengths

The quality of light (i.e., the wavelength of light received by plants) has a profound effect on secondary metabolite production. UV-B radiation, in particular, is a potent inducer of secondary metabolite biosynthesis. UV-B triggers the production of flavonoids, which act as UV shields, absorbing harmful radiation and protecting the plant from oxidative damage. In Arabidopsis thaliana, exposure to UV-B light induces the accumulation of anthocyanins and flavonols, both of which contribute to enhanced UV protection and defense (Kuhlmann & Müller, 2010). Additionally, blue and red light also influence the production of secondary metabolites. Blue light is known to stimulate the production of flavonoids, terpenoids, and phenolic acids, which enhance plant defense against pathogens and herbivores. In contrast, red light has been shown to affect the production of alkaloids and chlorophyll-related compounds, influencing plant growth and secondary metabolism (Zoratti et al., 2014).

2.1.3 Photoperiod and Metabolite Biosynthesis

The photoperiod (the duration of light exposure within 24 hours) also regulates secondary metabolite production. Plants exposed to longer daylight periods generally produce more terpenoids and phenolics, as these compounds are often synthesized in response to light-driven signals. Short-day plants, such as soybeans, tend to accumulate higher levels of isoflavonoids under shorter photoperiods, which enhances their defense against microbial pathogens (Koyama et al., 2012). In summary, light not only drives photosynthesis but also serves as a regulatory signal for the synthesis of a wide array of secondary metabolites. The intensity, wavelength, and duration of light exposure modulate the activity of biosynthetic enzymes and the expression of genes involved in metabolite production, optimizing plant defenses against environmental challenges.

2.2 The Role of Temperature in Metabolite Biosynthesis

Temperature is another key environmental factor that influences the synthesis of secondary metabolites. Both high and low temperatures can have significant effects on metabolic processes, altering the activity of enzymes involved in metabolite biosynthesis and affecting the overall levels of these compounds. The impact of temperature on secondary metabolism is species-specific and often depends on the plant's adaptive strategies for coping with heat stress or cold stress.

2.2.1 High Temperatures and Metabolite Production

High temperatures tend to accelerate metabolic processes, including those involved in the production of secondary metabolites such as terpenoids and alkaloids. In response to heat stress, plants increase the synthesis of these compounds to protect cellular structures and mitigate oxidative stress caused by elevated temperatures. For example, monoterpenes and sesquiterpenes are produced in greater quantities under high-temperature conditions due to the enhanced activity of terpene synthase enzymes (Sharkey et al., 2008). These terpenoids act as volatile compounds that repel herbivores and pathogens, while also functioning as antioxidants that protect cells from thermal damage. In medicinal plants like mint and basil, higher temperatures have been linked to increased production of essential oils, which are composed of various terpenoids. These oils not only serve as defense mechanisms but also enhance the medicinal properties of these plants (Sangwan et al., 2001). Similarly, in grapevine, high temperatures promote the production of phenolics and flavonoids, which contribute to the plant's antioxidant defenses and improve fruit quality (Mira de Orduña, 2010).

2.2.2 Low Temperatures and Metabolite Production

Low temperatures generally slow down metabolic processes but can also induce the production of certain secondary metabolites, particularly phenolics and alkaloids. Cold stress leads to the accumulation of anthocyanins, flavonoids, and other phenolic compounds that protect plants from cold-induced oxidative damage. These compounds act as antioxidants and stabilize cell membranes, helping plants survive in cold environments (Dixon & Paiva, 1995). For instance, in cold-tolerant species such as grapevine and strawberry, exposure to low temperatures leads to an increase in anthocyanin content, which enhances the plants' tolerance to freezing temperatures and UV radiation (Mancinelli, 1985). The enhanced production of flavonoids and phenolic acids under cold conditions is also associated with improved resistance to cold-related oxidative stress and a general enhancement of plant vigor (Kró et al., 2015).

2.2.3 Temperature Fluctuations and Metabolic Plasticity

Plants often encounter temperature fluctuations during the day or across seasons, which requires them to adjust their metabolic processes to maintain homeostasis. In response to such fluctuations, plants exhibit metabolic plasticity, altering the production of secondary metabolites to cope with varying environmental conditions. For example, in response to day-night temperature cycles, some plants adjust the levels of terpenoids and phenolics to optimize their defense against herbivores and pathogens during periods of peak activity (Holopainen & Gershenzon, 2010). Plants growing in regions with extreme temperature variation, such as desert environments, often produce high levels of terpenoids and alkaloids, which protect against both herbivores and the harsh environmental conditions (Gershenzon, 1984). This ability to modulate secondary metabolite production in response to temperature changes allows plants to maintain a balance between growth and defense, ensuring their survival in challenging environments.

2.3 Synergistic Effects of Light and Temperature on Metabolite Biosynthesis

While light and temperature individually influence metabolite biosynthesis, the combination of these factors can have synergistic effects on secondary metabolism. In many cases, the interaction between light intensity and temperature leads to enhanced production of terpenoids, flavonoids, and phenolics. For instance, in tomato plants, high light intensity combined with elevated temperatures results in increased production of carotenoids and terpenoids, both of which contribute to improved fruit quality and stress tolerance (Llorente et al., 2017). In grapevine, high temperatures and strong UV-B exposure synergistically boost the production of anthocyanins and flavonols, improving the plant's defense against UV radiation and heat stress while also enhancing the quality of the grapes for winemaking (Rouxinol et al., 2023). This interaction between light and temperature highlights the complexity of environmental regulation in secondary metabolite biosynthesis and underscores the importance of optimizing both factors to improve plant defense and metabolic output.

3. Nutrient Availability and Its Influence on Secondary Metabolite Pathways

Nutrient availability is a fundamental environmental factor that significantly influences the biosynthesis of secondary metabolites in plants. Nutrients, particularly nitrogen (N), phosphorus (P), and potassium (K), play a vital role in plant growth and development. However, when nutrient levels are limited or imbalanced, plants often shift their metabolic priorities from growth toward defense, resulting in changes in secondary metabolite production. This response is part of a survival strategy in which plants allocate more resources to the synthesis of defensive compounds, including alkaloids, phenolics, flavonoids, and terpenoids, that protect them from herbivores, pathogens, and abiotic stress. Understanding how nutrient availability influences secondary metabolite pathways is essential for optimizing agricultural practices and improving plant resilience in nutrient-poor soils.

3.1 Nitrogen Availability and Secondary Metabolite Biosynthesis

Nitrogen (N) is one of the most critical nutrients for plant metabolism, as it is a key component of amino acids, proteins, nucleic acids, and chlorophyll. Nitrogen availability has a profound effect on the biosynthesis of both primary and secondary metabolites, with a complex relationship between nitrogen levels and the production of nitrogen- and carbon-based secondary metabolites.

3.1.1 Nitrogen Deficiency and Carbon-Based Metabolites

Under conditions of nitrogen deficiency, plants tend to increase the production of carbon-based secondary metabolites, such as phenolics, flavonoids, and terpenoids. This shift occurs because nitrogen-limited plants reduce their growth and prioritize defense mechanisms by reallocating carbon from primary metabolism to secondary metabolism (Bryant et al., 1983). For example, phenylpropanoids, a group of carbon-based compounds, are often produced in higher quantities under nitrogen stress, as their biosynthesis does not require nitrogen and helps reinforce plant defenses against herbivores and pathogens. In species such as Medicago truncatula and Arabidopsis thaliana, nitrogen deficiency has been shown to stimulate the production of flavonoids and phenolic acids, which act as antioxidants and protect plants from oxidative stress. Similarly, terpenoid production, which includes essential oils and resins, is often upregulated under nitrogen-limited conditions. These compounds are known for their roles in deterring herbivores and inhibiting pathogen growth (Ormeño & Fernandez, 2012).

3.1.2 Nitrogen-Rich Conditions and Nitrogen-Based Metabolites

In contrast, under nitrogen-rich conditions, plants tend to produce more nitrogen-based secondary metabolites, such as alkaloids, cyanogenic glycosides, and glucosinolates. These compounds contain nitrogen in their structure and are often involved in direct defense against herbivores and pathogens. For example, alkaloids such as nicotine in Nicotiana tabacum (tobacco) and morphine in Papaver somniferum (opium poppy) are nitrogen-based compounds that increase in abundance when nitrogen is readily available (Lou & Baldwin, 2004). In Brassica species, nitrogen-rich soils lead to increased production of glucosinolates, sulfur-containing compounds that act as powerful defense chemicals against herbivores and pathogens. The myrosinase-glucosinolate system, in particular, is activated during herbivore attack, releasing toxic compounds that deter feeding (Bones & Rossiter, 2006). Nitrogen availability directly impacts the biosynthesis of glucosinolates by regulating key enzymes in the pathway, such as cytochrome P450 monooxygenases (Sønderby et al., 2010). However, excessive nitrogen levels can sometimes reduce the overall production of secondary metabolites by favoring primary metabolism and promoting growth. This trade-off occurs because high nitrogen availability encourages rapid growth and biomass accumulation, often at the expense of defense-related secondary metabolites (Koricheva et al., 1998).

3.2 Phosphorus Availability and Secondary Metabolite Biosynthesis

Phosphorus (P) is another essential nutrient that plays a crucial role in energy transfer, membrane synthesis, and signal transduction. Phosphorus availability significantly affects the biosynthesis of secondary metabolites, particularly in relation to alkaloids, flavonoids, and phenolics.

3.2.1 Phosphorus Deficiency and Secondary Metabolite Production

Under phosphorus deficiency, plants often exhibit an increase in the production of phenolic compounds and flavonoids, which serve as protective agents against phosphorus-related stress. For example, in white lupin (Lupinus albus), phosphorus deficiency has been shown to stimulate the production of isoflavonoids, which help protect the plant from oxidative stress and pathogen attack (Neumann et al., 1999). In addition to phenolics, phosphorus deficiency can also lead to an increase in terpenoid biosynthesis. In species like mint and rosemary, low phosphorus availability stimulates the production of essential oils rich in monoterpenes and sesquiterpenes, which play roles in defense against herbivores and microbial pathogens (Boncan et al., 2020).

3.2.2 Phosphorus-Rich Conditions and Secondary Metabolite Suppression

Conversely, when phosphorus is abundant, the production of secondary metabolites can be reduced as plants prioritize primary metabolic processes, such as growth and reproduction. In some cases, excessive phosphorus can lead to lower concentrations of phenolics and flavonoids, which weakens the plant's defense mechanisms and makes it more susceptible to pathogens and herbivores (Khan et al., 2023). This reduction in secondary metabolite production under phosphorus-rich conditions is part of a broader trade-off between growth and defense, where plants invest more resources in biomass production rather than in synthesizing defensive compounds.

3.3 Potassium and Its Influence on Metabolite Pathways

Potassium (K) is an essential macronutrient involved in various physiological processes, including osmotic regulation, enzyme activation, and stomatal function. Potassium availability also influences secondary metabolite production, although its effects are less well-studied compared to nitrogen and phosphorus.

3.3.1 Potassium Deficiency and Phenolic Accumulation

Under potassium deficiency, plants often increase the production of phenolic compounds and other antioxidants to mitigate the negative effects of potassium-related stress. Potassium deficiency can lead to the overproduction of reactive oxygen species (ROS), which cause cellular damage. To counteract this, plants upregulate the production of flavonoids, anthocyanins, and other phenolic compounds that act as ROS scavengers (Cakmak, 2005). In tomato plants, potassium deficiency has been shown to increase the concentration of phenolic acids and flavonoids, enhancing the plant's resistance to oxidative stress and improving its overall stress tolerance (Silva-Beltrán et al., 2015). Similarly, in grapevines, potassium-deficient plants produce higher levels of anthocyanins, which contribute to both antioxidant activity and the pigmentation of grape berries (Conde et al., 2006).

3.3.2 Potassium-Rich Conditions and Metabolite Production

On the other hand, potassium-rich conditions can lead to a reduction in the production of secondary metabolites, as plants allocate more resources to primary metabolic processes such as photosynthesis and protein synthesis. While adequate potassium levels are essential for maintaining plant growth and physiological functions, excessive potassium availability can shift metabolic priorities away from defense and toward growth, leading to reduced levels of phenolics and other defensive compounds (Koricheva et al., 1998).

3.4 Trade-offs Between Nutrient Availability and Defense

The relationship between nutrient availability and secondary metabolite production is governed by a series of trade-offs that plants must navigate to balance growth and defense. When nutrient levels are low, plants often invest more resources into the production of secondary metabolites to protect themselves from herbivores and pathogens. In contrast, when nutrients are abundant, plants prioritize primary metabolic processes, such as growth and reproduction, at the expense of defense. This trade-off is particularly evident in nitrogen- and phosphorus-limited environments, where plants allocate more carbon to secondary metabolite pathways to compensate for reduced growth. However, in nutrient-rich environments, the production of secondary metabolites may be reduced, making plants more vulnerable to biotic stressors. These trade-offs are critical for agricultural practices, as it allow for the optimization of nutrient management strategies to improve both crop yield and resistance to pests and diseases. By manipulating nutrient availability, farmers can influence the production of secondary metabolites and enhance the natural defenses of crops, reducing the need for chemical pesticides and fertilizers.

4. Genetic Regulation of Secondary Metabolite Biosynthesis

The production of secondary metabolites in plants is tightly regulated by a complex network of genetic controls. These metabolites, which include alkaloids, terpenoids, phenolics, flavonoids, and others, play vital roles in plant defense against herbivores, pathogens, and environmental stresses, as well as in ecological interactions with pollinators and other organisms. The biosynthesis of secondary metabolites is largely governed by the expression of genes involved in their biosynthetic pathways, many of which are regulated by transcription factors, gene clusters, and various signaling pathways that respond to internal and external stimuli. The genetic regulation of secondary metabolite biosynthesis is crucial for efforts to manipulate these pathways through genetic engineering, which can enhance the production of valuable compounds for agricultural, pharmaceutical, and industrial applications.

4.1 Role of Transcription Factors in Secondary Metabolite Regulation

Transcription factors (TFs) play a central role in the genetic regulation of secondary metabolite biosynthesis by modulating the expression of genes involved in the biosynthetic pathways. Transcription factors bind to specific promoter regions of target genes, thereby controlling their transcriptional activity. Several families of transcription factors, such as MYB, bHLH (basic helix-loop-helix), WRKY,

NAC, and ERF (ethylene response factors), have been shown to regulate the biosynthesis of various classes of secondary metabolites.

4.1.1 MYB Transcription Factors

The MYB transcription factor family is one of the largest and most well-characterized groups of transcription factors involved in regulating secondary metabolism. MYB transcription factors are particularly important in the regulation of flavonoid and phenolic biosynthesis, which are critical for plant defense and stress tolerance. For example, MYB75, also known as PRODUCTION OF ANTHOCYANIN PIGMENT 1 (PAP1), is a key regulator of anthocyanin biosynthesis in plants such as Arabidopsis thaliana. Overexpression of MYB75 leads to enhanced production of anthocyanins, which serve as antioxidants and provide protection against UV radiation and herbivory (Borevitz et al., 2000). Similarly, MYB transcription factors such as AtMYB12 in Arabidopsis regulate the production of flavonols, which are important for pathogen resistance and tolerance to abiotic stress (Stracke et al., 2007). In addition to anthocyanins and flavonols, MYB transcription factors also regulate the biosynthesis of phenolic acids and lignins. MYB46 and MYB83 are key regulators of lignin biosynthesis, which is crucial for cell wall reinforcement and defense against pathogens (Zhong et al., 2008). These transcription factors activate the expression of lignin biosynthetic genes such as PAL (phenylalanine ammonia-lyase), C4H (cinnamate-4-hydroxylase), and COMT (caffeic acid O-methyltransferase), leading to the accumulation of lignin in plant cell walls.

4.1.2 bHLH Transcription Factors

The bHLH (basic helix-loop-helix) transcription factors are also involved in the regulation of secondary metabolite biosynthesis, particularly in the flavonoid and alkaloid pathways. bHLH proteins often interact with MYB transcription factors to form regulatory complexes that control the expression of genes involved in secondary metabolite production. For instance, in Nicotiana tabacum (tobacco), bHLH transcription factors such as MYC2 play a key role in regulating the biosynthesis of nicotine, an important alkaloid involved in defense against herbivores. MYC2 is activated by jasmonic acid (JA) signaling and binds to the promoters of nicotine biosynthetic genes, such as PMT (putrescine N-methyltransferase), leading to increased nicotine production in response to herbivore attack (Shoji & Hashimoto, 2011). This demonstrates the importance of bHLH transcription factors in integrating hormonal signals with secondary metabolite biosynthesis.

4.1.3 WRKY and NAC Transcription Factors

WRKY and NAC transcription factors are well-known regulators of plant defense responses, including the biosynthesis of phenolics, terpenoids, and alkaloids. These transcription factors are often activated in response to biotic and abiotic stress, leading to the upregulation of defense-related secondary metabolites. WRKY transcription factors, such as WRKY33, are involved in the regulation of camalexin biosynthesis, a major phytoalexin in Arabidopsis that plays a critical role in defense against pathogens (Birkenbihl et al., 2012). Similarly, NAC transcription factors such as ANAC055 and ANAC072 regulate the production of phenolic acids and flavonoids in response to pathogen infection, enhancing the plant's resistance to biotic stress (Hoang et al., 2017).

4.2 Gene Clusters in Secondary Metabolite Pathways

In addition to transcription factors, the regulation of secondary metabolite biosynthesis often involves gene clusters, where multiple genes encoding enzymes for a specific biosynthetic pathway are co-located in the genome and coordinately regulated. Gene clusters are commonly found in the biosynthesis of alkaloids, terpenoids, and glucosinolates.

4.2.1 Alkaloid Biosynthetic Gene Clusters

Alkaloid biosynthetic pathways are often regulated by gene clusters that contain genes encoding the enzymes required for alkaloid production. In opium poppy (Papaver somniferum), the biosynthesis of morphine and other benzylisoquinoline alkaloids (BIAs) is controlled by a gene cluster that includes TYDC (tyrosine decarboxylase), CYP80A1 (cytochrome P450), and COR (codeinone reductase). These genes are co-regulated by transcription factors such as WRKY and bHLH, which modulate alkaloid production in response to environmental cues (Beaudoin & Facchini, 2014).

4.2.2 Terpenoid Biosynthetic Gene Clusters

Terpenoid biosynthesis is similarly regulated by gene clusters in several plant species. In maize (Zea mays), a gene cluster responsible for the biosynthesis of terpenoid volatiles is activated in response to herbivore damage, leading to the emission of volatile compounds that attract natural enemies of herbivores (Zerbe et al., 2013). This cluster includes genes encoding terpene synthases (TPS) and cytochrome P450 monooxygenases, which are involved in the conversion of simple terpenoid precursors into complex volatile terpenoids. The coordinated regulation of terpenoid biosynthetic gene clusters allows plants to produce these metabolites rapidly in response to herbivory or pathogen attack, enhancing their defense capabilities.

4.3 Hormonal Regulation of Secondary Metabolite Biosynthesis

Plant hormones such as jasmonic acid (JA), salicylic acid (SA), ethylene (ET), and abscisic acid (ABA) play essential roles in regulating the biosynthesis of secondary metabolites. These hormones often act as signaling molecules that activate transcription factors and gene clusters involved in secondary metabolite production.

4.3.1 Jasmonic Acid and Terpenoid Biosynthesis

Jasmonic acid (JA) is a key regulator of terpenoid biosynthesis, particularly in response to herbivore attack and mechanical wounding. JA signaling activates transcription factors such as MYC2, which regulate the expression of terpene synthase (TPS) genes involved in the biosynthesis of volatile terpenoids. These compounds act as defense molecules by deterring herbivores and attracting natural enemies (Pauwels & Goossens, 2011). In tomato (Solanum lycopersicum), JA signaling induces the production of monoterpenes and sesquiterpenes, which are emitted from damaged tissues to repel herbivores and attract predatory insects. JA also regulates the production of oxylipins, a group of secondary metabolites derived from fatty acids that play roles in defense against pathogens (Howe, 2004).

4.3.2 Salicylic Acid and Phenolic Biosynthesis

Salicylic acid (SA) is primarily involved in regulating the biosynthesis of phenolic compounds, which are crucial for plant defense against biotrophic pathogens. SA activates transcription factors such as WRKY and NAC, which upregulate the expression of genes involved in the phenylpropanoid pathway, leading to the accumulation of phenolic acids, flavonoids, and lignins (Pieterse et al., 2012). The SA signaling pathway also plays a key role in Systemic Acquired Resistance (SAR), where the production of secondary metabolites such as phytoalexins and pathogenesis-related (PR) proteins is enhanced to protect plants from subsequent pathogen attacks (Vlot et al., 2009).

4.4 Advances in Genetic Engineering for Enhanced Metabolite Production

The growing understanding of the genetic regulation of secondary metabolite biosynthesis has opened up new opportunities for **genetic engineering** aimed at enhancing metabolite production in plants. By manipulating transcription factors, gene clusters, and hormonal signaling pathways, researchers can increase the production of valuable secondary metabolites with applications in medicine, agriculture, and industry.

4.4.1 Overexpression of Key Transcription Factors

One common strategy for enhancing secondary metabolite production is the overexpression of key transcription factors involved in the regulation of biosynthetic pathways. For example, the overexpression of MYB and bHLH transcription factors has been used to increase the production of anthocyanins and flavonoids in crops such as tomato, grapevine, and apple, improving their nutritional and medicinal properties (Menconi et al., 2023). In Nicotiana tabacum, overexpression of MYC2 has been shown to increase the production of nicotine, enhancing the plant's defense against herbivores (Shoji & Hashimoto, 2011). Similarly, overexpression of WRKY transcription factors has been used to enhance the production of phytoalexins and other defense-related secondary metabolites in response to pathogen infection.

4.4.2 CRISPR-Cas9 for Targeted Gene Editing

The development of CRISPR-Cas9 technology has revolutionized plant biotechnology by enabling precise and targeted gene editing. Researchers have used CRISPR-Cas9 to knock out negative regulators of secondary metabolite biosynthesis, leading to enhanced production of valuable compounds. For instance, CRISPR-Cas9 has been used to knock out JAZ repressor proteins in the JA signaling pathway, resulting in constitutive activation of JA-responsive genes and increased production of terpenoids and alkaloids (Mipeshwaree Devi et al., 2023). This technology also allows for the modification of gene clusters involved in secondary metabolism, enabling the coordinated regulation of multiple genes in a pathway to enhance metabolite production. CRISPR-Cas9 has great potential for improving the biosynthesis of secondary metabolites in economically important crops and medicinal plants.

5. Transcription Factors in Controlling Metabolite Production

Transcription factors (TFs) are key regulatory proteins that control the expression of genes by binding to specific DNA sequences. They play a pivotal role in orchestrating the biosynthesis of secondary metabolites by activating or repressing genes involved in the metabolic pathways responsible for the production of compounds such as flavonoids, terpenoids, alkaloids, and phenolics. The regulation of these metabolic pathways is highly complex, often requiring the coordinated action of multiple transcription factors from different families to achieve precise control over secondary metabolite production. The ability of transcription factors to integrate signals from the environment and endogenous hormones makes them critical players in the plant's response to stress and defense mechanisms.

5.1 MYB Transcription Factors: Master Regulators of Flavonoid and Phenolic Biosynthesis

MYB transcription factors are one of the most extensively studied families of transcription factors involved in secondary metabolite production. Members of the MYB family are particularly important for the regulation of flavonoid and phenolic biosynthesis, which are crucial for plant defense against ultraviolet (UV) radiation, herbivores, and pathogens. MYB75, also known as PAP1 (Production of Anthocyanin Pigment 1), is a well-characterized transcription factor in Arabidopsis thaliana that controls the biosynthesis of anthocyanins, which are pigments with antioxidant properties. MYB75 binds to the promoters of genes in the phenylpropanoid pathway, such as chalcone synthase (CHS), dihydroflavonol 4-reductase (DFR), and flavonoid 3′-hydroxylase (F3′H), upregulating their expression and enhancing anthocyanin accumulation (Borevitz et al., 2000). Overexpression of MYB75 leads to a dramatic increase in anthocyanin content, providing enhanced protection against UV light and herbivores. Similarly, AtMYB12, another MYB transcription factor in Arabidopsis, regulates the biosynthesis of flavonols, which are important for plant resistance to both biotic and abiotic stresses (Stracke et al., 2007).

In addition to their role in flavonoid biosynthesis, MYB transcription factors also regulate lignin production. Lignin is a key component of the plant cell wall that provides structural integrity and protection against pathogens. MYB46 and MYB83 are master regulators of lignin biosynthesis in Arabidopsis, directly activating genes such as C4H (cinnamate-4-hydroxylase), 4CL (4-coumarate ligase), and COMT (caffeic acid O-methyltransferase), which are involved in the lignin biosynthetic pathway (Borevitz et al., 2000). These transcription factors play a crucial role in strengthening the cell wall and enhancing the plant's resistance to pathogens.

5.2 bHLH Transcription Factors: Coordinators of Alkaloid and Flavonoid Biosynthesis

Basic helix-loop-helix (bHLH) transcription factors are another important class of regulators involved in secondary metabolite biosynthesis, particularly in the production of alkaloids, terpenoids, and flavonoids. bHLH transcription factors often work in concert with MYB transcription factors to form regulatory complexes that control the expression of biosynthetic genes. In tobacco (*Nicotiana tabacum*), bHLH transcription factors such as MYC2 are key regulators of nicotine biosynthesis, a defense-related alkaloid that deters herbivores. MYC2 is activated by jasmonic acid (JA) signaling and binds to the promoters of nicotine biosynthetic genes, including putrescine N-methyltransferase (PMT) and quinolinate synthase (QPT), leading to enhanced nicotine production in response to herbivore attack (Shoji & Hashimoto, 2011). This example highlights the role of bHLH transcription factors in integrating hormonal signals with secondary metabolite production, ensuring that plants can rapidly produce defense compounds when needed. In addition to their role in alkaloid biosynthesis, bHLH transcription factors are also involved in regulating flavonoid biosynthesis. For instance, in maize (*Zea mays*), the bHLH transcription factor ZmBHLH interacts with MYB transcription factors to regulate the production of anthocyanins and other flavonoids in response to light and stress signals (Petroni & Tonelli, 2011). This cooperation between bHLH and MYB transcription factors allows for fine-tuned regulation of metabolite biosynthesis based on environmental conditions.

5.3 WRKY and NAC Transcription Factors: Defense-Associated Regulators of Phenolics and Terpenoids

WRKY transcription factors are central players in the regulation of plant defense mechanisms, including the biosynthesis of phenolic compounds and terpenoids. These transcription factors are often activated in response to biotic stress, such as pathogen infection or herbivore attack, and they regulate the expression of genes involved in defense-related secondary metabolism. For example, in Arabidopsis, WRKY33 is a key regulator of camalexin biosynthesis, a major phytoalexin that provides resistance against pathogens. WRKY33 directly activates genes in the camalexin biosynthetic pathway, such as PAD3 (phytoalexin-deficient 3), leading to the accumulation of camalexin during pathogen attack (Birkenbihl et al., 2012). Similarly, WRKY transcription factors regulate the biosynthesis of phenolic compounds such as salicylic acid (SA)-induced flavonoids and lignins, which strengthen the plant's defenses against biotrophic pathogens. NAC transcription factors also play a prominent role in regulating the biosynthesis of defense-related secondary metabolites. In rice (Oryza sativa), NAC transcription factors such as OsNAC have been shown to regulate the biosynthesis of terpenoid phytoalexins, which are produced in response to pathogen infection (Hoang et al., 2017). These terpenoids act as antimicrobial agents, inhibiting the growth of invading pathogens and protecting the plant from disease.

5.4 ERF Transcription Factors: Regulators of Terpenoid and Phenolic Biosynthesis

Ethylene response factor (ERF) transcription factors are involved in the regulation of both terpenoid and phenolic biosynthesis, particularly in response to ethylene (ET) and jasmonic acid (JA) signaling. ERF transcription factors play a crucial role in mediating plant responses to necrotrophic pathogens and herbivores by activating genes involved in the biosynthesis of defense-related secondary metabolites. In tomato (Solanum lycopersicum), the ERF transcription factor SlERF1 regulates the biosynthesis of monoterpenes and sesquiterpenes in response to herbivory. These terpenoids are emitted as volatile compounds that attract natural enemies of herbivores, thereby providing indirect defense to the plant (Berrocal-Lobo et al., 2002). Additionally, ERF transcription factors in Arabidopsis regulate the production of phenolic acids and lignins, which fortify the plant cell wall and enhance resistance to necrotrophic pathogens (Lorenzo et al., 2003).

5.5 Integration of Multiple Transcription Factors in Metabolite Regulation

The regulation of secondary metabolite biosynthesis often requires the coordinated action of multiple transcription factors. For example, in the regulation of flavonoid biosynthesis, MYB, bHLH, and WD40 transcription factors form a ternary complex known as the MBW complex, which controls the expression of key flavonoid biosynthetic genes (Zerbe et al., 2013). Similarly, in the regulation of terpenoid biosynthesis, MYC2 (a bHLH transcription factor), ERF, and WRKY transcription factors work together to regulate the production of volatile terpenoids in response to jasmonic acid and ethylene signaling. This integration of multiple transcription factors allows plants to fine-tune their secondary metabolite production based on environmental signals, ensuring that they can mount an effective defense against biotic and abiotic stress.

6. Advances in Genetic Engineering for Enhanced Metabolite Accumulation

Advances in genetic engineering have revolutionized the ability to manipulate the biosynthesis of secondary metabolites in plants, enabling the development of crops with enhanced resistance to pathogens, herbivores, and environmental stresses. By targeting the genetic pathways that regulate secondary metabolite biosynthesis, researchers can enhance the production of valuable compounds, including alkaloids, phenolics, terpenoids, and flavonoids, with important applications in medicine, agriculture, and industry. Several approaches, including the overexpression of transcription factors, gene editing technologies, and metabolic pathway engineering, have been employed to increase the accumulation of secondary metabolites in plants. These strategies have provided new insights into the regulation of metabolite biosynthesis and opened up possibilities for creating plants with improved stress tolerance and higher levels of valuable compounds.

6.1 Overexpression of Transcription Factors to Boost Metabolite Accumulation

One of the most effective strategies for enhancing secondary metabolite production is the overexpression of key transcription factors that regulate biosynthetic pathways. Transcription factors such as MYB, bHLH, WRKY, and ERF play central roles in controlling the expression of genes involved in metabolite biosynthesis, making them ideal targets for genetic manipulation. For example, the overexpression of MYB transcription factors has been used to enhance the production of anthocyanins and flavonoids in several crops. In tomato (Solanum lycopersicum), overexpression of the transcription factor AtMYB12 from Arabidopsis resulted in increased levels of flavonols, which improved the plant's antioxidant capacity and resistance to stress (Zhang et al., 2021). Similarly, overexpression of MYB75 in grapevine (Vitis vinifera) and apple (Malus domestica) has been shown to increase anthocyanin content, enhancing the fruit's color, flavor, and antioxidant properties (Espley et al., 2007). In Nicotiana tabacum, overexpression of the bHLH transcription factor MYC2 led to increased production of nicotine, an alkaloid that acts as a potent defense compound against herbivores (Shoji & Hashimoto, 2011). This demonstrates the potential of transcription factor overexpression to boost the production of defense-related secondary metabolites in response to environmental stress.

6.2 Metabolic Engineering for Enhanced Pathway Flux

Metabolic engineering aims to modify specific steps in the biosynthetic pathways of secondary metabolites to increase the flux of precursors through the pathway and enhance metabolite accumulation. This approach often involves the overexpression or suppression of key enzymes in the pathway, as well as the introduction of genes from other species to optimize metabolite production.

One successful example of metabolic engineering is the enhancement of terpenoid biosynthesis in mint (Mentha spicata), where researchers overexpressed terpene synthase (TPS) enzymes to increase the production of menthol and other essential oils. By optimizing the expression of enzymes such as limonene synthase and menthone reductase, researchers were able to significantly increase the yield of valuable terpenoids, which have applications in flavoring, fragrance, and pharmaceuticals (Lange & Ahkami, 2013). Similarly, metabolic engineering has been used to enhance the production of alkaloids in opium poppy (Papaver somniferum) by manipulating key enzymes in the biosynthetic pathway, such as codeinone reductase (COR) and salutaridine reductase (SalR). This approach has led to increased production of morphine and other benzylisoquinoline alkaloids, which are used in pain management and other medical applications (Beaudoin & Facchini, 2014).

6.3 CRISPR-Cas9 Gene Editing for Targeted Metabolite Enhancement

The development of CRISPR-Cas9 technology has provided researchers with an unprecedented level of precision in gene editing, allowing for the targeted modification of specific genes involved in secondary metabolite biosynthesis. By using CRISPR-Cas9 to knock out negative regulators or activate positive regulators of metabolite biosynthesis, researchers can enhance the accumulation of valuable compounds in plants. For instance, CRISPR-Cas9 has been used to knock out JAZ repressors in the jasmonic acid (JA) signaling pathway, leading to increased production of JA-responsive secondary metabolites, such as terpenoids and alkaloids (Zhang et al., 2021). This approach allows for the constitutive activation of defense-related pathways, resulting in enhanced metabolite accumulation without the need for external stress signals. In addition to knocking out genes, CRISPR-Cas9 can also be used to introduce new regulatory elements into biosynthetic pathways, enabling the fine-tuning of gene expression and metabolite production. For example, researchers have used CRISPR-Cas9 to modify the promoters of key biosynthetic genes, increasing their expression under specific environmental conditions and enhancing the production of phenolics and flavonoids in response to stress (Borevitz et al., 2000).

6.4 Challenges and Future Prospects in Genetic Engineering for Metabolite Accumulation

While significant progress has been made in enhancing secondary metabolite production through genetic engineering, several challenges remain. One of the major challenges is the complex regulation of secondary metabolite pathways, which often involves multiple transcription factors, enzymes, and signaling pathways. Manipulating one part of the pathway can have unintended effects on other metabolic processes, leading to trade-offs between growth and defense. Additionally, the cost of metabolite production can be a limiting factor, as secondary metabolites are often energetically expensive for plants to produce. Overexpression of biosynthetic genes can divert resources away from primary metabolism, reducing plant growth and yield. Finding the right balance between growth and defense remains a key challenge in the field of metabolic engineering. Despite these challenges, advances in synthetic biology and metabolic pathway modeling offer promising solutions for optimizing secondary metabolite production in plants. The integration of machine learning and genome editing technologies will likely lead to new strategies for fine-tuning metabolite biosynthesis, enabling the development of crops with enhanced resistance to stress and higher levels of valuable compounds.

Conclusion

The production of secondary metabolites in plants is intricately linked to a combination of environmental factors and genetic regulation, both of which play a vital role in determining the biosynthesis, accumulation, and function of these compounds. Environmental factors, such as light, temperature, water availability, and nutrient levels, are key drivers that influence the quantity and diversity of secondary metabolites in plants. These compounds, including flavonoids, terpenoids, alkaloids, and phenolics, are crucial for plant defense mechanisms, enabling plants to cope with biotic and abiotic stressors. The impact of environmental factors on secondary metabolite production is profound, as plants adjust their metabolic pathways to produce compounds that protect them from harmful environmental conditions. For instance, the exposure to ultraviolet (UV) light often leads to increased flavonoid biosynthesis, which helps mitigate UV-induced oxidative damage. Similarly, variations in temperature significantly affect the production of essential oils and terpenoids, which assist plants in maintaining homeostasis under thermal stress. This dynamic regulation highlights how plants rely on environmental cues to orchestrate their defense responses.

Light and temperature, in particular, play critical roles in shaping the biosynthesis of secondary metabolites. Light not only serves as an energy source for photosynthesis but also acts as a signal that activates the production of specific secondary metabolites. Different wavelengths of light, such as UV and blue light, stimulate the synthesis of phenolic compounds that act as antioxidants and photoprotective agents. Temperature, on the other hand, modulates the activity of enzymes involved in secondary metabolite biosynthesis. High temperatures often promote the accumulation of terpenoids and essential oils in aromatic plants, which can enhance their resistance to heat stress and reduce water loss. Conversely, low temperatures trigger the production of flavonoids and phenolic acids that protect plants from cold-induced oxidative damage. This complex relationship between light, temperature, and secondary metabolite production underscores the adaptive strategies plants employ to optimize their metabolism under varying environmental conditions.

Nutrient availability also plays a fundamental role in influencing secondary metabolite pathways. Plants experiencing nutrient deficiencies, such as nitrogen or phosphorus limitation, often redirect their metabolic resources toward the production of secondary metabolites rather than primary growth processes. This shift allows plants to enhance their stress tolerance by accumulating phenolic compounds and flavonoids, which help alleviate nutrient stress and promote survival under suboptimal conditions. Nitrogen deficiency, for instance, leads to an increase in phenolic compound production, as these metabolites are essential for strengthening plant cell walls and mitigating oxidative stress. Similarly, phosphorus limitation triggers flavonoid accumulation, which assists in nutrient scavenging and protection against oxidative damage. This nutrient-dependent regulation of secondary metabolites reveals how plants modulate their metabolic processes to balance growth and defense.

Genetic regulation is another critical aspect of secondary metabolite production, with numerous biosynthetic pathways governed by specific genes and enzymes. These biosynthetic pathways are controlled by transcription factors that respond to environmental stimuli and activate or repress the expression of key enzymes involved in metabolite production. Transcription factors such as MYB, WRKY, and bHLH are central to regulating the production of flavonoids, alkaloids, and terpenoids. MYB transcription factors, for example, are known to regulate the phenylpropanoid pathway, which is responsible for the biosynthesis of flavonoids and other phenolics. Similarly, WRKY transcription factors play a significant role in activating defense-related genes that control alkaloid and terpenoid biosynthesis. These regulatory proteins ensure that secondary metabolite production is tightly controlled and responsive to environmental changes, enabling plants to produce the necessary compounds when needed most.

The role of transcription factors in controlling secondary metabolite production cannot be overstated. These proteins act as molecular switches, determining the timing and intensity of metabolite synthesis. Under conditions of biotic stress, such as pathogen attack or herbivory, transcription factors quickly activate the expression of biosynthetic genes, leading to the accumulation of defense-related secondary metabolites. This rapid response is crucial for plant survival, as it enables plants to deploy their chemical defenses in a timely manner. Transcription factors not only regulate the production of antimicrobial compounds but also coordinate the synthesis of signaling molecules that trigger systemic resistance throughout the plant. This level of control ensures that secondary metabolite production is finely tuned to the specific needs of the plant, allowing for efficient resource allocation and effective defense.

Advances in genetic engineering have provided new tools for enhancing the production of secondary metabolites in plants. By manipulating key genes involved in biosynthetic pathways, scientists can increase the accumulation of valuable metabolites, improving plant resistance to environmental stress and pathogens. Techniques such as gene overexpression, gene silencing, and CRISPR-Cas9 genome editing have been used to modify the expression of biosynthetic enzymes and transcription factors, resulting in higher levels of secondary metabolites. For example, the overexpression of genes in the phenylpropanoid pathway has been shown to increase flavonoid production, enhancing plants' ability to withstand UV radiation and herbivorous pests. Similarly, genetic modifications aimed at boosting the production of terpenoids have led to improved resistance to heat and drought stress in several crop species. These genetic interventions offer promising avenues for developing crops with enhanced resilience, reduced reliance on chemical pesticides, and greater adaptability to changing environmental conditions.

However, while genetic engineering holds great potential, it must be approached with caution to avoid potential trade-offs. Enhancing secondary metabolite production can sometimes lead to unintended consequences, such as reduced plant growth or altered reproductive patterns. Therefore, it is essential to strike a balance between defense and growth, ensuring that the production of secondary metabolites does not compromise the plant's overall fitness. Additionally, the long-term ecological impacts of genetically engineered plants must be carefully evaluated, particularly in terms of their interactions with other organisms and their effects on ecosystems.

References

Al-Gabbiesh, A., Kleinwächter, M., & Selmar, D. (2015). Influencing the contents of secondary metabolites in spice and medicinal plants by deliberately applying drought stress during their cultivation. *Jordan Journal of Biological Sciences, 8*(1), 1-10.

Beaudoin, G.A.W., & Facchini, P.J. (2014). Benzylisoquinoline alkaloid biosynthesis in opium poppy. *Planta 240*, 19-32.

Berrocal-Lobo, M., Molina, A., & Solano, R. (2002). Constitutive expression of ETHYLENE-RESPONSE-FACTOR1 in Arabidopsis confers resistance to several necrotrophic fungi. *The Plant Journal, 29*(1), 23-32.

Bieza, K., & Lois, R. (2001). An Arabidopsis mutant tolerant to lethal ultraviolet-B levels shows constitutively elevated accumulation of flavonoids and other phenolics. *Plant Physiology, 126*(3), 1105-1115.

Birkenbihl, R. P., Diezel, C., & Somssich, I. E. (2012). Arabidopsis WRKY33 is a key transcriptional regulator of hormonal and metabolic responses toward Botrytis cinerea infection. *Plant Physiology, 159*(1), 266-285.

Boncan, D. A. T., Tsang, S. S. K., Li, C., Lee, I. H. T., Lam, H. M., Chan, T. F., & Hui, J. H. L. (2020). Terpenes and Terpenoids in Plants: Interactions with Environment and Insects. *International Journal of Molecular Science*, *21*(19), 7382.

Borevitz, J. O., Xia, Y., Blount, J., Dixon, R. A., & Lamb, C. (2000). Activation tagging identifies a conserved MYB regulator of phenylpropanoid biosynthesis. *The Plant Cell, 12*(12), 2383-2394.

Bryant, J. P., Chapin, F. S., & Klein, D. R. (1983). Carbon/nutrient balance of boreal plants in relation to vertebrate herbivory. *Oikos, 40*(3), 357-368.

Cakmak, I. (2005). The role of potassium in alleviating detrimental effects of abiotic stresses in plants. *Journal of Plant Nutrition and Soil Science, 168*(4), 521-530.

Conde, C., Silva, P., Fontes, N., Dias, A., Tavares, R., Sousa, M., Agasse, A., Delrot, S., Gerós, H. (2006). Biochemical changes throughout Grape Berry development and fruit and wine quality. *Food.* 1.

Das, S., Kwon, M., & Kim, J. Y. (2024) Enhancement of specialized metabolites using CRISPR/Cas gene editing technology in medicinal plants. *Frontiers in Plant Science*, *15*, 1279738.

Di Ferdinando, M., Brunetti, C., Fini, A., & Tattini, M. (2012). Flavonoids as Antioxidants in Plants Under Abiotic Stresses. In: Ahmad, P., Prasad, M. (eds) Abiotic Stress Responses in Plants. Springer, New York, NY.

Dixon, R. A. (2001). Natural products and plant disease resistance. *Nature, 411*(6839), 843-847.

Dixon, R. A., & Paiva, N. L. (1995). Stress-induced phenylpropanoid metabolism. *The Plant Cell, 7*(7), 1085-1097.

Dobhal, P., Purohit, V. K., Chandra, S., Rawat S., Prasad, P., Bhandari, U., Trivedi, V. L., & Nautiyal, M. C. (2024). Climate-induced changes in essential oil production and terpene composition in alpine aromatic plants, Plant Stress, Volume 12, 100445.

Espley, R. V., Hellens, R. P., Putterill, J., Stevenson, D. E., Kutty-Amma, S., Allan, A. C. (2007) Red colouration in apple fruit is due to the activity of the MYB transcription factor, MdMYB10. *Plant Journal*, *49*(3), 414-27.

Gershenzon, J. (1984). Changes in the Levels of Plant Secondary Metabolites Under Water and Nutrient Stress. In: Timmermann, B.N., Steelink, C., Loewus, F.A. (eds) Phytochemical Adaptations to Stress. Recent Advances in Phytochemistry, vol 18. Springer, Boston, MA.

Goufo, P., Pereira, J., Moutinho-Pereira, J., Correia, C. M., Figueiredo, N., Carranca, C., Rosa, E. A. S., & Trindade, H. (2014). Rice (*Oryza sativa* L.) phenolic compounds under elevated carbon dioxide (CO2) concentration. *Environmental and Experimental Botany*, *99*, 28-37.

Hoang, X. L. T., Nhi, D. N. H., Thu, N. B. A., Thao, N. P., & Tran, L. P. (2017). Transcription factors and their roles in signal transduction in plants under abiotic stresses. *Current Genomics*, *18*(6), 483-497.

Holopainen, J. K., & Gershenzon, J. (2010). Multiple stress factors and the emission of plant VOCs. *Trends in Plant Science, 15*(3), 176-184.

Howe, G. A. (2004). Jasmonates as signals in the wound response. *Journal of Plant Growth Regulation, 23*, 223-237.

Talha J., & San-Ji, G. (2023). WRKY transcription factors in plant defense. *Trends in Genetics, 39*(10), 787-801.

Khan, F., Siddique, A. B., Shabala, S., Zhou, M., & Zhao, C. (2023). Phosphorus plays key roles in regulating plants' physiological responses to abiotic stresses. *Plants (Basel), 12*(15), 2861.

Koricheva, J., Larsson, S., Haukioja, E., & Keinänen, M. (1998). Regulation of woody plant secondary metabolism by resource availability: hypothesis testing by means of meta-analysis. *Oikos, 83*(2), 212-226.

Koyama, K., Ikeda, H., Poudel, P. R., & Goto-Yamamoto, N. (2012). Light quality affects flavonoid biosynthesis in young berries of Cabernet Sauvignon grape. *Phytochemistry, 78*, 54-64.

Król, A., Amarowicz, R., & Weidner, S. (2015). The effects of cold stress on the phenolic compounds and antioxidant capacity of grapevine (*Vitis vinifera* L.) leaves. *Journal of Plant Physiology. 189*, 97-104.

Kuhlmann, F., & Müller, C. (2010). Impacts of ultraviolet radiation on interactions between plants and herbivorous insects: A chemo-ecological perspective. *Progress in Botany, 71*, 305-340.

Lange, B. M., & Ahkami, A. (2013). Metabolic engineering of plant monoterpenes, sesquiterpenes and diterpenes-current status and future opportunities. *Plant Biotechnol Journal*, 11, 169-196.

Llorente, B., Martinez-Garcia, J. F., Stange, C., & Rodriguez-Concepcion, M. (2017). Illuminating colors: Regulation of carotenoid biosynthesis and accumulation by light. *Current Opinion in Plant Biology, 37*, 49-55.

Lorenzo, O., Piqueras, R., Sánchez-Serrano, J. J., & Solano, R. (2003). Ethylene response factor1 integrates signals from ethylene and jasmonate pathways in plant defense. *The Plant Cell, 15*(1), 165-178.

Lou, Y., & Baldwin, I. T. (2004). Nitrogen supply influences herbivore-induced direct and indirect defenses and transcriptional responses in Nicotiana attenuata. *Plant Physiology, 135*(1), 496-506.

Mancinelli, A. L. (1985). Light-dependent anthocyanin synthesis: A model system for the study of plant photomorphogenesis. *Botanical Review, 51*(1), 107-157.

Menconi, J., Perata, P. & Gonzali, S. (2023). Novel R2R3 MYB transcription factors regulate anthocyanin synthesis in Aubergine tomato plants. *BMC Plant Biology* 23, 148.

Mipeshwaree Devi, A., Khedashwori Devi, K., Premi Devi, P., Lakshmipriyari Devi, M., Das, S. (2023). Metabolic engineering of plant secondary metabolites: prospects and its technological challenges. *Frontiers in Plant Science, 14*, 1171154.

Mira de Orduña, R. (2010). Climate change associated effects on grape and wine quality and production. *Food Research International, 43*(7), 1844-1855.

Neumann, G., Massonneau, A., & Martinoia, E. et al. (1999). Physiological adaptations to phosphorus deficiency during proteoid root development in white lupin. *Planta 208*, 373-382.

Nishad, R., Ahmed, T., Rahman, V. J., & Kareem, A. (2020). Modulation of plant defense system in response to microbial interactions. *Frontiers in Microbiology, 11*, 1298.

Ormeño, E., & Fernandez, C. (2012) Effect of Soil Nutrient on Production and Diversity of Volatile Terpenoids from Plants. *Current Bioactivity Compound*, *8*(1):71-79.

Pant P., Pandey S., & Dall'Acqua, S. (2021). The influence of environmental conditions on secondary metabolites in medicinal plants: A literature review. *Chemistry and Biodiversity*, *18*, e210034.

Pauwels, L., & Goossens, A. (2011). The JAZ proteins: A crucial interface in the jasmonate signaling cascade. *Plant Cell, 23*(9), 3089-3100.

Penuelas, J., & Estiarte, M. (1998). Can elevated CO2 affect secondary metabolism and ecosystem function? *Trends in Ecology & Evolution, 13*(1), 20-24.

Petroni, K., & Tonelli, C. (2011). Recent advances on the regulation of anthocyanin biosynthesis in reproductive organs. *Plant Science, 181*(3), 219-229.

Pieterse, C. M., Van der Does, D., Zamioudis, C., Leon-Reyes, A., & Van Wees, S. C. (2012). Hormonal modulation of plant immunity. *Annual Review of Cell and Developmental Biology, 28*, 489-521.

Qian, Z., He, L., & Li, F. (2024). Understanding cold stress response mechanisms in plants: an overview. *Frontiers in Plant Science*, *15*, 1443317.

Ramakrishna, A., & Ravishankar, G. A. (2011). Influence of abiotic stress signals on secondary metabolites in plants. *Plant Signaling & Behavior, 6*(11), 1720-1731.

Rouxinol, M. I., Martins, M. R., Barroso, J. M., & Rato, A. E. (2023). Wine grapes ripening: A review on climate effect and analytical approach to increase wine quality. *Applied Biosciences*, *2*(3), 347-372.

Sangwan, N. S., Farooqi, A. H. A., & Sangwan, R. S. (2001). Regulation of essential oil production in plants. *Plant Growth Regulation, 34*(1), 3-21.

Selmar, D. (2008). Potential of salt and drought stress to increase pharmaceutical significant secondary compounds in plants. *Landbauforschung Volkenrode, 58*(3), 139-144.

Sharkey, T. D., Wiberley, A. E., & Donohue, A. R. (2008). Isoprene emission from plants: why and how. *Annals of Botany*, *101*(1):5-18.

Shoji, T., & Hashimoto, T. (2011). Tobacco MYC2 regulates jasmonate-inducible Nicotine biosynthesis genes directly and by way of the NIC2-locus ERF genes. *Plant and Cell Physiology*, *52*(6), 1117-1130.

Silva-Beltrán, N. P., Ruiz-Cruz, S., Cira-Chávez, L. A., Estrada-Alvarado, M. I, Ornelas-Paz, J., de J., López-Mata, M. A., Del-Toro-Sánchez, C. L., Ayala-Zavala, J. F., & Márquez-Ríos, E. (2015). Total phenolic, flavonoid, tomatine, and tomatidine contents and antioxidant and antimicrobial activities of extracts of tomato plant. *International Journal of Analytical Chemistry*, 284071.

Sønderby, I. E., Geu-Flores, F., & Halkier, B. A. (2010). Biosynthesis of glucosinolates–gene discovery and beyond. *Trends in Plant Science, 15*(5), 283-290.

Song, C., Saloner, A., Fait, A., & Bernstein, N. (2023). Nitrogen deficiency stimulates cannabinoid biosynthesis in medical cannabis plants by inducing a metabolic shift towards production of low-N metabolites. *Industrial Crops and Products*, *202*, 116969.

Stracke, R., Ishihara, H., Huep, G., Barsch, A., Mehrtens, F., Niehaus, K., & Weisshaar, B. (2007).

Differential regulation of closely related R2R3-MYB transcription factors controls flavonol accumulation in different parts of the *Arabidopsis thaliana* seedling. *Plant Journal, 50*(4), 660-77.

Vlot, A. C., Dempsey, D. A., & Klessig, D. F. (2009). Salicylic acid, a multifaceted hormone to combat disease. *Annual Review of Phytopathology, 47*(1), 177-206.

Zerbe, P., Hamberger, B., Yuen, M. M. S., Chiang, A., Sandhu, H. K., Madilao, L. L., Nguyen, A., Hamberger, B., Bach, S. S., & Bohlmann, J. (2013) Gene discovery of modular diterpene metabolism in nonmodel systems. *Plant Physiology, 162*(2), 1073-1091.

Zhan, X., Chen, Z., Chen, R., & Shen, C. (2022). Environmental and genetic factors involved in plant protection-associated secondary metabolite biosynthesis pathways. *Frontiers in Plant Science, 13*, 877304.

Zhang, S., Zhang, L., Zou, H., Qiu, L., Zheng, Y., Yang, D., & Wang, Y. (2021). Effects of light on secondary metabolite biosynthesis in medicinal plants. *Frontiers in Plant Science, 12*, 781236

Zhong, R., Lee, C., Zhou, J., McCarthy, R. L., & Ye, Z. H. (2008). A battery of transcription factors involved in the regulation of secondary wall biosynthesis in Arabidopsis. *The Plant Cell, 20*(10), 2763-2782.

Zoratti, L., Karppinen, K., Luengo Escobar, A., Häggman, H., & Jaakola, L. (2014). Light-controlled flavonoid biosynthesis in fruits. *Frontiers in Plant Science, 5*, 534.

Chapter - 14

Secondary Metabolites in Rhizosphere Interactions and Disease Suppression

Sheetanshu Gupta[1], Pratima Raypa[2], Susheel Kumar[3], Zuby Gohar Ansari[4] and Vinod Chouhan[5]

[1]Department of Biochemistry, G.B. Pant University of Agriculture & Technology, Pantnagar, Uttarakhand, India

[2]Department of Biochemistry, H.N.B. Garhwal University, Srinagar, Uttarakhand, India

[3]Plant Molecular Virology Laboratory, CPMB Division, CSIR-National Botanical Research Institute, Rana Pratap Marg, Lucknow, Uttar Pradesh, India

[4]B.M. College of Agriculture, Jaswadi Road Khandwa, Madhya Pradesh, India

[5]Division of Plant Pathology, Indian Council of Agricultural Research-Indian Agricultural Research Institute (ICAR-IARI), New Delhi, India

Abstract

Secondary metabolites play a crucial role in the rhizosphere, the zone of soil that surrounds and is influenced by plant roots. In this dynamic environment, secondary metabolites serve as signaling molecules, mediators of plant-microbe interactions, and agents of disease suppression. These compounds, including phenolics, terpenoids, and flavonoids, are secreted by plant roots and influence microbial communities, enhancing beneficial associations with microorganisms like mycorrhizal fungi and rhizobacteria, while suppressing harmful pathogens. Secondary metabolites in the rhizosphere can modulate microbial populations by promoting beneficial organisms that support plant growth and health, while simultaneously inhibiting the proliferation of soil-borne pathogens. This dual function of secondary metabolites is critical for maintaining the delicate balance of rhizosphere microbial communities and protecting plants from root diseases. Furthermore, these compounds can alter the chemical environment of the soil, affecting nutrient availability and promoting allelopathic interactions that inhibit the growth of competing plant species. Recent research highlights the potential of using plant-derived secondary metabolites as natural biocontrol agents to suppress pathogens in agricultural systems, offering an eco-friendly alternative to chemical pesticides. Advances in genetic engineering have further explored the potential to enhance the production of these metabolites in plants, leading to increased disease resistance and improved plant-microbe interactions.

DOI: 10.1201/9781003717454-14

Keywords: Rhizosphere, Secondary Metabolites, Disease Suppression, Microbial Interactions, Allelopathy, Biocontrol

Introduction

Secondary metabolites are crucial in regulating interactions between plants and their surrounding environment, particularly within the rhizosphere, the soil region influenced by plant root exudates. These compounds, including flavonoids, terpenoids, alkaloids, and phenolics, play fundamental roles in structuring microbial communities, facilitating beneficial plant-microbe interactions, suppressing pathogens, and influencing the overall health and function of the rhizosphere. Through these complex interactions, secondary metabolites contribute to the enhancement of plant resilience against biotic and abiotic stresses, while also playing a significant part in nutrient cycling and soil health.

In the rhizosphere, secondary metabolites serve as signaling molecules that shape microbial populations, favoring beneficial organisms like mycorrhizal fungi and rhizobacteria while inhibiting harmful pathogens. These interactions not only influence plant health but also maintain the balance within the soil microbial community, contributing to the sustainability of the ecosystem. Research has shown that phenolic compounds secreted by plant roots can alter microbial diversity in the rhizosphere, promoting the growth of beneficial microbes that enhance nutrient uptake and protect against disease (Gargallo-Garriga et al., 2018). These microbes, in turn, assist plants by breaking down organic matter, solubilizing phosphates, and fixing nitrogen, which directly impacts plant growth and health. Flavonoids, for instance, are key mediators in the symbiosis between leguminous plants and nitrogen-fixing bacteria, such as *Rhizobium* species. By acting as chemoattractants and signaling molecules, flavonoids initiate the nodulation process, which leads to the formation of root nodules where nitrogen fixation occurs (Ghitti et al., 2024).

Plant-microbe interactions in the rhizosphere are often mediated by the secretion of secondary metabolites that facilitate communication between plants and beneficial microorganisms. These interactions can be mutualistic, where both plant and microbe benefit, or they can be antagonistic, where secondary metabolites inhibit harmful organisms. For example, flavonoids and terpenoids exuded by plant roots can promote the colonization of beneficial fungi, such as arbuscular mycorrhizal fungi (AMF), which enhance plant nutrient acquisition, particularly phosphorus. In return, AMF provide plants with improved access to nutrients and increased resistance to pathogens and environmental stresses. Similarly, root-secreted alkaloids have been shown to attract specific rhizobacteria that promote plant growth by producing phytohormones or by facilitating nutrient solubilization (Sasse et al., 2018). These intricate plant-microbe interactions underscore the role of secondary metabolites as pivotal regulators of the rhizosphere's microbial ecology.

The role of secondary metabolites in disease suppression within the rhizosphere is particularly well-documented. Root-secreted phenolics, terpenoids, and alkaloids have been shown to inhibit the growth of soil-borne pathogens, thereby reducing disease incidence in crops. These metabolites can directly target pathogen cell walls, disrupt their metabolic processes, or inhibit their reproductive abilities. Phenolics, for example, have been reported to have antifungal properties, effectively suppressing the growth of pathogenic fungi such as *Fusarium* and *Phytophthora* species (Yang et al., 2022). Similarly, terpenoids have been found to exhibit strong antimicrobial activities, particularly against bacterial pathogens, by disrupting cell membranes and interfering with bacterial enzyme functions. The production of these compounds is often upregulated in response to pathogen attack, highlighting their role in the plant's induced resistance mechanisms.

The suppression of pathogens through root-secreted secondary metabolites is not limited to direct antimicrobial action. These compounds also play a role in modulating the microbial community structure, promoting the proliferation of beneficial microorganisms that outcompete and inhibit harmful pathogens. This form of indirect disease suppression is a crucial aspect of plant defense strategies in the rhizosphere. For instance, certain phenolic compounds can enhance the activity of beneficial rhizobacteria, which produce antibiotics or siderophores that inhibit pathogen growth by depriving them of essential nutrients such as iron (Berendsen et al., 2012). This collaborative interaction between secondary metabolites and beneficial microbes creates a defensive barrier around the plant roots, protecting them from a wide range of soil-borne diseases.

Allelopathy, the chemical inhibition of one plant species by another through the release of secondary metabolites, is another important function of these compounds in the rhizosphere. Allelopathic secondary metabolites, such as phenolic acids, flavonoids, and terpenoids, are released into the soil where they can suppress the germination and growth of competing plant species. This phenomenon allows the allelopathic plant to dominate its environment by limiting competition for resources such as water, light, and nutrients. For example, compounds like p-hydroxybenzoic acid and ferulic acid, exuded by the roots of *Sorghum* species, have been shown to inhibit the growth of neighboring weeds, reducing competition, and enhancing the productivity of the crop (Zeng et al., 2008). Allelopathy is thus a key ecological strategy used by plants to modify the rhizosphere environment in their favor.

Environmental factors such as soil composition, moisture levels, temperature, and light availability significantly influence the production and activity of secondary metabolites in the rhizosphere. Plants adjust the quantity and types of metabolites they produce in response to these environmental conditions to optimize their defense and symbiotic strategies. For instance, drought stress has been shown to increase the production of certain flavonoids and phenolic compounds, which help plants mitigate oxidative stress and maintain root growth under water-limited conditions (Ramakrishna & Ravishankar, 2011). Similarly, nutrient deficiencies, particularly in nitrogen and phosphorus, can lead to an increase in the production of phenolics, which act as protective agents by modulating plant metabolism to conserve resources and enhance stress tolerance (Song et al., 2023).

Temperature and light availability also play a significant role in regulating the biosynthesis of secondary metabolites in the rhizosphere. High temperatures can enhance the production of terpenoids and alkaloids, which protect plants from heat stress and pathogen attack. Conversely, cooler temperatures have been shown to increase the accumulation of phenolic compounds that protect plants from cold-induced oxidative stress (Das et al., 2024). Light, particularly UV radiation, stimulates the production of flavonoids, which act as photoprotectants by absorbing harmful UV rays and reducing damage to plant tissues. These environmental influences underscore the adaptability of secondary metabolite production in response to external stressors, allowing plants to modify their rhizosphere environment to optimize survival and growth.

Recent advancements in genetic engineering have opened new avenues for enhancing the production of secondary metabolites in the rhizosphere to improve disease suppression and plant-microbe interactions. By manipulating key biosynthetic pathways, scientists have been able to increase the levels of specific secondary metabolites that enhance plant resistance to pathogens and improve beneficial microbial associations. Genetic modifications, such as overexpressing genes involved in the phenylpropanoid pathway, have been shown to increase the production of phenolic compounds, leading to enhanced disease resistance in crops such as wheat and rice (Bourgaud et al., 2006). Similarly, CRISPR/Cas9 gene-editing technology has enabled precise modifications to genes involved in the biosynthesis of flavonoids and terpenoids, resulting in plants with improved resilience to both biotic and abiotic stresses (Zhang et al., 2021).

The genetic engineering of secondary metabolite production also holds promise for improving the allelopathic potential of crops, allowing them to suppress weeds more effectively and reduce reliance on chemical herbicides. By enhancing the production of allelopathic compounds such as phenolic acids and flavonoids, researchers have been able to develop crop varieties that are more competitive against weeds, leading to higher yields and reduced environmental impact (Isah, 2019). This approach not only contributes to sustainable agricultural practices but also promotes healthier soil ecosystems by reducing the need for synthetic chemical inputs.

1. Role of Secondary Metabolites in Rhizosphere Microbial Communities

Secondary metabolites play a critical role in shaping the dynamics of rhizosphere microbial communities, which are essential for plant health, nutrient cycling, and disease suppression. The rhizosphere, the narrow zone of soil surrounding plant roots, is a hotspot for interactions between plants and microorganisms. Plants exude various compounds, including secondary metabolites, into the rhizosphere to influence the composition and activity of microbial communities. These metabolites not only act as defense agents against pathogenic organisms but also facilitate symbiotic relationships with beneficial microbes, such as mycorrhizal fungi and rhizobacteria. By modulating the chemical environment of the rhizosphere, secondary metabolites play a pivotal role in enhancing plant fitness and resilience to environmental stresses.

1.1 Root Exudates as Drivers of Microbial Community Structure

Plants release a wide array of secondary metabolites into the rhizosphere, often referred to as root exudates, which include flavonoids, alkaloids, terpenoids, phenolic compounds, and saponins. These compounds act as chemical signals that help shape the microbial composition of the rhizosphere by selectively promoting or inhibiting the growth of certain microbial species (Badri & Vivanco, 2009). For example, flavonoids secreted by legume roots are well-known to attract rhizobia, nitrogen-fixing bacteria that form beneficial symbiotic relationships with the plant. Flavonoids serve as signaling molecules that activate the Nod genes in rhizobia, initiating the formation of root nodules where nitrogen fixation occurs (Weston & Mathesius, 2013). In contrast, other secondary metabolites, such as saponins and terpenoids, have antimicrobial properties that can suppress the growth of pathogenic microorganisms in the rhizosphere, reducing the risk of root infections (Zaynab et al., 2018). By balancing these positive and negative interactions with microbes, plants can exert control over the rhizosphere microbiome, promoting beneficial relationships while inhibiting harmful organisms.

1.2 Defense Against Pathogens: Antimicrobial Effects of Secondary Metabolites

One of the primary roles of secondary metabolites in the rhizosphere is to protect plants from soil-borne pathogens. Various classes of secondary metabolites, including phenolics, alkaloids, and terpenoids, possess antimicrobial properties that inhibit the growth of bacteria, fungi, and other pathogens (Gorlenko et al., 2020). For example, phenolic acids such as caffeic acid and ferulic acid are commonly secreted by plant roots and can suppress the growth of pathogenic fungi like Fusarium oxysporum and Pythium spp., which cause root rot diseases (Lattanzio et al., 2006). Similarly, alkaloids such as berberine and quinoline derivatives are known for their broad-spectrum antimicrobial activity, targeting both bacterial and fungal pathogens in the rhizosphere. These compounds disrupt microbial cell membranes, interfere with enzymatic activity, and inhibit pathogen colonization on plant roots (Wink, 2015). By producing a diverse arsenal of antimicrobial secondary metabolites, plants can reduce pathogen pressure in the rhizosphere and maintain root health.

1.3 Role of Secondary Metabolites in Beneficial Plant-Microbe Interactions

While secondary metabolites act as defense compounds against pathogens, they also play an essential role in promoting beneficial interactions with symbiotic microorganisms in the rhizosphere. Mycorrhizal fungi and plant growth-promoting rhizobacteria (PGPR) are two groups of beneficial microbes that significantly enhance plant health by improving nutrient uptake, stress tolerance, and disease resistance. For instance, flavonoids are key signaling molecules in the establishment of symbiosis between plants and arbuscular mycorrhizal (AM) fungi. These compounds stimulate the germination of fungal spores and the growth of fungal hyphae toward plant roots, facilitating the colonization process (Lanfranco et al., 2018). Once established, AM fungi form a mutualistic relationship with the plant, where they enhance the uptake of phosphorus and other nutrients from the soil in exchange for carbon from the plant. PGPR also benefits from secondary metabolites in the rhizosphere. Compounds such as quorum sensing inhibitors can modulate microbial communication, influencing biofilm formation and microbial activity in the rhizosphere (Jamil et al., 2022). Additionally, certain terpenoids and alkaloids promote the colonization of beneficial bacteria, such as Bacillus and Pseudomonas species, which enhance plant growth by producing auxins, solubilizing phosphorus, and protecting against pathogens through induced systemic resistance (ISR) (Pieterse et al., 2012).

1.4 Indirect Effects of Secondary Metabolites on Rhizosphere Dynamics

Secondary metabolites also have indirect effects on rhizosphere microbial communities by altering the soil chemical environment and influencing nutrient availability. Phenolic compounds, for example, can chelate metal ions such as iron and aluminum, affecting microbial access to these essential nutrients. This, in turn, can influence the composition of microbial communities, as some microbes may be more adept at utilizing chelated nutrients than others (Cesco et al., 2012). Furthermore, the decomposition of secondary metabolites by rhizosphere microbes can generate new compounds with ecological relevance. For instance, the microbial breakdown of flavonoids and terpenoids can lead to the formation of allelochemicals that inhibit the germination and growth of competing plants, contributing to plant dominance in a given environment (Bais et al., 2006). These secondary interactions between plant metabolites and microbes add another layer of complexity to the role of secondary metabolites in shaping the rhizosphere ecosystem.

1.5 Co-Evolution of Plants and Rhizosphere Microbial Communities

The production of secondary metabolites in the rhizosphere is the result of millions of years of co-evolution between plants and microbes. Plants have evolved to secrete specific metabolites that shape the microbial community in ways that benefit plant health and growth. Conversely, microbes have adapted to these chemical environments by developing resistance to antimicrobial compounds or by evolving the ability to metabolize secondary metabolites for their own benefit (Kiers et al., 2010). This dynamic co-evolution has led to the establishment of highly specialized plant-microbe interactions, where both parties derive benefits from the exchange of chemical signals and nutrients. For example, the secretion of isoflavoids by legume roots not only attracts rhizobia but also influences the broader microbial community by enhancing the abundance of bacteria that promote nitrogen fixation and nutrient cycling. In return, rhizobia and other beneficial microbes produce secondary metabolites that induce systemic resistance in the host plant, further enhancing its ability to defend against pathogens (Zamioudis & Pieterse, 2012).

2. Plant-Microbe Interactions Mediated by Secondary Metabolites

Plant-microbe interactions in the rhizosphere are fundamental to plant health, nutrient acquisition, and overall ecosystem function. A critical aspect of these interactions is the mediation by secondary metabolites, a diverse group of compounds produced by plants that serve various ecological roles, including defense, signaling, and symbiosis. These metabolites, which include phenolics, flavonoids, alkaloids, terpenoids, and saponins, are secreted into the rhizosphere and influence the structure and function of microbial communities. By regulating microbial behavior and population dynamics, secondary metabolites mediate a complex network of interactions that either promote beneficial relationships or suppress harmful organisms. Understanding the role of secondary metabolites in plant-microbe interactions is essential for advancing sustainable agriculture and enhancing plant resilience to environmental stress.

2.1 Role of Secondary Metabolites in Symbiotic Interactions

Symbiotic relationships between plants and microbes are crucial for nutrient acquisition and stress tolerance. Secondary metabolites are key players in these interactions, acting as chemical signals that facilitate the establishment of symbioses with beneficial microorganisms such as mycorrhizal fungi and nitrogen-fixing bacteria.

2.1.1 Flavonoids in Legume-Rhizobia Symbiosis

One of the most well-studied examples of plant-microbe symbiosis mediated by secondary metabolites is the interaction between legumes and rhizobia, nitrogen-fixing bacteria. This mutualistic relationship is initiated by the secretion of flavonoids from legume roots into the rhizosphere. Flavonoids serve as signaling molecules, inducing the expression of Nod genes in rhizobia, which are responsible for the synthesis of Nod factors-key signaling molecules that trigger root nodule formation in legumes (Oldroyd & Downie, 2008). Once the rhizobia colonizes the root, it forms nodules where nitrogen fixation takes place. This interaction is crucial for converting atmospheric nitrogen into ammonia, which the plant can use for growth. In return, the plant supplies the rhizobia with carbohydrates. The specificity of this interaction is largely driven by the structure of the flavonoids produced by the plant, which determines the compatibility between the legume and its rhizobial partner (Wasson et al., 2009).

2.1.2 Terpenoids in Mycorrhizal Symbiosis

Terpenoids also play an essential role in mediating symbiotic interactions between plants and mycorrhizal fungi. These fungi form mutualistic associations with the roots of many plant species, enhancing the uptake of nutrients such as phosphorus and nitrogen. In return, the fungi receive carbon compounds from the plant. Terpenoids, such as strigolactones, are exuded by plant roots to stimulate the growth and branching of mycorrhizal fungal hyphae, facilitating root colonization (Akiyama et al., 2005). Strigolactones are also involved in the regulation of plant architecture and development, making them key molecules in balancing growth and symbiosis. In the rhizosphere, they act as signaling molecules that not only attract mycorrhizal fungi but also inhibit the germination of parasitic weeds such as Striga and Orobanche, demonstrating their dual role in beneficial symbiosis and defense (Wang et al., 2023).

2.2 Defense Against Pathogens and Herbivores

Secondary metabolites are crucial for plant defense, particularly in protecting against pathogenic microbes and herbivores. These compounds can have direct antimicrobial or antifeedant properties, or they may act indirectly by attracting beneficial microbes that enhance plant defense through induced systemic resistance (ISR).

2.2.1 Antimicrobial Properties of Phenolics and Alkaloids

Phenolic compounds such as phenolic acids, flavonoids, and tannins are well-known for their antimicrobial properties. In the rhizosphere, these compounds can inhibit the growth of pathogenic bacteria and fungi, preventing root infections and other soil-borne diseases. For example, ferulic acid and p-coumaric acid, two common phenolic acids found in root exudates, have been shown to suppress the growth of *Fusarium oxysporum* and *Rhizoctonia solani*, two major fungal pathogens that cause root rot in many crops (Harborne & Williams, 2000). Alkaloids such as quinolizidine alkaloids and berberine also exhibit potent antimicrobial activity. These compounds disrupt microbial cell walls, interfere with DNA replication, and inhibit enzyme activity, making them effective deterrents against both bacterial and fungal pathogens (Wink, 2015). In addition to their antimicrobial properties, alkaloids can act as deterrents to herbivores, reducing damage to roots and leaves.

2.2.2 Induced Systemic Resistance (ISR) via Plant Growth-Promoting Rhizobacteria (PGPR)

Secondary metabolites also play a role in the recruitment of beneficial microbes that enhance plant defense through induced systemic resistance (ISR). ISR is a plant defense mechanism that is activated by beneficial microbes, particularly plant growth-promoting rhizobacteria (PGPR), such as Pseudomonas and Bacillus species. These bacteria colonize plant roots and produce secondary metabolites that stimulate plant immune responses, priming the plant for future attacks by pathogens or herbivores (Pieterse et al., 2012). Phenolics and terpenoids secreted by plant roots can promote the colonization of beneficial microbes in the rhizosphere. Once established, these microbes enhance the production of defense-related metabolites in the plant, such as jasmonic acid (JA), salicylic acid (SA), and ethylene (ET), which are key signaling molecules in plant immunity (Zamioudis & Pieterse, 2012). By recruiting beneficial microbes through the secretion of secondary metabolites, plants can enhance their resistance to a wide range of biotic stresses.

2.3 Quorum Sensing and Microbial Communication

Secondary metabolites also influence quorum sensing, a form of microbial communication that regulates population density and behavior in bacterial communities. Quorum sensing inhibitors (QSIs), such as flavonoids and alkaloids, can disrupt microbial signaling, preventing the formation of harmful bacterial biofilms and reducing the virulence of pathogens (Jamil et al., 2022). By interfering with quorum sensing, plants can reduce the pathogenicity of bacteria in the rhizosphere and protect their roots from infection. For example, quorum-quenching flavonoids secreted by legume roots can inhibit the quorum-sensing systems of Agrobacterium tumefaciens, a bacterial pathogen that causes crown gall disease in many plants. By disrupting microbial communication, these flavonoids prevent the formation of galls on plant roots and stems (Majdura et al., 2023). This form of chemical defense allows plants to protect themselves from bacterial infections without directly killing the bacteria, reducing the selective pressure for resistance development.

2.4 Allelopathy and Competitive Interactions

Secondary metabolites can also mediate competitive interactions between plants and other organisms in the rhizosphere through a process known as allelopathy. Allelopathy refers to the release of chemical compounds by one plant that inhibits the growth or germination of neighboring plants. Many secondary metabolites, such as phenolics, terpenoids, and saponins, act as allelochemicals, giving the producing plant a competitive advantage in nutrient acquisition and space occupation. For example, juglone, a phenolic compound produced by black walnut (*Juglans nigra*), is a well-known allelochemical that inhibits the germination and growth of surrounding plants, reducing competition for resources (Bais et al., 2006). Similarly, saponins secreted by the roots of certain grasses can inhibit the growth of neighboring plants by disrupting cell membrane integrity and reducing nutrient uptake (Zaynab et al., 2018). These allelopathic interactions are a key aspect of plant-plant competition in the rhizosphere and contribute to the dominance of certain plant species in specific environments.

2.5 Co-Evolution of Plants and Microbes Mediated by Secondary Metabolites

The interactions between plants and microbes in the rhizosphere are shaped by co-evolutionary processes, where plants have evolved to produce secondary metabolites that modulate microbial behavior, and microbes have adapted to these chemical environments. This co-evolutionary arms race has led to the development of highly specialized interactions, where plants use secondary metabolites to recruit beneficial microbes and deter pathogens, while microbes evolve mechanisms to degrade or tolerate these compounds. For example, some beneficial microbes, such as Pseudomonas species, have evolved the ability to degrade plant-produced phenolic compounds and use them as a carbon source. This ability allows these microbes to thrive in the rhizosphere and establish beneficial relationships with the plant (Weller et al., 2002). On the other hand, pathogens such as Fusarium have developed resistance to certain antimicrobial secondary metabolites, highlighting the dynamic nature of plant-microbe co-evolution.

3. Disease Suppression through Root-Secreted Secondary Metabolites

Root-secreted secondary metabolites play a pivotal role in disease suppression within the rhizosphere, the soil zone surrounding plant roots. These compounds, which include phenolics, terpenoids, alkaloids, and flavonoids, serve as a plant's frontline defense against a wide array of soil-borne pathogens, including bacteria, fungi, oomycetes, and nematodes. By directly inhibiting pathogen growth or altering the microbial dynamics in the rhizosphere, root-exuded secondary metabolites help protect plants from root diseases, promoting plant health and resilience. Additionally, these compounds can induce systemic resistance in plants, preparing them to defend against future pathogen attacks. The mechanisms by which these metabolites suppress disease are crucial for advancing sustainable agricultural practices that minimize the reliance on chemical pesticides.

3.1 Direct Antimicrobial Effects of Secondary Metabolites

One of the most direct ways in which root-secreted secondary metabolites contribute to disease suppression is through their antimicrobial properties. Many secondary metabolites are toxic to a range of soil-borne pathogens, inhibiting their growth and preventing infections. For instance, phenolic acids such as ferulic acid, caffeic acid, and p-coumaric acid are commonly exuded by plant roots and can inhibit the growth of pathogenic fungi like *Fusarium oxysporum*, *Pythium spp.*, and *Rhizoctonia solani,* which cause diseases like root rot and damping-off in various crops (Lattanzio et al., 2006). These phenolic compounds disrupt fungal cell walls, interfere with enzyme activity, and inhibit spore germination, effectively reducing pathogen populations in the rhizosphere. In addition to phenolics, alkaloids also exhibit strong antimicrobial activity. Compounds like berberine and quinine, secreted by plant roots, are effective against both bacterial and fungal pathogens. Berberine, for example, is known to target the bacterial cell membrane, disrupting its integrity, and leading to cell death (Wink, 2015). The broad-spectrum antimicrobial effects of alkaloids make them valuable in suppressing soil-borne pathogens in the rhizosphere. Saponins, a group of steroidal and triterpenoidal glycosides, also contribute to disease suppression by exerting antifungal and antibacterial effects. These compounds have been shown to disrupt fungal cell membranes and inhibit the growth of pathogens like Phytophthora and Fusarium, which are responsible for significant crop losses worldwide (Zaynab et al., 2018). Through these direct antimicrobial actions, root-secreted secondary metabolites form an important part of the plant's defense strategy against soil-borne diseases.

3.2 Modulation of Microbial Communities for Disease Suppression

Beyond their direct antimicrobial effects, root-exuded secondary metabolites also modulate the microbial community in the rhizosphere, promoting beneficial microbes that contribute to disease suppression. By shaping the composition of the rhizosphere microbiome, plants can enhance the populations of plant growth-promoting rhizobacteria (PGPR) and mycorrhizal fungi, both of which play crucial roles in protecting plants from pathogens. For instance, flavonoids secreted by plant roots not only act as signaling molecules for beneficial microbes but also selectively inhibit pathogenic microorganisms. Flavonoids can promote the colonization of beneficial bacteria such as Pseudomonas fluorescens and Bacillus subtilis, which suppress soil-borne pathogens through the production of antimicrobial compounds like siderophores, antibiotics, and lipopeptides (Pieterse et al., 2012). These beneficial microbes also induce systemic resistance in the plant, priming it to defend against future pathogen attacks. Terpenoids, another group of secondary metabolites secreted by plant roots, also play a role in modulating microbial dynamics in the rhizosphere. Certain terpenoids, such as camalexin and glyceollin, are known to suppress pathogenic bacteria and fungi while promoting the growth of beneficial microbes that contribute to disease suppression (Tholl, 2015). By influencing the microbial composition of the rhizosphere, root-secreted terpenoids create an environment that is less conducive to pathogen establishment and more favorable to beneficial microbial activity.

3.3 Induced Systemic Resistance (ISR) and Disease Suppression

In addition to their antimicrobial and microbiome-modulating effects, root-secreted secondary metabolites play a crucial role in inducing systemic resistance in plants. Induced systemic resistance (ISR) is a defense mechanism in which beneficial microbes, particularly PGPR and mycorrhizal fungi, trigger a systemic immune response in the plant, enhancing its ability to resist a broad range of pathogens. The activation of ISR is often mediated by secondary metabolites secreted by the roots, which stimulate the colonization of beneficial microbes. These microbes, in turn, produce signaling molecules such as jasmonic acid (JA) and ethylene (ET) that activate the plant's defense responses. Once ISR is activated, plants produce defense-related proteins and secondary metabolites, such as phenolics, flavonoids, and alkaloids, in greater quantities, enhancing their resistance to both local and systemic pathogen attacks (Yu et al., 2019). For example, *Pseudomonas fluorescens*, a well-known PGPR, colonizes plant roots and induces ISR, leading to increased production of phenolic compounds in the shoots and leaves. These phenolics provide enhanced resistance against pathogens like *Botrytis cinerea* and *Pseudomonas syringae* (Pieterse et al., 2012). By leveraging the beneficial effects of root-secreted secondary metabolites, plants can activate a systemic defense network that offers long-lasting protection against a wide array of diseases.

3.4 Nematode Suppression by Root-Secreted Metabolites

Root-secreted secondary metabolites also play a critical role in suppressing nematode populations in the rhizosphere. Nematodes are a significant group of soil-dwelling parasites that infect plant roots, leading to reduced nutrient uptake, stunted growth, and yield losses. Certain secondary metabolites, particularly alkaloids, and terpenoids, are effective in deterring nematodes and reducing their populations around plant roots. For example, plants in the Solanaceae family, such as tomato and tobacco, secrete solanine, a glycoalkaloid that is toxic to nematodes. This compound disrupts the nematode's nervous system, preventing it from feeding on plant roots (Chitwood, 2002). Similarly, terpenoid compounds such as alpha-terpinene and beta-caryophyllene exuded by maize roots have been shown to repel root-knot nematodes (*Meloidogyne* spp.), reducing root infections and crop damage (Rasmann et al., 2005). By exuding these nematode-toxic compounds into the rhizosphere, plants can effectively suppress nematode populations and prevent root damage.

3.5 Allelopathic Effects on Pathogens

In addition to their antimicrobial and nematode-suppressing roles, root-secreted secondary metabolites can have allelopathic effects on soil-borne pathogens. Allelopathy refers to the release of chemical compounds by plants that inhibit the growth or germination of other organisms. Many secondary metabolites, such as phenolics and terpenoids, act as allelochemicals, suppressing the growth of pathogenic microbes in the rhizosphere. For example, juglone, a phenolic compound secreted by black walnut (*Juglans nigra*), has strong allelopathic effects on soil fungi, inhibiting the growth of pathogens such as *Verticillium dahliae*, which causes wilt diseases in many crops (Bais et al., 2006). Similarly, saponins secreted by leguminous plants have been shown to inhibit the growth of *Phytophthora* species, a group of oomycetes responsible for root rot diseases in a variety of plants (Zaynab et al., 2018).

Through these allelopathic interactions, root-secreted secondary metabolites reduce the pathogen load in the rhizosphere, enhancing plant health and productivity.

3.6 Co-Evolution of Plants and Pathogens in the Rhizosphere

The production of root-secreted secondary metabolites as a defense mechanism against pathogens is the result of millions of years of co-evolution between plants and their microbial adversaries. Plants have evolved the ability to synthesize and secrete a diverse array of antimicrobial compounds, while pathogens have developed mechanisms to evade or detoxify these metabolites. This ongoing evolutionary arms race has resulted in the fine-tuning of root exudation strategies that maximize plant defense while minimizing pathogen adaptation. For example, certain soil-borne pathogens, such as *Pythium spp.*, have evolved to detoxify phenolic compounds secreted by plant roots, allowing them to colonize plant tissues despite the presence of these defense compounds. In response, plants may alter the composition or concentration of their exudates to counteract the pathogen's detoxification mechanisms (Bakker et al., 2018). This dynamic co-evolutionary relationship highlights the complexity of disease suppression strategies mediated by root-secreted secondary metabolites.

4. Allelopathic Effects of Secondary Metabolites in the Rhizosphere

Allelopathy refers to the biological phenomenon by which plants release secondary metabolites into the surrounding environment to influence the growth, survival, or reproduction of other organisms, particularly neighboring plants, and microbes. In the rhizosphere, the narrow zone of soil influenced by plant roots, these allelochemicals play a critical role in shaping plant-plant interactions, microbial community composition, and overall soil health. By secreting allelochemicals, plants can gain a competitive advantage over neighboring species, suppress soil-borne pathogens, and modulate the dynamics of beneficial microbes. A variety of secondary metabolites, including phenolics, terpenoids, alkaloids, and saponins, have been identified as key players in allelopathic interactions, influencing the structure and function of the rhizosphere.

4.1 Phenolic Compounds and Allelopathy

Phenolic compounds are one of the most widely studied classes of allelochemicals and are known to significantly influence both plant-plant and plant-microbe interactions in the rhizosphere. These compounds are released into the soil via root exudates, leaf leachates, or decomposing plant material, and they have been shown to inhibit the germination and growth of competing plants as well as microbial pathogens. One prominent example of allelopathy mediated by phenolic compounds is the secretion of juglone by *Juglans nigra* (*black walnut*). Juglone, a quinone derivative, is toxic to many plant species, inhibiting seed germination and root elongation. It interferes with energy production by disrupting cellular respiration and inhibiting ATP synthesis, leading to reduced growth and death in sensitive plant species (Rietveld, 1983). This allelopathic effect allows *Juglans nigra* to dominate its growing area by suppressing competition from other plants.

Similarly, phenolic acids such as ferulic acid, caffeic acid, and p-coumaric acid, which are commonly exuded by plant roots, have been shown to exert strong allelopathic effects in the rhizosphere. These compounds can inhibit the growth of neighboring plant roots by disrupting cell wall formation and interfering with hormone signaling pathways. For instance, ferulic acid, a common phenolic exudate of grasses and cereals, inhibits the growth of neighboring weeds by blocking auxin transport, a key hormone involved in root elongation (Li et al., 2010). Phenolics also impact the microbial community in the rhizosphere. The allelopathic effect of phenolic acids can suppress soil-borne pathogens by inhibiting fungal and bacterial growth. Cinnamic acid and its derivatives, for example, are effective against a range of fungal pathogens, including *Fusarium oxysporum* and *Pythium ultimum*, which cause root rot and damping-off diseases (Inderjit, 1996). By shaping the microbial composition of the rhizosphere, phenolics help create a soil environment that is less conducive to pathogen colonization, providing plants with a defense mechanism against diseases.

4.2 Terpenoids as Allelochemicals

Terpenoids, a diverse group of secondary metabolites, are also involved in allelopathic interactions in the rhizosphere. These compounds, which are often volatile, are secreted by plant roots or released from decaying plant material and can have inhibitory effects on the growth of neighboring plants and microorganisms. One well-documented example of terpenoid-mediated allelopathy is the secretion of eugenol and camphor by *Salvia officinalis* (*sage*). These terpenoids inhibit the germination and growth of competing plant species in the rhizosphere, allowing *Salvia officinalis* to maintain dominance in its habitat. Eugenol has been shown to disrupt cellular membranes and interfere with the enzymatic processes of neighboring plants, leading to reduced growth (Lorenzo et al., 2008). Monoterpenes, such as alpha-pinene and beta-pinene, which are secreted by species like Pinus spp. (*pine trees*), are another example of allelopathic terpenoids. These compounds are highly volatile and can accumulate in the soil, where they inhibit the germination of competing plant seeds and disrupt root elongation. In particular, *Pinus* trees release terpenoids into the soil that reduce the growth of understory plants, giving *Pinus* species a competitive advantage in forest ecosystems (Cheng & Cheng, 2015). Monoterpenes can also have antimicrobial effects, inhibiting soil pathogens and promoting a healthier rhizosphere for the host plant.

4.3 Alkaloids and Saponins in Allelopathy

Alkaloids and saponins are other important classes of secondary metabolites involved in allelopathic interactions in the rhizosphere. These compounds, although primarily known for their roles in plant defense against herbivores and pathogens, also influence plant-plant competition and microbial dynamics in the soil. Alkaloids such as quinolizidine and morphine are secreted by certain leguminous plants, including Lupinus spp. (*lupins*) and *Papaver somniferum* (*opium poppy*), and have strong allelopathic effects on neighboring plants. These compounds interfere with cell division and protein synthesis, inhibiting the germination and growth of competing species (Wink, 2015). In addition to their phytotoxic effects, alkaloids can also impact soil microbial communities by inhibiting the growth of pathogenic bacteria and fungi. Saponins, a group of glycosides, are known for their detergent-like properties, which disrupt cell membranes. Plants such as *Trifolium spp.* (*clover*) and *Quillaja saponaria* (*soapbark tree*) release saponins into the rhizosphere, where they inhibit the growth of both neighboring plants and soil-borne pathogens (Zaynab et al., 2018). Saponins interfere with nutrient uptake and disrupt the structural integrity of plant cells, reducing the growth of nearby competing species. Their antimicrobial properties also help suppress fungal pathogens, reducing the incidence of root diseases in the rhizosphere.

4.4 Role of Secondary Metabolites in Microbial Allelopathy

Secondary metabolites also mediate microbial allelopathy, where plants suppress the growth of harmful soil microorganisms through the release of antimicrobial compounds. This type of allelopathy is critical in disease suppression, as it helps protect plants from infections caused by soil-borne pathogens. For example, the secretion of isothiocyanates by *Brassica spp.* (*mustard plants*) is a well-known mechanism of microbial allelopathy. These sulfur-containing compounds are derived from the hydrolysis of glucosinolates, a class of secondary metabolites present in many Brassicaceae species. Isothiocyanates are highly toxic to a wide range of soil pathogens, including *Rhizoctonia solani* and *Verticillium dahliae*, which cause severe root diseases in many crops (Halkier & Gershenzon, 2006). By releasing isothiocyanates into the rhizosphere, *Brassica* plants can suppress pathogen populations and reduce disease pressure. In addition to their antimicrobial effects, allelopathic secondary metabolites can also influence the competitive dynamics of microbial communities. Certain flavonoids and phenolics exuded by plant roots act as signaling molecules that selectively inhibit pathogenic microorganisms while promoting the growth of beneficial microbes, such as plant growth-promoting rhizobacteria (PGPR) and mycorrhizal fungi (Cesco et al., 2012). This selective recruitment of beneficial microbes further enhances plant health by creating a protective microbial barrier around the roots.

4.5 Ecological Implications of Allelopathy in the Rhizosphere

The allelopathic interactions mediated by secondary metabolites have significant ecological implications for plant communities and soil health. Allelopathy plays a key role in plant competition, allowing certain species to establish dominance by inhibiting the growth of competitors. For instance, invasive species such as *Centraurea maculosa* (*spotted knapweed*) release allelopathic compounds like catechin, which inhibit the growth of native plant species and facilitate the spread of the invader (Weir et al., 2004). Moreover, allelopathy can influence the succession of plant communities and the biodiversity of ecosystems. By altering the composition of plant species in a given area, allelopathic plants can reduce plant diversity and modify the structure of the ecosystem. In agricultural settings, allelopathic crops such as *Sorghum bicolor* (*sorghum*) and *Triticum aestivum* (*wheat*) are often used in crop rotation systems to suppress weeds and reduce the need for herbicides (Weston & Duke, 2003). These practices highlight the potential of allelopathy as a sustainable agricultural tool for weed management.

5. Environmental Factors Influencing Metabolite Production in the Rhizosphere

The production of secondary metabolites in the rhizosphere is highly influenced by various environmental factors. These metabolites, secreted by plant roots, play critical roles in plant defense, communication with microorganisms, and shaping the structure of the rhizosphere microbial communities. Environmental factors such as light, temperature, nutrient availability, water stress, and soil composition affect the biosynthesis, secretion, and activity of these secondary metabolites. As plants are sessile organisms, their ability to adjust to environmental changes is largely dependent on the modulation of their biochemical pathways, including the production of secondary metabolites. Understanding how environmental conditions impact metabolite production is essential for optimizing crop productivity and resilience, particularly in changing climates.

5.1 Light Intensity and Quality

Light is one of the most critical factors influencing the biosynthesis of secondary metabolites in plants. Light affects not only photosynthesis but also the production of a wide array of secondary metabolites, including flavonoids, phenolic acids, and terpenoids, which are then exuded into the rhizosphere. Light intensity plays a direct role in the accumulation of carbon-based secondary metabolites. For instance, flavonoid production in the roots of plants such as *Arabidopsis thaliana* and *Glycine max* (soybean) increases under high light intensity, as these compounds act as antioxidants that protect the plant from UV damage (Zoratti et al., 2014). Increased production of flavonoids in the roots can influence the microbial dynamics in the rhizosphere by promoting beneficial microorganisms, such as mycorrhizal fungi and plant growth-promoting rhizobacteria (PGPR), that rely on these compounds as signaling molecules or carbon sources (Badri & Vivanco, 2009). The quality of light referring to its wavelength also significantly impacts secondary metabolite production. Blue and UV-B light has been shown to enhance the production of flavonoids and phenolic acids in plant roots, compounds that not only act as UV protectants but also possess antimicrobial properties, suppressing the growth of soil-borne pathogens (Llorente et al., 2017). For example, plants exposed to higher levels of UV-B radiation often increase their secretion of phenolics and flavonoids into the rhizosphere, which in turn influences the composition of microbial communities by selectively inhibiting pathogenic fungi and bacteria.

5.2 Temperature

Temperature is another environmental factor that exerts a strong influence on secondary metabolite production in the rhizosphere. Both high and low temperatures can affect the enzymatic activities involved in the biosynthesis of these compounds, leading to changes in their secretion and concentration in the soil environment. At elevated temperatures, plants often increase the production of secondary metabolites such as terpenoids and alkaloids, which serve as defense compounds against herbivores and pathogens. For example, in Mentha spp. (mint), higher temperatures have been shown to enhance the production of essential oils, which include terpenoids with antimicrobial and nematicidal properties (Sangwan et al., 2001). These compounds, when secreted into the rhizosphere, can suppress pathogenic nematodes and bacteria, contributing to disease suppression in the root zone. Conversely, low temperatures can induce the accumulation of phenolic compounds such as anthocyanins and flavonoids, which help plants cope with cold-induced oxidative stress. In species like Vitis vinifera (grapevine), cold temperatures stimulate the production of these compounds, which can influence microbial activity in the rhizosphere by inhibiting the growth of cold-sensitive pathogens (Mancinelli, 1985). The production of root-secreted phenolics in response to cold stress can also enhance the recruitment of cold-tolerant microbial communities, which play a role in protecting the plant from freezing damage.

5.3 Nutrient Availability

Nutrient availability is a key determinant of secondary metabolite production in the rhizosphere. The levels of nitrogen (N), phosphorus (P), and potassium (K), among other essential nutrients, can influence both the quantity and quality of secondary metabolites secreted by plant roots. Under conditions of nitrogen deficiency, plants often increase the production of carbon-based secondary metabolites, such as phenolics and flavonoids, as a compensatory mechanism. This is because nitrogen-limited plants allocate more carbon to secondary metabolite production rather than to growth processes (Bryant et al., 1983). For example, in *Medicago sativa* (alfalfa), nitrogen deficiency has been shown to increase the secretion of isoflavonoids into the rhizosphere, enhancing the plant's defense against root pathogens such as Phytophthora and Fusarium species (Lillo et al., 2008). In contrast, phosphorus deficiency often leads to increased production of phenolics and terpenoids in plant roots. *Lupinus albus* (white lupin), for example, secretes higher levels of phenolic acids under phosphorus stress, which act as antimicrobial agents against soil pathogens and contribute to nutrient mobilization in the rhizosphere (Neumann et al., 2000). These compounds help to protect the plant from phosphorus-deficient stress while simultaneously reshaping the microbial community by promoting the growth of phosphorus-solubilizing bacteria.

5.4 Water Availability and Drought Stress

Water availability plays a critical role in the regulation of secondary metabolite production, particularly during periods of drought stress. When plants experience water scarcity, they often shift their metabolic priorities toward the production of defensive secondary metabolites that help mitigate stress and enhance resilience. During drought stress, plants increase the production of phenolic compounds, alkaloids, and terpenoids, which are secreted into the rhizosphere to defend against opportunistic pathogens that thrive under dry conditions. For example, in *Sorghum bicolor* (sorghum), drought stress induces the accumulation of sorgoleone, a phenolic compound with potent allelopathic and antimicrobial properties. Sorgoleone not only suppresses weed competition but also inhibits the growth of fungal pathogens like Aspergillus and Rhizoctonia, which cause root diseases during water-limited conditions. Root-secreted alkaloids also play a role in drought-induced defense. In species like *Nicotiana tabacum* (tobacco), drought stress increases the secretion of nicotine, an alkaloid that deters herbivores and inhibits pathogen growth in the rhizosphere (Lou & Baldwin, 2004). By modulating the production of these compounds in response to water availability, plants can enhance their survival and protect their root systems from biotic stressors.

5.5 Soil Composition and pH

The composition of the soil and its pH significantly affect the production and activity of secondary metabolites in the rhizosphere. The availability of nutrients, the presence of metal ions, and the soil's microbial activity are all influenced by its composition, which in turn impacts the secretion of root metabolites. Acidic soils, for instance, can lead to the increased production of phenolic acids and flavonoids in plant roots, as these compounds help in the chelation of metal ions like iron (Fe) and aluminum (Al), which become more soluble in acidic conditions. Plants such as *Oryza sativa* (rice) increase the secretion of flavonoids under acidic soil conditions to modulate microbial activity and mitigate the toxic effects of aluminum on root growth (Ku et al., 2020). Soil texture also plays a role in determining the retention and mobility of secondary metabolites in the rhizosphere. Sandy soils, which have low water-holding capacity, may lead to rapid leaching of root exudates, whereas clay-rich soils can retain secondary metabolites longer, allowing them to have more sustained effects on the microbial community (Cesco et al., 2012).

6. Genetic Engineering for Enhanced Rhizosphere Metabolite Production

Genetic engineering offers promising avenues for enhancing the production of secondary metabolites in the rhizosphere, which can improve plant defense, enhance beneficial microbial interactions, and contribute to overall plant health and productivity. By manipulating specific genes involved in the biosynthesis, regulation, and transport of secondary metabolites, it is possible to increase the levels of these compounds in plant root exudates, thus optimizing their roles in plant-microbe interactions and pathogen suppression. This approach has potential applications in sustainable agriculture, where genetically engineered crops can be developed to naturally produce higher amounts of beneficial compounds, reducing the need for chemical fertilizers and pesticides.

6.1 Genetic Engineering of Biosynthetic Pathways

One of the main targets for enhancing rhizosphere metabolite production is the biosynthetic pathways responsible for the production of secondary metabolites such as flavonoids, terpenoids, phenolics, and alkaloids. By overexpressing key enzymes or regulatory genes in these pathways, it is possible to significantly increase the levels of metabolites secreted by plant roots. For instance, the overexpression of genes involved in flavonoid biosynthesis, such as chalcone synthase (CHS) and flavonoid 3′-hydroxylase (F3′H), can lead to increased production of flavonoids in plants like Arabidopsis thaliana and *Glycine max* (soybean). These flavonoids play important roles in symbiotic interactions with rhizobia and mycorrhizal fungi, as well as in the suppression of soil-borne pathogens (Zamioudis & Pieterse, 2012). The enhanced production of flavonoids through genetic engineering could improve plant nutrient uptake, enhance disease resistance, and increase crop yields. In another example, the overexpression of terpene synthase (TPS) genes has been shown to increase the production of terpenoids in plant roots. Terpenoids such as squalene and beta-caryophyllene are involved in attracting beneficial microbes and deterring harmful soil pathogens (Tholl, 2006). By genetically engineering crops to overproduce these compounds, it is possible to enhance rhizosphere health and reduce the incidence of root diseases.

6.2 Transcription Factor Overexpression for Enhanced Metabolite Production

The regulation of secondary metabolite biosynthesis is often controlled by transcription factors (TFs), which bind to the promoter regions of biosynthetic genes and modulate their expression. Genetic engineering strategies that target these transcription factors can lead to increased metabolite production in the rhizosphere. For instance, overexpression of MYB transcription factors has been used to enhance the production of flavonoids and phenolic compounds in plants. In *Arabidopsis thaliana*, overexpression of AtMYB12 leads to increased flavonoid production, which enhances the plant's defense against pathogens and promotes beneficial microbial associations in the rhizosphere (Stracke et al., 2007). Similarly, the overexpression of WRKY and bHLH transcription factors has been shown to increase the production of phenolics and terpenoids, leading to improved resistance against root pathogens (Zhong et al., 2008). This approach has significant potential for crop improvement, as transcription factor overexpression can lead to coordinated upregulation of entire biosynthetic pathways, resulting in higher metabolite production. For example, genetically engineered rice (*Oryza sativa*) with overexpressed OsMYB transcription factors showed enhanced secretion of flavonoids into the rhizosphere, improving the plant's resistance to *Rhizoctonia solani*, a major fungal pathogen responsible for sheath blight disease (Peniche-Pavía et al., 2022).

6.3 CRISPR-Cas9 for Precise Gene Editing

The development of CRISPR-Cas9 gene-editing technology has revolutionized genetic engineering by allowing for precise modifications to be made to specific genes involved in secondary metabolite production. By knocking out negative regulators of metabolite biosynthesis or activating key genes, it is possible to enhance metabolite secretion into the rhizosphere. For example, using CRISPR-Cas9, researchers have successfully knocked out JAZ repressor proteins in the jasmonic acid (JA) signaling pathway, which negatively regulate the production of JA-responsive secondary metabolites. By removing these repressors, plants can constitutively produce higher levels of defense-related compounds such as terpenoids and alkaloids (Zhang et al., 2021). This leads to enhanced rhizosphere health, as these compounds not only deter pathogens but also attract beneficial microbes. CRISPR-Cas9 can also be used to modify the expression of genes encoding transport proteins responsible for exporting secondary metabolites from plant roots into the rhizosphere. By enhancing the expression of transporters such as ABC transporters, it is possible to increase the exudation of metabolites like flavonoids and phenolics, thereby improving plant-microbe interactions and pathogen resistance (Yazaki, 2005).

6.4 Engineering for Increased Metabolite Transport

The transport of secondary metabolites from plant roots into the rhizosphere is a critical step in their ecological function. Genetic engineering can be used to enhance the activity of metabolite transporters, thereby increasing the number of secondary metabolites secreted into the soil environment. For example, ATP-binding cassette (ABC) transporters are involved in the export of flavonoids, terpenoids, and phenolics from plant cells into the rhizosphere. By overexpressing genes encoding these transporters, researchers can increase the secretion of these metabolites, enhancing their roles in allelopathy, pathogen suppression, and symbiotic interactions (Ghitti et al., 2024). In crops like *Zea mays* (maize) and *Triticum aestivum* (wheat), increased transporter activity can lead to higher secretion of phenolic acids into the rhizosphere, which suppresses the growth of pathogenic fungi and promotes the colonization of beneficial microbes.

6.5 Engineering Plants for Enhanced Symbiotic Interactions

Genetic engineering can also be used to enhance symbiotic interactions between plants and beneficial microbes by increasing the production of root exudates that serve as signaling molecules. For example, flavonoids secreted by plant roots play a crucial role in the establishment of symbiotic relationships with rhizobia and mycorrhizal fungi. By genetically engineering plants to overproduce specific flavonoids, it is possible to enhance the efficiency of nitrogen fixation and phosphorus uptake, thereby improving plant growth and stress tolerance. In legumes such as Glycine max (soybean), overexpression of genes involved in flavonoid biosynthesis, such as chalcone synthase (CHS) and flavanone 3-hydroxylase (F3H), leads to increased nodulation by rhizobia, enhancing nitrogen fixation (Subramanian et al., 2007). Similarly, in cereals like *Triticum aestivum* (wheat), increasing the production of strigolactones a group of terpenoids that stimulate arbuscular mycorrhizal (AM) fungi can enhance phosphorus uptake and improve plant growth in nutrient-poor soils (Akiyama et al., 2005).

6.6 Potential Challenges and Future Prospects

While genetic engineering offers significant potential for enhancing secondary metabolite production in the rhizosphere, several challenges remain. One of the primary concerns is the unintended effects of increasing metabolite production on plant growth and development. For example, overproduction of certain secondary metabolites can lead to reduced growth or altered nutrient allocation, which may negatively impact crop yields. Additionally, the release of high concentrations of certain compounds into the rhizosphere may disrupt the balance of microbial communities, leading to unintended ecological consequences. Despite these challenges, the use of genetic engineering to enhance rhizosphere metabolite production holds great promise for improving crop resilience, disease suppression, and sustainable agriculture. Advances in synthetic biology, metabolic engineering, and gene-editing technologies will likely enable more precise control over secondary metabolite production, allowing researchers to optimize the roles of these compounds in plant-microbe interactions.

Conclusion

Secondary metabolites play a crucial role in shaping the microbial communities within the rhizosphere, the narrow region of soil influenced by root exudates and secretions. These bioactive compounds, produced by plants, influence the interactions between roots and the surrounding microbial community, including bacteria, fungi, and other soil organisms. The rhizosphere is a highly dynamic environment where plants and microbes engage in symbiotic, mutualistic, and antagonistic relationships. Secondary metabolites, including flavonoids, terpenoids, phenolics, and alkaloids, act as chemical signals and defense agents that mediate these interactions, promoting beneficial associations while suppressing pathogens. The role of secondary metabolites in the rhizosphere is a key factor in maintaining plant health, enhancing soil fertility, and influencing overall ecosystem function.

Secondary metabolites exert a profound influence on plant-microbe interactions in the rhizosphere. These compounds function as signaling molecules that promote the establishment of beneficial relationships with microorganisms such as mycorrhizal fungi and nitrogen-fixing bacteria. For example, flavonoids secreted by legume roots are crucial in initiating symbiotic interactions with *Rhizobium* bacteria, which fix atmospheric nitrogen into a form usable by plants. These compounds not only act as chemoattractants that guide bacteria to the root system but also trigger the expression of nodulation genes in the bacteria, leading to the formation of nitrogen-fixing nodules. Similarly, terpenoids and phenolics secreted by plant roots play a key role in establishing and maintaining symbiotic relationships with arbuscular mycorrhizal fungi, which enhance nutrient uptake, particularly phosphorus.

In addition to promoting beneficial interactions, secondary metabolites are integral to plant defense mechanisms in the rhizosphere, where they contribute to disease suppression. Root-secreted compounds such as phenolics, alkaloids, and saponins can inhibit the growth of pathogenic microorganisms and suppress soil-borne diseases. For example, phenolic acids like salicylic acid and coumarins have antimicrobial properties that target bacterial and fungal pathogens in the soil, limiting their ability to colonize the plant root system. These compounds can disrupt microbial cell walls, inhibit enzyme function, and interfere with DNA replication in pathogens. As a result, plants are better able to resist infections caused by pathogens such as *Pythium*, *Fusarium*, and *Rhizoctonia*, which can cause significant damage to root systems and reduce crop yields. By releasing secondary metabolites into the rhizosphere, plants create a protective chemical barrier that suppresses the proliferation of harmful microbes, thereby enhancing overall plant health.

One of the most significant roles of secondary metabolites in the rhizosphere is their allelopathic effects, where they inhibit the growth of neighboring plants or other competing organisms. Allelopathy refers to the chemical inhibition of one species by another, and it is an important strategy that plants use to compete for resources such as nutrients, water, and space. Many secondary metabolites secreted by plant roots act as allelopathic agents that can suppress the germination, growth, or establishment of competing plant species. For example, compounds such as juglone from black walnut trees (*Juglans nigra*) or benzoxazinoids from cereal crops like wheat and maize can significantly reduce the growth of neighboring plants by interfering with their root development and nutrient uptake. These allelopathic effects not only help the producing plant to dominate its immediate environment but also contribute to the regulation of plant diversity in ecosystems. By influencing the competitive dynamics of the rhizosphere, secondary metabolites play a crucial role in shaping the composition and structure of plant communities.

The production and release of secondary metabolites in the rhizosphere are highly influenced by environmental factors such as soil type, nutrient availability, temperature, and moisture levels. Environmental stressors, such as drought or nutrient deficiency, can stimulate the production of certain metabolites, as plants enhance their chemical defenses to cope with adverse conditions. For instance, under conditions of nutrient stress, particularly phosphorus or nitrogen limitation, plants tend to increase the exudation of organic acids and flavonoids, which not only assist in nutrient solubilization but also alter microbial activity in the rhizosphere. Similarly, abiotic stressors like drought or high salinity can trigger the production of defensive secondary metabolites, such as phenolics and terpenoids, which help the plant mitigate oxidative stress and maintain root function. Temperature fluctuations and seasonal changes can also affect the profile of secondary metabolites produced by plants, as different environmental conditions influence the metabolic pathways responsible for their synthesis. These environmental influences is crucial for optimizing agricultural practices, as manipulating environmental factors can enhance the production of beneficial secondary metabolites, thereby promoting plant health and increasing crop resilience.

Genetic engineering has opened up new possibilities for enhancing the production of secondary metabolites in the rhizosphere, thereby improving plant-microbe interactions and plant defense mechanisms. By targeting the genes responsible for the biosynthesis and regulation of secondary metabolites, scientists can increase the production of compounds that are beneficial for plant growth, disease resistance, and soil health. For example, genetic modification of plants to overexpress key enzymes in the phenylpropanoid pathway, which is responsible for the synthesis of many phenolics, can lead to increased production of these compounds in root exudates. This, in turn, enhances the plant's ability to suppress soil-borne pathogens and establish beneficial relationships with microorganisms.

Additionally, genetic engineering can be used to introduce new biosynthetic pathways into plants, enabling them to produce novel secondary metabolites that provide enhanced resistance to pests and diseases. For instance, the introduction of genes from other species that produce antimicrobial compounds, such as phytoalexins or alkaloids, can confer plants with greater resistance to a wider range of soil pathogens. This approach not only improves the resilience of crops but also reduces the need for chemical inputs like pesticides and fertilizers, promoting more sustainable agricultural practices. Another promising area of research involves the genetic manipulation of root exudation patterns to selectively promote beneficial microbial communities in the rhizosphere. By engineering plants to release specific secondary metabolites that favor the growth of plant growth-promoting rhizobacteria (PGPR) or mycorrhizal fungi, scientists can enhance nutrient uptake, improve soil structure, and increase plant productivity. This approach, often referred to as "rhizosphere engineering," holds great potential for optimizing plant-microbe interactions in agricultural systems, leading to more efficient and environmentally friendly farming practices.

References

Akiyama, K., Matsuzaki, K. I., & Hayashi, H. (2005). Plant sesquiterpenes induce hyphal branching in arbuscular mycorrhizal fungi. *Nature, 435*(7043), 824-827.

Badri, D. V., & Vivanco, J. M. (2009). Regulation and function of root exudates in plants. *Plant, Cell & Environment, 32*(6), 666-681.

Bais, H. P., Weir, T. L., Perry, L. G., Gilroy, S., & Vivanco, J. M. (2006). The role of root exudates in rhizosphere interactions with plants and other organisms. *Annual Review of Plant Biology, 57*, 233-266.

Bakker, M. G., Schlatter, D. C., Otto-Hanson, L., & Kinkel, L. L. (2014). Diffuse symbioses: Roles of plant secondary metabolites in structuring rhizosphere microbiomes. Molecular Ecology, *23*(6), 1571-1583.

Lou, Y., Baldwin, I. T. (2004). Nitrogen supply influences herbivore-induced direct and indirect defenses and transcriptional responses in *Nicotiana attenuata*. *Plant Physiology, 135*(1), 496-506.

Berendsen, R. L., Pieterse, C. M. J., & Bakker, P. A. H. M. (2012). The rhizosphere microbiome and plant health. *Trends in Plant Science, 17*(8), 478-486.

Bourgaud, F., Hehn, A., Larbat, R., Doerper, S., Gontier, E., Kellner, S., & Matern, U. (2006). Biosynthesis of coumarins in plants: A major pathway still to be unravelled for cytochrome P450 enzymes. *Phytochemistry Reviews, 5*(2-3), 293-308.

Wang, Y., Li, Z., Xu, M., Xiao, Z., Liu, C., Du, B., Xu, D., & Li, L. (2023). Signal molecules regulate the synthesis of secondary metabolites in the interaction between endophytes and medicinal plants. *Processes, 11*(3), 849.

Bryant, J. P., Chapin, F. S., & Klein, D. R. (1983). Carbon/nutrient balance of boreal plants in relation to vertebrate herbivory. *Oikos, 40*(3), 357-368.

Cesco, S., Mimmo, T., Tonon, G., Tomasi, N., & Pinton, R. (2012). Plant-borne flavonoids released into the rhizosphere: Impact on soil bio-activities related to plant nutrition. A review. *Biology and Fertility of Soils, 48*(2):123-149.

Cheng, F., & Cheng, Z. (2015). Research progress on the use of plant allelopathy in agriculture and the physiological and ecological mechanisms of allelopathy. *Frontiers in Plant Science, 6*, 1020.

Chitwood, D. J. (2002). Phytochemical-based strategies for nematode control. *Annual Review of Phytopathology, 40*(1), 221-249.

Gargallo-Garriga, A., Perez-Trujillo, M., Estiarte, M., Ogaya, R., Penuelas, J., & Sardans, J. (2018). Root exudate metabolomes change under drought and show limited capacity for recovery. *Scientific Reports, 8*(1), 12696.

Song, C., Saloner, A., Fait, A., & Bernstein, N. (2023). Nitrogen deficiency stimulates cannabinoid biosynthesis in medical cannabis plants by inducing a metabolic shift towards production of low-N metabolites. *Industrial Crops and Products*, *202*, 116969.

Ku, Y. S., Ng, M. S., Cheng, S. S., Lo, A. W., Xiao, Z., Shin, T. S., Chung, G., & Lam, H. M. (2020) Understanding the composition, biosynthesis, accumulation and transport of flavonoids in crops for the promotion of crops as healthy sources of flavonoids for human consumption. *Nutrients*, *12*(6), 1717.

Halkier, B. A., & Gershenzon, J. (2006). Biology and biochemistry of glucosinolates. *Annual Review of Plant Biology, 57*, 303-333.

Harborne, J. B., & Williams, C. A. (2000). Advances in flavonoid research since 1992. *Phytochemistry, 55*(6), 481-504.

Inderjit. (1996). Plant phenolics in allelopathy. *Botanical Review, 62*(2), 186-202.

Gorlenko, C. L., Kiselev, H. Y., Budanova, E. V., Zamyatnin, A. A., Jr., & Ikryannikova, L. N. (2020). Plant Secondary Metabolites in the Battle of Drugs and Drug-Resistant Bacteria: New Heroes or Worse Clones of Antibiotics? *Antibiotics*, *9*(4), 170.

Kiers, E. T., Rousseau, R. A., West, S. A., & Denison, R. F. (2003). Host sanctions and the legume-rhizobium mutualism. *Nature, 425*(6953), 78-81.

Lanfranco, L., Fiorilli, V., & Gutjahr, C. (2018). Partner communication and role of nutrients in the arbuscular mycorrhizal symbiosis. *New Phytologist, 220*(4), 1031-1046.

Lattanzio, V., Lattanzio, V. M. T., & Cardinali, A. (2006). Role of phenolics in the resistance mechanisms of plants against fungal pathogens and insects. In F. Imperato (Ed.), *Phytochemistry: Advances in Research* (pp. 23-67). Research Signpost.

Lillo, C., Lea, U. S., & Ruoff, P. (2008). Nutrient depletion as a key factor for manipulating gene expression and product formation in different branches of the flavonoid pathway. *Plant Cell & Environment*, *31*, 587-601.

Llorente, B., Martinez-Garcia, J. F., Stange, C., & Rodriguez-Concepcion, M. (2017). Illuminating colors: Regulation of carotenoid biosynthesis and accumulation by light. *Current Opinion in Plant Biology, 37*, 49-55.

Das, S., Kwon, M., & Kim, J.- Y. (2024). Enhancement of specialized metabolites using CRISPR/Cas gene editing technology in medicinal plants. *Frontiers in Plant Science*, *15*, 1279738.

Zaynab, M., Fatima, M., Abbas, S., Sharif, Y., Umair, M., Zafar, M. H., & Bahadar, K. (2018). Role of secondary metabolites in plant defense against pathogens. *Microbial Pathogenesis*, *124*, 198-202.

Mancinelli, A. L. (1985). Light-dependent anthocyanin synthesis: A model system for the study of plant photomorphogenesis. *Botanical Review, 51*(1), 107-157.

Majdura, J., Jankiewicz, U., Gałązka, A., & Orzechowski, S. (2023). The role of quorum sensing molecules in bacterial-plant interactions. *Metabolites*, *13*(1), 114.

Neumann, G., Massonneau, A., Langlade, N., Dinkelaker, B., Hengeler, C., Römheld, V., & Martinoia, E. (2000). Physiological aspects of cluster root function and development in phosphorus-deficient white lupin (*Lupinus albus* L.). *Annals of Botany*, *85*(6), 909-919.

Oldroyd, G. E. D., & Downie, J. A. (2008). Coordinating nodule morphogenesis with rhizobial infection in legumes. *Annual Review of Plant Biology, 59*, 519-546.

Yang, Y., Yang, X., Guo, X., Hu, X., Dong, D., Li, G., Xiong, X. (2022). Exogenously applied methyl Jasmonate induces early defense related genes in response to Phytophthora infestans infection in potato plants. *Horticultural Plant Journal, 8*(4), 511-526,

Peniche-Pavía, H. A., Guzmán, T. J., Magaña-Cerino, J. M., Gurrola-Díaz, C. M., & Tiessen, A. (2022). Maize flavonoid biosynthesis, regulation, and human health relevance: A review. *Molecules, 27*(16), 5166.

Pieterse, C. M., Van der Does, D., Zamioudis, C., Leon-Reyes, A., Van Wees, S. C. (2012). Hormonal modulation of plant immunity. *Annual Reviews in Cell Developmental Biology, 28*, 489-521.

Ramakrishna, A., & Ravishankar, G. A. (2011). Influence of abiotic stress signals on secondary metabolites in plants. *Plant Signaling & Behavior, 6*(11), 1720-1731.

Rasmann, S., Köllner, T. G., Degenhardt, J., Hiltpold, I., & Turlings, T. C. J. (2005). Recruitment of entomopathogenic nematodes by insect-damaged maize roots. *Nature, 434*(7034), 732-737.

Rietveld, W. J. (1983). Allelopathic effects of Juglans nigra on selected herbaceous plants. *Journal of Chemical Ecology, 9*(2), 295-308.

Sangwan, N. S., Farooqi, A. H. A., & Sangwan, R. S. (2001). Regulation of essential oil production in plants. *Plant Growth Regulation, 34*(1), 3-21.

Sasse, J., Martinoia, E., & Northen, T. (2018). Feed your friends: Do plant exudates shape the root microbiome? *Trends in Plant Science, 23*(1), 25-41.

Stracke, R., Ishihara, H., Barsch, G. A., Mehrtens, F., Niehaus, K., & Weisshaar, B. (2007). Differential regulation of flavonoid biosynthesis genes by R2R3-MYB transcription factors in Arabidopsis thaliana. *Planta, 50*(4), 660-677.

Subramanian, S., Stacey, G., & Yu, O. (2007). Distinct, crucial roles of flavonoids during legume nodulation. *Trends in Plant Science, 12*(7), 282-285.

Ghitti, E., Rolli, E., Vergani, L., & Borin, S. (2024). Flavonoids influence key rhizocompetence traits for early root colonization and PCB degradation potential of *Paraburkholderia xenovorans* LB400. *Frontiers in Plant Science, 15*, 1325048.

Tholl, D. (2006). Terpene synthases and the regulation, diversity, and biological roles of terpene metabolism. *Current Opinion in Plant Biology, 9*(3), 297-304.

Yu, K., Pieterse, C. M. J., Bakker, P. A. H. M., Berendsen, R. L. (2019). Beneficial microbes going underground of root immunity. *Plant Cell Environment, 42*(10), 2860-2870.

Jamil, F., Mukhtar, H., Fouillaud, M., & Dufossé, L. (2022). Rhizosphere signaling: Insights into plant–rhizomicrobiome interactions for sustainable agronomy. *Microorganisms, 10*(5), 899.

Wasson, A. P., Ramsay, K., Jones, M. G. K., & Mathesius, U. (2009). Differing requirements for flavonoids during the formation of lateral roots and nodules in *Medicago truncatula*. *New Phytologist, 183*(1), 167-179.

Weir, T. L., Park, S. W., & Vivanco, J. M. (2004). Biochemical and physiological mechanisms mediated by allelochemicals. *Current Opinion in Plant Biology, 7*(4), 472-479.

Weller, D. M., Raaijmakers, J. M., McSpadden Gardener, B. B., & Thomashow, L. S. (2002). Microbial populations responsible for specific soil suppressiveness to plant pathogens. *Annual Review of Phytopathology, 40*, 309-348.

Weston, L. A., & Duke, S. O. (2003). Weed and crop allelopathy. *Critical Reviews in Plant Sciences, 22*(3-4), 367-389.

Weston, L. A., & Mathesius, U. (2013) Flavonoids: their structure, biosynthesis and role in the rhizosphere, including allelopathy. Journal of Chemical Ecology, *39*(2), 283-97.

Wink, M. (2015). Modes of action of herbal medicines and plant secondary metabolites. *Medicines, 2*(3), 251-286.

Isah, T. (2019). Stress and defense responses in plant secondary metabolites production. *Biological Research, 52*, 39.

Yazaki, K. (2005). Transporters of secondary metabolites. *Current Opinion in Plant Biology, 8*(3), 301-307.

Zamioudis, C., & Pieterse, C. M. J. (2012). Modulation of host immunity by beneficial microbes. *Molecular Plant-Microbe Interactions, 25*(2), 139-150.

Zeng, R. S., Mallik, A. U., & Luo, S. M. (2008). Allelopathy in sustainable agriculture and forestry. *Springer Nature*.

Zhang, X., Xu, G., Cheng, C., Lei, L., Sun, J., & Xu, Y. (2021). Establishment of an *Agrobacterium*-mediated genetic transformation and CRISPR/Cas9-mediated targeted mutagenesis in hemp (*Cannabis sativa* l.). *Plant Biotechnology Journal, 19*, 1979-1987.

Zhong, R., Lee, C., Zhou, J., McCarthy, R. L., & Ye, Z. H. (2008). A battery of transcription factors involved in the regulation of secondary wall biosynthesis in Arabidopsis. *The Plant Cell, 20*(10), 2763-2782.

Zoratti, L., Karppinen, K., Luengo Escobar, A., Häggman, H., & Jaakola, L. (2014). Light-controlled flavonoid biosynthesis in fruits. *Frontiers in Plant Science, 5*, 534.

Chapter - 15

Induced Resistance Mechanisms: The Role of Secondary Metabolites

Pratima Raypa[1], Reshma Jadhav[2], Susheel Kumar[3], Praneet Chauhan[4] and Zuby Gohar Ansari[5]

[1]Department of Biochemistry, H.N.B. Garhwal University, Srinagar, Uttarakhand, India
[2]Department of Biochemistry, Mahatma Phule Krishi Vidyapeeth, Rahuri, Ahmednagar, Maharashtra, India
[3]Plant Molecular Virology Laboratory, CPMB Division, CSIR-National Botanical Research Institute, Rana Pratap Marg, Lucknow, Uttar Pradesh, India
[4]Department of Plant Pathology, Dr. Khem Singh Gill Akal College of Agriculture, Eternal University, Baru Sahib, Sirmour, Himachal Pradesh, India
[5]B.M. College of Agriculture, Jaswadi Road Khandwa, Madhya Pradesh, India

Abstract

Induced resistance mechanisms in plants are essential for their ability to defend against a variety of pathogens and pests. Secondary metabolites play a pivotal role in these mechanisms, acting as both direct defense compounds and signaling molecules that trigger broader resistance responses. When plants are exposed to pathogens, herbivores, or abiotic stressors, they can activate induced resistance pathways, including systemic acquired resistance (SAR) and induced systemic resistance (ISR). These pathways lead to the accumulation of secondary metabolites such as phenolics, flavonoids, alkaloids, and terpenoids, which help reinforce the plant's defense by directly inhibiting pathogen growth, deterring herbivores, or strengthening plant cell walls. In SAR, for example, salicylic acid acts as a critical signaling molecule, promoting the synthesis of defense-related proteins and phenolic compounds. In contrast, ISR, often mediated by beneficial rhizobacteria, triggers jasmonic acid and ethylene pathways, leading to the enhanced production of terpenoids and alkaloids. These metabolites are crucial for plant immunity, providing both localized and systemic protection. The synergistic interactions between different secondary metabolites and hormonal pathways further enhance the effectiveness of induced resistance. Recent advances in genetic engineering have enabled the modification of these pathways to boost secondary metabolite production, making crops more resistant to pathogens and reducing the need for chemical pesticides.

Keywords: Induced Resistance, Secondary Metabolites, Sar, Isr, Plant Immunity, Systemic Protection

DOI: 10.1201/9781003717454-15

Introduction

Induced resistance in plants is a critical aspect of their defense system, enabling them to respond effectively to pathogen attacks, herbivory, and various environmental stressors. This form of resistance allows plants to enhance their defensive capacity after initial exposure to a pathogen or stress signal, providing long-term protection against future attacks. Induced resistance is characterized by the activation of complex signaling networks and the accumulation of defense-related compounds, particularly secondary metabolites, which play a crucial role in reinforcing plant defenses. Two primary forms of induced resistance are Systemic Acquired Resistance (SAR) and Induced Systemic Resistance (ISR). Both mechanisms rely heavily on the production and regulation of secondary metabolites, which act as direct defense agents and as signaling molecules that modulate immune responses. By understanding the role of secondary metabolites in these processes, researchers and agriculturalists can harness the plant's natural defenses for more sustainable crop protection strategies.

Systemic Acquired Resistance (SAR) is a well-documented form of induced resistance that provides broad-spectrum defense against a wide range of pathogens, including bacteria, fungi, and viruses. SAR is often triggered by a localized infection that stimulates a systemic signal throughout the plant, leading to the activation of defense genes and the accumulation of defensive compounds in tissues far from the initial site of infection. Phenolic compounds, particularly salicylic acid (SA), play a central role in the SAR pathway. Salicylic acid acts as both a signaling molecule and a defense compound, inducing the expression of pathogenesis-related (PR) proteins and the synthesis of other phenolic metabolites that have antimicrobial properties. These phenolics, including flavonoids and lignins, reinforce plant cell walls, making it more difficult for pathogens to penetrate and establish infections (Durrant & Dong, 2004). Furthermore, SA-mediated SAR can provide long-lasting protection, often persisting for weeks or even months, making it a valuable tool for disease management in crops (Fu & Dong, 2013).

Induced Systemic Resistance (ISR) is another important form of plant defense, typically mediated by beneficial microorganisms such as plant growth-promoting rhizobacteria (PGPR). Unlike SAR, which is primarily driven by salicylic acid, ISR is regulated by the jasmonic acid (JA) and ethylene (ET) signaling pathways. These hormones stimulate the production of secondary metabolites like terpenoids and alkaloids, which have potent antimicrobial and anti-herbivore properties (Pieterse et al., 2014). ISR does not directly involve the activation of PR proteins; instead, it primes the plant to respond more rapidly and robustly to subsequent attacks. This priming effect enhances the plant's overall resistance, allowing it to deploy its defenses more efficiently when needed. The role of secondary metabolites in ISR is essential, as they contribute to both direct inhibition of pathogens and pests and the reinforcement of structural defenses, such as the fortification of cell walls with lignin and other phenolic compounds.

One of the key aspects of both SAR and ISR is the synergistic effect of secondary metabolites and other defense mechanisms. These metabolites often work in concert with each other and with other defense compounds, such as reactive oxygen species (ROS), to create a more effective defense response. For example, phenolics produced during SAR can enhance the production of ROS, which are toxic to pathogens and can trigger localized cell death to limit pathogen spread (Bailey-Serres & Mittler, 2016). Similarly, terpenoids produced during ISR can interact with phenolic compounds to disrupt pathogen cell membranes and inhibit their enzymatic functions, thereby enhancing the overall antimicrobial activity of the plant (Thomma et al., 1998). The ability of secondary metabolites to act synergistically with other defense mechanisms allows plants to mount a more comprehensive and robust defense, which is critical for their survival in pathogen-rich environments.

Hormonal signaling pathways play a crucial role in regulating the production of secondary metabolites during induced resistance. In SAR, salicylic acid is the primary hormone responsible for activating the synthesis of phenolic compounds, which are essential for both direct defense and signaling functions. In ISR, jasmonic acid and ethylene are the key hormones that regulate the production of terpenoids, alkaloids, and other secondary metabolites that contribute to resistance (Pieterse et al., 2012). These hormones not only activate the production of secondary metabolites but also modulate the expression of defense genes, ensuring that the plant's immune response is tailored to the specific type of threat it faces. Additionally, crosstalk between these hormonal pathways allows plants to fine-tune their defense responses, optimizing resource allocation and minimizing the negative impact on growth and development. For example, the interaction between SA and JA pathways ensures that plants can effectively balance their defenses against both biotrophic pathogens, which require living host cells, and necrotrophic pathogens, which kill host cells to feed on dead tissue (Koornneef & Pieterse, 2008).

Recent advancements in genetic engineering have opened up new possibilities for enhancing induced resistance in crops by manipulating the biosynthetic pathways involved in secondary metabolite production. Through techniques such as CRISPR/Cas9 gene editing and transgenic approaches, scientists have been able to increase the levels of key secondary metabolites involved in SAR and ISR, thereby boosting plant resistance to pathogens and pests. For example, overexpression of genes involved in the salicylic acid pathway has been shown to enhance SAR in crops such as rice and wheat, leading to increased resistance to fungal pathogens like *Magnaporthe oryzae* and *Puccinia graminis* (Shah & Zeier, 2013). Similarly, genetic modifications that enhance the production of jasmonic acid and terpenoids have been successful in improving ISR in crops, making them more resistant to insect herbivores and necrotrophic pathogens (Wasternack, 2007).

Another promising avenue of research involves the manipulation of transcription factors that regulate secondary metabolite biosynthesis. Transcription factors such as MYB, WRKY, and NAC play critical roles in controlling the expression of genes involved in the production of phenolics, terpenoids, and alkaloids (Pandey & Somssich, 2009). By overexpressing or silencing these transcription factors, researchers can increase or decrease the production of specific secondary metabolites, tailoring the plant's resistance mechanisms to the particular threats it faces. This approach has been particularly effective in enhancing the production of flavonoids and lignins, which contribute to both structural defenses and antimicrobial activity (Tong et al., 2024). Additionally, advances in metabolic engineering have allowed for the introduction of novel biosynthetic pathways into plants, enabling the production of secondary metabolites that are not normally present in certain crops. This technology has the potential to significantly enhance the disease resistance of staple crops, improving food security and reducing the need for chemical pesticides.

1. Overview of Induced Resistance in Plants

Induced resistance (IR) is a defense mechanism in plants that enhances their ability to resist or tolerate attacks by pathogens, insects, and environmental stressors. Unlike constitutive defenses, which are always present, induced resistance is activated in response to specific stimuli, either biotic or abiotic. This form of resistance allows plants to respond more dynamically to environmental threats by increasing the production of defensive compounds and proteins. The mechanisms of induced resistance are often broadly categorized into two major types: Systemic Acquired Resistance (SAR) and Induced Systemic Resistance (ISR), each triggered by different cues and signaling pathways. The phenomenon of induced resistance is widely recognized as an important tool for reducing crop losses and promoting sustainable agriculture by harnessing the plant's defense mechanisms to reduce the need for chemical pesticides. The ability to prime or enhance a plant's innate immune system through both biological and chemical inducers is a key area of research in crop protection.

1.1 Systemic Acquired Resistance (SAR)

Systemic Acquired Resistance (SAR) is triggered when a plant experiences a localized infection or stress, leading to the activation of defense responses in distal, uninfected parts of the plant. SAR is usually initiated by necrotrophic pathogens or mechanical damage and is often associated with the accumulation of salicylic acid (SA), a key signaling molecule in plant defense. This response is characterized by the expression of pathogenesis-related (PR) proteins, which include enzymes such as chitinases and glucanases that degrade the cell walls of fungal pathogens (Vlot et al., 2009). SAR provides broad-spectrum, long-lasting resistance against a wide range of pathogens, including viruses, bacteria, and fungi. Once activated, SAR primes the plant for enhanced defense, allowing it to mount a more robust response to subsequent infections. For example, in *Arabidopsis thaliana*, infection by *Pseudomonas syringae* triggers SAR, which leads to the upregulation of PR genes and enhanced resistance to future infections (Pieterse et al., 2014). In addition to salicylic acid, NPR1 (Nonexpressor of Pathogenesis-Related Genes 1) plays a central role in the SAR pathway. Upon activation by SA, NPR1 translocates to the nucleus, where it acts as a transcriptional co-activator of PR genes, thus amplifying the plant's immune response (Durrant & Dong, 2004).

1.2 Induced Systemic Resistance (ISR)

Induced Systemic Resistance (ISR) is a second form of induced resistance, typically triggered by beneficial microbes in the rhizosphere, such as plant growth-promoting rhizobacteria (PGPR) and mycorrhizal fungi. Unlike SAR, ISR is often associated with jasmonic acid (JA) and ethylene (ET) signaling pathways rather than salicylic acid. ISR does not involve the accumulation of PR proteins but rather primes the plant for a more effective response to pathogen or pest attacks when they occur (Van Loon et al., 1998). Rhizobacteria, such as *Pseudomonas fluorescens* and Bacillus subtilis, are well-known ISR inducers. These bacteria colonize plant roots and produce signaling compounds that activate the plant's defense pathways. For instance, *Pseudomonas fluorescens* triggers ISR in *Arabidopsis thaliana* and tomato (*Solanum lycopersicum*), providing enhanced resistance to bacterial pathogens such as *Pseudomonas syringae* and fungal pathogens like *Botrytis cinerea* (Pieterse et al., 2014). ISR also enhances the production of secondary metabolites such as phenolics, flavonoids, and terpenoids, which play a crucial role in plant defense against pathogens and herbivores. These metabolites are secreted into the rhizosphere and can inhibit the growth of soil-borne pathogens or deter herbivores from feeding on plant tissues. By leveraging ISR, plants can enhance their overall resilience to environmental challenges.

1.3 Priming and Defense Signaling Pathways

One of the most important aspects of induced resistance is the concept of priming, where plants are sensitized to respond more rapidly and robustly to subsequent pathogen or pest attacks. Primed plants can respond to external stressors by activating defense pathways more efficiently, often with a lower metabolic cost compared to constitutive defenses. The jasmonic acid (JA) and salicylic acid (SA) signaling pathways are the two primary routes through which induced resistance is regulated. Jasmonic acid is typically involved in responses to necrotrophic pathogens and herbivores, while salicylic acid is more active in defenses against biotrophic pathogens (Thaler et al., 2012). The cross-talk between these two pathways is essential for fine-tuning the plant's immune response. In many cases, these pathways interact antagonistically, meaning that the activation of one pathway can suppress the other. For example, high levels of salicylic acid can suppress jasmonic acid signaling, which may reduce the plant's ability to defend against certain types of pathogens or herbivores. Conversely, jasmonic acid can suppress salicylic acid signaling, potentially reducing the plant's resistance to biotrophic pathogens.

1.4 Biotechnological Approaches to Enhance Induced Resistance

Genetic engineering and biotechnology offer promising strategies to enhance induced resistance in crops. By manipulating key regulatory genes or introducing novel defense pathways, researchers can increase the plant's capacity to respond to environmental stress. For instance, overexpression of genes involved in the salicylic acid or jasmonic acid pathways can lead to enhanced resistance to pathogens. One example is the overexpression of NPR1 in rice (Oryza sativa), which enhances resistance to *Magnaporthe oryzae*, the causative agent of rice blast disease (Chern et al., 2005). Similarly, overexpression of ethylene response factors (ERFs), which are involved in the jasmonic acid signaling pathway, can enhance resistance to necrotrophic pathogens in crops such as tomatoes and tobacco (Shoresh et al., 2010). These advances highlight the potential for biotechnology to play a significant role in improving crop resilience through the activation of induced resistance mechanisms.

1.5 Role of Secondary Metabolites in Induced Resistance

Secondary metabolites, such as phenolics, flavonoids, alkaloids, and terpenoids, are integral components of the plant's induced defense responses. When induced resistance is activated, plants often increase the production of these compounds, which serve as antimicrobial agents, antifeedants, or signaling molecules that attract beneficial organisms or repel herbivores. For example, the production of phytoalexins, a group of antimicrobial secondary metabolites, is a key feature of both SAR and ISR. In grapevine (*Vitis vinifera*), the accumulation of resveratrol, a stilbene phytoalexin, is induced in response to infection by fungal pathogens such as *Botrytis cinerea*. This compound not only inhibits pathogen growth but also primes the plant for enhanced defense against future infections (Adrian et al., 2000). In addition to their direct antimicrobial effects, secondary metabolites also play a role in shaping the microbial community in the rhizosphere. By secreting compounds such as flavonoids and terpenoids, plants can attract beneficial microbes that help suppress pathogens or enhance nutrient uptake, further contributing to plant health and resilience (Badri & Vivanco, 2009).

2. Systemic Acquired Resistance and the Role of Phenolics

Systemic Acquired Resistance (SAR) is one of the most effective defense mechanisms in plants, providing long-lasting, broad-spectrum immunity against a wide array of pathogens, including bacteria, fungi, viruses, and even some insects. SAR is typically triggered by a localized infection or stress that activates defense responses in other, uninfected parts of the plant. The SAR mechanism involves several biochemical and molecular changes, many of which are mediated by salicylic acid (SA) and involve the production of pathogenesis-related (PR) proteins. An important group of compounds involved in SAR are phenolics, which play multiple roles, including acting as antioxidants, antimicrobials, and signaling molecules that enhance the plant's ability to resist future pathogen attacks.

2.1 Mechanism of Systemic Acquired Resistance

SAR is initiated when a plant is exposed to a pathogen or undergoes tissue damage, leading to the accumulation of salicylic acid at the infection site. This SA accumulation acts as a signal that triggers the activation of defense genes both locally and systemically throughout the plant. One of the key players in this pathway is the Nonexpressor of Pathogenesis-Related Genes 1 (NPR1), a transcription co-activator that plays a central role in the SAR signaling cascade. Once activated by SA, NPR1 moves to the nucleus, where it facilitates the transcription of pathogenesis-related (PR) genes (Durrant & Dong, 2004). PR proteins, such as chitinases and glucanases, are produced in response to SAR activation. These proteins degrade the cell walls of pathogens, thereby inhibiting their spread within the plant (Vlot et al., 2009). In addition to the production of PR proteins, SAR also primes the plant's immune system for faster and more robust responses to subsequent attacks, a phenomenon known as priming.

2.2 Phenolics as Key Players in SAR

Phenolics are a diverse group of secondary metabolites that play crucial roles in the plant's defense mechanisms, particularly in SAR. Phenolic compounds are involved in pathogen inhibition, cell wall fortification, and signaling during the defense response. Many phenolics are antioxidants, which help to neutralize reactive oxygen species (ROS) produced during pathogen attack, protecting plant cells from oxidative damage.

2.2.1 Antimicrobial Role of Phenolics

One of the primary roles of phenolics in SAR is their antimicrobial activity. Many phenolic compounds exhibit strong antifungal, antibacterial, and antiviral properties, which help prevent the spread of pathogens. For example, salicylic acid, itself a phenolic compound, not only functions as a signaling molecule in the SAR pathway but also has direct antimicrobial effects. Other phenolics, such as ferulic acid, caffeic acid, and coumaric acid, are exuded into the plant cell walls or rhizosphere to inhibit pathogen growth (Lattanzio et al., 2006). These compounds disrupt microbial membranes, interfere with enzyme activity, and inhibit spore germination, making them effective tools in pathogen resistance. For instance, in tomato (*Solanum lycopersicum*), the phenolic compound chlorogenic acid accumulates in response to infection by *Phytophthora infestans*, the causative agent of late blight. This phenolic compound has been shown to inhibit the growth of the pathogen, thereby contributing to disease resistance (Ngadze et al., 2012). Similarly, in grapevine (*Vitis vinifera*), the production of resveratrol, a stilbene phenolic, is increased during SAR activation, protecting *Botrytis cinerea* and other fungal pathogens (Adrian et al., 2000).

2.2.2 Phenolics in Cell Wall Strengthening

Another important role of phenolics in SAR is their contribution to cell wall fortification, which acts as a physical barrier to pathogen invasion. Lignin, a complex phenolic polymer, is deposited in plant cell walls during SAR, increasing the mechanical strength of the walls and making it more difficult for pathogens to penetrate the plant tissues. The production of lignin is often upregulated in response to pathogen attack, and enzymes involved in lignin biosynthesis, such as phenylalanine ammonia-lyase (PAL), are highly active during SAR. Ferulic acid and p-coumaric acid, two phenolic acids, are cross-linked into the plant cell wall during pathogen attack, further strengthening the cell wall and making it more resistant to enzymatic degradation by pathogens. These phenolic acids are involved in the formation of lignin precursors, which are essential for the structural reinforcement of the plant during SAR (Lattanzio et al., 2006).

2.2.3 Phenolics as Signaling Molecules

In addition to their antimicrobial and structural roles, phenolics also function as signaling molecules in the SAR pathway. Salicylic acid, a key phenolic in SAR, not only signals the activation of PR genes but also serves as a mobile signal that communicates the onset of SAR to distant tissues in the plant (Vlot et al., 2009). Once a pathogen is detected in one part of the plant, SA travels through the vascular system to uninfected tissues, where it primes the immune response and prepares the plant to fend off subsequent attacks. Other phenolics, such as flavonoids, also play signaling roles in plant defense. Flavonoids can modulate the activity of hormonal pathways, such as those involving jasmonic acid (JA) and ethylene (ET), which are critical for coordinating the plant's overall defense response. The cross-talk between salicylic acid (SA), jasmonic acid (JA), and ethylene (ET) is essential for fine-tuning the SAR response and ensuring that the plant mounts the appropriate defense strategy based on the nature of the threat (Thaler et al., 2012).

2.3 Phenolic Compounds in Priming and Memory of SAR

The priming effect of SAR is a phenomenon where plants exposed to initial stress or pathogen attack develop an enhanced capacity to respond to future challenges. Phenolic compounds play a crucial role in this process by maintaining the plant in a heightened state of alert, allowing for faster and more robust defense responses upon subsequent pathogen invasion. For instance, plants that have experienced SAR induction often accumulate phenolic precursors in their tissues, which can be rapidly converted into active defense compounds upon pathogen detection. This memory of prior infection is a critical feature of SAR, as it allows plants to respond more effectively to recurring stresses without the need for continuous activation of energy-intensive defense pathways (Conrath, 2009). This efficient use of resources is particularly important in agriculture, where maintaining crop productivity while minimizing pathogen damage is a constant challenge.

2.4 Application of SAR and Phenolics in Agriculture

Harnessing the power of SAR and phenolics for agricultural purposes is a promising approach to developing more sustainable crop protection strategies. Chemical inducers of SAR, such as acibenzolar-S-methyl (ASM), have been shown to enhance disease resistance in crops by stimulating the production of phenolics and other defense-related compounds (Walters et al., 2013). These inducers mimic the natural SAR response and activate the plant's immune system without causing damage or stress, reducing the need for chemical fungicides and pesticides. Additionally, breeding programs aimed at increasing the natural production of phenolics in crops, particularly in the rhizosphere, could lead to enhanced disease resistance and reduced pathogen pressure. For example, varieties of wheat (*Triticum aestivum*) and maize (*Zea mays*) that produce higher levels of root-secreted phenolic acids are more resistant to soil-borne pathogens like *Fusarium* and *Rhizoctonia*, contributing to better crop health and yield (Lanoue et al., 2010).

3. Induced Systemic Resistance and Secondary Metabolite Production

Induced Systemic Resistance (ISR) is a plant defense mechanism that enhances the plant's ability to resist pathogen attacks and herbivore damage. Unlike Systemic Acquired Resistance (SAR), which is primarily triggered by pathogen infection and relies on salicylic acid (SA) signaling, ISR is often activated by plant growth-promoting rhizobacteria (PGPR), such as *Pseudomonas* and *Bacillus* species, as well as mycorrhizal fungi. ISR operates mainly through jasmonic acid (JA) and ethylene (ET) signaling pathways and does not lead to the accumulation of pathogenesis-related (PR) proteins like SAR. However, ISR primes the plant for a more rapid and efficient defense response upon future pathogen or pest attacks, often involving the enhanced production of secondary metabolites. Secondary metabolites, including phenolics, flavonoids, alkaloids, and terpenoids, play key roles in ISR by acting as antimicrobial compounds, signaling molecules, and deterrents to herbivores. These metabolites are essential components of the plant's defense arsenal, providing both direct protection against invaders and facilitating beneficial interactions with symbiotic microbes.

3.1 Mechanism of Induced Systemic Resistance

ISR is triggered by beneficial microorganisms, primarily in the rhizosphere, where plant roots interact with plant growth-promoting rhizobacteria (PGPR) and mycorrhizal fungi. These microorganisms do not necessarily colonize plant tissues or cause infections, but they stimulate the plant's immune system, leading to an enhanced state of readiness to respond to future attacks. When a plant encounters PGPR or mycorrhizal fungi, the interaction triggers the jasmonic acid (JA) and ethylene (ET) signaling pathways, leading to the activation of defense responses. Unlike SAR, which induces the expression of PR genes, ISR enhances the plant's ability to produce defense-related secondary metabolites upon pathogen or pest attack (Pieterse et al., 2014). These secondary metabolites play a central role in ISR by acting as antimicrobial agents, insect deterrents, or even as signaling compounds that attract natural enemies of herbivores.

3.2 Role of Jasmonic Acid and Ethylene in ISR

The jasmonic acid (JA) and ethylene (ET) pathways are central to the activation of ISR. Jasmonic acid, a lipid-based signaling molecule, is involved in plant responses to necrotrophic pathogens and herbivores, while ethylene modulates plant defense responses to pathogens and mechanical stress. Both hormones act synergistically to prime plants for enhanced production of secondary metabolites, which are crucial for defense. Upon activation of ISR, jasmonic acid stimulates the production of phenolics, flavonoids, and terpenoids secondary metabolites that serve various protective functions. For example, phenolic compounds can fortify the plant cell wall, making it more difficult for pathogens to penetrate. Flavonoids have antimicrobial properties that help suppress pathogen growth, while terpenoids can repel herbivores and attract predators of herbivores (Lorenzo & Solano, 2005). In *Arabidopsis thaliana*, ISR induced by *Pseudomonas fluorescens* leads to the activation of the JA pathway, resulting in the accumulation of secondary metabolites such as camalexin, a phytoalexin with strong antifungal properties. This priming effect enhances the plant's ability to defend against *Botrytis cinerea*, a necrotrophic pathogen that causes gray mold in a variety of plants (Spoel et al., 2007).

3.3 Secondary Metabolite Production in ISR

The activation of ISR primes plants to produce secondary metabolites that play a critical role in defense against pathogens and herbivores. These metabolites include phenolics, terpenoids, and alkaloids, which provide both direct and indirect defense benefits.

3.3.1 Phenolics in ISR

Phenolic compounds are among the most important secondary metabolites involved in ISR. They function in both direct defense against pathogens and indirect defense by strengthening the plant's cell walls. Phenolic acids, such as ferulic acid and caffeic acid, are deposited in the cell walls during pathogen attack, providing physical reinforcement to prevent pathogen entry (Lattanzio et al., 2006). Moreover, phenolics can also act as antimicrobial agents, directly inhibiting the growth of pathogens. For instance, plants treated with *Bacillus subtilis* as an ISR-inducing agent show increased levels of chlorogenic acid and caffeic acid in the rhizosphere, which inhibit the growth of *Fusarium oxysporum*, a pathogen that causes wilt disease in many crops (Chen et al., 2013).

3.3.2 Flavonoids in ISR

Flavonoids are a diverse group of secondary metabolites involved in ISR that have both antimicrobial properties and signaling functions. When ISR is triggered, plants often increase the production of flavonoids in their roots and shoots. These compounds can inhibit pathogen growth, act as antioxidants, and even modulate the plant's hormonal balance during stress responses. In *Medicago truncatula*, flavonoid production is significantly enhanced following ISR activation by *Pseudomonas fluorescens*, leading to improved resistance against *Rhizoctonia solani*, a fungal pathogen. Flavonoids such as quercetin and kaempferol also play a role in attracting beneficial microbes, which further contributes to disease suppression (Mathesius, 2018).

3.3.3 Terpenoids and Alkaloids in ISR

Terpenoids are another group of secondary metabolites that are produced in response to ISR activation. Terpenoids serve as insect deterrents, antimicrobial agents, and herbivore repellents. For example, plants treated with ISR-inducing PGPR often produce higher levels of beta-caryophyllene, a volatile terpenoid that repels herbivores and attracts predators of herbivores, such as parasitic wasps (Dicke & Baldwin, 2010). Similarly, alkaloids, which include compounds like nicotine and quinine, are produced in higher quantities when ISR is activated. These compounds deter herbivores and inhibit the growth of pathogens by disrupting their cellular processes. For example, ISR activation in *Nicotiana tabacum* (tobacco) by *Bacillus subtilis* leads to increased nicotine production, which protects against herbivores such as caterpillars (Baldwin, 2001).

3.4 Indirect Defense Mechanisms Mediated by Secondary Metabolites

In addition to their direct defense roles, secondary metabolites also participate in indirect defense mechanisms. When ISR is triggered, plants may release volatile secondary metabolites, such as terpenoids, that attract the natural enemies of herbivores. This form of defense is known as induced indirect defense. For example, beta-caryophyllene, a volatile terpenoid produced by many plants, is released in response to herbivore damage following ISR activation. This compound attracts parasitic wasps that lay their eggs inside herbivorous caterpillars, leading to a natural reduction in herbivore populations (Degenhardt et al., 2009). Such indirect defenses are crucial for plants that cannot move or directly fend off herbivores. Moreover, ISR can enhance the production of volatile organic compounds (VOCs) that not only repel herbivores but also prime neighboring plants for defense. These VOCs serve as airborne signals, alerting nearby plants to impending attacks and triggering defense responses even before the herbivores arrive (Heil & Karban, 2010).

3.5 Engineering ISR for Enhanced Metabolite Production

Genetic engineering offers opportunities to enhance ISR and secondary metabolite production in crops, improving their resistance to pathogens and herbivores. By overexpressing genes involved in the jasmonic acid or ethylene pathways, researchers can enhance the plant's ability to produce defense-related secondary metabolites. For example, overexpression of jasmonate ZIM-domain (JAZ) repressors in *Arabidopsis thaliana* leads to increased production of secondary metabolites such as camalexin and glucosinolates, which are crucial for defense against necrotrophic pathogens (Jiang et al., 2021). Similarly, overexpression of ethylene response factors (ERFs) in crops like tomato (*Solanum lycopersicum*) and rice (*Oryza sativa*) leads to enhanced production of terpenoids and phenolics, providing increased resistance to fungal and bacterial pathogens (Shoresh et al., 2010).

4. Synergistic Effects of Secondary Metabolites in Resistance Mechanisms

Plants possess a diverse arsenal of secondary metabolites that play vital roles in their defense against pathogens, herbivores, and environmental stresses. These metabolites, including phenolics, flavonoids, terpenoids, alkaloids, and saponins, not only function as individual defense compounds but also work synergistically to enhance the overall resistance of plants. The synergistic effects of secondary metabolites can enhance the plant's ability to combat biotic stress by combining different defensive mechanisms, such as antimicrobial, antifeedant, or signaling actions, thereby providing a more robust defense strategy. Synergism between secondary metabolites occurs at multiple levels, including direct interactions between the compounds themselves, complementary modes of action against pathogens and pests, and coordinated signaling pathways that lead to the enhanced production of multiple metabolites in response to a threat. Understanding how these compounds work together can provide new insights into plant defense mechanisms and offer potential strategies for enhancing crop protection.

4.1 Complementary Modes of Action

One of the main ways secondary metabolites exhibit synergistic effects is through complementary modes of action. Different metabolites may target various aspects of a pathogen or pest's biology, making it more difficult for the invader to develop resistance. For example, phenolics and alkaloids can work together to inhibit pathogen growth by targeting different cellular processes. While phenolics disrupt cell wall integrity and membrane stability, alkaloids interfere with DNA replication and protein synthesis, creating a multi-targeted attack that overwhelms the pathogen's defenses (Wink, 2015). An example of such synergism can be seen in the interactions between phenolics and terpenoids. Phenolics, such as caffeic acid and ferulic acid, often function as cell wall fortifiers, providing a physical barrier to pathogen entry. Meanwhile, terpenoids, such as saponins and monoterpenes, disrupt cell membranes or interfere with microbial enzymatic activities. Together, these compounds provide both structural and biochemical protection against pathogens, making it harder for them to invade plant tissues (Lattanzio et al., 2006).

4.2 Synergistic Antimicrobial Properties

The antimicrobial activity of secondary metabolites is often enhanced when these compounds act synergistically. Plants produce a wide array of secondary metabolites that can inhibit bacterial, fungal, and viral pathogens. When multiple metabolites with different antimicrobial properties are produced in response to infection, their combined effects can be greater than the sum of their actions. For example, phenolic acids such as p-coumaric acid and ferulic acid work synergistically with flavonoids like quercetin and kaempferol to inhibit the growth of fungal pathogens such as *Fusarium oxysporum* and *Rhizoctonia solani*. While phenolic acids disrupt the pathogen's cell wall, flavonoids interfere with its metabolic pathways, leading to a more potent inhibition of pathogen growth (Ngadze et al., 2012). This type of synergy makes the plant's defense system more effective in preventing the spread of infection. In addition, saponins and alkaloids often show synergistic effects in their antimicrobial actions. Saponins, which have surfactant-like properties, disrupt the integrity of pathogen cell membranes, making the cells more permeable to other compounds, including alkaloids. Once the cell membrane is compromised, alkaloids such as berberine and quinine can enter the pathogen's cells more easily, disrupting critical cellular processes and leading to cell death (Zaynab et al., 2018). This cooperative interaction between different classes of secondary metabolites creates a powerful antimicrobial defense that can inhibit a broad range of pathogens.

4.3 Synergistic Role in Herbivore Deterrence

Secondary metabolites also work synergistically to deter herbivores. Plants produce a variety of antifeedant compounds that reduce the palatability of plant tissues and discourage herbivores from feeding. When different classes of metabolites are produced in concert, they can provide a more effective defense by targeting multiple aspects of the herbivore's physiology. For instance, alkaloids and terpenoids are often produced together in response to herbivore attacks. Alkaloids, such as nicotine and quinolizidine, act as neurotoxins that interfere with the nervous system of herbivores, while terpenoids, such as beta-caryophyllene and limonene, function as feeding deterrents by causing digestive disruption or repelling the herbivore through strong odors. This combination of neurotoxic and antifeedant effects makes the plant less attractive to herbivores, reducing the extent of damage (Baldwin, 2001). Moreover, flavonoids can enhance the deterrent effects of phenolics by altering the taste and digestibility of plant tissues, making them less appealing to herbivores. Tannins, a group of phenolic compounds, reduce the digestibility of plant proteins by binding to them, while flavonoids further inhibit the herbivore's digestive enzymes. Together, these compounds create a multifaceted defense that reduces the nutritional value of plant tissues, making them less desirable to herbivores (Rehman et al., 2012).

4.4 Synergy in Defense Signaling Pathways

In addition to their direct antimicrobial and antifeedant roles, secondary metabolites work synergistically within the plant's defense signaling pathways. The jasmonic acid (JA), salicylic acid (SA), and ethylene (ET) pathways are central to the plant's defense system, and secondary metabolites can modulate the cross-talk between these pathways to fine-tune the plant's response to different types of threats. For example, jasmonic acid is typically associated with defense against herbivores and necrotrophic pathogens, while salicylic acid is more involved in defense against biotrophic pathogens. When these pathways are activated together, they can lead to the production of a more diverse set of secondary metabolites. Phenolics and flavonoids, for instance, are produced in response to salicylic acid signaling, while terpenoids and alkaloids are often upregulated through jasmonic acid signaling (Thaler et al., 2012). The coordination between these pathways ensures that the plant produces the appropriate set of secondary metabolites in response to the specific nature of the attack. In some cases, secondary metabolites themselves act as signals to further enhance defense responses. Flavonoids and terpenoids can serve as signaling molecules that activate systemic acquired resistance (SAR) or induced systemic resistance (ISR), thereby priming the plant for more rapid and robust responses to subsequent attacks. By acting synergistically, these metabolites not only defend the plant directly but also amplify the plant's ability to resist future challenges (Pieterse et al., 2014).

4.5 Synergy in Plant-Microbe Interactions

Secondary metabolites also play a synergistic role in plant-microbe interactions, particularly in the rhizosphere, where plant roots interact with beneficial microbes. Flavonoids and terpenoids secreted by plant roots can work together to attract plant growth-promoting rhizobacteria (PGPR) and mycorrhizal fungi, which in turn enhance plant health by promoting nutrient uptake, improving stress tolerance, and inducing systemic resistance against pathogens. For example, in legumes, the secretion of flavonoids from the roots attracts rhizobia, which forms symbiotic relationships with the plant, fixing atmospheric nitrogen and improving the plant's nitrogen availability. At the same time, terpenoids secreted by the plant can attract mycorrhizal fungi, which enhance phosphorus uptake. The combined effects of flavonoids and terpenoids improve plant nutrition and health while also protecting against pathogens through the induced systemic resistance (ISR) pathway (Akiyama et al., 2005). In addition, secondary metabolites can work together to suppress soil-borne pathogens. Phenolics exuded from the roots can inhibit pathogen growth directly, while terpenoids and alkaloids may disrupt microbial communication or quorum sensing, preventing the formation of pathogenic biofilms. This multi-layered defense strategy creates a hostile environment for pathogens while promoting beneficial microbial communities (Badri & Vivanco, 2009).

5. Hormonal Pathways in Induced Resistance and Metabolite Synthesis

Hormonal pathways are crucial in regulating induced resistance (IR) in plants, governing the synthesis and accumulation of secondary metabolites that bolster plant defenses against pathogens, herbivores, and environmental stressors. The most significant hormonal pathways involved in IR include the salicylic acid (SA), jasmonic acid (JA), and ethylene (ET) signaling networks. Each of these pathways contributes to specific aspects of the plant's defense strategy, either individually or through cross-talk between them, enabling the plant to mount a robust response to various threats.

Secondary metabolites, such as phenolics, flavonoids, terpenoids, and alkaloids, are often produced in response to these hormonal signals and play key roles in both Systemic Acquired Resistance (SAR) and Induced Systemic Resistance (ISR). These compounds provide direct antimicrobial effects, deter herbivores, or serve as signaling molecules that modulate interactions between the plant and its surrounding microbial community. Understanding the interplay between hormonal pathways and secondary metabolite synthesis is critical for optimizing plant resistance mechanisms and enhancing crop resilience.

5.1 Salicylic Acid Pathway and Systemic Acquired Resistance (SAR)

The salicylic acid (SA) pathway is primarily associated with Systemic Acquired Resistance (SAR), a form of induced resistance that is activated in response to biotrophic pathogens, which feed on living plant tissues. Upon pathogen recognition, SA accumulates at the infection site and in distal tissues, triggering the expression of pathogenesis-related (PR) proteins and enhancing the production of secondary metabolites that deter pathogen spread (Vlot et al., 2009). This response is particularly effective against pathogens like *Pseudomonas syringae* and *Xanthomonas campestris*, which exploit living plant cells. Salicylic acid also plays a critical role in the synthesis of phenolic compounds, which are essential for cell wall strengthening and pathogen inhibition. For example, ferulic acid and caffeic acid are phenolic acids that accumulate in plant cell walls during SAR activation, increasing cell wall rigidity and making it more difficult for pathogens to penetrate. Additionally, SA modulates the production of flavonoids, such as quercetin and kaempferol, which act as antioxidants and antimicrobial agents. In crops like *Solanum lycopersicum* (*tomato*) and *Glycine max* (*soybean*), SA-induced flavonoid production enhances resistance to bacterial and fungal pathogens (Ngadze et al., 2012).

5.2 Jasmonic Acid Pathway and Induced Systemic Resistance (ISR)

Jasmonic acid (JA) is a key signaling molecule in the activation of Induced Systemic Resistance (ISR), which is typically triggered by plant growth-promoting rhizobacteria (PGPR) and is effective against necrotrophic pathogens and herbivores. The JA pathway regulates the synthesis of secondary metabolites that are critical for deterring herbivores and suppressing pathogen growth. JA-induced ISR does not rely on the accumulation of PR proteins like SAR but instead enhances the plant's ability to produce defense-related metabolites upon encountering a threat (Pieterse et al., 2014). In particular, JA signaling promotes the production of terpenoids and alkaloids, which serve as potent deterrents to herbivores. For example, in Nicotiana tabacum (*tobacco*), the activation of the JA pathway leads to increased synthesis of nicotine, an alkaloid that disrupts the nervous systems of herbivores, deterring them from feeding on the plant. Similarly, in Brassica napus (*canola*), JA signaling induces the production of glucosinolates, a group of sulfur-containing secondary metabolites that repel herbivores and inhibit fungal pathogens (Poelman et al., 2008). Additionally, JA plays a significant role in the regulation of phytoalexins, secondary metabolites produced in response to pathogen attack. Camalexin, a phytoalexin found in *Arabidopsis thaliana*, is synthesized in response to JA signaling and protects against necrotrophic pathogens like *Botrytis cinerea,* which causes gray mold in a variety of crops (Rehman et al., 2012). The production of these compounds enhances the plant's ability to resist pathogen infection and survive in hostile environments.

5.3 Ethylene Signaling in Defense

Ethylene (ET), a gaseous hormone, plays a synergistic role in defense, particularly in the regulation of ISR alongside jasmonic acid. While ethylene itself does not directly induce the production of secondary metabolites, it modulates the plant's response to biotic stress by enhancing the effects of JA signaling. Ethylene promotes the accumulation of secondary metabolites in response to both pathogen infection and herbivory, particularly in crops like Zea mays (*maize*) and *Triticum aestivum* (*wheat*), where it aids in deterring pests like the corn borer and aphids (Wasternack, 2007). The interaction between JA and ET pathways leads to the production of terpenoid volatiles, such as beta-caryophyllene and limonene, which not only deter herbivores but also attract natural enemies of pests. For example, beta-caryophyllene, produced in response to JA and ET signaling in *Zea mays*, attracts parasitic wasps that prey on herbivorous insects, providing an additional layer of defense (Degenhardt et al., 2009). This form of indirect defense is crucial for plants that cannot directly fend off herbivores.

5.4 Cross-Talk Between Hormonal Pathways and Secondary Metabolites

The interactions between the salicylic acid, jasmonic acid, and ethylene pathways allow for the fine-tuning of plant defense responses and the coordination of secondary metabolite production. Cross-talk between these pathways ensures that plants can prioritize defense strategies based on the nature of the threat. For instance, salicylic acid typically antagonizes jasmonic acid signaling, leading to the suppression of JA-mediated defenses when biotrophic pathogens are the primary threat (Thaler et al., 2012). However, in cases where plants face simultaneous attacks from both biotrophic and necrotrophic pathogens, such as *Pseudomonas syringae* and *Botrytis cinerea*, cross-talk between SA and JA/ET pathways can activate both defense mechanisms, leading to the production of a broader range of secondary metabolites. This coordinated defense response enhances the plant's ability to cope with multiple stresses by ensuring that the appropriate secondary metabolites are synthesized in response to specific threats.

6. Genetic Engineering to Enhance Induced Resistance in Crops

Genetic engineering has emerged as a powerful tool for enhancing induced resistance (IR) in crops by manipulating the underlying hormonal pathways and boosting the production of secondary metabolites. Advances in biotechnology allow for the precise modification of key genes involved in the salicylic acid (SA), jasmonic acid (JA), and ethylene (ET) pathways, which are critical for activating Systemic Acquired Resistance (SAR) and Induced Systemic Resistance (ISR). By engineering crops to overexpress genes associated with these pathways, it is possible to enhance their natural defense mechanisms, improve resistance to pathogens and herbivores, and reduce the reliance on chemical pesticides. Genetic engineering also allows for the introduction of novel traits that optimize the production and secretion of secondary metabolites, providing crops with enhanced protective capabilities against biotic stress. This approach has significant potential for improving crop resilience and sustainability, particularly in the context of climate change and increasing pathogen pressures.

6.1 Engineering the Salicylic Acid Pathway for Enhanced SAR

One of the primary targets for genetic engineering in crops is the salicylic acid (SA) pathway, which regulates Systemic Acquired Resistance (SAR). By manipulating genes involved in SA biosynthesis, researchers can increase the plant's capacity to mount an effective defense against biotrophic pathogens. For instance, overexpression of the NPR1 gene, which encodes the Nonexpressor of Pathogenesis-Related Genes 1 (NPR1) protein, has been shown to enhance SAR in crops like Oryza sativa (*rice*) and *Solanum tuberosum* (*potato*). NPR1 acts as a key regulator of SAR by modulating the transcription of pathogenesis-related (PR) genes in response to SA signaling (Chern et al., 2005). Transgenic plants overexpressing NPR1 exhibit increased resistance to biotrophic pathogens like *Magnaporthe oryzae*, the causative agent of rice blast disease, and *Phytophthora infestans*, which causes late blight in potatoes (Chern et al., 2005). In addition to NPR1, genes involved in salicylic acid biosynthesis, such as isochorismate synthase (ICS1), have been targeted for overexpression. In *Arabidopsis thaliana*, overexpression of ICS1 leads to increased SA production and enhanced resistance to *Pseudomonas syringae* (Wildermuth et al., 2001). Similar strategies can be applied to crops like Triticum aestivum (*wheat*) and Glycine max (*soybean*), where boosting SA biosynthesis could lead to increased resistance against a variety of biotrophic pathogens, such as *Puccinia striiformis*, the causative agent of wheat stripe rust, and *Phakopsora pachyrhizi*, which causes soybean rust. By enhancing the SA pathway, crops can sustain more robust Systemic Acquired Resistance (SAR), thereby reducing losses from major agricultural pathogens.

6.2 Engineering the Jasmonic Acid Pathway for Enhanced ISR

The jasmonic acid (JA) pathway is another critical target for genetic engineering, particularly in the context of Induced Systemic Resistance (ISR), which is often triggered by plant growth-promoting rhizobacteria (PGPR) and is effective against necrotrophic pathogens and herbivores. By manipulating genes that regulate JA biosynthesis and signaling, researchers can enhance the plant's production of secondary metabolites like terpenoids and alkaloids, which are vital for defense against herbivores and necrotrophic pathogens. One successful example of JA pathway engineering is the overexpression of LOX (lipoxygenase) genes, which are involved in the biosynthesis of jasmonic acid. In *Solanum lycopersicum* (*tomato*), overexpression of LOX3 results in higher JA levels and increased production of defense-related secondary metabolites, such as polyphenols and terpenoids. These compounds provide enhanced protection against *Botrytis cinerea*, a necrotrophic fungal pathogen, and herbivores like Spodoptera exigua (beet armyworm) (Howe & Schilmiller, 2002). This strategy has also been employed in Brassica napus (*canola*) to increase the synthesis of glucosinolates, which deter herbivores and suppress soil-borne pathogens like *Sclerotinia sclerotiorum*, the causative agent of white mold. Another approach involves the overexpression of JAZ (Jasmonate ZIM-domain) repressors, which negatively regulate the JA signaling pathway. By reducing the expression of these repressors using CRISPR-Cas9 or RNA interference (RNAi) technology, plants can sustain higher JA activity, leading to increased production of secondary metabolites involved in defense. For instance, in Nicotiana tabacum (*tobacco*), silencing of JAZ1 led to increased levels of nicotine, an alkaloid that deters herbivores such as *Manduca sexta* (tobacco hornworm) (Jiang et al., 2021). This enhanced JA pathway response strengthens ISR, enabling plants to mount a faster and more effective defense against both pathogens and herbivores.

6.3 Enhancing Ethylene-Mediated Resistance

The ethylene (ET) signaling pathway works synergistically with jasmonic acid in regulating Induced Systemic Resistance (ISR) and plays a role in defense against both necrotrophic pathogens and herbivores. Engineering ethylene signaling in crops can further enhance ISR by promoting the accumulation of defense-related secondary metabolites. In Zea mays (*maize*), overexpression of ERF (Ethylene Response Factor) genes, which are transcription factors activated by ethylene, has been shown to improve resistance to *Fusarium verticillioides*, the causative agent of maize ear rot. ERF transcription factors regulate the expression of genes involved in secondary metabolite biosynthesis, leading to increased production of beta-caryophyllene, a volatile terpenoid that repels herbivores and attracts natural enemies of pests, such as parasitic wasps (Degenhardt et al., 2009). This strategy not only improves direct plant defense but also enhances indirect defense, which involves attracting predatory organisms that help control pest populations. In addition, ethylene-mediated defenses can be enhanced through the overexpression of ethylene biosynthetic genes, such as ACO (1-aminocyclopropane-1-carboxylate oxidase), which catalyzes the final step in ethylene production. In crops like Triticum aestivum (*wheat*), increasing ethylene production leads to stronger defense responses against *Fusarium graminearum*, the pathogen responsible for Fusarium head blight (Bolt et al., 2017). By enhancing ethylene signaling, plants can more effectively activate ISR and produce the secondary metabolites needed for defense.

6.4 Engineering Cross-Talk Between Defense Pathways

A major focus of genetic engineering in crop protection is the cross-talk between salicylic acid (SA), jasmonic acid (JA), and ethylene (ET) signaling pathways. Fine-tuning this cross-talk allows plants to prioritize the appropriate defense response based on the type of threat they encounter. For instance, by manipulating key genes that modulate the antagonistic relationship between SA and JA pathways, researchers can enhance resistance to both biotrophic and necrotrophic pathogens simultaneously. For example, in *Arabidopsis thaliana*, the overexpression of NPR1, a master regulator of SA-mediated defense, has been combined with the overexpression of ERF1, a transcription factor that integrates JA and ET signals. This dual overexpression leads to a balanced activation of both SAR and ISR, resulting in increased resistance to both *Pseudomonas syringae* (biotrophic pathogen) and *Botrytis cinerea* (necrotrophic pathogen) (Spoel et al., 2007). Applying similar strategies in crops like *Glycine max* (*soybean*) and *Oryza sativa* (*rice*) could provide comprehensive protection against a wide range of pathogens.

6.5 Genetic Engineering for Enhanced Secondary Metabolite Production

In addition to manipulating hormonal pathways, genetic engineering can directly enhance the production of secondary metabolites that are critical for plant defense. By overexpressing genes involved in phenolic, flavonoid, and terpenoid biosynthesis, crops can be engineered to produce higher levels of these metabolites, improving their overall resistance to pathogens and pests. For example, in *Vitis vinifera* (*grapevine*), overexpression of stilbene synthase (STS), an enzyme involved in the biosynthesis of resveratrol, leads to increased production of this phytoalexin, which protects *Botrytis cinerea* and *Plasmopara viticola*, the causative agent of downy mildew (Adrian et al., 2000). Similarly, in *Medicago sativa* (*alfalfa*), the overexpression of chalcone synthase (CHS) and flavonoid 3'-hydroxylase (F3'H) enhances flavonoid production, improving resistance to root rot pathogens like *Phytophthora medicaginis* (Yu et al., 2009). The use of CRISPR-Cas9 and RNA interference (RNAi) technologies allows for even more precise editing of genes involved in secondary metabolite biosynthesis. For instance, CRISPR-mediated knockouts of negative regulators in the phenylpropanoid pathway can lead to increased production of lignin, flavonoids, and phenolic acids, which enhance cell wall integrity and provide broad-spectrum resistance to pathogens (Jiang et al., 2021).

Conclusion

Induced resistance in plants is a critical adaptive mechanism that enables plants to enhance their defense responses when exposed to external stressors, such as pathogens, pests, or environmental stresses. Unlike constitutive defenses, which are always active, induced resistance is activated in response to specific threats, allowing plants to conserve energy during times of peace and mount a strong defense when needed. This mechanism is highly complex and involves a network of signaling pathways, secondary metabolite production, and hormonal regulation. Induced resistance not only protects plants from immediate threats but also primes the plant's immune system, preparing it for future attacks. The two main forms of induced resistance are Systemic Acquired Resistance (SAR) and Induced Systemic Resistance (ISR), both of which rely heavily on the synthesis of secondary metabolites and the activation of specific hormonal pathways.

Systemic Acquired Resistance (SAR) is a well-characterized form of induced resistance that is triggered by localized infection or damage, leading to a whole-plant immune response. When a plant is attacked by a pathogen, it recognizes pathogen-associated molecular patterns (PAMPs), which trigger localized defense responses at the site of infection. Following this, signaling molecules, such as salicylic acid (SA), are produced and transported throughout the plant, activating defense genes in tissues distant from the infection site. SAR is often associated with the accumulation of pathogenesis-related (PR) proteins and phenolic compounds, which play a key role in strengthening the plant's defenses.

Phenolics are secondary metabolites that are crucial in the SAR response due to their antimicrobial properties and role in fortifying plant cell walls. When SAR is activated, phenolic compounds such as lignin and flavonoids accumulate in plant tissues, providing a physical barrier to invading pathogens. Lignin deposition strengthens the cell walls, making it more difficult for pathogens to penetrate, while flavonoids act as antioxidants and antimicrobial agents that neutralize reactive oxygen species (ROS) and inhibit pathogen growth. The role of phenolics in SAR underscores the importance of secondary metabolite production in plant immunity, as these compounds not only protect against immediate infections but also contribute to the long-term resilience of the plant.

Induced Systemic Resistance (ISR) is another form of induced resistance that is typically activated by beneficial microbes, such as plant growth-promoting rhizobacteria (PGPR) and certain mycorrhizal fungi. Unlike SAR, which is primarily mediated by salicylic acid, ISR is regulated by the plant hormones jasmonic acid (JA) and ethylene. When plants interact with beneficial microbes in the soil, these microbes trigger the production of signaling molecules in the roots that activate defense responses throughout the plant. This leads to the production of secondary metabolites, including alkaloids, terpenoids, and phenolics, which enhance the plant's ability to defend against a wide range of pathogens and pests.

The production of secondary metabolites in ISR plays a crucial role in plant defense. For example, terpenoids and alkaloids, which are often produced in response to jasmonic acid signaling, serve as potent anti-insect compounds, deterring herbivory and inhibiting insect growth. Phenolic compounds, which are also produced during ISR, contribute to both antimicrobial and antioxidant defenses, helping the plant to withstand both biotic and abiotic stresses. The activation of ISR is often more generalized than SAR, providing broad-spectrum resistance that is effective against multiple types of pathogens, including bacteria, fungi, and insects. This broad-spectrum resistance makes ISR particularly valuable in agricultural settings, where crops are exposed to a variety of stressors.

One of the most interesting aspects of induced resistance is the synergistic effect of secondary metabolites in enhancing plant defense mechanisms. Secondary metabolites rarely act in isolation; instead, they work together to create a multilayered defense system that protects the plant from a wide range of threats. For instance, phenolics can act synergistically with terpenoids to enhance antimicrobial activity, while alkaloids may work in concert with other compounds to deter herbivores more effectively. This synergy is often mediated by the hormonal signaling pathways that regulate induced resistance, ensuring that the right combination of metabolites is produced in response to specific threats.

The interplay between different secondary metabolites and hormonal pathways in induced resistance is crucial for optimizing the plant's defense strategy. Salicylic acid, jasmonic acid, and ethylene are the primary hormones involved in regulating SAR and ISR, and they often interact to fine-tune the plant's response to specific stressors. For example, salicylic acid is typically associated with defense against biotrophic pathogens (which require living host cells), while jasmonic acid and ethylene are more effective against necrotrophic pathogens (which kill host cells) and herbivores. The cross-talk between these pathways ensures that the plant can mount a defense that is tailored to the type of threat it faces.

The synthesis of secondary metabolites is tightly linked to the hormonal pathways that regulate induced resistance. For example, the production of alkaloids and terpenoids in response to herbivory is often driven by jasmonic acid signaling. Jasmonic acid activates a cascade of transcription factors that regulate the expression of genes involved in secondary metabolite biosynthesis, leading to the accumulation of defense compounds in the plant tissues most vulnerable to attack. Similarly, phenolic production in response to pathogen infection is often regulated by salicylic acid, which induces the expression of genes involved in phenolic biosynthesis and transport. This hormonal regulation ensures that secondary metabolites are produced in the right amounts and in the right tissues, optimizing the plant's defense strategy while minimizing unnecessary energy expenditure.

In recent years, genetic engineering has emerged as a powerful tool for enhancing induced resistance in crops. By manipulating the genes involved in secondary metabolite biosynthesis and hormonal signaling, researchers can develop plants that are more resistant to pests, diseases, and environmental stresses. For example, crops can be engineered to overexpress key enzymes in the jasmonic acid or salicylic acid pathways, leading to increased production of secondary metabolites that enhance resistance. Genetic engineering can also be used to introduce new defense genes from other species, expanding the range of metabolites that a plant can produce in response to stress.

One of the most successful applications of genetic engineering in enhancing induced resistance has been the development of crops that produce higher levels of phytoalexins—antimicrobial compounds that are produced in response to pathogen attacks. By introducing genes from naturally resistant plants, scientists have been able to create crops that are more resilient to fungal and bacterial infections. For instance, the introduction of genes that regulate the production of resveratrol, a phenolic compound with strong antimicrobial properties, has been shown to increase resistance to fungal pathogens in genetically modified plants. In addition to enhancing pathogen resistance, genetic engineering can also be used to improve a plant's ability to resist herbivores. By manipulating the biosynthetic pathways responsible for producing terpenoids and alkaloids, researchers can create plants that are less palatable to insects, reducing the need for chemical insecticides. This approach not only improves crop yields but also contributes to more sustainable agricultural practices by reducing the environmental impact of pesticide use.

References

Adrian, M., Jeandet, P., Douillet-Breuil, A. C., Tesson, L., & Bessis, R. (2000). Stilbene content of mature *Vitis vinifera* berries in response to UV-C elicitation. *Journal of Agricultural and Food Chemistry, 48*(12), 6103-6105.

Akiyama, K., Matsuzaki, K. I., & Hayashi, H. (2005). Plant sesquiterpenes induce hyphal branching in arbuscular mycorrhizal fungi. *Nature, 435*(7043), 824-827.

Badri, D. V., & Vivanco, J. M. (2009). Regulation and function of root exudates in plants. *Plant, Cell & Environment, 32*(6), 666-681.

Bailey-Serres, J., & Mittler, R. (2006). The roles of reactive oxygen species in plant cells. *Plant Physiology, 141*(2), 311.

Baldwin, I. T. (2001). An ecologically motivated analysis of plant-herbivore interactions in native tobacco. *Plant Physiology, 127*(4), 1449-1458.

Bolt, S., Zuther, E., Zintl, S., Hincha, D. K., & Schmülling, T. (2017). ERF105 is a transcription factor gene of *Arabidopsis thaliana* required for freezing tolerance and cold acclimation. *Plant, Cell & Environment, 40*, 108-120.

Chen, X. H., Koumoutsi, A., Scholz, R., Eisenreich, A., Schneider, K., Heinemeyer, I., & Borriss, R. (2007). Comparative analysis of the complete genome sequence of the plant growth–promoting bacterium *Bacillus amyloliquefaciens* FZB42. *Nature Biotechnology, 25*(9), 1007-1014.

Chern, M., Fitzgerald, H. A., Canlas, P. E., Navarre, D. A., & Ronald, P. C. (2005). Overexpression of a rice NPR1 homolog leads to constitutive activation of defense response and hypersensitivity to light. *Molecular Plant Microbe Interaction, 18*(6), 511-20.

Conrath, U. (2009). Priming of Induced Plant Defense Responses. *Advances in Botanical Research*, Academic Press, *51*, 361-395

Degenhardt, J., Hiltpold, I., Köllner, T. G., Frey, M., Gierl, A., Gershenzon, J., & Turlings, T. C. J. (2009). Restoring a maize root signal that attracts insect-killing nematodes to control a major pest. *Proceedings of the National Academy of Sciences, 106*(32), 13213-13218.

Dicke, M., & Baldwin, I. T. (2010). The evolutionary context for herbivore-induced plant volatiles: Beyond the "cry for help". *Trends in Plant Science, 15*(3), 167-175.

Durrant, W. E., & Dong, X. (2004). Systemic acquired resistance. *Annual Review of Phytopathology, 42*, 185-209.

Fu, Z. Q., & Dong, X. (2013). Systemic acquired resistance: Turning local infection into global defense. *Annual Review of Plant Biology, 64*, 839-863.

Heil, M., & Karban, R. (2010). Explaining evolution of plant communication by airborne signals. *Trends in Ecology & Evolution, 25*(3), 137-144.

Howe, G. A., & Schilmiller, A. L. (2002). Oxylipin metabolism in response to stress. *Current Opinion in Plant Biology, 5*(3), 230-236.

Jiang, C., Lv, G., Tu, Y., Cheng, X., Duan, Y., Zeng, B., & He, B. (2021). Applications of CRISPR/Cas9 in the synthesis of secondary metabolites in filamentous fungi. *Frontiers in Microbiology, 12*, 638096.

Koornneef, A., & Pieterse, C. M. J. (2008). Cross talk in defense signaling. *Plant Physiology, 146*(3), 839-844.

Lanoue, A., Burlat, V., Henkes, G. J., Koch, I., Schurr, U., & Rose, U. S. (2010). De novo biosynthesis of defense root exudates in response to Fusarium attack in barley. *New Phytologist, 185*, 577-588.

Lattanzio, V., Lattanzio, V. M. T., & Cardinali, A. (2006). Role of phenolics in the resistance mechanisms of plants against fungal pathogens and insects. In F. Imperato (Ed.), *Phytochemistry: Advances in Research* (pp. 23-67). Research Signpost.

Lorenzo, O., & Solano, R. (2005). Molecular players regulating the jasmonate signalling network. *Current Opinion in Plant Biology, 8*(5), 532-540.

Mathesius U. (2018). Flavonoid Functions in Plants and Their Interactions with Other Organisms. *Plants (Basel), 7*(2), 30.

Ngadze, E., Icishahayo, D., Coutinho, T. A., & van der Waals, J. E. (2012). Role of polyphenol oxidase, phenylalanine ammonia-lyase, total phenolic compounds, and chlorogenic acid in resistance of potato to soft rot. *Plant Disease, 96*(2), 186-192.

Pandey, S. P., & Somssich, I. E. (2009). The WRKY transcription factor family in plant immunity. *Plant Physiology, 150*(4), 1648-1655.

Pieterse, C. M. J., Van der Does, D., Zamioudis, C., Leon-Reyes, A., & Van Wees, S. C. M. (2012). Hormonal modulation of plant immunity. *Annual Review of Cell and Developmental Biology, 28*, 489-521.

Pieterse, C. M. J., Zamioudis, C., Berendsen, R. L., Weller, D. M., Van Wees, S. C. M., & Bakker, P. A. H. M. (2014). Induced systemic resistance by beneficial microbes. *Annual Review of Phytopathology, 52*(1), 347-375.

Poelman, E. H., Broekgaarden, C., Van Loon, J. J. A., & Dicke, M. (2008). Early season herbivore differentially affects plant defence responses to subsequently colonizing herbivores and their abundance in the field. *Molecular Ecology, 17*(14), 3352-3365.

Rehman, F., Khan, F., & Badruddin, S. (2012). Role of Phenolics in Plant Defense Against Insect Herbivory. In: Khemani, L., Srivastava, M., Srivastava, S. (eds) Chemistry of Phytopotentials: Health, Energy and Environmental Perspectives. Springer, Berlin, Heidelberg.

Shah, J., & Zeier, J. (2013). Long-distance communication and signal amplification in systemic acquired resistance. *Frontiers in Plant Science, 4*, 30.

Shoresh, M., Harman, G. E., & Mastouri, F. (2010). Induced systemic resistance and plant responses to fungal biocontrol agents. *Annual Review of Phytopathology, 48*, 21-43.

Spoel, S. H., Johnson, J. S., & Dong, X. (2007). Regulation of tradeoffs between plant defenses against pathogens with different lifestyles. *Proceedings of the National Academy of Sciences, 104*(47), 18842-18847.

Thaler, J. S., Humphrey, P. T., & Whiteman, N. K. (2012). Evolution of jasmonate and salicylate signal crosstalk. *Trends in Plant Science, 17*(5), 260-270.

Thomma B. P. H. J., Eggermont K., Penninckx I. A. M. A., Mauch-Mani B., Vogelsang R., Cammue B. P. A., & Broekaert W. F. (1998). Separate jasmonate-dependent and salicylate-dependent defense-response pathways in Arabidopsis are essential for resistance to distinct microbial pathogens. *Proceedings of National Academy of Sciences USA, 95*, 15107-15111.

Tong, Y., Xue, J., Li, Q., & Zhang, L. (2024). A generalist regulator: MYB transcription factors regulate the biosynthesis of active compounds in medicinal plants. *Journal of Experimental Botany*, 75(16), 4729-4744.

Van Loon, L. C., Bakker, P. A. H. M., & Pieterse, C. M. J. (1998). Systemic resistance induced by rhizosphere bacteria. *Annual Review of Phytopathology, 36*, 453-483.

Vlot, A. C., Dempsey, D. A., & Klessig, D. F. (2009). Salicylic acid, a multifaceted hormone to combat disease. *Annual Review of Phytopathology, 47*(1), 177-206.

Walters, D. R., Ratsep, J., & Havis, N. D. (2013). Controlling crop diseases using induced resistance: Challenges for the future. *Journal of Experimental Botany, 64*(5), 1263-1280.

Wasternack C. (2007). Jasmonates: An Update on biosynthesis, signal transduction and action in plant stress response, growth and development. *Annals of Botany*, *100*(4), 681-697.

Wildermuth, M. C., Dewdney, J., Wu, G., & Ausubel, F. M. (2001). Isochorismate synthase is required to synthesize salicylic acid for plant defence. *Nature, 414*(6863), 562-565.

Wink, M. (2015). Modes of action of herbal medicines and plant secondary metabolites. *Medicines, 2*(3), 251-286.

Yu, O., Shi, J., Hession, A. O., Maxwell, C. A., McGonigle, B., & Odell, J. T. (2003). Metabolic engineering to increase isoflavone biosynthesis in soybean seeds. *Phytochemistry, 63*, 753-763.

Zaynab, M., Fatima, M., Abbas, S., Sharif, Y., Umair, M., Zafar, M. H., & Bahadar, K. (2018). Role of secondary metabolites in plant defense against pathogens. *Microbial Pathogenesis, 124*, 198-202.

Chapter - 16

Applications of Plant Secondary Metabolites in Sustainable Agriculture

Krishna Chaitanya Tirunagaru[1], Zuby Gohar Ansari[2], Reshma Jadhav[3] and Susheel Kumar[4]

[1]Department of Agricultural Entomology, School of Agricultural Sciences & Technology (SAST), Narsee Monjee Institute of Management Studies (NMIMS), Shirpur, Dhule District, Maharashtra, India
[2]B.M. College of Agriculture, Jaswadi Road Khandwa, Madhya Pradesh, India
[3]Department of Biochemistry, Mahatma Phule Krishi Vidyapeeth, Rahuri, Ahmednagar, Maharashtra, India
[4]Plant Molecular Virology Laboratory, CPMB Division, CSIR-National Botanical Research Institute, Rana Pratap Marg, Lucknow, Uttar Pradesh, India

Abstract

Plant secondary metabolites have emerged as essential tools for promoting sustainability in agriculture by offering eco-friendly alternatives to synthetic chemicals for pest and disease control. These metabolites, which include phenolics, flavonoids, alkaloids, and terpenoids, serve as natural defense compounds that can be harnessed to protect crops from pathogens and pests, reducing the reliance on chemical pesticides. In sustainable agriculture, secondary metabolites are increasingly used as biopesticides, bioherbicides, and soil health enhancers, contributing to the development of integrated pest management (IPM) systems. These natural compounds have the potential to control a wide range of plant diseases and pests while minimizing environmental harm. For example, phenolics and flavonoids can disrupt the growth and reproduction of harmful insects, while terpenoids and alkaloids can inhibit the growth of fungal and bacterial pathogens. Additionally, the allelopathic properties of certain secondary metabolites can be used to suppress weed growth, promoting healthier crop competition and reducing the need for synthetic herbicides. Furthermore, secondary metabolites play a critical role in enhancing soil health by fostering beneficial microbial communities in the rhizosphere, which can improve nutrient uptake and plant resilience. The use of plant-derived metabolites in sustainable agriculture not only improves crop yield and quality but also promotes environmental sustainability by reducing chemical inputs, enhancing biodiversity, and improving soil health. Advances in genetic engineering have further expanded the potential of secondary metabolites in agriculture by allowing

DOI: 10.1201/9781003717454-16

the modification of plants to produce higher levels of these compounds, leading to more resilient and productive crops.

Keywords: Sustainable Agriculture, Secondary Metabolites, Biopesticides, Bioherbicides, Integrated Pest Management, Soil Health

Introduction

Secondary metabolites play a critical role in integrated pest management (IPM) strategies, offering natural, eco-friendly alternatives to synthetic chemicals in agricultural practices. These compounds, which include flavonoids, alkaloids, terpenoids, and phenolics, are naturally produced by plants to defend against pests, pathogens, and competitors. The incorporation of secondary metabolites into IPM approaches helps reduce the reliance on chemical pesticides, thus minimizing environmental damage and fostering sustainable agricultural practices. Secondary metabolites are highly versatile in their function, acting as direct deterrents to pests, inhibiting pathogen growth, and even modulating the behavior and development of herbivores. This natural defense system provides a foundation for more sustainable and targeted pest control strategies in modern agriculture. One of the main advantages of using secondary metabolites in IPM is their ability to target specific pests while maintaining the ecological balance, allowing beneficial insects and organisms to thrive. This selective action is key to reducing pesticide resistance in pests, which is a major problem associated with synthetic chemical use (Divekar et al., 2022).

The use of secondary metabolites as biopesticides and bioherbicides is increasingly gaining attention due to their potential in sustainable pest management. Biopesticides derived from secondary metabolites, such as neem oil (rich in terpenoids) or pyrethrins (derived from *Chrysanthemum* flowers), have demonstrated strong efficacy in controlling a wide variety of agricultural pests. These natural products work through multiple modes of action, including insect growth regulation, disruption of feeding behavior, and inhibition of reproductive processes (Isman, 2017). Unlike synthetic pesticides, which often target a single biochemical pathway, secondary metabolites typically act on multiple biological targets, making it more difficult for pests to develop resistance. This multi-target approach also reduces the environmental persistence of these compounds, as they are more easily degraded in soil and water, minimizing their ecological footprint. Similarly, bioherbicides, such as those derived from allelopathic secondary metabolites, can be used to suppress weed growth, offering an alternative to synthetic herbicides like glyphosate, which has raised environmental and health concerns (Duke et al., 2018).

The allelopathic properties of secondary metabolites are particularly valuable for weed control. Allelopathy refers to the chemical inhibition of one plant by another, often mediated by secondary metabolites secreted into the soil through root exudates or leaf litter. These allelopathic compounds can suppress the germination, growth, and development of competing plant species, reducing the need for synthetic herbicides. For example, *Sorghum bicolor* produces sorgoleone, a potent allelopathic compound that inhibits the growth of weeds by disrupting their photosynthetic processes (Weston & Duke, 2003). By harnessing the allelopathic potential of plants, farmers can reduce the chemical load on the environment while promoting crop productivity. Furthermore, allelopathic secondary metabolites are often species-specific, which allows for more targeted weed suppression without affecting neighboring non-target plant species. This specificity is a significant advantage over broad-spectrum chemical herbicides that can harm beneficial plants and reduce biodiversity in agricultural systems (Inderjit & Duke, 2003).

In addition to their direct effects on pests and weeds, secondary metabolites contribute to enhancing soil health by promoting beneficial microbial communities in the rhizosphere. These metabolites serve as signaling molecules that facilitate plant-microbe interactions, such as symbiotic relationships with mycorrhizal fungi and nitrogen-fixing bacteria. The presence of phenolics and flavonoids in root exudates has been shown to enhance the colonization of beneficial microbes, which in turn improve nutrient availability and uptake by plants (Badri et al., 2013). For instance, flavonoids play a critical role in establishing symbiotic interactions between leguminous plants and *Rhizobium* bacteria, which fix atmospheric nitrogen into a form that plants can readily absorb. This symbiotic relationship reduces the need for nitrogen fertilizers, promoting more sustainable and environmentally friendly agricultural practices. Additionally, secondary metabolites such as terpenoids and alkaloids can suppress soil-borne pathogens, further contributing to plant health and productivity by reducing disease incidence (Walker et al., 2003).

The environmental benefits of using plant-derived compounds in agriculture are numerous. Unlike synthetic chemicals, secondary metabolites are biodegradable and less likely to accumulate in the environment, reducing the risks of soil and water contamination. Their natural origin means they are often less toxic to non-target organisms, including pollinators, beneficial insects, and aquatic life. Moreover, the use of plant-based compounds supports biodiversity in agricultural ecosystems by fostering a balance between pest control and the conservation of beneficial species (Isman & Grieneisen, 2014). The reduced environmental impact of secondary metabolites makes them a vital component of sustainable agriculture, where the goal is to minimize the use of harmful chemicals while maintaining crop productivity. Furthermore, plant-derived compounds offer a renewable resource for pest and weed management, as they can be sustainably sourced from crops or produced through biotechnological means, reducing the reliance on non-renewable petrochemicals used in synthetic pesticide production.

Advances in genetic engineering have opened up new possibilities for increasing the production of secondary metabolites in crops, further enhancing their utility in sustainable agriculture. By manipulating the biosynthetic pathways responsible for secondary metabolite production, scientists can create crop varieties that produce higher levels of these compounds, thereby improving their natural defenses against pests and diseases. For example, overexpression of genes involved in the phenylpropanoid pathway has been shown to increase the production of phenolic compounds, such as flavonoids, which enhance resistance to insect herbivores and fungal pathogens (Jiang et al., 2023). Similarly, the genetic modification of terpenoid biosynthesis pathways has led to crops with increased levels of terpenoids, which are known to deter pests and protect against abiotic stressors such as drought and UV radiation (Tholl, 2015).

CRISPR/Cas9 gene editing technology has provided a precise and efficient tool for modifying secondary metabolite pathways in crops. This technology allows scientists to target specific genes involved in metabolite production, resulting in crops with enhanced pest resistance without affecting overall growth and development (Zhou et al., 2014). By increasing the levels of specific secondary metabolites, genetically engineered crops can reduce the need for chemical inputs, leading to more sustainable and resilient agricultural systems. Additionally, metabolic engineering approaches have been used to introduce novel secondary metabolite pathways into plants that do not naturally produce certain defense compounds. This strategy has been particularly successful in improving the pest and disease resistance of staple crops, such as rice and maize, which are traditionally more vulnerable to biotic stressors (Paddon & Keasling, 2014).

1. Role of Secondary Metabolites in Integrated Pest Management (IPM)

1.1 Overview of Secondary Metabolites in Pest Control

Secondary metabolites, often referred to as "plant defensive compounds," are organic compounds produced by plants that are not essential for their basic metabolic functions but play critical roles in defense mechanisms. These compounds are integral to the strategy of Integrated Pest Management (IPM) due to their ability to naturally deter or kill pest organisms. IPM aims to reduce the excessive use of synthetic pesticides by integrating multiple approaches to pest control, with secondary metabolites being one of the most promising strategies. Secondary metabolites are highly diverse, and their complex chemical structures are often specific to certain plant species. These compounds act against a variety of pests, including insects, fungi, bacteria, nematodes, and even other plants. What makes secondary metabolites particularly suitable for IPM is their ability to be highly selective, affecting only specific pest organisms while being less toxic to humans and non-target species. This feature is crucial in maintaining ecological balance and preventing the widespread harm often associated with synthetic chemical pesticides (Wink, 2018).

In the context of pest control, secondary metabolites have various biological activities, such as direct toxicity, disruption of reproduction, inhibition of feeding, and repelling pests. This versatility allows for more nuanced pest management strategies, where different metabolites can be used in combination to target various stages of a pest's life cycle. Unlike synthetic pesticides, which may target a single physiological pathway, secondary metabolites tend to interact with multiple biological processes, reducing the likelihood of pests developing resistance (Tariq et al., 2023). The growing need for sustainable agriculture and environmental preservation has heightened interest in the use of secondary metabolites in IPM. Farmers are increasingly seeking alternatives to synthetic chemicals, which often lead to pest resistance, environmental degradation, and health issues. In this context, secondary metabolites represent a sustainable, eco-friendly solution. Additionally, their use aligns with organic farming principles, which prohibit or limit the use of synthetic inputs (Isman, 2020).

1.2 Mechanism of Action of Secondary Metabolites in IPM

Secondary metabolites exhibit a wide range of mechanisms that make them highly effective in IPM. These mechanisms ensure that pests are controlled without causing undue harm to the surrounding environment or non-target organisms.

1.2.1 Toxicity to Pests

Toxicity is one of the most direct ways secondary metabolites affect pest populations. Many plants produce toxic compounds that can interfere with the basic physiological functions of pests. Alkaloids are a prominent example of this, with compounds like nicotine from *Nicotiana tabacum* and morphine from *Papaver somniferum* having profound neurotoxic effects on insects. Nicotine, in particular, interferes with neurotransmitter function by binding to nicotinic acetylcholine receptors, causing paralysis and eventual death of the pest. This neurotoxic effect makes alkaloid-based pesticides potent options for controlling a wide range of insect pests (Casida, 2018). Another example of toxic secondary metabolites is the group of compounds known as pyrethrins, derived from *Chrysanthemum* flowers. Pyrethrins act on the nervous systems of insects by targeting sodium channels, leading to hyper-excitation, paralysis, and death. This compound is so effective that synthetic derivatives, known as pyrethroids, have been developed and widely used as commercial insecticides. However, natural pyrethrins are favored in IPM because of their rapid biodegradability and lower toxicity to mammals, birds, and beneficial insects (Schmutterer, 1995). Phenolics, another class of secondary metabolites, also exhibit toxic effects. They inhibit the activity of key enzymes in pests, interfere with cell wall biosynthesis in fungi, and create oxidative stress. For example, tannins, a type of phenolic compound, bind to proteins in the digestive tracts of herbivores, reducing the efficiency of digestion and deterring further feeding. Such compounds are invaluable in managing insect herbivory without the use of synthetic pesticides (Lattanzio, 2013).

1.2.2 Feeding Deterrence

Apart from direct toxicity, many secondary metabolites function as feeding deterrents, making plants less palatable or entirely inedible to pests. Terpenoids, such as limonoids and saponins, are among the most effective in this regard. Terpenoids often have a bitter taste or a strong odor that repels herbivores. For instance, azadirachtin, a triterpenoid found in the neem tree (*Azadirachta indica*), acts as both a feeding deterrent and a growth disruptor. When insects come into contact with azadirachtin, it interferes with their hormonal system, specifically ecdysone, a hormone critical for molting. This dual effect not only discourages feeding but also prevents insects from maturing and reproducing, thus reducing the pest population over time (Isman, 2020). Essential oils, rich in terpenoids, are another group of secondary metabolites that serve as feeding deterrents. Oils from plants like lavender (*Lavandula angustifolia*), peppermint (*Mentha piperita*), and clove (*Syzygium aromaticum*) have been shown to repel a variety of insect pests. These oils are volatile and evaporate quickly, making them ideal for repelling insects in a natural and non-toxic manner. By using essential oils as part of an IPM strategy, farmers can protect crops without relying on synthetic chemicals that often have negative environmental impacts (Bakkali et al., 2008).

1.2.3 Interference with Pest Growth and Reproduction

Another important mechanism by which secondary metabolites contribute to IPM is through interference with the growth and reproduction of pests. Compounds like azadirachtin and other limonoids disrupt the hormonal balance in insects, preventing them from successfully molting, pupating, or reproducing. This class of compounds acts as insect growth regulators (IGRs), and their use in IPM is particularly beneficial because they provide long-term control without killing adult pests outright. Instead, by preventing the next generation of pests from developing, these compounds ensure a gradual decline in pest populations (Schmutterer, 1995). Similarly, flavonoids, another group of phenolic compounds, have been found to inhibit egg-laying in certain insect species. By making plants less attractive for oviposition, flavonoids can effectively reduce pest populations without causing immediate mortality. This selective pressure on pest behavior is crucial for maintaining ecological balance, as it allows for the natural regulation of pest populations without the drastic consequences of widespread pesticide use (Wink, 2018).

1.3 Applications of Secondary Metabolites in Modern IPM Strategies

1.3.1 Biological Control Using Secondary Metabolites

The use of secondary metabolites in biological control is a cornerstone of IPM. Biological control refers to the use of living organisms or their by-products (such as secondary metabolites) to manage pest populations. Plants that produce high levels of defensive compounds are often used in companion planting or as part of crop rotation strategies to naturally repel pests. For example, marigolds (*Tagetes* species) are known for producing thiophenes, which are highly effective against nematodes. By planting marigolds alongside susceptible crops, farmers can reduce nematode infestations without relying on synthetic nematicides (Hooks & Johnson, 2003). Another strategy involves the use of cover crops or trap crops that produce secondary metabolites to attract pests away from the main crop. These trap crops can then be treated with natural pesticides derived from secondary metabolites, providing a dual-layer of protection (Sarkar et al., 2018).

1.3.2 Formulation of Botanical Pesticides

Secondary metabolites are increasingly being used to develop botanical pesticides pesticides derived directly from plant extracts. Neem-based products, which contain azadirachtin, have been particularly successful as broad-spectrum insecticides. These products are effective against a variety of pests, including aphids, caterpillars, whiteflies, and mites, while remaining safe for beneficial insects like bees and ladybugs. Botanical pesticides break down rapidly in the environment, reducing the risk of contamination of soil and water systems (Isman, 2020). Other successful examples include rotenone, an isoflavonoid used to control leaf-eating insects, and pyrethrins, which are effective against a wide range of insect pests. These botanical pesticides are integral to IPM because they offer a sustainable alternative to synthetic chemicals, reducing the risk of pesticide resistance and environmental damage (Casida, 2018).

1.3.3 Enhancing Plant Resistance through Breeding and Biotechnology

One of the most promising advancements in the use of secondary metabolites within IPM is the ability to enhance plant resistance to pests through breeding and biotechnology. Plant breeding techniques can be used to select traits that increase the concentration or potency of secondary metabolites, thereby naturally fortifying the plant's defenses. This method is particularly important for crop plants that do not naturally produce high levels of these protective compounds or for which synthetic pesticide use is restricted. For example, wild relatives of crop plants often possess higher levels of secondary metabolites due to the selective pressures of natural environments. By cross-breeding these wild relatives with commercial varieties, it is possible to introduce these defensive traits into economically important crops. For instance, in the case of tomato plants, wild relatives have been found to produce high levels of acyl sugars that deter pests such as aphids and whiteflies. Through selective breeding, these traits can be introduced into domesticated tomato varieties, providing them with enhanced natural defenses (Pandey et al., 2023). Beyond traditional breeding methods, advances in biotechnology have allowed for the genetic modification of plants to overexpress secondary metabolites or produce entirely new compounds. Genetic engineering tools such as CRISPR-Cas9 and transgenic techniques have enabled scientists to directly manipulate the genes responsible for secondary metabolite production. For example, researchers have successfully inserted genes from neem (*Azadirachta indica*) into tobacco plants to allow them to produce azadirachtin, an effective insect deterrent and growth regulator. This biotechnological approach not only increases the efficacy of IPM strategies but also reduces the need for external inputs like chemical pesticides (Schmutterer, 1995).

The use of biotechnology also extends to the development of microbial biofactories for the production of secondary metabolites. Microbes can be genetically engineered to produce plant secondary metabolites in large quantities, which can then be formulated into biopesticides. This approach is both scalable and sustainable, providing a constant supply of bioactive compounds that can be used in IPM. For instance, microbes engineered to produce alkaloids, terpenoids, and phenolics have been developed, offering a promising alternative to the extraction of these compounds directly from plants, which can be resource-intensive (Pyne et al., 2019).

1.4 Challenges and Future Directions in Using Secondary Metabolites for IPM

Despite the many benefits of secondary metabolites in IPM, there are several challenges that need to be addressed to fully realize their potential.

1.4.1 Variability in Secondary Metabolite Production

One of the main challenges in utilizing secondary metabolites is the variability in their production, both within individual plants and across different environments. Secondary metabolite production is often influenced by external factors such as soil nutrients, water availability, temperature, and pest pressure. For instance, some plants may produce higher levels of defensive compounds under stress conditions, while others may decrease their production due to resource limitations. This variability can make it difficult to consistently rely on plants as a source of pest control in IPM programs. To overcome this challenge, ongoing research is focused on identifying the specific environmental triggers that enhance secondary metabolite production. This includes studying the impact of abiotic stressors such as drought, salinity, and nutrient deficiency on metabolite synthesis. By understanding these triggers, agricultural practices can be optimized to create conditions that promote the production of bioactive compounds, ensuring a more reliable source of pest resistance (Wink, 2018).

1.4.2 Pest Resistance to Secondary Metabolites

Another challenge in the use of secondary metabolites is the potential for pests to develop resistance over time. Just as pests have developed resistance to synthetic pesticides, there is a risk that continuous exposure to certain secondary metabolites may lead to the selection of resistant pest populations. This is particularly true for monogenic resistance, where a single gene mutation in the pest may render it immune to the effects of the metabolite. To mitigate the risk of resistance development, IPM strategies must emphasize the importance of rotating different types of secondary metabolites with varying modes of action. This prevents pests from being continuously exposed to the same compounds and reduces the likelihood of resistance building up. Moreover, combining secondary metabolites with other pest control strategies, such as biological control using predators or parasitoids, can create a multi-layered defense that is harder for pests to overcome (Bajwa & Kogan, 2002).

1.4.3 Scaling Up the Production of Secondary Metabolites

While secondary metabolites hold great promise for IPM, one of the major hurdles is the large-scale production of these compounds. Many secondary metabolites are produced in small quantities by plants, making it challenging to harvest enough material for widespread agricultural use. For example, neem trees produce azadirachtin in relatively low concentrations, and large-scale harvesting of neem seeds is not always practical or sustainable. As a result, finding cost-effective and scalable methods for producing secondary metabolites is a key area of research. Biotechnological solutions, such as microbial biofactories and plant cell cultures, offer promising alternatives for the mass production of secondary metabolites. By engineering microorganisms to produce these compounds in controlled environments, it is possible to bypass the limitations of plant-based production. Additionally, advancements in synthetic biology may allow for the creation of entirely new compounds with enhanced pesticidal properties, tailored specifically for IPM applications (Pyne et al., 2019).

1.4.4 Regulatory and Market Challenges

The regulatory landscape for biopesticides derived from secondary metabolites is often complex and varies between countries. In some regions, the registration process for new biopesticides can be lengthy and costly, creating a barrier to their adoption by farmers. Moreover, the market for secondary metabolite-based pesticides is still relatively small compared to the market for synthetic pesticides, which are often cheaper and more widely available. Overcoming these regulatory and market challenges will require concerted efforts from governments, industry stakeholders, and researchers to promote the benefits of these eco-friendly solutions. Education and outreach will also play a critical role in increasing the adoption of secondary metabolites in IPM. Many farmers may be unfamiliar with the advantages of using biopesticides or may lack the technical knowledge to implement these strategies effectively. Training programs that demonstrate the efficacy and economic benefits of secondary metabolite-based products can help bridge this gap and encourage wider adoption (Isman, 2020).

2. Use of Secondary Metabolites as Biopesticides and Bioherbicides

2.1 Secondary Metabolites as Biopesticides and Bioherbicides

Secondary metabolites, the diverse group of organic compounds synthesized by plants, have shown great promise in agricultural systems, particularly in their application as biopesticides and bioherbicides. These naturally occurring chemicals, which include alkaloids, phenolics, terpenoids, and flavonoids, serve as plant defense mechanisms against a variety of biotic stresses, such as insect herbivores, pathogens, and competing plants (allelopathy). The use of secondary metabolites in the form of biopesticides and bioherbicides offers an eco-friendly alternative to synthetic chemicals, which often have deleterious effects on non-target organisms, lead to environmental contamination, and contribute to the development of resistance in pests and weeds (Isman, 2020). Biopesticides and bioherbicides derived from secondary metabolites align with the principles of sustainable agriculture, as they are biodegradable, target-specific, and generally safer for humans, animals, and beneficial organisms. As interest in organic farming and environmentally conscious pest management grows, secondary metabolites are increasingly being explored for their role in the development of biologically based pesticides and herbicides that mitigate the negative impacts of conventional agrochemicals (Schmutterer, 1995).

2.2 Mechanisms of Secondary Metabolites as Biopesticides

The use of secondary metabolites as biopesticides revolves around their various modes of action against pests. These compounds can be toxic, repellent, or disrupt critical physiological functions in insects and other pests, making them highly effective in managing pest populations without resorting to synthetic chemical pesticides.

2.2.1 Toxicity to Insects and Pathogens

Many secondary metabolites are toxic to insects and other pests, making them ideal candidates for use as biopesticides. One notable example is pyrethrins, a class of terpenoids found in the flowers of *Chrysanthemum* species. Pyrethrins attack the nervous system of insects by interfering with sodium channels in nerve cells, leading to paralysis and death. These compounds are highly effective against a broad range of insect pests, including mosquitoes, flies, and aphids. Unlike synthetic pyrethroids, natural pyrethrins break down quickly in the environment, reducing the risk of ecological harm (Casida, 2018). Azadirachtin, a limonoid isolated from the neem tree (*Azadirachta indica*), is another widely used biopesticide that exhibits multiple modes of action. Azadirachtin acts as an insect growth regulator by disrupting the hormonal system of insects, inhibiting molting and reproduction. It also has antifeedant properties, deterring insects from feeding on treated plants. The combination of these effects makes azadirachtin an excellent biopesticide, particularly for controlling pests such as whiteflies, aphids, and caterpillars in both field crops and greenhouses (Schmutterer, 1995). Additionally, secondary metabolites like flavonoids and saponins have been shown to inhibit the growth of fungal pathogens. For instance, saponins disrupt the cell membranes of fungi, causing cell lysis and death. This antifungal property makes saponin-rich plants, such as quinoa (*Chenopodium quinoa*), valuable in protecting crops from fungal infections, particularly in humid climates where fungal diseases are prevalent (Morrissey & Osbourn, 1999).

2.2.2 Repellent and Antifeedant Effects

Another important mechanism through which secondary metabolites function as biopesticides is by repelling pests or preventing them from feeding on plants. Terpenoids and alkaloids are particularly effective as repellents and antifeedants. For example, limonene, a monoterpenoid found in citrus peels, is known for its strong insect-repellent properties. Limonene is used in several commercial insect-repellent products and is effective against mosquitoes, fleas, and flies. The repellent action of secondary metabolites is an important tool in IPM, as it prevents pest infestations without the need for lethal measures (Wink, 2018). Other secondary metabolites, such as phenolics and tannins, interfere with the digestive processes of herbivores. Tannins, which are found in high concentrations in plants like oak (*Quercus* species), bind to proteins in the digestive tracts of herbivores, reducing their ability to absorb nutrients. This not only deters feeding but also weakens the pest over time, reducing its ability to reproduce and survive. Such antifeedant effects are particularly useful in controlling leaf-chewing pests like caterpillars and beetles (Lattanzio, 2013).

2.2.3 Disruption of Reproduction and Growth

Some secondary metabolites act as insect growth regulators (IGRs), disrupting the normal growth and development of insect pests. Azadirachtin from Neem is one of the most well-known natural IGRs. It interferes with the production of ecdysone, the hormone responsible for molting in insects. Without this hormone, insects are unable to shed their exoskeletons and transition to the next stage of development, leading to death. This mode of action is particularly useful for managing insect populations over the long term, as it targets the pest's life cycle rather than simply killing adults (Schmutterer, 1995). Similarly, phytoecdysteroids, plant analogs of insect molting hormones, have been shown to disrupt the growth and development of various insect pests. These compounds are structurally similar to the ecdysteroids produced by insects, and when ingested, they interfere with the insect's hormonal balance, leading to abnormal development and death. Phytoecdysteroids are found in several plant species, including spinach (*Spinacia oleracea*) and quinoa (*Chenopodium quinoa*), and offer a natural alternative to synthetic insecticides (Lafont et al., 2021).

2.3 Mechanisms of Secondary Metabolites as Bioherbicides

In addition to their role as biopesticides, secondary metabolites are also important in the development of bioherbicides herbicides derived from natural sources that inhibit the growth of unwanted plants (weeds). Weeds are a major problem in agricultural systems, competing with crops for resources such as light, water, and nutrients. Conventional herbicides, while effective, often lead to issues such as herbicide resistance and environmental contamination. Secondary metabolites offer a more sustainable solution to weed control, as they are biodegradable and pose less risk to non-target species.

2.3.1 Allelopathy: Inhibition of Weed Growth

Allelopathy refers to the chemical inhibition of one plant species by another through the release of secondary metabolites into the environment. Many plants produce allelopathic compounds that inhibit the germination or growth of nearby plants, making them natural bioherbicides. Phenolic acids, terpenoids, and flavonoids are some of the most common allelopathic compounds found in plants (Weston & Duke, 2003). For example, juglone, a naphthoquinone produced by black walnut trees (*Juglans nigra*), is a potent allelochemical that inhibits the growth of many herbaceous plants. Juglone is released into the soil through the tree's roots, leaves, and husks, where it affects the respiration and photosynthesis of nearby plants, ultimately reducing their growth and survival. This natural herbicidal effect has been exploited in the development of bioherbicides for use in organic farming systems (Melander et al., 2017). Similarly, sorgoleone, a lipid benzoquinone produced by sorghum (*Sorghum bicolor*), is a powerful allelochemical that inhibits the germination of weed seeds. Sorgoleone disrupts the electron transport chain in the chloroplasts of susceptible plants, preventing photosynthesis and leading to the death of the weed. This compound is particularly effective against grass weeds in cereal crops and is being explored as a natural herbicide for sustainable agriculture (Czarnota et al., 2001).

2.3.2 Direct Herbicidal Action

In addition to their allelopathic effects, some secondary metabolites exhibit direct herbicidal activity. For instance, the compound leptospermone, found in the essential oils of *Leptospermum scoparium* (Manuka) and *Callistemon citrinus* (bottlebrush), inhibits the enzyme 4-hydroxyphenylpyruvate dioxygenase (HPPD) in plants. This enzyme is essential for the biosynthesis of carotenoids, which protect chlorophyll from oxidative damage during photosynthesis. By inhibiting HPPD, leptospermone causes the rapid breakdown of chlorophyll, leading to the death of the weed. This mechanism of action is similar to that of synthetic HPPD-inhibiting herbicides, but leptospermone offers a natural alternative with lower environmental impact (Dayan et al., 2011). Another example of a direct-acting bioherbicide is the compound catechol, which is found in several plant species, including buckwheat (*Fagopyrum esculentum*) and tea (*Camellia sinensis*). Catechol disrupts cell membrane integrity in plants, causing leakage of cellular contents and ultimately leading to cell death. This phytotoxic effect makes catechol a valuable tool in controlling broadleaf weeds, particularly in organic farming systems where synthetic herbicides are not permitted (Inderjit & Duke, 2003).

2.4 Applications and Future Prospects of Biopesticides and Bioherbicides

The use of secondary metabolites as biopesticides and bioherbicides is rapidly gaining momentum as a key component of sustainable agriculture. These natural compounds offer an environmentally friendly alternative to synthetic chemicals, reducing the risk of pollution, pesticide resistance, and harm to non-target organisms. Biopesticides derived from secondary metabolites are already in use in several agricultural systems. Neem-based products, which contain azadirachtin, are widely used to control insect pests in both conventional and organic farming. Similarly, essential oils rich in terpenoids are used as natural insect repellents and fungicides. Bioherbicides, though still in the early stages of development, show great promise for sustainably controlling weeds. Compounds like sorgoleone and leptospermone are being actively researched for their potential to replace synthetic herbicides in weed management programs (Isman, 2020). However, there are challenges to the widespread adoption of biopesticides and bioherbicides. These include the variability in efficacy due to environmental conditions, the need for large-scale production of secondary metabolites, and regulatory hurdles related to the approval of new biopesticide products. Despite these challenges, advancements in biotechnology and plant breeding offer new opportunities to enhance the production and effectiveness of secondary metabolites for use in pest and weed management. Genetically engineered plants that overproduce bioactive compounds, as well as microbial biofactories for the large-scale production of secondary metabolites, represent exciting future directions in this field (Pyne et al., 2019).

3. Allelopathic Properties of Secondary Metabolites for Weed Control

3.1 Allelopathy in Weed Control

Allelopathy is a biological phenomenon in which plants release chemicals into the environment that affect the growth, survival, and reproduction of neighboring plants. These chemicals, often referred to as allelochemicals, are a subset of secondary metabolites that plants produce as a defense mechanism to reduce competition from other plant species. The use of allelopathic properties for weed control offers a natural and sustainable alternative to synthetic herbicides, which are associated with environmental contamination, human health risks, and the development of herbicide-resistant weed populations (Melanderet al., 2017). Secondary metabolites with allelopathic effects include a wide range of chemical classes such as phenolics, terpenoids, flavonoids, alkaloids, and glycosides. These compounds can be released into the environment through root exudation, leaching from leaves, decomposition of plant residues, or volatilization from plant tissues. Once released, these metabolites can inhibit seed germination, disrupt cell division, or interfere with physiological processes in weeds, effectively reducing their ability to compete with crops. Understanding the allelopathic potential of these secondary metabolites has led to significant interest in their application as bioherbicides, particularly in sustainable and organic farming systems (Inderjit & Duke, 2003).

3.2 Mechanisms of Allelopathic Action in Weed Control

The mechanisms by which secondary metabolites exert allelopathic effects on weeds are varied and depend on the specific chemical composition of the allelochemicals involved. Some of the key mechanisms include the inhibition of seed germination, disruption of photosynthesis, interference with nutrient uptake, and oxidative stress induction.

3.2.1 Inhibition of Seed Germination and Early Growth

One of the primary ways secondary metabolites exhibit allelopathic properties is by inhibiting seed germination and early seedling growth. Phenolic compounds, such as ferulic acid, p-coumaric acid, and vanillic acid, are known to inhibit the germination of weed seeds by disrupting cellular respiration and enzyme activity necessary for the germination process. For example, studies have shown that the presence of high levels of phenolics in the soil can delay or prevent the germination of common weeds such as barnyard grass (*Echinochloa crus-galli*) and crabgrass (*Digitaria sanguinalis*) (Weston & Duke, 2003). Terpenoids also play a significant role in the inhibition of seed germination. Sorgoleone, a potent benzoquinone exuded from the roots of sorghum (*Sorghum bicolor*), is a well-known allelochemical that suppresses the germination of a wide range of weed species. It inhibits photosystem II (PSII) in the chloroplasts of susceptible plants, effectively preventing them from conducting photosynthesis during the critical early stages of development (Czarnota et al., 2001). The use of sorgoleone-producing plants or extracts in crop rotations or intercropping systems can thus reduce weed pressure without the need for synthetic herbicides.

3.2.2 Disruption of Photosynthesis and Respiration

Some allelopathic secondary metabolites directly disrupt photosynthesis, the process by which plants convert light energy into chemical energy. Terpenoids, such as monoterpenes and sesquiterpenes, are particularly effective at interfering with photosynthetic processes in weed plants. For instance, cineole, a monoterpene found in eucalyptus (*Eucalyptus globulus*), inhibits electron transport in the chloroplasts of weed species, leading to a reduction in photosynthetic efficiency. This disruption ultimately results in the depletion of energy reserves necessary for weed growth and survival (Ranganathan & Kandasamy, 2008). In addition to disrupting photosynthesis, secondary metabolites can interfere with respiration by inhibiting key enzymes involved in cellular respiration. Quinones, such as juglone from black walnut (*Juglans nigra*), inhibit cytochrome oxidase in the mitochondrial electron transport chain, reducing the production of ATP, the energy currency of the cell. This reduction in energy availability impairs the growth and development of weed species, allowing crops to outcompete weeds for resources (Melander et al., 2017).

3.2.3 Induction of Oxidative Stress in Weeds

Many secondary metabolites induce oxidative stress in weed plants by generating reactive oxygen species (ROS), which cause damage to cellular structures such as membranes, proteins, and DNA.

For example, phenolic acids, including caffeic acid and gallic acid, are known to generate ROS in plant tissues, leading to lipid peroxidation, protein degradation, and ultimately, cell death. The ability of these allelochemicals to induce oxidative stress in weeds without affecting crop plants makes them particularly valuable in selective weed control (Bais et al., 2006). Flavonoids, another class of secondary metabolites, also play a role in inducing oxidative stress in weeds. These compounds can chelate essential metal ions such as iron and copper, which are necessary cofactors for various antioxidant enzymes. By depleting these ions, flavonoids reduce the ability of weed plants to detoxify ROS, leading to oxidative damage and growth inhibition. Flavonoid-rich plants, such as buckwheat (*Fagopyrum esculentum*), have been used in crop rotations to suppress the growth of problematic weeds through this mechanism (Weston, 2005).

3.3 Examples of Allelopathic Secondary Metabolites for Weed Control

There are numerous examples of plants that produce allelopathic secondary metabolites capable of controlling weeds in agricultural systems. Below are some of the most well-studied examples:

3.3.1 Juglone from Black Walnut (*Juglans nigra*)

Juglone, a naphthoquinone produced by the black walnut tree, is one of the most well-known allelochemicals. Juglone is released into the soil through the roots, leaves, and fruit husks of the walnut tree and inhibits the growth of many herbaceous plants, including common agricultural weeds. Juglone affects multiple physiological processes in weeds, including respiration and photosynthesis, by inhibiting key enzymes in these pathways. Its broad-spectrum herbicidal properties make it a valuable tool in integrated weed management, particularly in organic farming systems where synthetic herbicides are not permitted (Melander et al., 2017).

3.3.2 Sorgoleone from Sorghum (*Sorghum bicolor*)

Sorgoleone, a lipid benzoquinone exuded by the roots of sorghum, is another potent allelochemical used for weed control. It inhibits photosystem II in the chloroplasts of susceptible plants, preventing them from conducting photosynthesis and ultimately leading to their death. Sorgoleone has been shown to be effective against a wide range of weed species, including pigweed (*Amaranthus retroflexus*) and velvetleaf (*Abutilon theophrasti*). The ability of sorghum to produce this compound makes it an ideal candidate for use in crop rotations or cover cropping systems to suppress weed populations naturally (Czarnota et al., 2001).

3.3.3 Leptospermone from Manuka (*Leptospermum scoparium*)

Leptospermone, a natural herbicide found in the essential oils of manuka and bottlebrush plants (*Callistemon citrinus*), inhibits the enzyme 4-hydroxyphenylpyruvate dioxygenase (HPPD), which is essential for the biosynthesis of carotenoids in plants. Carotenoids protect chlorophyll from oxidative damage, and without them, plants cannot conduct photosynthesis efficiently. Leptospermone causes the rapid degradation of chlorophyll in weeds, leading to their death. This allelopathic compound is being studied as a potential bioherbicide for controlling broadleaf weeds in cereal crops (Dayan et al., 2011).

3.3.4 Phenolic Acids from Rye (*Secale cereale*)

Rye is well known for its allelopathic properties, primarily due to the presence of phenolic acids such as ferulic acid, p-coumaric acid, and vanillic acid in its roots and residues. These phenolic acids inhibit the germination and growth of many weed species, including lambsquarters (*Chenopodium album*) and pigweed (*Amaranthus retroflexus*). Rye is commonly used as a cover crop in conservation tillage systems, where it suppresses weed growth through the release of allelopathic compounds during decomposition. This strategy reduces the need for chemical herbicides, making rye an essential component of sustainable agriculture (Weston & Duke, 2003).

3.4 Applications and Challenges of Using Allelopathy for Weed Control

The use of allelopathic secondary metabolites in weed control offers several advantages over traditional herbicides. Allelochemicals are biodegradable, reducing the risk of environmental contamination. They are also often more selective, targeting specific weed species while leaving crops and non-target organisms unharmed. This selectivity makes allelopathy an attractive option for integrated weed management, particularly in organic and conservation agriculture (Inderjit & Duke, 2003). However, there are also challenges associated with the use of allelopathy for weed control. One of the main limitations is the variability in allelopathic activity, which can be influenced by environmental factors such as soil type, moisture, temperature, and microbial activity. The concentration of allelochemicals in plant tissues can also vary depending on the growth stage and health of the plant, making it difficult to predict the consistency and effectiveness of allelopathic weed suppression in field conditions (Weston, 2005). Another challenge is the potential for crop allelopathy, where allelochemicals released by the crop may inadvertently inhibit the growth of the crop itself or subsequent crops in a rotation. Careful selection of crop varieties and timing of planting can help mitigate this risk, but further research is needed to fully understand the interactions between allelochemicals and crop plants (Melander et al., 2017). Despite these challenges, ongoing research in plant breeding, biotechnology, and agronomic practices holds promise for enhancing the use of allelopathy in weed management. Genetic engineering techniques, for example, could be used to increase the production of specific allelochemicals in crop plants, making them more effective at suppressing weeds. Additionally, the development of microbial biofactories for the production of allelopathic compounds could provide a scalable and sustainable source of bioherbicides for large-scale agriculture (Pyne et al., 2019).

4. Enhancing Soil Health Through Secondary Metabolite Production

4.1 Soil Health and Secondary Metabolites

Soil health is a critical component of sustainable agriculture, as healthy soils support plant growth, enhance biodiversity, regulate water cycles, and help in the sequestration of carbon. Soil health is often evaluated based on physical, chemical, and biological factors, including soil structure, nutrient availability, water infiltration, microbial activity, and organic matter content. The role of secondary metabolites in enhancing soil health is increasingly recognized, as these compounds influence a variety of soil processes, from nutrient cycling to microbial interactions. Secondary metabolites, produced by plants, microbes, and soil organisms, act as chemical signals and defense compounds that regulate interactions between organisms, mediate plant growth, and contribute to the overall balance of soil ecosystems (Bais et al., 2006). In this context, secondary metabolites function beyond plant defense mechanisms; they play a pivotal role in enhancing soil health by influencing microbial communities, facilitating nutrient cycling, and promoting organic matter decomposition. The strategic use of plants and microorganisms that produce beneficial secondary metabolites can lead to healthier soils and more sustainable farming practices. The ways secondary metabolites contribute to soil health can inform agricultural practices that reduce the need for synthetic inputs, promote biodiversity, and improve long-term soil fertility (Bakker et al., 2020).

4.2 Role of Secondary Metabolites in Nutrient Cycling

One of the key functions of secondary metabolites in soil ecosystems is their role in nutrient cycling. Nutrient cycling is a vital process where organic matter decomposes and nutrients, such as nitrogen, phosphorus, and potassium, are made available for plant uptake. Secondary metabolites influence this process by interacting with soil microorganisms, including bacteria and fungi, which are responsible for breaking down organic matter and releasing nutrients.

4.2.1 Decomposition and Organic Matter Turnover

Secondary metabolites can either promote or inhibit the decomposition of organic matter in soils, depending on their chemical nature and the organisms they affect. For example, phenolic compounds, such as lignin, are known to slow down the decomposition of plant residues because they form complex polymers that are difficult for microbes to break down. This reduced decomposition rate leads to the accumulation of organic matter in the soil, enhancing soil structure and increasing its capacity to retain water and nutrients. However, other secondary metabolites, such as terpenoids and alkaloids, may stimulate the activity of certain microbial communities that accelerate the breakdown of less complex organic compounds, thus facilitating nutrient release (Lattanzio, 2013). Microbial populations, particularly decomposers such as fungi and bacteria, play a significant role in this process. Plants that produce secondary metabolites like saponins and flavonoids can stimulate the growth of beneficial decomposer microbes. For example, saprotrophic fungi, which decompose dead plant material, are often stimulated by plant exudates rich in secondary metabolites. This activity enhances the breakdown of organic material into humus, improving the soil's organic matter content and nutrient availability (Smith & Read, 2008).

4.2.2 Nitrogen Fixation and Mineralization

Secondary metabolites also influence the processes of nitrogen fixation and mineralization in soils. Nitrogen fixation is carried out by symbiotic bacteria, such as *Rhizobium* species, which form associations with leguminous plants. Certain secondary metabolites produced by these plants, including flavonoids, act as signaling molecules that initiate the symbiotic relationship between the plants and nitrogen-fixing bacteria. These flavonoids activate nod genes in bacteria, triggering the formation of root nodules where nitrogen fixation occurs. This process enriches the soil with biologically available nitrogen, reducing the need for synthetic nitrogen fertilizers (Masson-Boivin & Sachs, 2018). In addition to nitrogen fixation, secondary metabolites play a role in nitrogen mineralization, the process by which organic nitrogen is converted into inorganic forms, such as ammonium and nitrate, which plants can absorb. Phenolic acids, such as ferulic acid and vanillic acid, can affect the activity of nitrifying and denitrifying bacteria, which are responsible for converting organic nitrogen into plant-available forms. By modulating microbial activity, secondary metabolites ensure that nitrogen is released at a rate that aligns with plant uptake, thus improving nitrogen use efficiency in soils (Kuzyakov, 2010).

4.3 Impact of Secondary Metabolites on Soil Microbial Communities

Soil microbial communities are essential for maintaining soil health, as they participate in nutrient cycling, organic matter decomposition, and the suppression of soil-borne diseases. Secondary metabolites can significantly alter the composition and function of these microbial communities, promoting beneficial microorganisms while suppressing harmful ones. This balance is crucial for maintaining soil fertility and plant health.

4.3.1 Promotion of Beneficial Microbes

Many secondary metabolites promote the growth and activity of beneficial soil microbes, such as mycorrhizal fungi, rhizobacteria, and decomposer organisms. Mycorrhizal fungi, for instance, form symbiotic associations with plant roots and enhance the plant's ability to absorb water and nutrients, particularly phosphorus. Certain secondary metabolites, like terpenoids and flavonoids, enhance the colonization of plant roots by mycorrhizal fungi, facilitating this mutualistic relationship. These compounds act as chemical signals that attract mycorrhizal fungi to plant roots, promoting root-fungal interactions that improve nutrient uptake and plant resilience (Smith & Read, 2008). Similarly, rhizobacteria that promote plant growth (PGPR) are also influenced by plant-produced secondary metabolites. Compounds like phenolics, exuded by plant roots, create a favorable environment for PGPR, which, in turn, improve plant health by producing growth-promoting substances, such as indole-3-acetic acid (IAA), and by enhancing nutrient availability. The presence of these beneficial bacteria reduces the need for chemical fertilizers, contributing to the long-term sustainability of agricultural systems (Lugtenberg & Kamilova, 2009).

4.3.2 Suppression of Soil Pathogens

Secondary metabolites also play a crucial role in suppressing soil-borne pathogens, which can cause significant damage to crops. These metabolites act as natural biocontrol agents, limiting the growth of harmful microbes and reducing the incidence of plant diseases. For example, alkaloids and saponins produced by certain plants have antimicrobial properties that inhibit the growth of pathogenic fungi and bacteria. The production of these compounds in the rhizosphere creates a protective barrier around plant roots, reducing the likelihood of infection (Bais et al., 2006). A well-known example of secondary metabolites suppressing soil pathogens is the production of glucosinolates in plants of the Brassicaceae family. When plant tissues containing glucosinolates are incorporated into the soil, they break down into isothiocyanates, which have strong antifungal and antibacterial properties. This process, known as biofumigation, is used as a natural method to control soil-borne pathogens, such as *Fusarium* and *Rhizoctonia*, which cause root rots and wilts in crops. By reducing the reliance on chemical fungicides, biofumigation helps maintain soil health and biodiversity (Sarwar et al., 1998).

4.4 Secondary Metabolites in Organic Matter Formation and Soil Structure

Secondary metabolites contribute to the formation of stable organic matter, also known as humus, which is vital for maintaining soil structure and fertility. Humus improves soil aeration, water retention, and nutrient availability, all of which are essential for plant growth. The interaction between secondary metabolites and microbial communities plays a key role in the formation and stabilization of humus.

4.4.1 Soil Aggregation and Structure

Soil aggregation refers to the clumping together of soil particles, which improves the soil's ability to retain water and resist erosion. Secondary metabolites, particularly polysaccharides and lignins, contribute to the formation of soil aggregates by providing binding agents that hold soil particles together. These compounds interact with soil microbes to form a stable matrix that enhances soil structure. A well-aggregated soil has better water infiltration, reduced compaction, and improved root penetration, all of which are critical for crop productivity and soil health (Frey et al., 2006). Polysaccharides produced by soil microorganisms in response to secondary metabolites help glue soil particles together, forming stable aggregates. These microbial exudates act as a carbon source for soil microorganisms, promoting the formation of humus. Humus-rich soils have a higher capacity to store nutrients and water, improving the resilience of agricultural systems to climate variability and reducing the need for irrigation and fertilizers (Lehmann & Kleber, 2015).

4.4.2 Carbon Sequestration and Organic Matter Stabilization

Secondary metabolites also play an essential role in carbon sequestration, the process by which carbon is captured and stored in soils as organic matter. Lignin, a phenolic polymer found in plant cell walls, is particularly important in this regard. Due to its complex and recalcitrant nature, lignin resists microbial degradation and contributes to long-term carbon storage in soils. The presence of lignin and other resistant phenolics in plant residues slows down the decomposition process, leading to the accumulation of stable organic matter, which helps sequester carbon from the atmosphere and mitigates climate change (Thevenot et al., 2010). The stabilization of organic matter by secondary metabolites also enhances the soil's ability to act as a carbon sink. Soils with high levels of stabilized organic matter not only improve soil health but also contribute to reducing greenhouse gas emissions by storing carbon that would otherwise be released into the atmosphere. This process is particularly important in the context of climate-smart agriculture, where the dual goals of improving soil fertility and mitigating climate change are paramount (Lal, 2011).

4.5 Challenges and Future Prospects of Enhancing Soil Health Through Secondary Metabolites

While the benefits of secondary metabolites in enhancing soil health are clear, there are challenges to their widespread adoption and use in agricultural systems. One of the main challenges is the variability in secondary metabolite production, which is influenced by factors such as plant species, environmental conditions, and soil type. Plants may produce different quantities and types of secondary metabolites depending on stress factors like drought, nutrient availability, and pest pressure. This variability makes it difficult to predict the consistency of soil health benefits from secondary metabolites across different agricultural systems (Wink, 2018). Furthermore, research into the specific roles of secondary metabolites in soil ecosystems is still in its early stages. While significant progress has been made in understanding the interactions between secondary metabolites and soil microorganisms, more research is needed to identify the best management practices for harnessing these compounds to improve soil health. Integrating secondary metabolite production into crop management strategies, such as cover cropping, crop rotations, and organic amendments, offers a promising pathway for enhancing soil fertility and sustainability (Bakker et al., 2012). In the future, advancements in plant breeding and biotechnology could enhance the production of beneficial secondary metabolites in crops, making it easier to integrate these compounds into soil health management practices. Genetic engineering and microbial inoculation strategies may allow for the targeted manipulation of soil microbiomes to optimize the production and effects of secondary metabolites on soil health. As agriculture shifts towards more sustainable practices, the role of secondary metabolites in maintaining healthy soils will become increasingly important.

5. Environmental Benefits of Using Plant-Derived Compounds in Agriculture

5.1 Plant-Derived Compounds in Agriculture

Agriculture is a sector that significantly impacts the environment, particularly due to the widespread use of synthetic fertilizers, herbicides, and pesticides. These chemicals often contribute to soil degradation, water pollution, and loss of biodiversity, as well as the development of resistant pest species. The growing awareness of these environmental challenges has led to a renewed interest in plant-derived compounds as a sustainable alternative to synthetic agrochemicals. Plant-derived compounds, including secondary metabolites such as alkaloids, terpenoids, phenolics, and flavonoids, offer an eco-friendly solution for pest management, weed control, and soil health improvement. These compounds, being naturally biodegradable and selective in their action, have the potential to reduce the harmful impacts associated with conventional agricultural practices. The use of these compounds aligns with the principles of agroecology and organic farming, which aim to create agricultural systems that are resilient, sustainable, and environmentally benign. In addition to reducing chemical inputs, plant-derived compounds enhance biodiversity, improve soil health, and support ecosystem services such as pollination and natural pest control (Isman, 2020).

5.2 Reduction of Chemical Inputs and Pollution

One of the most significant environmental benefits of using plant-derived compounds in agriculture is the reduction in the use of synthetic chemicals. Synthetic pesticides and herbicides, while effective in controlling pests and weeds, often have unintended consequences for non-target species, including beneficial insects, soil microbes, and wildlife. They also persist in the environment, leading to soil and water contamination. In contrast, plant-derived compounds such as neem-based biopesticides and essential oils break down more rapidly in the environment, reducing the risk of long-term pollution. For example, azadirachtin, a limonoid extracted from neem trees (*Azadirachta indica*), is a widely used biopesticide that disrupts the growth and reproduction of insect pests while posing minimal risk to non-target organisms. It degrades quickly in sunlight and does not accumulate in the soil or water, making it an ideal alternative to persistent synthetic pesticides (Schmutterer, 1995). Similarly, pyrethrins, derived from *Chrysanthemum* species, are natural insecticides that degrade rapidly in the environment, offering effective pest control without the toxic residues associated with synthetic pyrethroids (Casida, 2018). The use of plant-derived compounds also reduces the reliance on chemical fertilizers. For example, leguminous plants that produce secondary metabolites like flavonoids play a crucial role in promoting symbiotic nitrogen fixation, reducing the need for synthetic nitrogen fertilizers. This not only decreases the energy required for fertilizer production but also mitigates the environmental issues associated with nitrogen runoff, such as eutrophication and waterway pollution (Masson-Boivin & Sachs, 2018).

5.3 Enhancing Biodiversity and Ecosystem Services

Plant-derived compounds contribute to the enhancement of biodiversity in agricultural ecosystems. Conventional agricultural practices, particularly monocultures and the heavy use of synthetic pesticides, often lead to the simplification of ecosystems, reducing species richness and disrupting ecosystem services such as pollination, natural pest control, and nutrient cycling. By contrast, the use of plant-derived compounds encourages a more diverse and balanced ecosystem. For instance, many plant-derived pesticides are highly selective, targeting specific pests while leaving beneficial insects such as pollinators and natural predators unharmed. Essential oils from plants like lavender (*Lavandula angustifolia*) and peppermint (*Mentha piperita*) are effective against insect pests but do not harm pollinators such as bees and butterflies. By protecting beneficial organisms, plant-derived compounds help maintain the ecological balance necessary for natural pest control and pollination services, which are critical for crop production and ecosystem resilience (Bakkali et al., 2008). Moreover, the use of allelopathic plants—those that release biochemicals to inhibit the growth of weeds—reduces the need for chemical herbicides. Allelopathic plants such as rye (*Secale cereale*) and sorghum (*Sorghum bicolor*) produce compounds that suppress weed growth, thus enhancing biodiversity by reducing the dominance of invasive species without negatively affecting native flora. This promotes a more diverse plant community, which in turn supports a wider range of insects, birds, and other wildlife (Weston & Duke, 2003).

5.4 Improvement of Soil Health and Carbon Sequestration

Plant-derived compounds play an important role in improving soil health and enhancing the soil's ability to sequester carbon. Healthy soils are rich in organic matter and microbial diversity, both of which are essential for sustainable agriculture. Secondary metabolites produced by plants can enhance soil microbial activity, promote organic matter decomposition, and improve nutrient cycling. For example, phenolic compounds exuded by plant roots can stimulate the activity of beneficial soil microbes, such as mycorrhizal fungi and nitrogen-fixing bacteria. These microbes improve soil structure, increase nutrient availability, and enhance plant growth. By promoting healthy microbial communities, plant-derived compounds reduce the need for chemical fertilizers and soil amendments, which can degrade soil quality over time (Kuzyakov, 2010). In addition to promoting microbial activity, plant-derived compounds such as lignin contribute to carbon sequestration in soils. Lignin, a complex phenolic polymer found in plant cell walls, is resistant to microbial decomposition, allowing it to persist in the soil and store carbon for long periods. This process helps mitigate climate change by reducing the amount of carbon dioxide in the atmosphere. The accumulation of organic matter also improves soil fertility, water retention, and resistance to erosion, creating more resilient agricultural systems (Lehmann & Kleber, 2015).

5.5 Reducing Pest and Disease Resistance

One of the major challenges in modern agriculture is the development of resistance to pests and diseases to synthetic pesticides and fungicides. Over-reliance on chemical control methods has led to the evolution of resistant pest populations, making it increasingly difficult to manage crop diseases and infestations. Plant-derived compounds offer a solution to this problem by providing alternative modes of action that pests and pathogens are less likely to develop resistance to. Many secondary metabolites act on multiple physiological pathways in pests, making it harder for them to evolve resistance. For example, azadirachtin from neem trees disrupts the hormonal system of insects, affecting their molting and reproduction. This multi-target mode of action makes it more difficult for pests to develop resistance compared to single-target synthetic insecticides. Similarly, compounds like saponins and flavonoids have broad-spectrum antimicrobial properties that make it challenging for pathogens to develop resistance (Schmutterer, 1995). By rotating or combining plant-derived compounds with other pest management strategies, farmers can reduce the likelihood of resistance development and ensure more sustainable pest control. This approach, known as integrated pest management (IPM), is crucial for maintaining the effectiveness of biopesticides and reducing the need for synthetic inputs.

6. Genetic Engineering for Increased Production of Secondary Metabolites in Crops

6.1 Introduction to Genetic Engineering and Secondary Metabolites

Secondary metabolites are critical for plant defense and play significant roles in agriculture, particularly in pest and disease control, weed management, and soil health. However, many economically important crops produce these compounds in limited quantities, which can restrict their effectiveness in agricultural applications. Genetic engineering has emerged as a powerful tool to enhance the production of secondary metabolites in crops, allowing for increased concentrations of these bioactive compounds to be used for sustainable agricultural practices. By manipulating the biosynthetic pathways that lead to secondary metabolite production, scientists can create genetically engineered plants that produce higher levels of these compounds. This approach has the potential to improve crop resilience, reduce the need for synthetic agrochemicals, and increase the availability of plant-derived compounds for biopesticides, bioherbicides, and other applications in agriculture. The use of genetic engineering for the production of secondary metabolites is a promising area of research that could significantly enhance the sustainability and efficiency of modern agricultural systems (Degenhardt et al., 2009).

6.2 Strategies for Increasing Secondary Metabolite Production in Crops

Genetic engineering approaches to enhance the production of secondary metabolites in crops typically involve the manipulation of key genes involved in biosynthetic pathways. These strategies can be categorized into several broad approaches, including the overexpression of biosynthetic genes, the introduction of novel biosynthetic pathways, and the manipulation of regulatory genes that control secondary metabolite production.

6.2.1 Overexpression of Biosynthetic Genes

One of the most straightforward strategies for increasing secondary metabolite production in crops is the overexpression of genes that encode enzymes in the biosynthetic pathways of these compounds. For example, genes involved in the production of flavonoids, alkaloids, or terpenoids can be introduced into crops to boost the synthesis of these compounds. In the case of flavonoids, overexpression of genes such as *chalcone synthase* (CHS), which catalyzes the first step in the flavonoid biosynthetic pathway, has been shown to significantly increase flavonoid production in plants. This increased production can enhance the plant's resistance to pests and pathogens, as flavonoids are known to have antimicrobial and antifeedant properties (Tariq et al., 2023). Similarly, the overexpression of genes involved in the biosynthesis of alkaloids, such as those encoding *strictosidine synthase* in the production of vinblastine, has been used to increase alkaloid content in medicinal plants and crops (O'Connor & Maresh, 2006).

6.2.2 Introduction of Novel Biosynthetic Pathways

In addition to overexpressing native biosynthetic genes, genetic engineering can be used to introduce novel biosynthetic pathways into crops that do not naturally produce certain secondary metabolites. This strategy allows for the production of valuable bioactive compounds in plants that are otherwise not capable of synthesizing them. For example, the introduction of genes from the neem tree (*Azadirachta indica*) into other crops could enable these plants to produce azadirachtin, a potent insect growth regulator and antifeedant. This would enhance the plant's natural defenses against insect pests, reducing the need for chemical pesticides (Schmutterer, 1995). Similarly, the transfer of genes involved in the biosynthesis of taxol, a terpenoid produced by the Pacific yew tree (*Taxus brevifolia*), into crop plants could provide a sustainable source of this valuable compound, which is used as an anticancer drug.

6.2.3 Manipulation of Regulatory Genes

Another approach to enhancing secondary metabolite production is the manipulation of regulatory genes that control the expression of entire biosynthetic pathways. By targeting transcription factors that regulate secondary metabolite biosynthesis, scientists can increase the production of these compounds in a more controlled and coordinated manner. For instance, transcription factors like MYB and bHLH are known to regulate the expression of genes involved in flavonoid and anthocyanin biosynthesis. By engineering crops to overexpress these transcription factors, researchers can enhance the accumulation of flavonoids, which protect against UV radiation, pests, and pathogens. This strategy has been successfully applied in several crop species, including tomatoes and grapes, to improve their resistance to environmental stressors (Grotewold, 2006).

6.3 Applications of Genetically Engineered Crops for Secondary Metabolite Production

The use of genetically engineered crops that produce higher levels of secondary metabolites has numerous applications in agriculture, ranging from pest management to improving crop resilience and enhancing the nutritional value of food products.

6.3.1 Pest and Disease Management

One of the most important applications of genetically engineered crops that produce increased levels of secondary metabolites is in pest and disease management. By enhancing the production of compounds such as alkaloids, flavonoids, and terpenoids, crops can be made more resistant to insect herbivores, fungal pathogens, and bacterial diseases. This reduces the need for synthetic pesticides and fungicides, leading to more sustainable farming practices (Isman, 2020). For example, genetically engineered cotton plants that overexpress gossypol, a terpenoid with insecticidal properties, have been developed to protect against bollworms and other insect pests. This approach reduces the need for chemical insecticides, which can have harmful environmental effects and contribute to the development of resistant pest populations (Rathore et al., 2020).

6.3.2 Improving Crop Resilience to Environmental Stress

Genetic engineering of secondary metabolite pathways can also improve crop resilience to environmental stressors such as drought, salinity, and UV radiation. Many secondary metabolites, particularly flavonoids and terpenoids, play a role in protecting plants from oxidative stress caused by environmental challenges. By increasing the production of these compounds, crops can be made more tolerant to harsh growing conditions, reducing crop losses and improving food security. For example, plants that produce high levels of anthocyanins, a type of flavonoid, are more resistant to UV radiation and drought stress. Engineering crops to accumulate higher levels of anthocyanins can help protect them from the damaging effects of climate change, making agriculture more resilient in the face of increasing environmental variability (Tariq et al., 2023).

6.3.3 Enhancing Nutritional and Medicinal Value of Crops

In addition to their role in pest management and stress tolerance, secondary metabolites contribute to the nutritional and medicinal value of crops. Many secondary metabolites, such as flavonoids, carotenoids, and glucosinolates, have antioxidant, anti-inflammatory, and anticancer properties that make them valuable for human health. Genetic engineering can be used to enhance the production of these compounds in crops, improving their nutritional quality and health benefits. For instance, the genetic modification of crops to increase the production of carotenoids, such as beta-carotene in golden rice, has been a successful example of biofortification. This approach addresses vitamin A deficiency in populations that rely on rice as a staple food. Similarly, increasing the production of glucosinolates in Brassica vegetables, such as broccoli and kale, enhances their cancer-preventive properties, providing additional health benefits to consumers (González-Peña et al., 2023).

6.4 Challenges and Future Directions in Genetic Engineering for Secondary Metabolite Production

While genetic engineering offers significant potential for increasing the production of secondary metabolites in crops, several challenges need to be addressed. One of the primary challenges is the complexity of secondary metabolite biosynthetic pathways, which often involve multiple enzymes and regulatory steps. Manipulating these pathways without disrupting other aspects of plant growth and development requires a detailed understanding of plant metabolism and gene regulation. Additionally, there are concerns about the environmental and ecological risks associated with genetically engineered crops, particularly in terms of gene flow to wild relatives and the potential impact on non-target organisms. These concerns must be carefully considered when developing genetically engineered plants for agricultural use (Raman, 2017). Despite these challenges, advancements in synthetic biology and CRISPR-Cas9 genome editing hold promise for overcoming the technical barriers to enhancing secondary metabolite production in crops. These technologies allow for more precise and targeted genetic modifications, reducing the likelihood of unintended effects and improving the efficiency of secondary metabolite biosynthesis.

Conclusion

Secondary metabolites produced by plants play a pivotal role in agriculture, particularly in the context of Integrated Pest Management (IPM) and sustainable farming practices. These naturally occurring compounds serve multiple ecological functions in plants, acting as defense mechanisms against pests, pathogens, and competition from other plants. By harnessing the power of secondary metabolites, modern agriculture is increasingly shifting toward more sustainable practices, reducing the reliance on synthetic chemicals and exploring plant-derived alternatives for pest control, weed management, and soil health enhancement. The use of secondary metabolites as biopesticides, bioherbicides, and in soil management not only helps mitigate environmental harm but also promotes biodiversity and ecosystem balance. Innovations in genetic engineering further amplify the potential for enhanced production of these compounds, creating new possibilities for environmentally friendly agricultural solutions.

One of the most significant roles of secondary metabolites in agriculture is their application in Integrated Pest Management (IPM). IPM aims to control pest populations in a way that minimizes risks to human health, beneficial organisms, and the environment. In this context, secondary metabolites offer a natural, non-toxic alternative to conventional pesticides. These compounds can act as insect repellents, antifeedants, growth inhibitors, and even toxins for specific pest species. Alkaloids, terpenoids, and flavonoids, for example, are known to exhibit strong insecticidal properties, making them highly effective in controlling a range of agricultural pests. By using plants or plant-derived extracts containing these metabolites, farmers can reduce their dependence on synthetic pesticides, which are often associated with environmental contamination and the development of pesticide-resistant pests.

In IPM, the use of secondary metabolites is not only beneficial for pest control but also supports the preservation of beneficial insects such as pollinators and natural predators of pests. Since secondary metabolites are often species-specific or act on a limited range of pests, their application can help maintain the balance of the agroecosystem without disrupting non-target organisms. Neem (*Azadirachta indica*), for example, produces azadirachtin, a limonoid that interferes with the growth and reproduction of insects. This compound has been widely used in IPM programs due to its selective toxicity and minimal environmental impact. By incorporating plant-derived compounds into IPM strategies, agriculture can move toward more sustainable and ecologically sound practices.

Secondary metabolites also play a key role as biopesticides and bioherbicides, offering natural alternatives to synthetic agrochemicals. Biopesticides derived from secondary metabolites are typically less harmful to the environment and non-target species, making them an attractive option for organic and sustainable farming systems. Alkaloids, such as nicotine and pyrethrins, derived from plants like tobacco and chrysanthemums, are well-known for their insecticidal properties. These compounds disrupt the nervous systems of insects, leading to paralysis and death. Pyrethrins, in particular, have been used extensively as natural insecticides in organic farming due to their broad-spectrum activity and low toxicity to mammals.

In addition to their use as insecticides, secondary metabolites also function as bioherbicides for weed control. Weeds represent a major challenge in agriculture, competing with crops for nutrients, water, and light. Many secondary metabolites possess allelopathic properties, meaning they can inhibit the growth of neighboring plants. Allelopathy refers to the chemical warfare between plants, where one plant releases compounds that suppress the germination and growth of others. For example, phenolic compounds produced by species such as black walnut (*Juglans nigra*) can inhibit the growth of surrounding plants by interfering with root development and nutrient uptake. Harnessing allelopathic compounds from plants offers a natural solution for weed management, reducing the need for chemical herbicides that can contaminate soil and water sources.

Furthermore, secondary metabolites contribute to enhancing soil health by fostering beneficial microbial activity and improving nutrient cycling. Soil health is a critical component of sustainable agriculture, and secondary metabolites can play a direct role in maintaining and restoring soil quality. Certain plant compounds, such as flavonoids, serve as signaling molecules in plant-microbe interactions, particularly in symbiotic relationships with nitrogen-fixing bacteria. For instance, flavonoids released by legume roots stimulate the growth of *Rhizobium* bacteria, which in turn fix atmospheric nitrogen into a form that plants can use for growth. This process not only enhances soil fertility but also reduces the need for chemical fertilizers, which can lead to nutrient runoff and pollution.

Additionally, secondary metabolites can influence the composition and activity of soil microbial communities, promoting the growth of beneficial microbes that contribute to nutrient cycling, pathogen suppression, and organic matter decomposition. Terpenoids and phenolics, for example, can act as antimicrobial agents that selectively suppress harmful soil pathogens while promoting the growth of beneficial microorganisms. This microbial balance is essential for maintaining healthy soils and ensuring long-term agricultural productivity.

The environmental benefits of using plant-derived compounds in agriculture are substantial. One of the most critical advantages is the reduction of chemical inputs such as synthetic pesticides, herbicides, and fertilizers. The overuse of these chemicals has been linked to soil degradation, water contamination, and loss of biodiversity. By replacing or complementing synthetic chemicals with secondary metabolites, agriculture can mitigate these environmental impacts and promote more sustainable practices. Plant-derived biopesticides and bioherbicides are biodegradable and generally have lower persistence in the environment compared to synthetic chemicals, reducing the risk of soil and water contamination. Moreover, they tend to be less toxic to non-target species, including pollinators, aquatic organisms, and soil fauna, thus preserving ecosystem health. Additionally, the use of plant-derived compounds can help reduce the problem of chemical resistance in pests and weeds. Many synthetic pesticides and herbicides target specific biochemical pathways, and their repeated use can lead to the evolution of resistant populations. Secondary metabolites, on the other hand, often act through multiple mechanisms or target several physiological processes in pests, making it more difficult for resistance to develop. The diverse modes of action provided by these compounds can therefore contribute to more effective and sustainable pest and weed management strategies.

Advances in genetic engineering have further expanded the potential for increasing the production of secondary metabolites in crops. By manipulating the genes responsible for the biosynthesis of these compounds, researchers can enhance the natural defense mechanisms of plants and improve their ability to resist pests, diseases, and environmental stresses. Genetic engineering enables the overexpression of key enzymes in biosynthetic pathways or the introduction of genes from other species to boost the production of target metabolites. For example, genetically modified plants that overproduce terpenoids or alkaloids can offer enhanced resistance to insect pests, reducing the need for external pesticide applications. In addition to enhancing pest resistance, genetic engineering can also be used to increase the production of allelopathic compounds, thereby improving the crop's ability to suppress weeds naturally. Crops that are engineered to produce higher levels of allelopathic chemicals could offer an environmentally friendly solution for weed management, reducing the need for herbicides and improving overall crop yield. Genetic engineering has also been applied to improve the nutritional content and resilience of crops, further contributing to sustainable agriculture by reducing the need for chemical inputs and supporting food security.

References

Badri, D. V., Zolla, G., Bakker, M. G., Manter, D. K., & Vivanco, J. M. (2013). Potential impact of soil microbiomes on the leaf metabolome and on herbivore feeding behavior. *New Phytologist, 198*(1), 264-273.

Bais, H. P., Weir, T. L., Perry, L. G., Gilroy, S., & Vivanco, J. M. (2006). The role of root exudates in rhizosphere interactions with plants and other organisms. *Annual Review of Plant Biology, 57*, 233-266.

Bajwa, W. I., & Kogan, M. (2002). Compendium of IPM Definitions (CID)–What is IPM and how is it defined in the worldwide literature? *IPM Practitioner, 24*(5-6), 1-10.

Bakkali, F., Averbeck, S., Averbeck, D., & Idaomar, M. (2008). Biological effects of essential oils – A review. *Food and Chemical Toxicology, 46*(2), 446-475.

Bakker, M. G., Manter, D. K., Sheflin, A. M., Weir, T. L., & Vivanco, J. M. (2012). Harnessing the rhizosphere microbiome through plant breeding and agricultural management. *Plant and Soil, 360*(1-2), 1-13.

Casida, J. E. (2018). Neonicotinoids and other insect nicotinic receptor competitive modulators: Progress and prospects. *Annual Review of Entomology, 63*, 125-144.

Czarnota, M. A., Paul, R. N., Dayan, F. E., & Weston, L. A. (2001). Mode of action, localization of production, chemical nature, and activity of sorgoleone: A potent PSII inhibitor in sorghum root exudates. *Weed Technology, 15*(3), 813-825.

Dayan, F. E., Cantrell, C. L., & Duke, S. O. (2011). Natural products in crop protection. *Bioorganic & Medicinal Chemistry, 17*(12), 4022-4034.

Degenhardt, J., Kollner, T. G., & Gershenzon, J. (2009). Monoterpene and sesquiterpene synthases and the origin of terpene skeletal diversity in plants. *Phytochemistry*, 70(15-16), 1621-1637.

Divekar, P. A., Narayana, S., Divekar, B. A., Kumar, R., Gadratagi, B. G., Ray, A., Singh, A. K., Rani, V., Singh, V., Singh, A. K., Kumar, A., Singh, R. P., Meena, R. S., & Behera, T. K.. (2022). Plant secondary metabolites as defense tools against herbivores for sustainable crop protection. *International Journal of Molecular Science, 23*(5), 2690.

Duke, S. O., Cantrell, C. L., Meepagala, K. M., Wedge, D. E., Tabanca, N., & Schrader, K. K. (2018). Natural toxins for use in pest management. *Toxins, 2*(8), 1943-1962.

Frey, S. D., Thiet, R., & Batten, K. M. (2006). Bacterial and fungal contributions to carbon sequestration in agroecosystems. *Soil Science Society of America Journal*, 70.

González-Peña, M. A., Ortega-Regules, A. E., Anaya de Parrodi, C., & Lozada-Ramírez, J. D. (2023). Chemistry, occurrence, properties, applications, and encapsulation of carotenoids-a review. *Plants (Basel)*, *12*(2), 313.

Grotewold, E. (2006). The genetics and biochemistry of floral pigments. *Annual Review of Plant Biology, 57*, 761-780.

Hooks, C. R., & Johnson, M. W. (2003). Impact of agricultural diversification on the insect community of cruciferous crops. *Crop Protection, 22*(2), 223-238.

Inderjit, & Duke, S. O. (2003). Ecophysiological aspects of allelopathy. *Planta, 217*(4), 529-539.

Isman, M. B. (2017). Bridging the gap: Moving botanical insecticides from the laboratory to the farm. *Industrial Crops and Products, 110*, 10-14.

Isman, M. B. (2020). Botanical insecticides in the twenty-first century-Fulfilling their promise? *Annual Review of Entomology, 65*(1), 233-249.

Isman, M. B., & Grieneisen, M. L. (2014). Botanical insecticide research: Many publications, limited useful data. *Trends in Plant Science, 19*(3), 140-145.

Jiang, L., Gao, Y., Han, L., Zhang, W., & Fan, P. (2023). Designing plant flavonoids: harnessing transcriptional regulation and enzyme variation to enhance yield and diversity. *Frontiers in Plant Science, 14*, 1220062.

Kuzyakov, Y. (2010). Priming effects: Interactions between living and dead organic matter. *Soil Biology & Biochemistry, 42*(9), 1363-1371.

Lafont, R., Balducci, C., & Dinan, L. (2021). Ecdysteroids. *Encyclopedia, 1*(4), 1267-130.

Lal, R. (2011). Sequestering carbon in soils of agro-ecosystems. *Food Policy, 36*(1), S33-S39.

Lattanzio, V. (2013). Phenolic compounds: Introduction. In P. A. Michael (Ed.), *Natural Products* (1543-1580). Springer.

Lehmann, J., & Kleber, M. (2015). The contentious nature of soil organic matter. *Nature, 528*(7580), 60-68.

Lugtenberg, B., & Kamilova, F. (2009). Plant-growth-promoting rhizobacteria. *Annual Review of Microbiology, 63*, 541-556.

Masson-Boivin C., Sachs J. L. (2018). Symbiotic nitrogen fixation by rhizobia-the roots of a success story. *Current Opinion in Plant Biology, 44*, 7-15.

Melander, Bo & Liebman, Matt & Davis, Adam & Gallandt, Eric & Bàrberi, Paolo & Moonen, A.C. & Rasmussen, Jesper & Weide, R. & Vidotto, Francesco. (2017). Non-Chemical Weed Management: Expanding Horizons.

Morrissey, J. P., & Osbourn, A. E. (1999). Fungal resistance to plant antibiotics as a mechanism of pathogenesis. *Microbiology and Molecular Biology Reviews, 63*(3), 708-724.

O'Connor, S. E., & Maresh, J. J. (2006). Chemistry and biology of monoterpene indole alkaloid biosynthesis. *Natural Product Reports, 23*(4), 532-547.

Paddon, C. J., & Keasling, J. D. (2014). Semi-synthetic artemisinin: A model for the use of synthetic biology in pharmaceutical development. *Nature Reviews Microbiology, 12*(5), 355-367.

Pandey, S., da Silva, A. L. B. R., Dutta, B., Chong, J. H., Mutschler, M. A., & Schmidt, J. M. (2023). Acylsugar tomato lines suppress whiteflies and *Amblyseius swirs*kii establishment. *Entomologia Experimentalis et Applicata, 171*, 745-753.

Pyne, M. E., Narcross, L., Martin, V. J. J. (2019). Engineering plant secondary metabolism in microbial systems. *Plant Physiology, 179*(3), 844-861.

Raman, R. (2017). The impact of genetically modified (GM) crops in modern agriculture: A review. *GM Crops & Food, 8*(4), 195-208.

Ranganathan, C., & Kandasamy, O. (2008). Allelopathic Effect of Eucalyptus globulus Labill. on Cyperus rotundus L. and Cynodon dactylon L. Pers. *Journal of Agronomy and Crop Science*, 179, 123-126.

Rathore, K. S., Pandeya, D., Campbell, L. M., Wedegaertner, T. C., Puckhaber, L., Stipanovic, R. D., & Hake, K. (2020). Ultra-low gossypol cottonseed: Selective gene silencing opens up a vast resource of plant-based protein to improve human nutrition. *Critical Reviews in Plant Sciences, 39*(1), 1-29.

Sarkar, S. C., Wang, E., Wu, S., & Lei, Z. (2018). Application of Trap Cropping as Companion Plants for the Management of Agricultural Pests: A Review. *Insects, 9*(4), 128.

Sarwar, M., Kirkegaard, J., & Wong, P. et al. (1998). Biofumigation potential of brassicas. *Plant and Soil*, 201, 103-112.

Schmutterer, H. (1995). *The Neem Tree, Azadirachta indica, and other Meliaceous Plants: Sources of Unique Natural Products for Integrated Pest Management, Medicine, Industry, and Other Purposes.* VCH Publishers.

Smith, S. E., & Read, D. J. (2008). *Mycorrhizal Symbiosis* (3rd ed.). Academic Press, London.

Tariq, H., Asif, S., Andleeb, A., Hano, C., & Abbasi, B. H. (2023). Flavonoid production: Current trends in plant metabolic engineering and de novo microbial production. *Metabolites, 13*(1), 124.

Thevenot, M., Dignac, M. F., & Rumpel, C. (2010). Fate of lignins in soils: A review. *Soil Biology & Biochemistry, 42*(8), 1200-1211.

Tholl, D. (2015). Biosynthesis and biological functions of terpenoids in plants. *Advances in Biochemical Engineering/Biotechnology, 148*, 63-106.

Walker, T. S., Bais, H. P., Grotewold, E., & Vivanco, J. M. (2003). Root exudation and rhizosphere biology. *Plant Physiology, 132*(1), 44-51.

Weston, L. A. (2005). History and current trends in the use of allelopathy for weed management. *HortTechnology, 15*(3), 529–534.

Weston, L. A., & Duke, S. O. (2003). Weed and crop allelopathy. *Critical Reviews in Plant Sciences, 22*(3-4), 367-389.

Wink, M. (2018). Plant secondary metabolites modulate insect behavior-Steps toward biocontrol ?. *Journal of Chemical Ecology, 44*(9), 914-925. *Frontiers in Physiology, 9*, 364.

Zhou, H., Liu, B., Weeks, D. P., Spalding, M. H., & Yang, B. (2014). Large chromosomal deletions and heritable small genetic changes induced by CRISPR/Cas9 in rice. *Nucleic Acids Research, 42*(17), 10903-10914.

Chapter - 17

Biotechnological Advances in Enhancing Secondary Metabolite Production

Vinod Chouhan[1], Susheel Kumar[2], Pratima Raypa[3], Zuby Gohar Ansari[4] and Praneet Chauhan[5]

[1]Division of Plant Pathology, Indian Council of Agricultural Research-Indian Agricultural Research Institute (ICAR-IARI), New Delhi, India

[2]Plant Molecular Virology Laboratory, CPMB Division, CSIR-National Botanical Research Institute, Rana Pratap Marg, Lucknow, Uttar Pradesh, India

[3]Department of Biochemistry, H.N.B. Garhwal University, Srinagar, Uttarakhand, India

[4]B.M. College of Agriculture, Jaswadi Road Khandwa, Madhya Pradesh, India

[5]Department of Plant Pathology, Dr. Khem Singh Gill Akal College of Agriculture, Eternal University, Baru Sahib, Sirmour, Himachal Pradesh, India

Abstract

The advancement of biotechnology has revolutionized the field of plant secondary metabolite production, offering new techniques to enhance the quantity and quality of these vital compounds for use in agriculture, medicine, and industry. Secondary metabolites, including flavonoids, alkaloids, terpenoids, and phenolics, play essential roles in plant defense mechanisms, stress responses, and ecological interactions. Through biotechnological interventions such as genetic engineering, metabolic pathway manipulation, tissue culture techniques, and synthetic biology, scientists have been able to increase the biosynthesis of these valuable compounds in plants. Gene editing technologies like CRISPR/Cas9 allow precise modifications to key biosynthetic genes, leading to improved metabolite production without compromising plant health or growth. Additionally, metabolic engineering has enabled the reprogramming of plant cells to overproduce specific secondary metabolites, offering enhanced resistance to diseases and pests and improving overall crop resilience. Tissue culture methods, such as hairy root cultures and cell suspension cultures, have also provided controlled environments for large-scale production of secondary metabolites, bypassing the limitations of field-grown plants. The use of omics technologies, including genomics, transcriptomics, and proteomics, has further facilitated the identification of regulatory genes and pathways involved in secondary metabolite biosynthesis, enabling targeted biotechnological interventions. Moreover, advances in synthetic biology have led

DOI: 10.1201/9781003717454-17

to the development of novel biofactories, where microorganisms are engineered to produce plant secondary metabolites on an industrial scale, reducing reliance on traditional agriculture.

Keywords: Biotechnology, Secondary Metabolites, Genetic Engineering, Metabolic Engineering, Crispr/Cas9, Synthetic Biology

Introduction

The role of genetic engineering in enhancing secondary metabolite production in plants has gained significant attention due to its potential applications in agriculture, pharmaceuticals, and biotechnology. Secondary metabolites such as flavonoids, alkaloids, terpenoids, and phenolics play vital roles in plant defense mechanisms, stress responses, and ecological interactions, while also serving as valuable bioactive compounds in medicine and industry. By manipulating biosynthetic pathways through genetic engineering, researchers have been able to increase the production of these metabolites, optimizing their yield for various commercial applications. Genetic engineering approaches, including gene editing technologies like CRISPR/Cas9, metabolic engineering, tissue culture techniques, and synthetic biology, have opened up new possibilities for the targeted production of secondary metabolites, bypassing the limitations of traditional breeding methods. These advances in biotechnology have not only improved the efficiency of secondary metabolite production but also allowed for the creation of novel compounds with enhanced bioactivity, thereby expanding the potential of these metabolites in diverse sectors (Basu et al., 2023).

CRISPR/Cas9 and other gene editing technologies have revolutionized the field of secondary metabolite biosynthesis by enabling precise modifications of key genes involved in metabolite production. CRISPR/Cas9, in particular, allows for targeted alterations to the plant genome, enabling the enhancement or suppression of specific genes that regulate secondary metabolite pathways. This precise editing tool has facilitated the overproduction of valuable metabolites in plants, such as increasing the levels of alkaloids or phenolic compounds, which are critical for both plant defense and their use in pharmaceuticals (Gan & Ling, 2022). Through gene editing, researchers can not only increase the concentration of these metabolites but also modify their structure to improve their efficacy. For instance, CRISPR/Cas9 has been successfully used to upregulate genes in the flavonoid biosynthesis pathway, leading to enhanced production of flavonoids with stronger antioxidant properties. This technique has proven particularly effective in fine-tuning metabolic pathways without compromising plant growth and development, making it a powerful tool for both research and commercial applications (Yin et al., 2017).

Metabolic engineering is another vital approach used to overproduce targeted secondary metabolites by manipulating the enzymes and genes involved in their biosynthetic pathways. This method involves altering the expression of rate-limiting enzymes or introducing new biosynthetic routes to increase the flux towards the desired metabolite. Metabolic engineering has been particularly successful in enhancing the production of terpenoids, alkaloids, and phenolics, which are commonly used in pharmaceuticals and agrochemicals. By redirecting metabolic flux and optimizing the efficiency of biosynthetic pathways, scientists have been able to significantly increase the yield of these compounds in plants (Pickens et al., 2011). For example, the overexpression of key enzymes in the terpenoid biosynthesis pathway has led to the production of higher levels of terpenoids in plants like *Artemisia annua*, which is used to produce the antimalarial compound artemisinin. Similarly, the introduction of genes from other organisms, such as bacteria or fungi, into plant metabolic pathways has expanded the range of secondary metabolites that plants can produce, offering new possibilities for drug development and pest control (Mipeshwaree Devi et al., 2023).

Tissue culture techniques have also been widely used to enhance secondary metabolite production in controlled environments. These techniques, such as hairy root cultures and cell suspension cultures, provide an alternative to field-grown plants by allowing for the large-scale production of secondary metabolites in vitro. Hairy root cultures, induced by the infection of plants with *Agrobacterium rhizogenes*, are particularly effective for the production of root-derived secondary metabolites, including alkaloids and terpenoids. These cultures offer several advantages, including faster growth rates, genetic stability, and the ability to produce high levels of metabolites without the need for environmental stressors (Murthy et al., 2014). In addition to hairy root cultures, cell suspension cultures have been employed to produce metabolites like flavonoids and phenolics in a controlled, sterile environment. These techniques allow for the precise control of nutrient availability, light, temperature, and other factors that influence metabolite production, ensuring consistent yields of high-quality compounds. Moreover, tissue culture techniques are valuable for the production of rare or endangered medicinal plants, enabling the conservation of these species while still providing access to their valuable bioactive compounds (Morey & Peebles, 2022)..

Omics technologies, including genomics, transcriptomics, proteomics, and metabolomics, have played a crucial role in identifying the biosynthetic pathways involved in secondary metabolite production. These high-throughput technologies provide comprehensive insights into the genetic and molecular networks that regulate metabolite biosynthesis, allowing researchers to pinpoint key genes and enzymes that can be targeted for enhancement. Genomics and transcriptomics have been instrumental in uncovering the complex regulatory networks that control secondary metabolite pathways, particularly in non-model plants with significant medicinal or agricultural value (Chen et al., 2011). Proteomics, on the other hand, offers a deeper understanding of the post-translational modifications and protein-protein interactions that influence enzyme activity in these pathways. Metabolomics complements these approaches by providing a detailed profile of the metabolites produced under different environmental conditions or genetic manipulations, enabling researchers to monitor the effects of metabolic engineering or gene editing on secondary metabolite production (Palazzotto & Weber, 2018). Together, these omics technologies provide a powerful toolkit for optimizing secondary metabolite biosynthesis, facilitating the discovery of new bioactive compounds and enhancing the production of existing ones.

Synthetic biology has further expanded the potential for secondary metabolite production by enabling the microbial production of plant-derived compounds. By transferring plant biosynthetic pathways into microbial hosts like yeast or bacteria, synthetic biology allows for the large-scale, cost-effective production of secondary metabolites in fermentation systems. This approach has been particularly successful in producing high-value metabolites like artemisinin, paclitaxel, and vinblastine, which are used in cancer therapy and other medical treatments (Paddon & Keasling, 2014). Microbial production offers several advantages over traditional plant-based extraction methods, including higher yields, faster production times, and the ability to produce complex compounds that are difficult to synthesize chemically. Moreover, synthetic biology enables the engineering of new biosynthetic pathways that combine enzymes from different organisms, creating novel metabolites with enhanced bioactivity or improved pharmacokinetic properties. This approach has the potential to revolutionize the production of secondary metabolites, providing a sustainable and scalable solution for the pharmaceutical and agricultural industries (Smanski et al., 2016).

1. Genetic Engineering for Enhanced Secondary Metabolite Production

1.1 Secondary Metabolites and Genetic Engineering

Secondary metabolites are diverse organic compounds produced by plants, fungi, and microorganisms that are not directly involved in their growth, development, or reproduction but play vital roles in defense mechanisms and ecological interactions. These compounds include alkaloids, flavonoids, terpenoids, phenolics, and glycosides, and they serve crucial roles in pest resistance, pathogen defense, and overall plant survival. Additionally, many secondary metabolites have pharmaceutical and industrial applications due to their bioactive properties, such as anti-inflammatory, antimicrobial, and antioxidant effects. While these metabolites are essential for plants' ecological interactions, many crop species produce these compounds in insufficient quantities for effective large-scale agricultural or pharmaceutical use. The advent of genetic engineering has opened new avenues for enhancing the production of secondary metabolites in plants. Genetic engineering allows for the manipulation of biosynthetic pathways to increase the concentration of these valuable compounds, either by overexpressing key biosynthetic enzymes, introducing novel pathways, or modulating regulatory networks that control metabolite synthesis. These approaches are particularly important for creating crops with enhanced pest resistance, improved stress tolerance, and higher yields of bioactive compounds for pharmaceutical purposes (De Luca & St-Pierre, 2000).

1.2 Mechanisms for Enhancing Secondary Metabolite Production via Genetic Engineering

Genetic engineering strategies for increasing secondary metabolite production typically target specific enzymes within the biosynthetic pathways or the transcriptional regulators that control these pathways. The most common strategies include gene overexpression, pathway rerouting, and gene silencing.

1.2.1 Overexpression of Biosynthetic Enzymes

A straightforward approach to enhancing secondary metabolite production is the overexpression of genes that encode key enzymes in biosynthetic pathways. By increasing the levels of these enzymes, the overall flux through the pathway is elevated, leading to higher production of the desired metabolite. For example, the overexpression of chalcone synthase (CHS), which catalyzes the first step in the flavonoid biosynthesis pathway, has been shown to significantly increase flavonoid content in transgenic plants (Winkel-Shirley, 2001). Flavonoids are well-known for their antioxidant properties and their role in plant defense, and their increased production through genetic engineering has been explored in crops like tomato and soybean to enhance pest resistance and nutritional value. Similarly, the production of alkaloids, which serve as potent defense compounds against herbivores and pathogens, can be increased by overexpressing key enzymes in the alkaloid biosynthetic pathway. For example, in *Catharanthus roseus*, the overexpression of strictosidine synthase has been used to boost the production of the alkaloid vinblastine, a compound with significant anti-cancer properties (Srivastava & Srivastava, 2013).

1.2.2 Introduction of Novel Biosynthetic Pathways

Another approach to enhancing secondary metabolite production is the introduction of entire biosynthetic pathways into plants that do not naturally produce the desired compound. This strategy has been particularly useful in expanding the range of bioactive compounds that can be harvested from crop plants. For instance, genes responsible for the biosynthesis of resveratrol, a polyphenolic compound with antioxidant and anti-cancer properties, have been successfully introduced into crops like tomato and grapevine, increasing their medicinal and nutritional value (Huccetogullari et al., 2019). The production of taxol, a valuable terpenoid used in cancer treatment, has also been attempted through genetic engineering. Genes involved in the taxol biosynthetic pathway from the Pacific yew tree (*Taxus brevifolia*) have been transferred into other plants, providing a sustainable source for this medically important compound (Jennewein & Croteau, 2001). This highlights the potential of synthetic biology to create novel pathways in plants, allowing for the production of high-value compounds in a more sustainable and scalable manner than harvesting from wild plant populations.

1.2.3 Regulation of Transcription Factors

In addition to manipulating the enzymes directly involved in secondary metabolite biosynthesis, genetic engineering can target the transcription factors that regulate entire metabolic pathways. Transcription factors such as MYB, bHLH, and WRKY are known to control the expression of multiple genes within secondary metabolite pathways. By overexpressing or modulating these transcription factors, it is possible to simultaneously activate the expression of several biosynthetic genes, leading to a coordinated increase in metabolite production (Li et al., 2019). For example, MYB transcription factors have been used to enhance anthocyanin production in crops such as maize and Arabidopsis, leading to higher levels of these flavonoids, which are known for their antioxidant properties and potential health benefits (Zhao et al., 2023). Manipulating these transcription factors can also enhance a plant's defense capabilities, as many secondary metabolites serve as deterrents against herbivores and pathogens.

1.2.4 Gene Silencing and Metabolic Flux Redistribution

In some cases, genetic engineering strategies focus on silencing or knocking out genes that compete for precursor molecules in biosynthetic pathways. By knocking out competing pathways, metabolic flux can be redirected towards the desired secondary metabolite, increasing its production. For example, silencing genes involved in the production of unwanted flavonoid derivatives has been used to enhance the production of specific anthocyanins in genetically modified plants (Gutterson & Zhang, 2004). Additionally, CRISPR-Cas9 gene-editing technology allows for the precise silencing of unwanted pathways or the modification of regulatory elements to enhance secondary metabolite production. This technology has made it easier to fine-tune metabolic pathways in plants, offering more precise control over the levels of specific compounds (Mishra & Pandey, 2024).

1.3 Applications in Agriculture and Industry

The use of genetic engineering to enhance secondary metabolite production has far-reaching applications in agriculture and industry. By increasing the levels of defense-related secondary metabolites, crops can be made more resistant to pests, diseases, and environmental stresses, reducing the need for chemical inputs such as pesticides and fertilizers. This aligns with the goals of sustainable agriculture, which seeks to minimize the environmental impact of farming while maintaining high productivity.

1.3.1 Pest and Disease Resistance

One of the most promising applications of enhanced secondary metabolite production is in improving plant resistance to pests and diseases. Many secondary metabolites, such as alkaloids, terpenoids, and phenolics, have strong antimicrobial, antifungal, and insecticidal properties. For example, crops that produce high levels of alkaloids or saponins can deter herbivores and protect themselves from insect attacks. Genetic engineering can enhance these natural defense mechanisms by increasing the production of these compounds in key tissues such as leaves and stems, reducing the need for external pesticide applications (Srivastava &Srivastava, 2013). Similarly, the overproduction of phytoalexins, a group of antimicrobial compounds, can enhance a plant's resistance to fungal and bacterial pathogens. Genetic modifications that increase the production of phytoalexins like camalexin in Arabidopsis or glyceollins in soybean have been shown to improve disease resistance, leading to healthier crops and higher yields (Bednarek, 2012).

1.3.2 Pharmaceutical and Nutraceutical Production

In addition to agricultural benefits, genetic engineering for enhanced secondary metabolite production has significant implications for the pharmaceutical and nutraceutical industries. Many secondary metabolites are valued for their medicinal properties, including alkaloids like vincristine (used in cancer treatment) and terpenoids like artemisinin (used in malaria treatment). By engineering crop plants to produce these high-value compounds, it is possible to create more sustainable and scalable sources of these pharmaceuticals. For instance, the production of artemisinin, a sesquiterpene lactone used in antimalarial drugs, has been successfully increased in genetically engineered *Artemisia annua* plants by manipulating the enzymes in its biosynthetic pathway. This has the potential to lower the cost of artemisinin production and increase the availability of this life-saving drug (Peplow, 2013). Similarly, crops engineered to produce higher levels of polyphenols and flavonoids, such as resveratrol, quercetin, and catechins, have been explored for their potential to improve human health. These compounds are known for their antioxidant, anti-inflammatory, and cardioprotective properties, and increasing their content in food crops can provide nutraceutical benefits to consumers (Huccetogullari et al., 2019).

While genetic engineering offers significant promise for enhancing secondary metabolite production, several challenges need to be addressed. One of the primary challenges is the complexity of secondary metabolite biosynthetic pathways, which often involve multiple enzymes and regulatory steps. Manipulating one part of the pathway can have unintended effects on other aspects of plant metabolism, leading to reduced growth or fitness. Additionally, there are regulatory and consumer acceptance issues surrounding genetically modified (GM) crops. Public concerns about the safety and environmental impact of GM crops remain a barrier to their widespread adoption in many regions, despite the potential benefits. It is essential to continue research into the safety and efficacy of genetically engineered crops to address these concerns and ensure that the benefits outweigh the risks (Gelinsky & Hilbeck, 2018). Despite these challenges, the future of genetic engineering for secondary metabolite production looks promising. Advances in genome editing technologies like CRISPR-Cas9 offer more precise and efficient methods for manipulating biosynthetic pathways, reducing the risk of off-target effects and improving the scalability of these modifications. Additionally, synthetic biology approaches that combine multiple pathways into a single organism could lead to the production of entirely new compounds with applications in agriculture, medicine, and industry (Pyne et al., 2019).

2. CRISPR/Cas9 and Gene Editing in Metabolite Biosynthesis

2.1 CRISPR/Cas9 and Its Role in Metabolite Biosynthesis

Clustered Regularly Interspaced Short Palindromic Repeats (CRISPR) and CRISPR-associated protein 9 (Cas9) technology have revolutionized genetic engineering due to their simplicity, precision, and versatility in editing specific DNA sequences. This genome-editing tool, originally derived from a bacterial immune system, allows for precise modifications of genes in a wide range of organisms, including plants, animals, and microorganisms. Since its discovery, CRISPR/Cas9 has been applied extensively in agriculture to improve crop traits, increase resistance to diseases, enhance stress tolerance, and, more recently, manipulate the production of secondary metabolites. Secondary metabolites are valuable for their ecological roles in plants, including defense against pests and pathogens, and for their applications in medicine, agriculture, and industry. These metabolites, such as alkaloids, flavonoids, terpenoids, and phenolics, often have complex biosynthetic pathways controlled by multiple genes. CRISPR/Cas9 offers a powerful means to precisely edit these pathways, thereby enhancing the production of desirable compounds or reducing the synthesis of unwanted by-products (Basu et al., 2023). By enabling targeted modifications in key genes, CRISPR/Cas9 accelerates the production of high-value secondary metabolites, reduces the reliance on synthetic inputs, and creates opportunities for sustainable production systems in agriculture and industry.

2.2 Mechanisms of CRISPR/Cas9 in Gene Editing for Metabolite Biosynthesis

CRISPR/Cas9 operates by utilizing a guide RNA (gRNA) that is designed to bind to a specific sequence of DNA within the target genome. The Cas9 enzyme, directed by the gRNA, creates a double-stranded break at the target site. The cell then attempts to repair this break through one of two mechanisms: non-homologous end joining (NHEJ), which often results in small insertions or deletions (indels), or homology-directed repair (HDR), which can introduce precise changes when a donor template is provided. In the context of secondary metabolite biosynthesis, CRISPR/Cas9 can be used to enhance, suppress, or alter the production of specific metabolites by editing the genes encoding biosynthetic enzymes or regulatory factors that control these pathways. By selectively knocking out or modifying genes, scientists can direct the metabolic flux towards desired compounds or eliminate the synthesis of harmful or undesirable by-products. This level of precision was difficult to achieve with earlier gene-editing techniques, making CRISPR/Cas9 an essential tool for advancing metabolic engineering (Das et al., 2024).

2.2.1 Targeting Key Enzymes in Biosynthetic Pathways

One of the most common uses of CRISPR/Cas9 in metabolite biosynthesis is the direct editing of genes encoding key enzymes in biosynthetic pathways. Many secondary metabolite pathways involve several enzymatic steps, each catalyzed by a specific enzyme. By knocking out or modifying these enzymes, it is possible to alter the flow of metabolites through the pathway, thereby increasing or decreasing the production of specific compounds. For example, in the case of alkaloid biosynthesis,

CRISPR/Cas9 has been used to knock out genes encoding strictosidine synthase, an enzyme crucial for the production of monoterpene indole alkaloids like vinblastine and vincristine. These compounds are important for their anticancer properties. Modifying strictosidine synthase genes can increase the yield of these valuable compounds, making the production of these pharmaceuticals more efficient and cost-effective (Sato et al., 2001). Similarly, CRISPR/Cas9 has been applied to modify genes involved in flavonoid biosynthesis. By editing the gene for chalcone synthase (CHS), a key enzyme in the early steps of flavonoid production, researchers can enhance or suppress the production of specific flavonoids, which are important for plant defense, pigmentation, and human health (Huccetogullari et al., 2019). Such edits can be used to boost the production of anthocyanins in crops, increasing their nutritional value and market appeal.

2.2.2 Regulation of Transcription Factors

Another powerful application of CRISPR/Cas9 in metabolite biosynthesis involves the regulation of transcription factors, which control the expression of entire sets of biosynthetic genes. Transcription factors such as MYB, WRKY, and bHLH play central roles in regulating the synthesis of secondary metabolites by activating or repressing specific pathways in response to environmental cues or developmental signals. By editing these transcription factors, researchers can coordinately regulate the expression of multiple genes involved in secondary metabolite production. For instance, CRISPR/Cas9-mediated knockouts or overexpression of MYB transcription factors have been shown to enhance anthocyanin biosynthesis in plants. Anthocyanins, which belong to the flavonoid class of compounds, have antioxidant properties and contribute to the coloration of fruits, vegetables, and flowers. Modulating the activity of MYB transcription factors can lead to increased accumulation of anthocyanins, which can improve the nutritional value of crops like tomatoes, blueberries, and grapes (Oudelaar & Higgs, 2021).

2.2.3 Genome-Wide Metabolic Engineering

CRISPR/Cas9 also enables genome-wide metabolic engineering by allowing for the simultaneous editing of multiple genes. This is particularly important for secondary metabolite pathways that involve multiple genes spread across the genome. Through multiplexed CRISPR/Cas9 approaches, researchers can target several genes at once, either knocking them out or introducing specific modifications to fine-tune the metabolic network. For example, in terpenoid biosynthesis, multiple enzymes and regulatory proteins are involved in the production of essential oils, resins, and other valuable compounds. By using CRISPR/Cas9 to simultaneously edit several genes in the terpenoid pathway, scientists can significantly increase the yield of specific terpenoids, such as limonene or taxol, which are used in perfumes, insect repellents, and pharmaceuticals (Kumari et al., 2024).

2.3 Applications of CRISPR/Cas9 in Secondary Metabolite Production

The use of CRISPR/Cas9 in secondary metabolite biosynthesis has led to numerous practical applications in agriculture, pharmaceuticals, and industrial biotechnology. By modifying the biosynthetic capabilities of plants and microorganisms, CRISPR/Cas9 enables the large-scale production of high-value compounds that were previously difficult to synthesize or extract.

2.3.1 Agricultural Applications

In agriculture, CRISPR/Cas9 has been used to enhance the production of secondary metabolites that provide resistance to pests, diseases, and environmental stress. For example, researchers have used CRISPR/Cas9 to edit genes in rice to increase the production of phytoalexins, antimicrobial compounds that protect plants from fungal infections. By enhancing phytoalexin production, these genetically edited rice plants are more resistant to diseases like rice blast, reducing the need for chemical fungicides and improving crop yields (Shan et al., 2013). CRISPR/Cas9 has also been applied to improve the production of terpenoids in crops. Terpenoids, such as saponins and gossypol, are known for their insecticidal properties. By editing genes involved in terpenoid biosynthesis, scientists can create crops that are more resistant to herbivores, reducing the need for synthetic pesticides and promoting more sustainable farming practices (Chen et al., 2017).

2.3.2 Pharmaceutical and Nutraceutical Production

Secondary metabolites are a rich source of pharmaceuticals and nutraceuticals, and CRISPR/Cas9 is being used to enhance the production of these compounds in both plants and microorganisms. For instance, artemisinin, a sesquiterpene lactone used as a treatment for malaria, is naturally produced in *Artemisia annua*. Through CRISPR/Cas9-mediated editing, scientists have increased the production of artemisinin by enhancing the expression of key enzymes in its biosynthetic pathway (Liu et al., 2011). This has significant implications for improving the availability and reducing the cost of this life-saving drug. Similarly, CRISPR/Cas9 has been used to enhance the production of resveratrol in grapes. Resveratrol is a polyphenolic compound with antioxidant and anti-inflammatory properties, making it a valuable nutraceutical. By editing genes involved in resveratrol biosynthesis, researchers have developed grape varieties with higher levels of this compound, which could lead to more health-promoting foods (Ren et al., 2023).

2.3.3 Industrial Biotechnology

In industrial biotechnology, CRISPR/Cas9 is being applied to optimize the production of secondary metabolites in microorganisms, such as bacteria and yeast. Microbial systems are often used as "biofactories" for producing high-value compounds because they can be easily cultured and genetically manipulated. CRISPR/Cas9 has been used to enhance the production of antibiotics, biofuels, and other industrial products by editing metabolic pathways in microorganisms. For example, in *Escherichia coli*, CRISPR/Cas9 has been used to increase the production of polyketides, a class of secondary metabolites with antibiotic and anticancer properties. By editing genes involved in polyketide biosynthesis, researchers have optimized the production of these compounds in microbial cultures, providing a scalable and cost-effective method for producing pharmaceuticals (Zhao et al., 2023). Despite the many advantages of CRISPR/Cas9, there are challenges associated with its use in metabolite biosynthesis. One major challenge is the potential for off-target effects, where unintended regions of the genome are edited, leading to unexpected phenotypes or metabolic disruptions. Advances in CRISPR technology, such as the development of high-fidelity Cas9 variants, are helping to mitigate this issue by improving the specificity of genome editing (Basu et al., 2023). Another challenge is the complexity of secondary metabolite biosynthetic pathways, which often involve feedback loops, regulatory networks, and interactions between different metabolic processes. These interactions are crucial for ensuring that editing one part of the pathway does not negatively impact other aspects of plant or microbial growth and development. Looking ahead, CRISPR/Cas9 holds immense potential for further advancements in secondary metabolite production. The development of more precise gene-editing tools, coupled with advances in synthetic biology, could lead to the creation of entirely new biosynthetic pathways, enabling the production of novel compounds with applications in agriculture, medicine, and industry. Furthermore, the integration of CRISPR/Cas9 with omics technologies, such as genomics, transcriptomics, and metabolomics, will allow for a more comprehensive understanding of metabolite biosynthesis and its regulation, paving the way for more efficient and sustainable production systems (Das et al., 2024).

3. Metabolic Engineering to Overproduce Targeted Secondary Metabolites

3.1 Metabolic Engineering

Metabolic engineering refers to the deliberate modification of cellular metabolic pathways to enhance the production of desired compounds or to introduce new biosynthetic capabilities into an organism. This field draws on the principles of molecular biology, biochemistry, and systems biology to optimize the metabolic flux toward the synthesis of specific target molecules. In the context of secondary metabolites, metabolic engineering aims to increase the yield of bioactive compounds that have agricultural, pharmaceutical, and industrial importance. Secondary metabolites, including alkaloids, terpenoids, flavonoids, and phenolics, are often produced in low quantities in their native hosts, making large-scale production challenging. Through metabolic engineering, researchers can redirect the organism's metabolic resources to overproduce these valuable compounds (Stephanopoulos, 2012). Over the past few decades, advances in genetic engineering, synthetic biology, and omics technologies have dramatically expanded the potential for metabolic engineering to manipulate secondary metabolite production in plants, microbes, and other organisms. With the integration of genome-editing tools such as CRISPR/Cas9, metabolic pathways can be restructured with precision, leading to enhanced yields of desired metabolites. Metabolic engineering offers a sustainable alternative to the extraction of secondary metabolites from natural sources, which is often resource-intensive and environmentally damaging. This approach also opens the door to producing novel compounds that are difficult or impossible to obtain from nature (Zhao et al., 2023).

3.2 Key Strategies for Overproducing Targeted Secondary Metabolites

Several strategies are employed in metabolic engineering to increase the production of targeted secondary metabolites. These strategies include enhancing precursor availability, redirecting metabolic flux, overexpressing key enzymes, and eliminating competitive pathways. Additionally, the regulation of transcription factors that control entire biosynthetic pathways and the introduction of novel pathways from other organisms are essential strategies to achieve high yields of the desired metabolites.

3.2.1 Enhancing Precursor Availability

One of the most fundamental strategies in metabolic engineering is to increase the availability of precursors required for the biosynthesis of secondary metabolites. Secondary metabolites are often synthesized from primary metabolites such as amino acids, fatty acids, or simple sugars. By increasing the supply of these precursors, the metabolic pathway leading to the production of the desired secondary metabolite can be driven more efficiently. For instance, in the production of flavonoids, metabolic engineers often focus on increasing the availability of phenylalanine, the precursor for many flavonoids. Overexpressing the genes involved in the shikimate pathway, which produces phenylalanine, can increase the flux of precursors into flavonoid biosynthesis. This strategy has been successfully used in transgenic plants such as tomato and Arabidopsis to boost the production of flavonoids, which are valued for their antioxidant properties (Liu et al., 2021). Similarly, the terpenoid pathway can be engineered to increase the production of geranyl pyrophosphate (GPP), a common precursor for many terpenoids, including monoterpenes and diterpenes. By manipulating the mevalonate (MVA) pathway in plants or microbes, researchers can enhance the availability of GPP, leading to increased production of valuable terpenoids such as limonene and taxol (Ajikumar et al., 2010).

3.2.2 Redirecting Metabolic Flux

Another critical aspect of metabolic engineering is the redirection of metabolic flux toward the desired product. In many organisms, metabolic pathways are highly branched, leading to the production of multiple metabolites from the same precursors. To maximize the yield of the target metabolite, it is necessary to minimize the production of by-products and channel metabolic resources specifically toward the desired biosynthetic route. In alkaloid production, for example, the metabolic flux can be redirected by silencing or knocking out genes involved in the production of unwanted alkaloids, thus focusing the metabolic effort on the synthesis of the desired compound. This has been demonstrated in the production of monoterpene indole alkaloids, where genes involved in competing pathways were downregulated to boost the yield of anti-cancer alkaloids such as vinblastine and vincristine in *Catharanthus roseus* (Sears & Boger, 2015). CRISPR/Cas9 technology has facilitated the precise editing of metabolic pathways by allowing for the targeted knockdown of genes in competing pathways. For example, in *Saccharomyces cerevisiae* (yeast), CRISPR/Cas9 was used to delete genes involved in ethanol production, redirecting the carbon flux toward the production of carotenoids, which are used as natural pigments and antioxidants in the food and pharmaceutical industries (Jin et al., 2017).

3.2.3 Overexpression of Key Enzymes

One of the most widely used strategies in metabolic engineering is the overexpression of enzymes that catalyze key steps in secondary metabolite biosynthesis. By increasing the activity of these enzymes, the overall flux through the pathway can be enhanced, leading to higher yields of the target compound.

For instance, the overexpression of strictosidine synthase, a key enzyme in the biosynthesis of indole alkaloids, has been employed to increase the production of vinblastine and other medically valuable alkaloids in *Catharanthus roseus* (Srivastava & Srivastava, 2013). Similarly, overexpression of chalcone synthase (CHS) in plants has been used to increase the production of flavonoids, which have important health benefits due to their antioxidant, anti-inflammatory, and anti-cancer properties (Winkel-Shirley, 2001). In microbes, overexpression of enzymes in the mevalonate pathway has been used to boost the production of terpenoids, such as artemisinin and taxol, which are important for pharmaceutical applications. By introducing multiple copies of genes encoding enzymes such as HMGR (3-hydroxy-3-methylglutaryl-CoA reductase), researchers can increase the metabolic flux through the terpenoid pathway, leading to higher yields of these valuable compounds (Martin et al., 2003).

3.2.4 Eliminating Competitive Pathways

Eliminating or reducing the activity of competing metabolic pathways is a common strategy used to focus the metabolic resources on the desired biosynthetic pathway. This approach is particularly important when multiple pathways compete for the same precursor molecules. For example, in the production of anthocyanins (a class of flavonoids), the accumulation of other flavonoid derivatives such as flavonols can reduce the overall yield of anthocyanins. By knocking out or silencing the genes responsible for flavonol production, metabolic engineers can ensure that more precursors are available for anthocyanin biosynthesis. This strategy has been successfully employed in several plant species to increase anthocyanin content, which is valuable for its health benefits and as a natural colorant in the food industry (Gutterson & Zhang, 2004). In microbial systems, gene deletion or knockout strategies are commonly used to eliminate by-product formation. For example, in the microbial production of polyketides, which are used as antibiotics and anticancer agents, the deletion of competing pathways that produce unwanted polyketide derivatives can significantly increase the yield of the target compound (Fischbach & Walsh, 2006).

3.2.5 Introduction of Novel Biosynthetic Pathways

In some cases, metabolic engineering involves the introduction of entirely new biosynthetic pathways into an organism to produce compounds that are not naturally synthesized by the host. This is achieved through the transfer of genes from other organisms, often involving complex biosynthetic pathways with multiple enzymes. A well-known example is the introduction of the artemisinin biosynthetic pathway from *Artemisia annua* into *Saccharomyces cerevisiae* (yeast). By introducing the genes responsible for artemisinin biosynthesis into yeast, researchers were able to create a microbial system capable of producing artemisinic acid, a precursor to the antimalarial drug artemisinin. This achievement provides a scalable and cost-effective alternative to the extraction of artemisinin from plants, which is resource-intensive and subject to fluctuations in supply (Ro et al., 2006). The introduction of novel biosynthetic pathways also enables the production of non-native compounds in crops. For instance, the biosynthesis of resveratrol, a polyphenol with antioxidant properties, has been successfully introduced into tomato plants, increasing their health benefits and market value (Huccetogullari et al., 2019).

3.3 Applications of Metabolic Engineering in Agriculture, Pharmaceuticals, and Industry

Metabolic engineering for the overproduction of secondary metabolites has numerous applications across agriculture, pharmaceuticals, and industrial biotechnology. By enhancing the production of high-value compounds, metabolic engineering contributes to sustainable agriculture, improved healthcare, and the development of bio-based industrial products.

3.3.1 Agricultural Applications

In agriculture, metabolic engineering has been applied to improve the production of secondary metabolites that provide resistance to pests, diseases, and environmental stress. For instance, crops engineered to produce higher levels of phytoalexins and terpenoids are more resistant to fungal pathogens and insect herbivores. This reduces the need for chemical pesticides and promotes more sustainable farming practices. The production of flavonoids and other antioxidant compounds in crops has also been enhanced through metabolic engineering to improve crop quality and nutritional value. For example, metabolic engineering has been used to increase the production of anthocyanins in fruits and vegetables, providing health benefits to consumers and improving the visual appeal of food products (Lila, 2004).

3.3.2 Pharmaceutical and Nutraceutical Production

In the pharmaceutical industry, metabolic engineering plays a crucial role in producing secondary metabolites that serve as active ingredients in drugs. The production of plant-derived alkaloids, terpenoids, and phenolics, which are used in cancer treatment, pain management, and antimicrobial therapies, has been significantly enhanced through metabolic engineering. For example, the metabolic engineering of *Catharanthus roseus* has led to the overproduction of vinblastine and vincristine, which are used as chemotherapeutic agents (Srivastava & Srivastava, 2013). Similarly, the production of nutraceuticals, such as resveratrol and other polyphenols, has been increased in crops like grapes and tomatoes through metabolic engineering. These compounds are valued for their antioxidant and anti-inflammatory properties and are used in dietary supplements and functional foods (Huccetogullari et al., 2019).

3.3.3 Industrial Biotechnology

In industrial biotechnology, metabolic engineering is used to produce bio-based chemicals, fuels, and materials from renewable resources. For example, the production of biofuels such as biodiesel and bioethanol has been enhanced through the metabolic engineering of microorganisms to overproduce fatty acids and alcohols. Similarly, metabolic engineering has been applied to increase the production of bioplastics and other biodegradable materials from secondary metabolites such as polylactic acid (Fischer et al., 2008). The microbial production of terpenoids, polyketides, and other secondary metabolites with industrial applications has also been improved through metabolic engineering. These compounds are used in the manufacture of fragrances, flavors, and cosmetics, as well as in the development of bio-based insecticides and fungicides (Jennewein & Croteau, 2001). While metabolic engineering holds great promise for the overproduction of targeted secondary metabolites, there are several challenges that must be addressed. One of the main challenges is the complexity of secondary metabolite biosynthetic pathways, which often involve multiple enzymes, cofactors, and regulatory mechanisms. Understanding and manipulating these pathways without causing unintended metabolic imbalances or reducing organismal fitness requires a deep understanding of metabolic networks and regulatory systems (Nielsen & Keasling, 2016). Another challenge is the potential for metabolic burden, where the overproduction of secondary metabolites diverts resources away from essential cellular functions, leading to reduced growth or productivity. Strategies such as dynamic pathway regulation and the use of synthetic biology tools to optimize metabolic flux can help address this challenge. Looking forward, advances in synthetic biology, CRISPR/Cas9 gene editing, and systems biology are expected to further enhance the capabilities of metabolic engineering. The development of novel biosynthetic pathways, the integration of multi-omics data, and the use of computational models to predict metabolic outcomes will enable more precise and efficient manipulation of secondary metabolite production. These advances will contribute to more sustainable agricultural practices, improved healthcare, and the development of bio-based industries (Lee et al., 2024).

4. Tissue Culture Techniques for Secondary Metabolite Production

4.1 Tissue Culture and Secondary Metabolite Production

Tissue culture, or in vitro culture, is the technique of growing plant cells, tissues, or organs in a controlled environment, typically on a nutrient medium under sterile conditions. This technique has become an invaluable tool for the production of secondary metabolites, which are bioactive compounds that plants produce as part of their defense mechanisms or ecological interactions. These metabolites—such as alkaloids, terpenoids, flavonoids, and phenolics—are widely used in pharmaceuticals, cosmetics, agriculture, and food industries due to their medicinal, pesticidal, and nutritional properties. The challenge with extracting secondary metabolites from plants in the field is that they are often produced in low quantities, may be seasonally available, or require the destruction of the plant, leading to sustainability issues. Tissue culture techniques offer a solution to these limitations by providing an alternative method to produce secondary metabolites in a controlled environment, which ensures a consistent and scalable supply of these valuable compounds. In vitro techniques can also help enhance the yield of secondary metabolites by manipulating growth conditions or by using biotechnological tools such as elicitors, precursor feeding, or genetic transformation (Verpoorte et al., 1991).

4.2 Types of Tissue Culture Techniques for Secondary Metabolite Production

Several tissue culture techniques have been developed to produce secondary metabolites, each with its advantages and applications. The most common methods include callus culture, cell suspension culture, organ culture (such as root and shoot cultures), hairy root culture, and bioreactor-based systems. These techniques enable the production of secondary metabolites without the need for whole plants and allow for the precise control of environmental factors to optimize metabolite synthesis.

4.2.1 Callus Culture

Callus culture is one of the most widely used tissue culture techniques for secondary metabolite production. A callus is an unorganized mass of cells that forms when plant tissues are grown on a nutrient medium supplemented with growth regulators, such as auxins and cytokinins. These undifferentiated cells have the potential to produce a variety of secondary metabolites, depending on the plant species and the culture conditions. One of the main advantages of callus culture is that it can be initiated from small amounts of plant material, such as leaf, stem, or root explants, making it possible to produce secondary metabolites without destroying the entire plant. Additionally, the growth and metabolic activity of callus cells can be easily manipulated by adjusting the nutrient composition, light conditions, and the type and concentration of growth regulators. For example, callus cultures of *Catharanthus roseus* have been used to produce valuable indole alkaloids such as vinblastine and vincristine, which are used in cancer treatment (Gutierrez-Valdes et al., 2020).

4.2.2 Cell Suspension Culture

Cell suspension culture is an extension of callus culture, where cells from a callus are dispersed in a liquid medium, allowing them to grow and divide as single cells or small cell clusters in a continuously agitated system. This technique is particularly advantageous for large-scale production of secondary metabolites because it allows for the cultivation of plant cells in bioreactors, where the culture conditions can be optimized for maximum metabolite production. In cell suspension cultures, the production of secondary metabolites can be enhanced through the use of elicitors substances that stimulate the production of bioactive compounds. Elicitors can be biotic (derived from pathogens, fungi, or other organisms) or abiotic (such as heavy metals, UV light, or physical stress). For example, elicitors such as jasmonic acid and salicylic acid have been used to increase the production of secondary metabolites like anthraquinones in cell suspension cultures of *Rubia tinctorum* (Zhao et al., 2005). Cell suspension cultures have been successfully used for the production of a wide range of secondary metabolites, including paclitaxel (Taxol), an anticancer compound produced by *Taxus* species. By optimizing the nutrient medium, elicitor treatments, and bioreactor conditions, researchers have significantly increased the yield of paclitaxel in suspension cultures, providing an alternative to the extraction of this compound from yew trees (Tabata, 2006).

4.2.3 Organ Culture (Root and Shoot Cultures)

Organ culture involves the cultivation of specific plant organs, such as roots, shoots, or leaves, in vitro. This technique is particularly useful for the production of secondary metabolites that are organ-specific, meaning that they are synthesized predominantly in certain plant tissues. For example, many root-derived secondary metabolites, such as alkaloids, are produced at higher concentrations in root cultures than in other plant parts. Root culture is often used for the production of tropane alkaloids, which are found in plants like *Atropa belladonna* and *Datura* species. These alkaloids, including hyoscyamine and scopolamine, are important for their medicinal properties as anticholinergic drugs. Root cultures provide a stable and sustainable system for the production of these compounds without the need to harvest entire plants (Palazon et al., 2003). Shoot cultures, on the other hand, are used for the production of metabolites like flavonoids, which are often synthesized in aerial parts of plants. Shoot cultures of *Hypericum perforatum* (St. John's Wort) have been used to produce hypericin, a secondary metabolite with antidepressant and antiviral properties. By manipulating the culture conditions, researchers can optimize the production of hypericin in shoot cultures (Karppinen et al., 2016).

4.2.4 Hairy Root Culture

Hairy root culture is a specialized form of root culture that is induced by infection with the bacterium *Agrobacterium rhizogenes*. This bacterium transfers part of its DNA into the plant cells, causing them to form highly branched, fast-growing "hairy roots" that are capable of synthesizing high levels of secondary metabolites. Hairy root cultures are particularly advantageous because they are genetically stable and can grow without the need for added hormones, making them ideal for the long-term production of metabolites. Hairy root cultures have been extensively used for the production of a variety of secondary metabolites. For example, hairy root cultures of *Catharanthus roseus* have been used to produce the anti-cancer alkaloids vinblastine and vincristine. Additionally, hairy root cultures of *Withania somnifera* (ashwagandha) have been used to enhance the production of withanolides, which have immunomodulatory and anti-inflammatory properties (Tabata, 2006).

4.2.5 Bioreactor-Based Systems

To scale up the production of secondary metabolites from tissue cultures, bioreactor-based systems are used. Bioreactors are vessels designed to provide the optimal conditions for the growth of plant cells, tissues, or organs in large volumes. These systems allow for the precise control of environmental factors such as temperature, pH, oxygen concentration, and nutrient supply, making them ideal for industrial-scale production of secondary metabolites. There are several types of bioreactors used in plant tissue culture, including stirred-tank bioreactors, airlift bioreactors, and bubble column bioreactors. Each type has its advantages, depending on the type of tissue culture being used. For example, airlift bioreactors are commonly used for hairy root cultures, as they provide efficient oxygenation without damaging the delicate root structures. In contrast, stirred-tank bioreactors are often used for cell suspension cultures, as they provide continuous mixing and nutrient distribution (Pyne et al., 2019). Bioreactor-based systems have been successfully employed for the large-scale production of secondary metabolites such as shikonin from *Lithospermum erythrorhizon* and ginsenosides from *Panax ginseng*. These systems not only increase the yield of secondary metabolites but also reduce the production costs, making them suitable for commercial applications in the pharmaceutical and nutraceutical industries (Nielsen & Keasling, 2016).

4.3 Enhancing Secondary Metabolite Production in Tissue Culture

Several strategies can be used to enhance the production of secondary metabolites in tissue culture systems. These strategies include the use of elicitors, precursor feeding, optimization of culture conditions, and genetic transformation.

4.3.1 Elicitation

Elicitation is a widely used technique in tissue culture to enhance the production of secondary metabolites. Elicitors are compounds that stimulate the plant's defense mechanisms, leading to an increase in the production of secondary metabolites. Elicitors can be of biotic origin (such as pathogen-derived molecules) or abiotic origin (such as heavy metals, UV light, or physical stress). For example, jasmonic acid, a plant hormone involved in stress responses, has been used as an elicitor to increase the production of secondary metabolites in cell cultures. Jasmonic acid has been shown to enhance the production of paclitaxel in *Taxus* cell cultures and ginsenosides in *Panax ginseng* cultures (Sohn et al., 2022). Similarly, salicylic acid has been used to increase the production of phenolic compounds in several plant species.

4.3.2 Precursor Feeding

Precursor feeding involves the addition of metabolic precursors to the culture medium to increase the production of secondary metabolites. By supplying the cells with key intermediates of the biosynthetic pathway, researchers can enhance the flux of metabolites through the pathway, leading to higher yields of the desired product. For example, feeding cell cultures of *Catharanthus roseus* with tryptophan, a precursor of the indole alkaloid biosynthetic pathway, has been shown to increase the production of alkaloids such as ajmalicine and serpentine (Murthy et al., 2014). Similarly, the addition of mevalonic acid, a precursor in the terpenoid biosynthetic pathway, has been used to enhance the production of terpenoids in cell suspension cultures of *Taxus* species.

4.3.3 Optimization of Culture Conditions

The production of secondary metabolites in tissue cultures can be significantly influenced by the culture conditions, including the composition of the nutrient medium, the concentration of growth regulators, light, temperature, and pH. By optimizing these conditions, researchers can maximize the yield of secondary metabolites. For example, altering the ratio of auxins to cytokinins in the culture medium can influence the differentiation of plant cells and the production of secondary metabolites. In *Hypericum perforatum* shoot cultures, adjusting the concentration of cytokinins has been shown to increase the production of hypericin, a secondary metabolite with antidepressant properties (Karppinen et al., 2018).

4.3.4 Genetic Transformation

Genetic transformation involves the introduction of foreign genes into plant cells to enhance the production of secondary metabolites. This can be achieved through the use of *Agrobacterium tumefaciens*-mediated transformation or other genetic engineering techniques. By introducing genes encoding key enzymes in the biosynthetic pathway of a desired secondary metabolite, researchers can increase the metabolic flux toward the production of that compound. For example, the overexpression of genes involved in the production of tropane alkaloids has been used to enhance the yield of these compounds in root cultures of *Atropa belladonna* (Yun et al., 1992). Similarly, genetic transformation has been used to increase the production of anthocyanins in cell cultures of *Vitis vinifera* (grape) (Kobayashi et al., 2005).

Tissue culture techniques for secondary metabolite production have a wide range of applications in the pharmaceutical, nutraceutical, cosmetic, and agricultural industries. These techniques offer a sustainable and scalable alternative to traditional methods of secondary metabolite extraction, which often rely on the harvesting of endangered or slow-growing plants. In the pharmaceutical industry, tissue culture techniques are used to produce valuable compounds such as paclitaxel, vinblastine, and scopolamine, which are used in cancer treatment and other medical applications. In the nutraceutical and cosmetic industries, tissue culture is used to produce compounds like resveratrol, ginsenosides, and flavonoids, which have antioxidant and anti-aging properties. Looking forward, the integration of tissue culture with modern biotechnological tools such as CRISPR/Cas9 gene editing and synthetic biology offers exciting prospects for the enhanced production of secondary metabolites. By combining tissue culture with these advanced techniques, it may be possible to create plant cell factories that produce high yields of novel or rare secondary metabolites with important medicinal properties (Ren et al., 2023).

5. Omics Technologies in Identifying Biosynthetic Pathways

5.1 Omics Technologies

The advent of omics technologies has revolutionized the way biological research is conducted, allowing scientists to explore the complex networks of genes, proteins, metabolites, and other molecules that govern cellular functions. In the context of plant biology and secondary metabolite production, omics approaches, including genomics, transcriptomics, proteomics, and metabolomics, have provided invaluable insights into the identification and characterization of biosynthetic pathways. These technologies offer a systems-level understanding of the processes involved in the production of secondary metabolites, which are often controlled by multiple genes, enzymes, and regulatory factors. By applying omics tools, researchers can identify key genes and enzymes in biosynthetic pathways, elucidate regulatory mechanisms, and discover novel metabolites. This knowledge is crucial for optimizing the production of valuable bioactive compounds through genetic engineering, metabolic engineering, and synthetic biology (Gutterson & Zhang, 2004).

5.2 Genomics: Mapping Biosynthetic Genes

Genomics is the study of an organism's complete set of DNA, including all of its genes. In the context of secondary metabolite biosynthesis, genomics provides a foundation for identifying genes involved in the production of these compounds. Advances in next-generation sequencing (NGS) technologies have made it possible to rapidly sequence the entire genomes of plants and microbes, allowing for the identification of biosynthetic gene clusters (BGCs) responsible for the production of secondary metabolites.

5.2.1 Genome Sequencing and Annotation

One of the primary applications of genomics in secondary metabolite research is the sequencing and annotation of genomes to identify biosynthetic genes. Genome sequencing allows researchers to assemble the complete genetic code of an organism and locate the genes encoding key enzymes in secondary metabolite biosynthesis. For example, the sequencing of the *Catharanthus roseus* genome led to the identification of key genes involved in the biosynthesis of the anti-cancer alkaloids vinblastine and vincristine (Pyne et al., 2019). Genome annotation involves the identification and functional characterization of genes within the genome. Through annotation, researchers can link specific genes to their roles in biosynthetic pathways and predict the functions of unknown genes. Tools such as BLAST, Pfam, and KEGG are commonly used to annotate genomes and assign functions to biosynthetic genes.

5.2.2 Biosynthetic Gene Cluster Discovery

In many cases, genes involved in the production of secondary metabolites are organized into clusters known as biosynthetic gene clusters (BGCs). These clusters often contain all of the genes necessary for the synthesis, modification, and transport of a specific secondary metabolite. Genomic analysis has been instrumental in discovering BGCs in plants, fungi, and bacteria. For example, the identification of the taxol biosynthetic gene cluster in *Taxus* species was a key breakthrough in understanding how this valuable anticancer compound is produced. By sequencing the genome and identifying the BGC responsible for taxol biosynthesis, researchers have been able to explore new ways of producing this compound using metabolic engineering and synthetic biology (Kobayashi et al., 2005).

5.3 Transcriptomics: Decoding Gene Expression

While genomics provides a static map of an organism's genetic blueprint, transcriptomics reveals the dynamic expression of genes under different conditions. Transcriptomics involves the study of the complete set of RNA transcripts produced by an organism at a given time, providing insights into which genes are being actively expressed. This is particularly important in secondary metabolite research, where gene expression can be influenced by environmental factors, developmental stages, and biotic or abiotic stress.

5.3.1 RNA Sequencing for Pathway Identification

RNA sequencing (RNA-seq) is a powerful transcriptomics tool used to quantify gene expression levels and identify differentially expressed genes in various conditions. In secondary metabolite research, RNA-seq is used to compare gene expression profiles in tissues or cells that produce high levels of metabolites with those that produce low levels. This helps identify key regulatory genes and enzymes involved in biosynthetic pathways. For example, RNA-seq analysis of *Salvia miltiorrhiza*, a medicinal plant known for producing tanshinones (diterpenoids with therapeutic properties), revealed differential expression of genes involved in tanshinone biosynthesis in roots compared to leaves. This allowed researchers to pinpoint key enzymes and transcription factors regulating tanshinone production (Hou et al., 2021).

5.3.2 Co-expression Network Analysis

Co-expression network analysis is a transcriptomics technique used to identify groups of genes that are co-regulated and expressed together. By constructing gene co-expression networks, researchers can identify potential regulatory genes that control the expression of entire biosynthetic pathways. These networks help reveal how secondary metabolite production is coordinated at the molecular level. For instance, co-expression analysis in *Arabidopsis thaliana* has been used to identify regulatory genes controlling the production of flavonoids, phenolic compounds that protect plants from UV radiation and pathogens. Through this approach, researchers discovered novel transcription factors that modulate the expression of multiple flavonoid biosynthetic genes, providing new targets for genetic engineering to enhance flavonoid production (Yin et al., 2017).

5.4 Proteomics: Understanding Enzymatic Machinery

Proteomics is the large-scale study of proteins, the molecules that carry out most cellular functions, including enzymatic reactions involved in biosynthetic pathways. While transcriptomics reveals which genes are expressed, proteomics focuses on identifying the proteins encoded by these genes and understanding their functional roles. In the context of secondary metabolite production, proteomics helps elucidate the enzymatic machinery responsible for converting precursors into bioactive compounds.

5.4.1 Protein Identification and Characterization

Proteomics techniques, such as mass spectrometry (MS) and two-dimensional gel electrophoresis, are used to identify and characterize the proteins involved in secondary metabolite biosynthesis. By analyzing protein extracts from tissues producing high levels of secondary metabolites, researchers can identify key enzymes in the biosynthetic pathway and assess their abundance, activity, and post-translational modifications. For example, proteomics analysis in *Camptotheca acuminata*, the source of the anti-cancer alkaloid camptothecin, led to the identification of several enzymes involved in camptothecin biosynthesis, including strictosidine synthase and camptothecin synthase. Understanding the role of these enzymes has facilitated the metabolic engineering of camptothecin production in microbial systems (Lee et al., 2024).

5.4.2 Post-Translational Modifications and Enzyme Regulation

Many biosynthetic enzymes undergo post-translational modifications (PTMs) that affect their activity, stability, or localization within the cell. Proteomics allows researchers to study these modifications, such as phosphorylation, glycosylation, and ubiquitination, which can regulate enzyme function. Understanding how PTMs influence biosynthetic enzymes provides insights into how secondary metabolite production can be fine-tuned in response to environmental or developmental cues. For instance, proteomic studies on flavonoid biosynthesis in *Glycine max* (soybean) have shown that key enzymes in the pathway, such as chalcone isomerase, undergo phosphorylation, which regulates their activity. By manipulating these PTMs, researchers can potentially increase the flux through the flavonoid biosynthetic pathway, enhancing the production of these valuable compounds (Dastmalchi & Dhaubhadel, 2015).

5.5 Metabolomics: Profiling Metabolic Products

Metabolomics is the comprehensive study of metabolites, the small molecules that are the end products of cellular processes. In the context of secondary metabolite production, metabolomics provides a direct measure of the metabolites being synthesized by an organism, allowing researchers to map biosynthetic pathways and quantify the levels of bioactive compounds. Metabolomics is often used in combination with genomics, transcriptomics, and proteomics to provide a complete picture of secondary metabolite biosynthesis.

5.5.1 Metabolite Profiling for Pathway Elucidation

Metabolomics techniques, such as gas chromatography-mass spectrometry (GC-MS) and liquid chromatography-mass spectrometry (LC-MS), are used to profile the metabolites present in plant tissues, microbial cultures, or cell extracts. By comparing the metabolite profiles of different samples, researchers can identify which metabolites are produced under specific conditions and link them to their corresponding biosynthetic pathways. For example, metabolomics analysis in *Hypericum perforatum* (St. John's Wort) has been used to profile the production of hypericin, a secondary metabolite with antidepressant properties. By correlating metabolite levels with gene expression data, researchers have been able to identify key enzymes and regulatory genes involved in hypericin biosynthesis (Karppinen et al., 2018).

5.5.2 Flux Analysis for Metabolic Pathway Mapping

Metabolomics also enables the measurement of metabolic flux, which refers to the rates at which metabolites are produced and consumed in a biosynthetic pathway. Flux analysis is critical for understanding how metabolic resources are allocated within a cell and how they can be redirected toward the production of target compounds. This information is particularly valuable for optimizing the production of secondary metabolites through metabolic engineering. For instance, flux analysis in *Escherichia coli* engineered to produce flavonoids has provided insights into how carbon flux can be redirected from primary metabolism to flavonoid biosynthesis. By optimizing key steps in the pathway, researchers have been able to increase the production of flavonoids in microbial systems, providing a scalable platform for producing these compounds (Lee et al., 2024).

5.6 Integrating Omics Data: Systems Biology Approaches

One of the key strengths of omics technologies is their ability to generate large datasets that can be integrated to provide a holistic understanding of biological systems. In the context of secondary metabolite production, integrating data from genomics, transcriptomics, proteomics, and metabolomics is crucial for fully elucidating biosynthetic pathways and identifying regulatory mechanisms that control metabolite synthesis. This approach, known as systems biology, enables researchers to model entire biological processes and predict how changes at the genetic, enzymatic, or environmental levels will affect metabolite production.

5.6 Integrating Omics Data: Systems Biology Approaches

5.6.1 Data Integration and Pathway Modeling

Systems biology integrates diverse omics data to create comprehensive models of metabolic pathways. By combining genomic data (which provides the genetic blueprint) with transcriptomic data (which reveals gene expression patterns), proteomic data (which identifies the functional proteins), and metabolomic data (which profiles the end products), researchers can reconstruct entire biosynthetic pathways and understand how they are regulated. This integrated approach allows for the identification of bottlenecks in pathways, the discovery of new enzymes or regulatory elements, and prediction of how environmental factors influence metabolite production. For example, in the production of taxol (paclitaxel), a complex diterpenoid used in cancer treatment, systems biology approaches have been employed to integrate transcriptomic, proteomic, and metabolomic data. By modeling the taxol biosynthetic pathway in *Taxus* cells, researchers identified key rate-limiting steps in the pathway and were able to manipulate culture conditions to enhance taxol production (Tabata, 2006).

5.6.2 Metabolic Flux Balance Analysis

Flux balance analysis (FBA) is a computational method used in systems biology to model the flow of metabolites through a metabolic network. By integrating metabolomics data with enzyme kinetics and stoichiometric models, FBA helps predict how changes in gene expression or enzyme activity will affect the overall production of secondary metabolites. This approach is particularly useful for optimizing metabolic pathways in engineered organisms, where maximizing the yield of a target compound is the goal. In microbial systems, FBA has been applied to optimize the production of secondary metabolites like polyketides, flavonoids, and terpenoids. For instance, FBA was used to model the production of resveratrol in engineered *Saccharomyces cerevisiae* and identified metabolic bottlenecks that, when alleviated through genetic engineering, resulted in higher resveratrol yields (Jennewein & Croteau, 2001).

5.6.3 Omics-Assisted Metabolic Engineering

The integration of omics data not only advances our understanding of biosynthetic pathways but also supports metabolic engineering efforts aimed at overproducing secondary metabolites. By using omics-assisted metabolic engineering, researchers can systematically identify the key genes, enzymes, and regulatory elements that need to be modified to optimize metabolite production. This approach has been particularly effective in microbial systems, where genetic tools like CRISPR/Cas9 can be used to edit multiple genes simultaneously. For example, in the microbial production of flavonoids, transcriptomic and metabolomic data were used to identify key bottlenecks in the biosynthesis of naringenin, a precursor to many flavonoids. By engineering these bottlenecks in *Escherichia coli* using CRISPR/Cas9, researchers were able to significantly increase naringenin production (Ro et al., 2006).

5.7 Applications of Omics in Secondary Metabolite Research

The application of omics technologies in secondary metabolite research has opened up new possibilities for drug discovery, agriculture, and biotechnology. By identifying novel biosynthetic pathways and optimizing existing ones, omics technologies are paving the way for the sustainable production of valuable bioactive compounds.

5.7.1 Pharmaceutical Applications

Many secondary metabolites are used as active pharmaceutical ingredients (APIs) in drugs. Omics technologies have been instrumental in identifying the biosynthetic pathways for important pharmaceuticals such as paclitaxel, vincristine, artemisinin, and camptothecin. These discoveries have enabled the development of biotechnological methods for producing these drugs more sustainably, often through microbial fermentation or plant cell culture systems. For example, the integration of genomics and metabolomics in *Artemisia annua* led to the discovery of key enzymes in the artemisinin biosynthetic pathway, enabling the production of this antimalarial drug in engineered yeast cells (Gelinsky & Hilbeck, 2018).

5.7.2 Agricultural Applications

In agriculture, secondary metabolites play a crucial role in plant defense against pests and pathogens. Omics technologies have been used to identify the biosynthetic pathways for natural pesticides and herbicides, such as alkaloids, terpenoids, and phenolics. By understanding these pathways, researchers can breed or engineer crops to produce higher levels of these natural defense compounds, reducing the need for synthetic agrochemicals. For instance, transcriptomic and metabolomic studies of maize (corn) identified key genes involved in the biosynthesis of benzoxazinoids, which are natural insect repellents. By selecting for maize varieties with higher expression of these genes, breeders have developed pest-resistant maize that requires fewer chemical inputs (Huccetogullari, et al., 2019). In industrial biotechnology, secondary metabolites are used as raw materials for the production of biofuels, bioplastics, and other biochemicals. Omics technologies have helped identify and optimize the biosynthetic pathways for these compounds, enabling their production in microbial or plant-based systems. For example, genomics and metabolomics have been used to enhance the production of terpenoids, which are used in the manufacture of fragrances, flavors, and biofuels. By engineering the terpenoid biosynthetic pathway in *Escherichia coli*, researchers were able to increase the production of limonene, a monoterpene used as a biofuel precursor (Winkel-Shirley, 2001).

6. Synthetic Biology and Microbial Production of Plant Metabolites

6.1 Synthetic Biology and Microbial Production

Synthetic biology is a rapidly growing field that combines principles of engineering, biology, and computer science to design and construct new biological systems or redesign existing ones for useful purposes. It involves the reprogramming of organisms to perform specific tasks or produce valuable compounds. In recent years, synthetic biology has been increasingly applied to the production of plant metabolites, particularly those with pharmaceutical, nutraceutical, and industrial value. By leveraging microbial systems such as *Escherichia coli* and yeast (*Saccharomyces cerevisiae*), scientists can produce plant metabolites in a scalable, sustainable, and controlled environment. Microbial production of plant metabolites through synthetic biology offers several advantages over traditional extraction methods, which often require large quantities of plant material, are labor-intensive, and are environmentally unsustainable. Microbes can be engineered to produce large amounts of plant secondary metabolites by introducing the necessary biosynthetic genes and pathways into their genomes. This approach is especially useful for producing complex compounds like alkaloids, terpenoids, flavonoids, and phenolics, which are often difficult to synthesize chemically or extract from plants in sufficient quantities (Nielsen & Keasling, 2016).

6.2 Key Approaches in Synthetic Biology for Metabolite Production

Synthetic biology enables the microbial production of plant metabolites through a variety of strategies, including pathway reconstruction, modular design, gene optimization, and metabolic engineering. These approaches are used to reconstitute the plant metabolic pathways in microbial hosts, allowing for efficient and high-yield production of target compounds.

6.2.1 Pathway Reconstruction in Microbes

One of the foundational approaches in synthetic biology is the reconstruction of plant biosynthetic pathways in microbial hosts. Plant secondary metabolites are often produced through complex biosynthetic pathways involving multiple enzymes. Synthetic biologists can reconstruct these pathways by identifying and transferring the necessary genes from plants to microbial systems such as *E. coli* or yeast. For example, the production of artemisinin, an antimalarial drug originally derived from *Artemisia annua*, was successfully transferred into yeast through the reconstruction of its biosynthetic pathway. By introducing plant genes responsible for artemisinic acid production into yeast, researchers were able to develop a microbial system capable of producing artemisinic acid, a precursor to artemisinin. This microbial production system has significantly reduced the cost of artemisinin, making it more accessible for malaria treatment (Paddon & Keasling, 2014). Pathway reconstruction can be applied to various other metabolites, such as taxol (paclitaxel), a diterpenoid used in cancer therapy, and vinblastine, an alkaloid used in chemotherapy. By transferring the necessary enzymes and optimizing the pathways, microbes can be turned into "cell factories" that produce these valuable compounds at a commercial scale.

6.2.2 Modular Design and Standardization

Synthetic biology emphasizes the modular design of biological systems, where metabolic pathways are constructed from standardized genetic parts or modules that can be easily assembled, modified, and optimized. These modules include promoters, ribosome binding sites, coding sequences, and terminators, all of which can be tailored to control the expression of specific genes within a biosynthetic pathway. For example, a modular approach was used to produce flavonoids in *Escherichia coli*. Researchers designed a modular system to express the flavonoid biosynthetic pathway by introducing genes encoding key enzymes such as chalcone synthase, chalcone isomerase, and flavanone synthase. By optimizing the expression levels of each enzyme through the use of modular promoters and ribosome binding sites, they were able to maximize flavonoid production in the microbial host (Lee et al., 2024). The modular nature of synthetic biology also allows for the rapid assembly of different metabolic pathways, enabling researchers to mix and match genes from various sources to produce novel compounds. This approach is particularly useful for combinatorial biosynthesis, where multiple enzymes from different species are combined to generate new chemical structures with potentially valuable bioactive properties.

6.2.3 Gene Optimization and Codon Usage

Gene optimization is a critical aspect of synthetic biology, especially when transferring plant genes into microbial systems. Plant genes often have different codon usage preferences compared to microbes, which can lead to inefficient gene expression and low yields of the desired metabolite. To address this issue, synthetic biologists optimize plant genes for expression in microbial hosts by altering the codon sequences to match the preferred codon usage of the host organism. For instance, in the microbial production of taxol precursors, the plant genes involved in taxol biosynthesis were codon-optimized for expression in *Escherichia coli*. This optimization improved the expression of the enzymes in the microbial host, resulting in higher production levels of taxadiene, a key intermediate in the taxol biosynthetic pathway (Ajikumar et al., 2010). Codon optimization is a widely used strategy to enhance the efficiency of synthetic biology systems and improve the yield of target metabolites.

6.2.4 Metabolic Engineering for Flux Optimization

Once a biosynthetic pathway is introduced into a microbial host, it is essential to optimize the metabolic flux toward the production of the desired metabolite. This is achieved through metabolic engineering, where the expression levels of key enzymes are fine-tuned, competing pathways are knocked out, and precursor availability is increased. The goal is to maximize the flow of metabolites through the desired pathway while minimizing the production of unwanted by-products. For example, in the production of resveratrol, a polyphenol with antioxidant properties, *Saccharomyces cerevisiae* was engineered to enhance the availability of malonyl-CoA, a key precursor in resveratrol biosynthesis. By increasing the flux through the malonyl-CoA pathway and optimizing the expression of resveratrol biosynthetic enzymes, researchers were able to achieve high titers of resveratrol in yeast (Misirli et al., 2015). Metabolic engineering is a key strategy for improving the efficiency and scalability of microbial production systems.

6.3 Microbial Hosts for Plant Metabolite Production

Several microbial systems are commonly used for the production of plant metabolites through synthetic biology, each with its own advantages and limitations. The choice of host depends on factors such as the complexity of the biosynthetic pathway, the yield requirements, and the downstream processing needs.

6.3.1 Escherichia coli

Escherichia coli is one of the most widely used microbial hosts in synthetic biology due to its rapid growth rate, ease of genetic manipulation, and well-characterized metabolic network. It is commonly used for the production of simple plant metabolites, such as flavonoids, phenolics, and terpenoids, which require relatively short and straightforward biosynthetic pathways. For example, *E. coli* has been engineered to produce naringenin, a precursor to flavonoids, by introducing genes from plants and optimizing the expression of key enzymes in the pathway. The microbial production of naringenin has potential applications in the nutraceutical industry, where flavonoids are valued for their antioxidant and anti-inflammatory properties (Huccetogullari et al., 2019). However, *E. coli* is not always the ideal host for producing complex or highly modified metabolites. Its lack of organelles such as the endoplasmic reticulum (ER) and Golgi apparatus, which are involved in the modification and transport of certain metabolites in plants, can limit its ability to produce certain compounds that require complex post-translational modifications.

6.3.2 Saccharomyces cerevisiae

Saccharomyces cerevisiae, or baker's yeast, is another popular microbial host for synthetic biology applications. Yeast has several advantages over *E. coli*, including the ability to carry out post-translational modifications such as glycosylation, which is required for the production of many plant secondary metabolites. Yeast also has a robust metabolic network and is well-suited for the production of terpenoids, alkaloids, and other complex compounds. The microbial production of artemisinin, one of the most well-known successes of synthetic biology, was achieved using yeast. By engineering yeast to express the enzymes required for artemisinic acid production, researchers were able to develop a scalable system for producing this valuable antimalarial drug (Ro et al., 2006). Yeast is also used for the production of other high-value metabolites, such as cannabinoids and essential oils.

6.3.3 Streptomyces and Other Actinomycetes

Streptomyces and other actinomycetes are filamentous bacteria that have been widely used in natural product discovery and synthetic biology. These microbes are particularly well-suited for producing polyketides, alkaloids, and other secondary metabolites with complex structures. Actinomycetes naturally produce a wide range of bioactive compounds, including antibiotics, antifungals, and anticancer agents, making them an ideal platform for synthetic biology applications. In synthetic biology, *Streptomyces* species have been engineered to produce plant-derived metabolites such as alkaloids and terpenoids. For example, *Streptomyces* was engineered to produce the alkaloid noscapine, a cough suppressant and potential anticancer agent, by introducing plant-derived genes involved in noscapine biosynthesis (Mipeshwaree Devi et al., 2023). The use of actinomycetes for synthetic biology offers great potential for producing complex and novel compounds.

6.4 Applications of Synthetic Biology in Plant Metabolite Production

The microbial production of plant metabolites through synthetic biology has wide-ranging applications in the pharmaceutical, agricultural, and industrial sectors. By producing these metabolites in microbial systems, industries can reduce their reliance on traditional plant extraction methods, which are often inefficient, costly, and unsustainable.

6.4.1 Pharmaceutical Applications

Many plant secondary metabolites are used as active pharmaceutical ingredients (APIs) in drugs. Synthetic biology enables the microbial production of these APIs at a commercial scale, providing a more sustainable and cost-effective alternative to traditional methods. Examples include the production of artemisinin for malaria treatment, taxol for cancer therapy, and cannabinoids for pain management and neurological disorders (Nielsen & Keasling, 2016). Synthetic biology also offers the potential to create novel drug candidates by combining biosynthetic enzymes from different plants or modifying existing pathways to produce new chemical structures. This approach, known as combinatorial biosynthesis, allows for the creation of new compounds with potential therapeutic properties.

6.4.2 Agricultural Applications

In agriculture, plant secondary metabolites play a crucial role in plant defense against pests and diseases. Synthetic biology allows for the microbial production of these metabolites, which can be used as natural pesticides, herbicides, or fungicides. For example, flavonoids and terpenoids produced through microbial fermentation can be applied as biopesticides, reducing the need for synthetic chemical inputs in agriculture (Jennewein & Croteau, 2001). Additionally, synthetic biology enables the production of plant hormones and growth regulators, such as gibberellins and auxins, which can be used to enhance crop growth and yield. By producing these compounds in microbial systems, agricultural industries can reduce their reliance on chemical fertilizers and growth enhancers. Plant metabolites are used in a wide range of industrial applications, including the production of biofuels, bioplastics, fragrances, and flavors. Synthetic biology enables the microbial production of these metabolites in a scalable and sustainable manner. For example, terpenoids such as limonene and pinene, which are used in the production of biofuels and fragrances, can be produced in engineered microbes at high yields (Liu et al., 2011). Synthetic biology also enables the production of bioplastics from plant-derived metabolites. For example, polyhydroxyalkanoates (PHAs), a class of biodegradable plastics, can be produced from plant sugars using engineered microbes. These bioplastics offer a sustainable alternative to petroleum-based plastics, reducing the environmental impact of plastic production.

Conclusion

The production of secondary metabolites—bioactive compounds with significant therapeutic potential—has been a focal point in both pharmaceutical and biotechnological research. These metabolites, often derived from plants, fungi, and other organisms, play crucial roles in medicine, providing the foundation for treatments that address a wide array of health conditions, including cancer, infectious diseases, and inflammation. As the global demand for these compounds increases, particularly for sustainable and scalable production, modern biotechnology has introduced several advanced methods to enhance the production of secondary metabolites. Key among these are genetic engineering, CRISPR/Cas9 gene editing, metabolic engineering, tissue culture techniques, omics technologies, and synthetic biology. These approaches offer innovative ways to increase the yield, efficiency, and diversity of secondary metabolite production, thereby facilitating the development of new drugs and therapeutic agents.

Genetic engineering for enhanced secondary metabolite production has revolutionized the field of plant biotechnology. By manipulating the genetic makeup of plants and microorganisms, scientists can increase the production of valuable secondary metabolites or introduce new biosynthetic pathways into these organisms. For example, transgenic plants have been developed to overproduce specific compounds by introducing or upregulating genes involved in their biosynthesis. *Catharanthus roseus*, a plant known for producing anticancer alkaloids, has been genetically modified to increase the yield of vinblastine and vincristine, two critical compounds used in chemotherapy. Similarly, metabolic pathways responsible for producing terpenoids, flavonoids, and alkaloids have been enhanced in transgenic plants through the introduction of regulatory genes or transcription factors that promote secondary metabolite synthesis.

In microbial systems, genetic engineering has facilitated the transfer of plant biosynthetic pathways into organisms like *Escherichia coli* and *Saccharomyces cerevisiae*. These microbes, when genetically modified, can produce plant-derived compounds more efficiently and sustainably than traditional plant extraction methods. For instance, the production of artemisinin, an antimalarial compound, has been significantly enhanced through genetic engineering, where the biosynthetic pathway of artemisinin was transferred into yeast, leading to large-scale production in bioreactors. This approach not only makes production more scalable but also reduces dependency on plant cultivation, which can be subject to environmental and economic challenges.

CRISPR/Cas9 and other gene-editing technologies have further refined our ability to manipulate the biosynthesis of secondary metabolites. CRISPR/Cas9, in particular, has emerged as a powerful tool for precise gene editing, allowing for targeted modifications in both plants and microorganisms. This technology enables scientists to knock out genes that negatively regulate metabolite production or insert new genes that enhance biosynthetic pathways. By using CRISPR/Cas9, researchers can modify key enzymes in the biosynthesis of alkaloids, flavonoids, or terpenoids, leading to increased production of these metabolites. For example, the knockout of specific repressor genes in *Catharanthus roseus* has led to enhanced production of valuable alkaloids used in cancer treatment.

In microbial systems, CRISPR/Cas9 allows for the precise modification of metabolic pathways to optimize the production of plant-derived compounds. By editing genes that control precursor availability or pathway flux, scientists can increase the yield of target compounds. The precision of CRISPR/Cas9 makes it particularly valuable for metabolic engineering in microbes, where small changes in enzyme activity or gene expression can significantly impact the overall production of secondary metabolites. This technology has opened new possibilities for the large-scale production of complex bioactive compounds, including cannabinoids, opioids, and antibiotics, by engineering microbial "cell factories."

Metabolic engineering, closely related to genetic engineering and gene editing, focuses on optimizing the entire metabolic network within an organism to increase the production of desired secondary metabolites. This approach often involves modifying multiple genes and enzymes within a biosynthetic pathway to improve the efficiency of metabolite production. By overexpressing certain enzymes, downregulating competing pathways, and enhancing precursor availability, scientists can push more metabolic flux toward the production of specific compounds.

For instance, in the production of resveratrol, a polyphenolic compound with antioxidant and anti-inflammatory properties, *Saccharomyces cerevisiae* has been engineered to increase the availability of precursors such as tyrosine and malonyl-CoA. By optimizing the expression of enzymes involved in the resveratrol biosynthetic pathway, researchers have achieved higher yields of this valuable compound, which has applications in nutraceuticals and pharmaceuticals. Metabolic engineering has also been applied to produce other plant-derived compounds, such as flavonoids, anthocyanins, and alkaloids, in microbial systems, offering a sustainable alternative to traditional plant extraction. Tissue culture techniques represent another important advancement in the production of secondary metabolites, particularly for compounds that are difficult to produce through conventional agriculture or microbial systems. Plant tissue culture involves growing plant cells, tissues, or organs in a controlled environment, where they can be induced to produce secondary metabolites. Callus cultures, cell suspension cultures, and hairy root cultures are among the most widely used tissue culture techniques for this purpose.

Hairy root cultures, in particular, have proven to be highly effective for the production of root-derived secondary metabolites such as alkaloids, terpenoids, and phenolics. By infecting plant tissues with *Agrobacterium rhizogenes*, scientists can induce the formation of hairy roots, which exhibit rapid growth and stable metabolite production. These cultures can be maintained in bioreactors, allowing for continuous and large-scale production of bioactive compounds without the need for whole plants. This technique has been successfully applied to produce compounds such as withanolides from *Withania somnifera* and ginsenosides from *Panax ginseng*, both of which have significant medicinal value.

Omics technologies—encompassing genomics, transcriptomics, proteomics, and metabolomics—have become essential tools for identifying the biosynthetic pathways responsible for secondary metabolite production. These high-throughput techniques allow scientists to map out the entire set of genes, transcripts, proteins, and metabolites involved in the synthesis of bioactive compounds, providing a comprehensive understanding of how these compounds are produced and regulated. Genomics, for example, can reveal the full complement of biosynthetic gene clusters in an organism, while transcriptomics can show which genes are being expressed under specific conditions. Metabolomics, meanwhile, provides insights into the diversity and abundance of metabolites within a given organism or tissue. By integrating these datasets, researchers can identify key genes and enzymes involved in secondary metabolite production and manipulate them to increase yield. Omics technologies have been instrumental in the discovery of new biosynthetic pathways in plants, fungi, and bacteria, leading to the identification of novel secondary metabolites with potential therapeutic applications.

Synthetic biology, an interdisciplinary field that combines biology and engineering, has further expanded the possibilities for secondary metabolite production by enabling the design and construction of entirely new biosynthetic pathways. Using synthetic biology techniques, researchers can reassemble metabolic pathways from multiple organisms to create "synthetic" organisms capable of producing novel or improved bioactive compounds. One of the most notable examples of synthetic biology in secondary metabolite production is the microbial synthesis of artemisinin. By reconstructing the artemisinin biosynthetic pathway in yeast, scientists have developed a scalable and cost-effective method for producing this critical antimalarial compound. Similarly, synthetic biology has been used to produce other complex secondary metabolites, such as taxol, opioids, and cannabinoids, in microbial systems. This approach not only reduces the environmental impact of plant cultivation but also enables the production of compounds that are difficult to extract from their natural sources.

References

Ajikumar, P. K., Xiao, W. H., Tyo, K. E., Wang, Y., Simeon, F., Leonard, E., & Stephanopoulos, G. (2010). Isoprenoid pathway optimization for taxol precursor overproduction in *Escherichia coli*. *Science, 330*(6000), 70-74.

Pickens, L. B., Tang, Y., Chooi, Y. H. (2011). Metabolic engineering for the production of natural products. *Annual Reviews in Chemical Biomolecular Engineering, 2*, 211-36.

Palazzotto, E., & Weber, T. (2018). Omics and multi-omics approaches to study the biosynthesis of secondary metabolites in microorganisms. *Current Opinion in Microbiology, 45*, 109-116.

Bednarek, P. (2012). Chemical warfare or modulators of defence responses-The function of secondary metabolites in plant immunity. *Current Opinion in Plant Biology, 15*(4), 407-414.

Chen, F., Tholl, D., Bohlmann, J., & Pichersky, E. (2011). The family of terpene synthases in plants: a mid-size family of genes for specialized metabolism that is highly diversified throughout the kingdom. *The Plant Journal, 66*, 212-229.

Chen, X., Lu, X., & Shu, N. et al. (2017). Targeted mutagenesis in cotton (*Gossypium hirsutum* L.) using the CRISPR/Cas9 system. *Scientific Reports, 7*, 44304.

Sato, F., Hashimoto, T., Hachiya, A., Tamura, K., Choi, K.-B., Morishige, T., Fujimoto, H., & Yamada,

Y. (2001). Metabolic engineering of plant alkaloid biosynthesis. *Proceedings of the National Academy of Sciences USA, 98*, 367-372.

De Luca, V., & St-Pierre, B. (2000). The cell and developmental biology of alkaloid biosynthesis. *Trends in Plant Science, 5*(4), 168-173.

Sears, J. E., & Boger, D. L. (2015). Total synthesis of vinblastine, related natural products, and key analogues and development of inspired methodology suitable for the systematic study of their structure-function properties. *Accounts of Chemical Research, 48*(3), 653-662.

Fischbach, M. A., & Walsh, C. T. (2006). Assembly-line enzymology for polyketide and nonribosomal peptide antibiotics: Logic, machinery, and mechanisms. *Chemical Reviews, 106*(8), 3468-3496.

Fischer, C. R., Klein-Marcuschamer, D., & Stephanopoulos, G. (2008). Selection and optimization of microbial hosts for biofuels production. *Metabolic Engineering, 10*(6), 295-304.

Liu, B., Wang, H., Du, Z., Li, G., & Ye, H. (2011). Metabolic engineering of artemisinin biosynthesis in Artemisia annua L. *Plant Cell Reports, 30*(5), 689-694.

Sohn, S. I., Pandian, S., Rakkammal, K., Largia, M. J. V., Thamilarasan, S. K., Balaji, S., Zoclanclounon, Y. A. B., Shilpha, J., & Ramesh, M. (2022). Jasmonates in plant growth and development and elicitation of secondary metabolites: An updated overview. *Frontiers in Plant Science, 13*, 942789.

Morey, K. J., & Peebles, C. A. M. (2022). Hairy roots: An untapped potential for production of plant products. *Frontiers in Plant Science, 13*, 937095.

Pyne, M. E., Narcross, L., & Martin, V. J. J. (2019). Engineering Plant secondary metabolism in microbial systems. *Plant Physiology, 179*(3), 844-861.

Gutterson, N., & Zhang, J. Z. (2004). Genomics applications to biotech traits: A revolution in progress?. *Current Opinion in Plant Biology, 7*(2), 226-230.

Jennewein, S., & Croteau, R. (2001). Taxol: Biosynthesis, molecular genetics, and biotechnological applications. *Applied Microbiology and Biotechnology, 57*(1-2), 13-19.

Dastmalchi, M., & Dhaubhadel, S. (2015). Soybean chalcone isomerase: evolution of the fold, and the differential expression and localization of the gene family. Planta, *241*(2), 507-523.

Jin, Y. S., & Cate, J. H. (2017). Metabolic engineering of yeast for lignocellulosic biofuel production. *Current Opinion in Chemical Biology, 41*, 99-106.

Karppinen, K., Tegelberg, P., Häggman, H., & Jaakola, L. (2018). Abscisic acid regulates anthocyanin biosynthesis and gene expression associated with cell wall modification in ripening bilberry (*Vaccinium myrtillus* L.) fruits. *Frontiers in Plant Science, 9*, 1259.

Mishra, A., & Pandey, V. P. (2024). CRISPR/Cas system: A revolutionary tool for crop improvement. *Biotechnology Journal, 19*(2), e2300298.

Tabata, H. (2006). Production of paclitaxel and the related taxanes by cell suspension cultures of Taxus species. *Currents Drug Targets, 7*(4), 453-461.

Kobayashi, S., Goto-Yamamoto, N., & Hirochika, H. (2005). Retrotransposon-induced mutations in grape skin color. *Science, 304*(5673), 982.

Misirli, G., Hallinan, J., & Wipat, A. (2015). Composable Modular Models for Synthetic Biology. *Journal of Emerging Technology and Computational Systematics*, 11(3), 1-19.

Zhao, Y., Liu, G., & Yang, F. et al. (2023). Multilayered regulation of secondary metabolism in medicinal plants. *Molecular Horticulture*, *3*, 11.

Lee, Y., Hwang, S., Kim, W., Kim, J. H., Palsson, B. O., & Cho, B.-K. (2024). CRISPR-aided genome engineering for secondary metabolite biosynthesis in Streptomyces.*Journal of Industrial Microbiology and Biotechnology*, *51*, kuae009.

Lila, M. A. (2004). Anthocyanins and human health: An in vitro investigative approach. *Journal of Biomedicine and Biotechnology, 2004*(5), 306-313.

Ren, C., Gathunga, E.K., & Li, X. et al. (2023). Efficient genome editing in grapevine using CRISPR/LbCas12a system. *Molecular Horticulture*, *3*, 21.

Martin, V. J., Pitera, D. J., Withers, S. T., Newman, J. D., & Keasling, J. D. (2003). Engineering a mevalonate pathway in *Escherichia coli* for production of terpenoids. *Nature Biotechnology, 21*(7), 796-802.

Murthy, H. N., Lee, E. J., & Paek, K. Y. (2014). Production of secondary metabolites from cell and organ cultures: Strategies and approaches for biomass improvement and metabolite accumulation. *Plant Cell Tissue and Organ Culture*, *118*, 1-6.

Nielsen, J., & Keasling, J. D. (2016). Engineering cellular metabolism. *Cell, 164*(6), 1185-1197.

Kumari, M., Checker, V. G., Kathpalia, R., Srivastava, V., Singh, I. K., & Singh, A. (2024). Metabolic engineering for enhanced terpenoid production: Leveraging new horizons with an old technique. *Plant Physiology and Biochemistry*, *210*, 108511

Oudelaar, A. M., & Higgs, D. R. (2021). The relationship between genome structure and function. *Nature Reviews Genetics, 22*(12, 808.

Paddon, C. J., & Keasling, J. D. (2014). Semi-synthetic artemisinin: A model for the use of synthetic biology in pharmaceutical development. *Nature Reviews Microbiology, 12*(5), 355-367.

Yun, D. J., Hashimoto, T., & Yamada, Y. (1992). Metabolic engineering of medicinal plants: transgenic Atropa belladonna with an improved alkaloid composition. *Proceedings of National Academy of Sciences USA*, 89(24), 11799-803.

Gelinsky, E., & Hilbeck, A. (2018). European Court of Justice ruling regarding new genetic engineering methods scientifically justified: a commentary on the biased reporting about the recent ruling. *Environmental Sciences Europe*, *30*, 52.

Ro, D. K., Paradise, E. M., Ouellet, M., Fisher, K. J., Newman, K. L., Ndungu, J. M., & Keasling, J. D. (2006). Production of the antimalarial drug precursor artemisinic acid in engineered yeast. *Nature, 440*(7086), 940-943.

Gan, W. C., & Ling, A. P. K. (2022). CRISPR/Cas9 in plant biotechnology: applications and challenges. *BioTechnologia (Pozn)*, *103*(1), 81-93.

Shan, Q., Wang, Y., Li, J., Zhang, Y., Chen, K., Liang, Z., & Gao, C. (2013). Targeted genome modification of crop plants using a CRISPR/Cas9 system. *Nature Biotechnology, 31,* 686–688.

Gutierrez-Valdes, N., Häkkinen, S. T., Lemasson, C., Guillet, M., Oksman-Caldentey, K.-M., Ritala, A., & Cardon, F. (2020). Hairy root cultures-A versatile tool with multiple applications. *Frontiers in Plant Science*, *11*, 33.

Mipeshwaree Devi, A., Khedashwori Devi, K., Premi Devi, P., Lakshmipriyari Devi, M., & Das, S. (2023). Metabolic engineering of plant secondary metabolites: prospects and its technological challenges. *Frontiers in Plant Science*, *14*, 1171154.

Smanski, M. J., Bhatia, S., Zhao, D., & Park, Y. J. (2016). Synthetic biology to access and expand nature's chemical diversity. *Nature Reviews Microbiology, 14*(3), 135-149.

Stephanopoulos, G. (2012). Synthetic biology and metabolic engineering. *ACS Synthetic Biology, 1*(11), 514-525.

Srivastava, S., Srivastava, A.K. (2013). Biotechnology and Genetic Engineering for Alkaloid Production. In: Ramawat, K., Mérillon, JM. (eds) Natural Products. Springer, Berlin, Heidelberg.

Verpoorte, R., van der Heijden, R., & ten Hoopen, H. J. G. (1991). Plant cell biotechnology for the production of alkaloids: Present status and prospects. *The Alkaloids: Chemistry and Pharmacology, 40*, 1-187.

Winkel-Shirley, B. (2001). Flavonoid biosynthesis: A colorful model for genetics, biochemistry, cell biology, and biotechnology. *Plant Physiology, 126*(2), 485-493.

Hou, M., Wang, R., Zhao, S., & Wang, Z. (2021). Ginsenosides in Panax genus and their biosynthesis. *Acta Pharmaceutica Sinica B*, *11*(7), 1813-1834.

Yin, K., Gao, C., & Qiu, J. L. (2017). Progress and prospects in plant genome editing. *Nature Plants*, 3, 17107.

Das, S., Kwon, M., & Kim, J. -Y. (2024). Enhancement of specialized metabolites using CRISPR/Cas gene editing technology in medicinal plants. *Frontiers in Plant Science*, *15*, 1279738.

Huccetogullari, D., Luo, Z.W. & Lee, S.Y. (2019). Metabolic engineering of microorganisms for production of aromatic compounds. *Microbial Cell Factories*, *18*, 41.

Basu, U., Riaz Ahmed, S., Bhat, B. A., Anwar, Z., Ali, A., Ijaz, A., Gulzar, A., Bibi, A., Tyagi, A., Nebapure, S. M., Goud, C. A., Ahanger, S. A., Ali, S., & Mushtaq, M. (2023). A CRISPR way for accelerating cereal crop improvement: Progress and challenges. Frontiers in Genetics, *13*, 866976.

Zhao, J., Davis, L. C., & Verpoorte, R. (2005). Elicitor signal transduction leading to production of plant secondary metabolites. *Biotechnology Advances, 23*(4), 283-333.

Liu, W., Feng, Y., Yu, S., Fan, Z., Li, X., Li, J., & Yin, H. (2021). The flavonoid biosynthesis network in plants. *International Journal of Molecular Science*, *22*(23), 12824.

Chapter - 18

Exploring Secondary Metabolites in Medicinal Plants for Disease Control

Pratima Raypa[1], Reshma Jadhav[2], Vinod Chouhan[3], Susheel Kumar[4] and Zuby Gohar Ansari[5]

[1]Department of Biochemistry, H.N.B. Garhwal University, Srinagar, Uttarakhand, India
[2]Department of Biochemistry, Mahatma Phule Krishi Vidyapeeth, Rahuri, Ahmednagar, Maharashtra, India
[3]Division of Plant Pathology, Indian Council of Agricultural Research-Indian Agricultural Research Institute (ICAR-IARI), New Delhi, India
[4]Plant Molecular Virology Laboratory, CPMB Division, CSIR-National Botanical Research Institute, Rana Pratap Marg, Lucknow, Uttar Pradesh, India
[5]B.M. College of Agriculture, Jaswadi Road Khandwa, Madhya Pradesh, India
[5]Department of Plant Pathology, Dr. Khem Singh Gill Akal College of Agriculture, Eternal University, Baru Sahib, Sirmour, Himachal Pradesh, India

Abstract

Secondary metabolites found in medicinal plants have long been recognized for their potent therapeutic properties and their crucial role in disease control. These compounds, including alkaloids, flavonoids, terpenoids, and phenolics, are responsible for a wide range of bioactivities such as antimicrobial, anti-inflammatory, antiviral, and anticancer effects. Medicinal plants, widely used in traditional medicine systems, owe much of their therapeutic efficacy to the presence of these bioactive secondary metabolites. Alkaloids like morphine, quinine, and berberine have been pivotal in treating pain, malaria, and bacterial infections, respectively, while flavonoids exhibit strong antioxidant properties, helping to reduce inflammation and prevent chronic diseases. Terpenoids and phenolics, too, have demonstrated significant antimicrobial and anticancer properties, making them valuable candidates for drug development. The ability of secondary metabolites to inhibit pathogen growth and modulate immune responses has led to their increasing utilization in modern pharmaceuticals, offering natural alternatives to synthetic drugs. Furthermore, advances in biotechnology and plant tissue culture have facilitated the mass production of these compounds, ensuring a sustainable supply for medicinal use. Ongoing research continues to explore novel secondary metabolites with potential applications in treating emerging diseases, to develop new, effective, and safe therapeutic agents.

DOI: 10.1201/9781003717454-18

Keywords: Medicinal Plants, Secondary Metabolites, Disease Control, Antimicrobial, Bioactive Compounds, Drug Development

Introduction

The therapeutic roles of secondary metabolites in medicinal plants are well-established, as these compounds have been utilized for centuries in traditional medicine systems and have contributed significantly to modern pharmacology. Secondary metabolites such as alkaloids, flavonoids, terpenoids, and phenolics possess a wide range of bioactivities, including antimicrobial, anti-inflammatory, antioxidant, and anticancer properties. These compounds are often found in high concentrations in medicinal plants, where they serve as a defense mechanism against herbivores, pathogens, and environmental stressors. In the context of human health, secondary metabolites from medicinal plants offer potent therapeutic benefits that have been validated by scientific research and clinical applications. As natural products with complex chemical structures, these compounds have become indispensable in drug discovery and development, providing templates for synthetic drugs and serving as direct therapeutic agents in treating a variety of diseases (Atanasov et al., 2021).

Alkaloids, flavonoids, and terpenoids are three major classes of secondary metabolites with well-documented antimicrobial properties. Alkaloids, such as morphine, quinine, and berberine, are known for their potent activity against bacteria, viruses, and parasites. These nitrogen-containing compounds often target specific biochemical pathways in pathogens, disrupting essential processes such as DNA replication or protein synthesis, leading to cell death or growth inhibition (Wink, 2010). Flavonoids, which are polyphenolic compounds widely distributed in plants, exhibit strong antimicrobial activity against a broad range of pathogens, including Gram-positive and Gram-negative bacteria. Their mechanism of action often involves disrupting microbial cell membranes, inhibiting enzyme activity, and scavenging reactive oxygen species (ROS) that are essential for pathogen survival (Cushnie & Lamb, 2011). Terpenoids, a large and diverse class of compounds, have demonstrated significant antimicrobial activity through mechanisms such as disrupting microbial cell walls and membranes, interfering with quorum sensing, and inhibiting biofilm formation. These terpenoid compounds, such as artemisinin and thymol, are widely used in traditional medicine and have been developed into modern drugs to treat diseases like malaria and bacterial infections (Mipeshwaree Devi et al., 2023).

Phenolics, another important class of secondary metabolites, play a crucial role in inflammation modulation and disease prevention. These compounds exhibit strong antioxidant activity, which allows them to neutralize free radicals and reduce oxidative stress, a key factor in the development of chronic diseases such as cardiovascular diseases, cancer, and neurodegenerative disorders. Phenolic acids, flavonoids, and tannins are known for their anti-inflammatory properties, as they inhibit pro-inflammatory enzymes such as cyclooxygenase (COX) and lipoxygenase (LOX), as well as downregulate the expression of inflammatory mediators like cytokines and chemokines (Pandey & Rizvi, 2009). The ability of phenolics to modulate inflammatory pathways has made them valuable in both preventive healthcare and as therapeutic agents for managing inflammatory diseases. For example, resveratrol, a well-known phenolic found in grapes and red wine, has shown promising results in reducing inflammation and oxidative stress in patients with cardiovascular diseases (Singh et al., 2015).

Recent biotechnological advances have significantly improved the production of medicinal compounds derived from secondary metabolites. Biotechnology has enabled the large-scale production of these valuable compounds through techniques such as plant cell culture, metabolic engineering, and synthetic biology. Plant cell culture systems, including hairy root cultures and suspension cultures, offer a sustainable and controlled environment for producing secondary metabolites without the need for harvesting wild plants, which can be endangered or difficult to cultivate. These systems have been particularly successful in producing high-value compounds like paclitaxel, an anti-cancer drug derived from *Taxus* species, and artemisinin, an anti-malarial compound from *Artemisia annua* (Smetanska, 2008). Metabolic engineering and synthetic biology have further enhanced the production of medicinal compounds by allowing the modification of biosynthetic pathways to increase the yield of desired metabolites or produce novel compounds with enhanced therapeutic properties. By introducing new genes or optimizing existing pathways, researchers can manipulate plant metabolic networks to produce higher levels of alkaloids, flavonoids, and terpenoids, thus improving the availability of these compounds for pharmaceutical use (Thirumurugan et al., 2018).

Exploring new secondary metabolites for drug development is a promising area of research, as many medicinal plants remain underexplored in terms of their chemical composition and bioactivity. Advances in analytical techniques, such as high-performance liquid chromatography (HPLC) and mass spectrometry (MS), have enabled the rapid identification and characterization of novel secondary metabolites with potential therapeutic applications (Seca & Pinto, 2018). These technologies, combined with bioassays to evaluate the biological activity of isolated compounds, have facilitated the discovery of new drug candidates from plants. For example, recent studies on the Amazonian rainforest have led to the identification of several new alkaloids and terpenoids with potent anticancer and antimicrobial properties. By exploring the vast chemical diversity of plants, researchers can uncover new molecules that could serve as the basis for future drugs, particularly in areas where existing treatments are limited or ineffective.

The integration of plant-based medicines into modern healthcare systems has gained significant momentum, as there is increasing recognition of the value of natural products in treating chronic diseases and addressing the limitations of synthetic drugs. Plant-based medicines offer several advantages, including lower toxicity, fewer side effects, and the potential for synergistic effects when used in combination with other therapies (Saklani & Kutty, 2008). Herbal medicines and supplements, which are often based on traditional knowledge, are becoming more widely accepted and used in both preventive and therapeutic healthcare settings. For instance, the use of plant-derived compounds like curcumin (from turmeric) and ginseng has gained popularity in complementary and alternative medicine for managing inflammatory diseases, boosting immunity, and promoting overall well-being (Zhao et al., 2024). However, integrating plant-based medicines into modern healthcare systems also requires rigorous scientific validation, including clinical trials to assess safety, efficacy, and dosage. Regulatory frameworks must be established to ensure the quality and standardization of herbal products, as variability in plant sources, extraction methods, and compound concentrations can affect the consistency and effectiveness of these medicines.

1. Therapeutic Roles of Secondary Metabolites in Medicinal Plants

1.1 Medicinal Plants and Secondary Metabolites

Medicinal plants have been a cornerstone of healthcare systems for millennia, playing critical roles in various traditional medicinal practices such as Ayurveda, Traditional Chinese Medicine (TCM), and indigenous healing systems. The bioactive compounds responsible for the therapeutic properties of these plants are known as secondary metabolites—organic molecules that do not participate directly in basic metabolic processes like growth or reproduction but are crucial for a plant's interaction with its environment. These compounds are typically produced as part of the plant's defense mechanisms against herbivores, pathogens, and environmental stressors. Secondary metabolites encompass a wide variety of chemical classes, including alkaloids, terpenoids, flavonoids, phenolics, and glycosides. They are notable for their wide range of pharmacological activities, such as antimicrobial, anticancer, anti-inflammatory, antioxidant, and neuroprotective properties. This diverse functionality underpins their significance in modern medicine, where many of these metabolites serve as direct or precursor compounds for pharmaceuticals.

1.2 Alkaloids: Powerful Agents in Cancer Treatment and Pain Management

Alkaloids, a class of nitrogen-containing compounds, are among the most pharmacologically potent secondary metabolites found in medicinal plants. Alkaloids are known for their wide-ranging effects on human physiology, including anticancer, analgesic, and antimalarial properties. Due to their complex chemical structures and potent bioactivity, alkaloids have been extensively studied and utilized in modern medicine.

1.2.1 Anticancer Alkaloids

Several plant-derived alkaloids are used as chemotherapy agents due to their ability to disrupt cell division, making them highly effective against cancer. Notably, *Catharanthus roseus* (Madagascar periwinkle) produces vinblastine and vincristine, two potent alkaloids that have revolutionized the treatment of cancers such as Hodgkin's lymphoma, leukemia, and breast cancer. These compounds work by binding to tubulin, a protein essential for mitotic spindle formation, thereby inhibiting cell division and triggering apoptosis in rapidly proliferating cancer cells (Jordan & Wilson, 2004). Another prominent example is paclitaxel (commercially known as Taxol), which is derived from the bark of the Pacific yew tree (*Taxus brevifolia*). Paclitaxel stabilizes microtubules, preventing their depolymerization during mitosis, which halts cancer cell proliferation. Paclitaxel has proven effective in treating a variety of cancers, including ovarian, breast, and non-small cell lung cancer, making it one of the most valuable plant-derived compounds in oncology (Wani et al., 1971).

1.2.2 Analgesic Alkaloids

Alkaloids also play a critical role in pain management, particularly in the form of opioid analgesics. Morphine and codeine, derived from *Papaver somniferum* (opium poppy), are powerful analgesics that have been used for centuries to relieve severe pain. These compounds work by binding to opioid receptors in the central nervous system, where they inhibit the transmission of pain signals (Harrison et al., 1999). Morphine, while highly effective for pain relief, is also associated with the risk of addiction and tolerance, necessitating careful clinical use. In addition to pain management, certain alkaloids have been utilized as antimalarial agents. Quinine, extracted from the bark of the *Cinchona* tree, was historically one of the first effective treatments for malaria. Quinine interferes with the malaria parasite's ability to degrade hemoglobin within red blood cells, leading to the parasite's death. Today, synthetic derivatives like chloroquine remain important in the treatment of malaria, especially in regions with chloroquine-sensitive *Plasmodium* strains (Achan et al., 2011).

1.3 Terpenoids: Anti-inflammatory, Antimicrobial, and Antiviral Potentials

Terpenoids, also referred to as isoprenoids, are another large class of secondary metabolites, consisting of molecules built from five-carbon isoprene units. They are produced by plants as part of their defense mechanisms and have been extensively used in medicine due to their anti-inflammatory, antimicrobial, antiviral, and anticancer properties.

1.3.1 Anti-inflammatory Terpenoids

Terpenoids such as curcumin, a compound found in *Curcuma longa* (turmeric), have garnered significant attention for their potent anti-inflammatory effects. Curcumin acts by modulating various signaling molecules, including nuclear factor kappa B (NF-κB) and cyclooxygenase-2 (COX-2), which play key roles in inflammation. This modulation makes curcumin a promising therapeutic agent for managing chronic inflammatory conditions like arthritis, cardiovascular diseases, and neurodegenerative disorders (Aggarwal & Harikumar, 2009). Another group of anti-inflammatory terpenoids includes boswellic acids, extracted from the resin of *Boswellia serrata* (Indian frankincense). Boswellic acids are 5-lipoxygenase inhibitors, meaning they prevent the formation of leukotrienes, which are potent pro-inflammatory mediators. Due to their efficacy in reducing inflammation and pain, boswellic acids have been used in the treatment of osteoarthritis, rheumatoid arthritis, and other inflammatory diseases (Siddiqui, 2011).

1.3.2 Antimicrobial and Antiviral Terpenoids

Terpenoids also exhibit strong antimicrobial and antiviral activities, making them useful for treating infectious diseases. Thymol and carvacrol, terpenoids found in thyme (*Thymus vulgaris*) and oregano (*Origanum vulgare*), have shown efficacy against a wide range of bacterial and fungal pathogens. These compounds disrupt microbial cell membranes, leading to the leakage of essential cellular components and the eventual death of the pathogens. Their broad-spectrum antimicrobial action makes them valuable in the development of natural disinfectants and preservatives (Marchese et al., 2016). Artemisinin, a sesquiterpene lactone isolated from *Artemisia annua* (sweet wormwood), is another noteworthy example of a terpenoid with potent antimalarial activity. Artemisinin and its derivatives are currently the gold standard for treating multidrug-resistant strains of *Plasmodium falciparum*, the parasite responsible for the most severe form of malaria. Artemisinin works by generating reactive oxygen species (ROS) within the parasite cells, which leads to cell damage and death (Tu, 2016).

1.4 Flavonoids: Potent Antioxidants and Cardioprotective Agents

Flavonoids are polyphenolic compounds commonly found in fruits, vegetables, and medicinal plants. They are well-known for their antioxidant properties, which play a vital role in protecting cells from oxidative stress—a key contributor to chronic diseases such as cardiovascular disorders, diabetes, and neurodegenerative conditions.

1.4.1 Cardioprotective Flavonoids

Flavonoids such as quercetin and rutin, found in onions (*Allium cepa*) and citrus fruits, have been shown to improve cardiovascular health. These compounds exert their protective effects by enhancing endothelial function, reducing blood pressure, and inhibiting platelet aggregation. Quercetin's antioxidant properties help neutralize free radicals and prevent the oxidative modification of low-density lipoprotein (LDL) cholesterol, a major contributor to atherosclerosis. In addition, epigallocatechin gallate (EGCG), a flavonoid found in green tea (*Camellia sinensis*), has been extensively studied for its cardioprotective properties. EGCG reduces LDL cholesterol levels and improves vascular endothelial function, making it effective in reducing the risk of heart disease (Hodgson et al., 2015).

1.4.2 Neuroprotective Flavonoids

The neuroprotective potential of flavonoids is also gaining attention, particularly in the context of age-related neurodegenerative diseases like Alzheimer's and Parkinson's disease. Kaempferol, a flavonoid found in various green leafy vegetables, has been shown to protect neurons by reducing neuroinflammation and oxidative damage, both of which contribute to the progression of neurodegenerative diseases. Resveratrol, another flavonoid found in grapes and red wine, has garnered significant interest due to its ability to activate sirtuin proteins, which are involved in cellular repair and longevity. Resveratrol's neuroprotective effects include reducing oxidative stress and inflammation, potentially slowing the onset of neurodegenerative diseases (Baur & Sinclair, 2006).

1.5 Phenolics: Multifaceted Compounds for Antimicrobial and Antidiabetic Uses

Phenolic compounds are known for their broad spectrum of bioactivities, including antimicrobial, antioxidant, and antidiabetic properties. These metabolites are found in various medicinal plants and are used to treat infections, manage oxidative stress, and control blood sugar levels in diabetic patients.

1.5.1 Antimicrobial Phenolics

Eugenol, a phenolic compound found in clove (*Syzygium aromaticum*), exhibits strong antimicrobial activity against a range of pathogens, particularly those involved in dental infections. Eugenol's ability to disrupt bacterial and fungal cell membranes makes it an effective natural antiseptic, widely used in dentistry for treating toothaches and as an ingredient in mouthwashes (Pramod et al., 2010).

1.5.2 Antidiabetic Phenolics

Phenolic compounds such as ferulic acid and caffeic acid, found in grains, coffee, and vegetables, have shown promise in managing diabetes. Ferulic acid enhances insulin sensitivity and promotes glucose uptake in cells, helping to regulate blood sugar levels. Caffeic acid, with its antioxidant properties, reduces oxidative stress, a common complication in diabetes, thereby improving overall glycemic control (Delgado et al., 2023).

1.6 Glycosides: Cardiac and Respiratory Health Support

Glycosides are compounds composed of a sugar molecule linked to a non-sugar entity (aglycone), often contributing to cardiac and respiratory health. Cardiac glycosides, in particular, have a long history of use in treating heart conditions.

1.6.1 Cardiotonic Glycosides

Cardiac glycosides, such as digoxin and digitoxin from *Digitalis purpurea* (foxglove), are well-known for their ability to increase the force of heart contractions. These compounds inhibit the sodium-potassium ATPase pump, leading to an increase in intracellular calcium concentration, which strengthens heart contractions and improves cardiac output. This makes them essential for treating heart failure and certain arrhythmias (Roberts et al., 2016).

1.6.2 Respiratory Glycosides

Some glycosides also support respiratory health by acting as expectorants and bronchodilators. Glycyrrhizin, found in licorice root (*Glycyrrhiza glabra*), has been used traditionally to treat respiratory conditions such as bronchitis and asthma. Glycyrrhizin's anti-inflammatory and antiviral properties also make it effective in treating respiratory infections, including those caused by viruses (Fiore et al., 2005).

2. Antimicrobial Properties of Alkaloids, Flavonoids, and Terpenoids

2.1 Plant-Derived Antimicrobials

Plants produce a vast array of secondary metabolites, including alkaloids, flavonoids, and terpenoids, which serve as their primary defense mechanisms against microbial pathogens. These natural compounds exhibit significant antimicrobial activity against bacteria, fungi, viruses, and parasites, making them important components in the development of new therapeutic agents. With the growing global concern over antibiotic resistance, the antimicrobial properties of plant-derived compounds are being explored more intensively as potential alternatives or complements to synthetic antimicrobial drugs. The antimicrobial action of these secondary metabolites arises from their ability to disrupt vital processes in microbial cells, such as membrane integrity, protein function, and genetic material replication. Alkaloids, flavonoids, and terpenoids represent three major classes of plant secondary metabolites that have demonstrated broad-spectrum antimicrobial properties. These compounds are not only key to plant survival in hostile environments but are also vital in human medicine, where they offer alternative strategies for combating resistant pathogens.

2.2 Alkaloids: Disruptors of Bacterial and Fungal Pathogens

Alkaloids are nitrogen-containing compounds found in many medicinal plants. They are particularly noted for their diverse pharmacological activities, including antimicrobial properties. Alkaloids exert their antimicrobial effects through a variety of mechanisms, including inhibition of DNA and protein synthesis, disruption of membrane integrity, and interference with enzymatic functions in microbes.

2.2.1 Antibacterial Activity of Alkaloids

Alkaloids such as berberine, found in plants like *Berberis vulgaris* (barberry), exhibit potent antibacterial activity against both Gram-positive and Gram-negative bacteria. Berberine's antibacterial effects are primarily attributed to its ability to intercalate with microbial DNA, thereby inhibiting DNA replication and transcription. This disruption halts bacterial proliferation and induces cell death (Sun et al., 2009). Another example is the alkaloid sanguinarine, derived from *Sanguinaria canadensis* (bloodroot). Sanguinarine is known for its ability to disrupt bacterial membranes, causing the leakage of cellular contents and ultimately leading to bacterial cell death. Its broad-spectrum activity against oral pathogens makes it a key ingredient in oral hygiene products designed to prevent gum disease and tooth decay (Senchina et al., 2009).

2.2.2 Antifungal Properties of Alkaloids

Alkaloids have also been shown to possess significant antifungal activity. For instance, the alkaloid solamargine, found in *Solanum* species, exhibits strong antifungal effects by disrupting the cell membranes of fungi, leading to increased membrane permeability and cell lysis. Solamargine has been particularly effective against fungal pathogens such as *Candida albicans*, which causes candidiasis in humans (Man et al., 2010). The antifungal alkaloid tomatidine, derived from the tomato plant (*Solanum lycopersicum*), has been shown to inhibit the growth of *Candida* species by interfering with ergosterol synthesis, a vital component of fungal cell membranes. This disruption of the membrane structure leads to the collapse of the fungal cells, making tomatidine a promising candidate for antifungal drug development (Dorsaz et al., 2017).

2.3 Flavonoids: Potent Antimicrobial and Anti-Inflammatory Agents

Flavonoids, a large class of polyphenolic compounds, are widespread in fruits, vegetables, and medicinal plants. These compounds are well-known for their antioxidant and anti-inflammatory properties, but they also exhibit significant antimicrobial activity. Flavonoids act against a variety of pathogens by disrupting microbial membranes, inhibiting enzyme function, and interfering with nucleic acid synthesis.

2.3.1 Antibacterial Activity of Flavonoids

Quercetin, a flavonoid found in apples, onions, and berries, has demonstrated considerable antibacterial effects against both Gram-positive and Gram-negative bacteria. Quercetin's antimicrobial action is multifaceted, including the inhibition of DNA gyrase, an essential enzyme in bacterial DNA replication, as well as the disruption of bacterial cell membranes. Quercetin is effective against antibiotic-resistant strains of *Staphylococcus aureus*, making it a potential candidate for combating resistant bacterial infections (Wu et al., 2013). Catechins, particularly epigallocatechin gallate (EGCG), found in green tea (*Camellia sinensis*), are another class of flavonoids with potent antibacterial properties. EGCG disrupts the lipid bilayers of bacterial membranes and inhibits the efflux pumps that bacteria use to expel antibiotics from their cells. This dual action not only directly kills bacteria but also enhances the efficacy of conventional antibiotics against resistant strains, such as methicillin-resistant *Staphylococcus aureus* (MRSA) (Friedman, 2007).

2.3.2 Antiviral and Antifungal Properties of Flavonoids

Flavonoids also possess antiviral and antifungal properties, making them versatile antimicrobial agents. For example, hesperidin, a flavonoid found in citrus fruits, has shown activity against viruses such as the influenza virus and herpes simplex virus (HSV). Hesperidin inhibits viral replication by blocking the entry of the virus into host cells and interfering with viral DNA replication. Flavonoids like luteolin and apigenin, found in celery, parsley, and chamomile, exhibit antifungal activity by inhibiting the synthesis of ergosterol in fungal membranes, similar to the mechanism of action of many commercial antifungal drugs. This makes flavonoids promising candidates for the development of new antifungal therapies, particularly in the face of rising antifungal resistance (Li et al., 2016).

2.4 Terpenoids: Multifaceted Antimicrobial Compounds

Terpenoids, also known as isoprenoids, represent the largest class of plant secondary metabolites, and they play a significant role in protecting plants from microbial infections. These compounds have been widely studied for their antimicrobial properties, which are attributed to their ability to disrupt microbial cell membranes, inhibit protein synthesis, and impair microbial respiration.

2.4.1 Antibacterial and Antifungal Terpenoids

Thymol and carvacrol, two terpenoids found in essential oils of thyme (*Thymus vulgaris*) and oregano (*Origanum vulgare*), have strong antibacterial properties. These compounds destabilize bacterial membranes, leading to increased permeability and the leakage of vital intracellular components. Thymol and carvacrol are effective against a wide range of bacterial pathogens, including *Escherichia coli*, *Salmonella*, and *Listeria monocytogenes*, making them valuable for use in food preservation and as natural disinfectants (Marchese et al., 2016). Another notable terpenoid is menthol, derived from peppermint (*Mentha piperita*), which exhibits both antibacterial and antifungal properties. Menthol disrupts microbial cell membranes and inhibits bacterial respiration, leading to cell death. Its broad-spectrum antimicrobial activity makes it a common ingredient in over-the-counter topical antiseptics and oral hygiene products.

2.4.2 Antiviral Activity of Terpenoids

In addition to their antibacterial and antifungal properties, terpenoids also exhibit antiviral activity. For example, the terpenoid glycyrrhizin, derived from licorice root (*Glycyrrhiza glabra*), has been shown to inhibit the replication of several viruses, including hepatitis C virus (HCV) and human immunodeficiency virus (HIV). Glycyrrhizin interferes with viral attachment to host cells and inhibits viral gene expression, making it a potential candidate for antiviral drug development (Fiore et al., 2008). Eucalyptol, a terpenoid found in eucalyptus oil, also demonstrates antiviral activity. Eucalyptol inhibits the replication of respiratory viruses, such as influenza and respiratory syncytial virus (RSV), by modulating the host immune response and reducing inflammation in the respiratory tract (Shao et al., 2020).

2.5 Mechanisms of Antimicrobial Action

The antimicrobial activities of alkaloids, flavonoids, and terpenoids are diverse, involving several mechanisms that target critical microbial processes. These mechanisms include:

» **Membrane Disruption**: Many of these compounds disrupt microbial cell membranes, increasing permeability and leading to the leakage of essential cellular contents. This is a common mechanism of action for terpenoids like thymol and carvacrol, as well as alkaloids like sanguinarine.

» **Inhibition of DNA and Protein Synthesis**: Alkaloids such as berberine and flavonoids like quercetin inhibit the synthesis of DNA, RNA, and proteins in microbial cells, preventing their replication and growth.

» **Interference with Metabolic Pathways**: Terpenoids and flavonoids often target key enzymes involved in microbial metabolism. For example, flavonoids like luteolin inhibit enzymes in fungal ergosterol synthesis, while terpenoids like eucalyptol modulate viral gene expression.

» **Efflux Pump Inhibition**: Flavonoids like catechins inhibit bacterial efflux pumps, which bacteria use to expel antibiotics and resist treatment. This action not only kills bacteria but also restores the efficacy of existing antibiotics.

The antimicrobial properties of alkaloids, flavonoids, and terpenoids have immense potential in various fields, including medicine, agriculture, and food preservation. These natural compounds provide a promising alternative to synthetic antibiotics and fungicides, especially in the face of rising antimicrobial resistance. In agriculture, plant-derived antimicrobials can be used to protect crops from bacterial and fungal infections, reducing the need for chemical pesticides. In the food industry, these compounds can be employed as natural preservatives to prevent spoilage and extend shelf life. In medicine, the future development of plant-based antimicrobial drugs offers a pathway toward combating resistant pathogens. The combination of plant-derived antimicrobials with conventional antibiotics holds the potential for synergistic therapies that could overcome resistance mechanisms in bacteria and fungi. Additionally, ongoing research into the mechanisms of action of these secondary metabolites will provide valuable insights into how they can be optimized for therapeutic use.

3. The Role of Phenolics in Inflammation and Disease Prevention

3.1 Introduction to Phenolic Compounds and Their Health Benefits

Phenolic compounds, often referred to as polyphenols, are a large group of secondary metabolites produced by plants. These compounds are recognized for their significant health-promoting properties, particularly in the prevention and management of chronic diseases. Phenolics are widely distributed in fruits, vegetables, grains, tea, wine, and medicinal plants, making them a regular part of the human diet. Structurally, phenolics are characterized by one or more hydroxyl groups attached to aromatic rings, which contribute to their potent antioxidant activity. This antioxidant property enables phenolics to neutralize free radicals, thereby preventing oxidative stress a key factor in the development of chronic diseases such as cardiovascular disease, diabetes, neurodegenerative disorders, and certain cancers. Beyond their antioxidant capacity, phenolic compounds are noted for their anti-inflammatory, antimicrobial, and anticarcinogenic properties. Their ability to modulate inflammation and influence cellular pathways associated with chronic diseases positions phenolics as essential compounds in the field of nutraceuticals and preventive medicine.

3.2 Mechanisms of Action in Inflammation Modulation

Inflammation is a natural immune response to infection or injury, but chronic inflammation is linked to the progression of many diseases, including heart disease, diabetes, and cancer. Phenolic compounds have been shown to play a critical role in modulating the inflammatory process by interfering with various molecular pathways that lead to inflammation. These compounds inhibit the activity of enzymes, transcription factors, and signaling molecules involved in inflammatory responses.

3.2.1 Inhibition of Pro-Inflammatory Enzymes

One of the primary mechanisms through which phenolic compounds exert anti-inflammatory effects is by inhibiting enzymes that are involved in the production of pro-inflammatory mediators.

Cyclooxygenase (COX) and lipoxygenase (LOX) are key enzymes responsible for the production of prostaglandins and leukotrienes, which are potent inflammatory mediators. Many phenolic compounds, including flavonoids and phenolic acids, inhibit the activity of these enzymes, thereby reducing the production of pro-inflammatory molecules. For instance, resveratrol, a stilbenoid phenolic found in grapes and red wine, has been shown to inhibit COX-2, an enzyme that promotes inflammation in conditions such as arthritis and cardiovascular disease (Jang et al., 1997). Similarly, caffeic acid, a phenolic acid found in coffee and various fruits, inhibits LOX activity, thus reducing the formation of leukotrienes that contribute to inflammatory conditions such as asthma and arthritis (Semis et al., 2021).

3.2.2 Modulation of NF-κB and MAPK Signaling Pathways

Another key mechanism by which phenolics reduce inflammation is through the modulation of transcription factors and signaling pathways, particularly the nuclear factor-kappa B (NF-κB) and mitogen-activated protein kinase (MAPK) pathways. NF-κB is a major transcription factor that regulates the expression of pro-inflammatory genes, including cytokines (such as TNF-α and IL-6), chemokines, and adhesion molecules. Phenolic compounds have been shown to inhibit the activation of NF-κB, thus preventing the transcription of pro-inflammatory genes. Curcumin, the main phenolic compound in turmeric (*Curcuma longa*), is one of the most studied natural inhibitors of the NF-κB pathway. Curcumin suppresses NF-κB activation by inhibiting the phosphorylation and degradation of IκBα, a protein that inhibits NF-κB's translocation to the nucleus (Aggarwal & Harikumar, 2009). By blocking this pathway, curcumin reduces the production of inflammatory cytokines, making it effective in managing chronic inflammatory diseases such as inflammatory bowel disease and rheumatoid arthritis. Additionally, phenolic compounds like quercetin and epigallocatechin gallate (EGCG) from green tea inhibit the MAPK signaling pathway, which is involved in cellular responses to stress, inflammation, and apoptosis. By modulating these pathways, phenolics can attenuate the production of pro-inflammatory cytokines and reduce inflammation in various disease states (Cheng et al., 2019).

3.3 Antioxidant Properties and Disease Prevention

The antioxidant activity of phenolic compounds plays a crucial role in preventing the onset and progression of oxidative stress-related diseases. Oxidative stress occurs when there is an imbalance between reactive oxygen species (ROS) and the body's ability to neutralize them with antioxidants. ROS can damage cellular components, including lipids, proteins, and DNA, leading to chronic diseases such as cancer, cardiovascular disease, and neurodegenerative disorders.

3.3.1 Neutralization of Free Radicals

Phenolics neutralize free radicals by donating electrons, stabilizing them, and preventing further oxidative damage. Flavonoids, phenolic acids, and tannins are particularly effective at scavenging ROS, including superoxide anions, hydroxyl radicals, and peroxynitrite. Catechins, found in green tea, and anthocyanins, present in berries, are two classes of phenolics that have demonstrated strong antioxidant capacity. For example, catechins protect endothelial cells from oxidative damage, thereby reducing the risk of atherosclerosis and other cardiovascular diseases. The antioxidant effects of catechins have been linked to improved blood vessel function and reduced arterial stiffness, which are key factors in preventing heart disease (Hodgson et al., 2015). Similarly, anthocyanins help protect cells from oxidative DNA damage, which is a precursor to cancer development (He & Giusti, 2010).

3.3.2 Protection Against Neurodegenerative Diseases

Oxidative stress is a major contributing factor to the development of neurodegenerative diseases such as Alzheimer's and Parkinson's disease. The brain is particularly vulnerable to oxidative damage due to its high oxygen consumption and lipid content. Phenolic compounds have shown promise in protecting neurons from oxidative stress and improving cognitive function in aging populations. Resveratrol, for instance, crosses the blood-brain barrier and exhibits neuroprotective effects by activating sirtuins, a family of proteins involved in cellular repair and longevity. Studies have shown that resveratrol reduces the accumulation of amyloid-beta plaques, a hallmark of Alzheimer's disease, and enhances memory and cognitive function in animal models (Baur & Sinclair, 2006). Similarly, curcumin's ability to reduce oxidative damage and inflammation in the brain has made it a subject of interest in the prevention and treatment of neurodegenerative diseases (Ringman et al., 2005).

3.4 Phenolics and Cardiovascular Health

Cardiovascular disease (CVD) is one of the leading causes of death worldwide, and phenolic compounds have been shown to have protective effects against several risk factors associated with CVD, including high blood pressure, cholesterol levels, and arterial inflammation.

3.4.1 Reduction of Oxidized LDL

One of the primary mechanisms by which phenolics protect against CVD is by preventing the oxidation of low-density lipoprotein (LDL) cholesterol. Oxidized LDL is a major contributor to the formation of atherosclerotic plaques, which can block blood vessels and lead to heart attacks and strokes. Phenolic compounds such as quercetin, resveratrol, and catechins reduce the oxidation of LDL cholesterol by neutralizing free radicals and enhancing the activity of antioxidant enzymes such as superoxide dismutase and glutathione peroxidase (Del Rio et al., 2010). Red wine polyphenols, particularly resveratrol, have been associated with a reduced risk of heart disease due to their ability to inhibit LDL oxidation and promote vasodilation. This phenomenon, known as the "French Paradox," refers to the lower incidence of cardiovascular disease observed in populations that consume moderate amounts of red wine despite a high-fat diet (Renaud & de Lorgeril, 1992).

3.4.2 Improvement of Endothelial Function

Phenolics also play a critical role in improving endothelial function, which is essential for maintaining vascular health. Endothelial dysfunction, characterized by impaired vasodilation and increased inflammation, is a precursor to atherosclerosis and other cardiovascular conditions. Flavonoids such as epicatechin, found in cocoa and dark chocolate, have been shown to improve endothelial function by enhancing nitric oxide (NO) production, which relaxes blood vessels and lowers blood pressure (Grassi et al., 2011). In addition to their direct effects on blood vessels, phenolics reduce inflammation in the vascular system by inhibiting the expression of adhesion molecules that promote the recruitment of inflammatory cells to the endothelium. This anti-inflammatory effect reduces the risk of plaque formation and arterial blockages, which are key contributors to heart attacks and strokes (González et al., 2011).

3.5 Phenolics in Cancer Prevention

Phenolic compounds have been widely studied for their anticancer properties. These compounds can inhibit the initiation, promotion, and progression of cancer through multiple mechanisms, including the induction of apoptosis, inhibition of cell proliferation, and suppression of angiogenesis (the formation of new blood vessels that feed tumors).

3.5.1 Induction of Apoptosis

Phenolics such as resveratrol, curcumin, and epigallocatechin gallate (EGCG) have been shown to induce apoptosis (programmed cell death) in cancer cells. Resveratrol activates the p53 pathway, which is a critical regulator of apoptosis, leading to the selective death of cancer cells while sparing healthy cells (Athar et al., 2007). Similarly, curcumin induces apoptosis in various cancer cell lines by modulating multiple signaling pathways, including NF-κB, Akt, and MAPK, which are often dysregulated in cancer (Shishodia et al., 2005).

3.5.2 Inhibition of Angiogenesis and Metastasis

Angiogenesis, the formation of new blood vessels, is a crucial process in tumor growth and metastasis. Phenolic compounds can inhibit angiogenesis by downregulating vascular endothelial growth factor (VEGF) and other pro-angiogenic factors. For example, quercetin has been shown to inhibit VEGF expression in cancer cells, thereby preventing the development of new blood vessels that supply nutrients to tumors (Athar et al., 2007). In addition to inhibiting angiogenesis, phenolics such as curcumin and EGCG can suppress metastasis, and the spread of cancer cells to distant organs. These compounds inhibit the activity of matrix metalloproteinases (MMPs), enzymes that degrade the extracellular matrix and allow cancer cells to invade surrounding tissues (Singh & Aggarwal, 1995).

4. Biotechnological Advances in the Production of Medicinal Compounds

4.1 Biotechnology and Medicinal Compound Production

Biotechnology, a field at the intersection of biology and technology, has revolutionized the production of medicinal compounds by providing innovative approaches to harness biological systems for therapeutic purposes. Traditional methods of producing medicinal compounds such as plant extraction, chemical synthesis, or fermentation often face limitations in scalability, cost, and environmental sustainability. Biotechnological advances, however, have overcome many of these challenges, enabling more efficient, sustainable, and cost-effective production of a wide variety of bioactive compounds. One of the primary goals of biotechnology in medicine is to enhance the production of secondary metabolites bioactive compounds that serve as pharmaceuticals or precursors to drugs. These compounds, often found in low concentrations in their native sources, can be produced in larger quantities through genetic manipulation, metabolic engineering, tissue culture techniques, and microbial systems. The development of tools like genetic engineering, synthetic biology, and CRISPR/Cas9 gene editing has significantly advanced the ability to manipulate biological pathways and organisms for the production of high-value medicinal compounds.

4.2 Genetic Engineering for Enhanced Compound Production

Genetic engineering involves the modification of an organism's genome to introduce, enhance, or suppress the production of specific compounds. In the context of medicinal compound production, genetic engineering allows for the optimization of biosynthetic pathways in plants, microbes, and even animals to increase the yield of valuable metabolites.

4.2.1 Transgenic Plants for High-Yield Production

One of the most significant biotechnological advances has been the development of transgenic plants that are engineered to produce higher yields of medicinal compounds. By introducing genes from other species or by overexpressing native biosynthetic genes, scientists can enhance the production of specific metabolites within plant systems. For instance, *Catharanthus roseus*, a plant known for producing anticancer alkaloids such as vincristine and vinblastine, has been genetically modified to overproduce these valuable compounds. By manipulating key enzymes in the alkaloid biosynthetic pathway, researchers have significantly increased the yield of these anticancer agents (Glenn et al., 2013). Additionally, transgenic approaches have been applied to increase the production of artemisinin, an antimalarial compound derived from *Artemisia annua*. By introducing genes involved in artemisinin biosynthesis into other high-biomass plants, researchers have created alternative systems capable of producing large quantities of the compound, helping to meet global demand for malaria treatment (Ro et al., 2006).

4.2.2 Metabolic Engineering in Microbial Systems

Microbial systems such as bacteria, yeast, and fungi have been extensively engineered to produce medicinal compounds through the introduction of plant biosynthetic pathways. Microbes offer several advantages, including rapid growth rates, well-characterized genetic systems, and ease of cultivation in bioreactors, making them ideal platforms for industrial-scale production of pharmaceuticals. For example, *Saccharomyces cerevisiae* (yeast) has been engineered to produce artemisinic acid, a precursor to artemisinin. This biotechnological achievement has significantly reduced the cost of artemisinin production and made it more accessible for malaria treatment, especially in regions where the disease is endemic (Paddon et al., 2013). Similar approaches have been used to engineer yeast and bacteria to produce compounds such as opioids, cannabinoids, and taxanes, offering sustainable alternatives to plant extraction and chemical synthesis.

4.3 Tissue Culture Techniques for Medicinal Compound Production

Plant tissue culture has emerged as a powerful tool for the production of medicinal compounds, particularly for compounds that are produced in low quantities or are difficult to synthesize chemically. Tissue culture techniques allow for the in vitro cultivation of plant cells, tissues, or organs under controlled conditions, enabling the large-scale production of bioactive compounds without the need for whole plants.

4.3.1 Callus and Cell Suspension Cultures

Callus culture, where undifferentiated plant cells are grown on nutrient media, is a widely used method for producing medicinal compounds. Callus cultures can be initiated from small explants of plant tissue and used to produce secondary metabolites in a controlled environment. For example, callus cultures of *Taxus* species have been used to produce paclitaxel (taxol), a powerful anticancer compound. By optimizing culture conditions and using elicitors to enhance secondary metabolite production, researchers have significantly increased the yield of paclitaxel in tissue cultures (Kolewe et al., 2008). Cell suspension cultures, which involve growing plant cells in liquid media, offer another method for large-scale production of medicinal compounds. These cultures allow for the continuous production of secondary metabolites, which can be harvested from the culture medium. Cell suspension cultures have been successfully used for the production of compounds such as ginsenosides (from *Panax ginseng*) and anthraquinones (from *Rubia tinctorum*), both of which have medicinal applications (Zhao et al., 2005).

4.3.2 Hairy Root Cultures

Hairy root cultures are a specialized form of tissue culture that is initiated by infecting plant tissues with the bacterium *Agrobacterium rhizogenes*. This bacterium transfers part of its DNA into the plant cells, causing them to produce rapidly growing, genetically stable "hairy roots." Hairy root cultures are particularly effective for the production of root-derived medicinal compounds, such as alkaloids and terpenoids. For example, hairy root cultures of *Withania somnifera* (ashwagandha) have been used to enhance the production of withanolides, a group of bioactive compounds with immunomodulatory and anticancer properties. The ability of hairy root cultures to continuously produce high levels of secondary metabolites without genetic instability makes them a valuable tool in biotechnological production systems (Saleem et al., 2020).

4.4 Synthetic Biology and the Reprogramming of Metabolic Pathways

Synthetic biology, an emerging field that combines engineering principles with biology, is pushing the boundaries of biotechnological production of medicinal compounds. Through the reprogramming of metabolic pathways in microbial and plant systems, synthetic biology enables the design and construction of novel biosynthetic routes that can produce existing and new bioactive compounds with greater efficiency.

4.4.1 Pathway Reconstruction in Microbes

One of the key strategies in synthetic biology is the reconstruction of plant biosynthetic pathways in microbial hosts such as yeast or bacteria. By identifying and transferring the genes responsible for the biosynthesis of a medicinal compound from plants to microbes, researchers can create microbial "cell factories" that produce the desired compounds in large quantities. For instance, the production of opioids such as morphine and codeine has been transferred from the opium poppy (*Papaver somniferum*) to yeast. This involved the introduction of more than 20 genes from the poppy plant into yeast, enabling the microbial production of these pain-relieving compounds. Such biotechnological advances offer a sustainable alternative to traditional opioid production, which relies on the cultivation and extraction of alkaloids from poppy plants (Harrison et al., 1999).

4.4.2 Combinatorial Biosynthesis for Novel Compounds

Synthetic biology also allows for the combinatorial assembly of biosynthetic pathways from different species, enabling the production of novel compounds that do not exist in nature. By mixing and matching genes from different plants, microbes, or even animals, researchers can create new metabolic pathways that lead to the synthesis of novel bioactive compounds with potential therapeutic applications. For example, synthetic biology approaches have been used to create new cannabinoids by combining enzymes from the cannabis plant with those from other species. These novel cannabinoids may have unique pharmacological properties that differ from those of naturally occurring cannabinoids like THC and CBD, opening up new avenues for drug development (Gülck & Møller, 2020).

4.5 CRISPR/Cas9 and Genome Editing in Medicinal Compound Production

The CRISPR/Cas9 gene-editing system has revolutionized the field of biotechnology by enabling precise and efficient manipulation of genetic material. In the context of medicinal compound production, CRISPR/Cas9 allows for the targeted modification of biosynthetic pathways in plants, microbes, and even animals to enhance the production of valuable secondary metabolites.

4.5.1 Gene Editing in Plants

CRISPR/Cas9 has been successfully used to enhance the production of medicinal compounds in plants by editing key genes involved in biosynthetic pathways. For example, CRISPR/Cas9 was used to knock out genes involved in the production of unwanted alkaloids in *Catharanthus roseus*, thereby increasing the yield of desired anticancer alkaloids such as vinblastine (Sun et al., 2009). In addition to increasing compound yields, CRISPR/Cas9 can be used to create plants with improved resistance to pathogens or environmental stresses, further enhancing their suitability for large-scale production of medicinal compounds.

4.5.2 Gene Editing in Microbes

CRISPR/Cas9 has also been applied in microbial systems to optimize the production of medicinal compounds. By editing genes involved in precursor synthesis, enzyme expression, and pathway regulation, researchers can improve the efficiency of microbial production systems. For instance, CRISPR/Cas9 has been used to enhance the production of resveratrol in *Escherichia coli* by optimizing the expression of key enzymes in the resveratrol biosynthetic pathway (Lee et al., 2024). This powerful tool allows for the fine-tuning of metabolic pathways, enabling the production of high-value compounds with greater precision and reduced production costs.

4.6 Bioreactor Technology and Industrial-Scale Production

To translate biotechnological advances into industrial-scale production, bioreactor technology plays a critical role. Bioreactors are vessels that provide optimal conditions for the growth of microorganisms or plant cells in large volumes, enabling the commercial production of medicinal compounds.

4.6.1 Plant Cell Bioreactors

Plant cell bioreactors are used to culture plant cells in a controlled environment for the production of secondary metabolites. These systems allow for the continuous production of compounds such as paclitaxel, ginsenosides, and shikonin, which are difficult to produce in large quantities through traditional cultivation methods. Bioreactors offer the advantage of scalability, allowing for the production of medicinal compounds on an industrial scale (González et al., 2011).

4.6.2 Microbial Fermentation Systems

Microbial fermentation systems, such as those used for yeast or bacterial cultures, are widely employed in the pharmaceutical industry for the production of antibiotics, vaccines, and bioactive compounds. These systems provide a cost-effective and sustainable method for producing large quantities of medicinal compounds through the cultivation of genetically engineered microbes. For example, microbial fermentation has been used to produce insulin, human growth hormone, and various antibiotics, revolutionizing the pharmaceutical industry (Wu et al., 2013).

5. Exploring New Secondary Metabolites for Drug Development

5.1 Introduction to Secondary Metabolites in Drug Discovery

Secondary metabolites, produced by plants, fungi, bacteria, and marine organisms, represent a vast and largely untapped reservoir of bioactive compounds with immense therapeutic potential. Unlike primary metabolites, which are essential for basic cellular functions, secondary metabolites are not directly involved in growth, development, or reproduction. Instead, they serve specialized functions such as defense against herbivores, competition with other microorganisms, and adaptation to environmental stresses. These compounds, due to their chemical diversity and biological activities, have historically been a rich source of pharmaceuticals, providing the foundations for many of the drugs used today. Penicillin, for example, was one of the earliest breakthroughs derived from a secondary metabolite produced by *Penicillium* fungi, revolutionizing the treatment of bacterial infections. Similarly, other natural products, including paclitaxel (Taxol) from the Pacific yew tree and artemisinin from *Artemisia annua*, have proven to be critical in cancer therapy and malaria treatment, respectively. Today, the exploration of new secondary metabolites is more critical than ever due to the rising challenge of antimicrobial resistance, the need for novel cancer therapies, and the demand for drugs to treat chronic diseases. Advances in biotechnology, synthetic biology, and analytical techniques have accelerated the discovery of new secondary metabolites. Researchers are now able to explore previously inaccessible organisms, such as deep-sea microbes, and leverage cutting-edge tools like genome mining and metabolomics to identify novel bioactive compounds with the potential to become the next generation of therapeutics.

5.2 The Importance of Biodiversity in Secondary Metabolite Discovery

Biodiversity plays a crucial role in the discovery of new secondary metabolites. Each species, whether plant, microbe, or marine organism, has evolved unique metabolic pathways to produce secondary metabolites that help them survive in their particular environments. These compounds, shaped by millions of years of evolution, often possess potent bioactive properties that can be harnessed for drug development.

5.2.1 Terrestrial Plants and Microorganisms

Plants have historically been the most prolific producers of secondary metabolites, with compounds such as alkaloids, flavonoids, and terpenoids being widely explored for medicinal purposes. For instance, the indole alkaloids from *Catharanthus roseus* have led to the development of the anticancer drugs vinblastine and vincristine. Similarly, the steroidal alkaloids from *Solanum* species have shown promise as anti-inflammatory agents. However, microorganisms especially bacteria and fungi are increasingly recognized for their capacity to produce bioactive secondary metabolites. Soil-dwelling bacteria like *Streptomyces* are well-known for producing antibiotics such as streptomycin and tetracycline. The exploration of these microorganisms from diverse environments, such as rainforests and deserts, has led to the discovery of new compounds with antibacterial, antifungal, and anticancer activities (Zhao et al., 2024).

5.2.2 Marine Ecosystems: A New Frontier

The oceans, covering more than 70% of the Earth's surface, represent a largely untapped reservoir of biodiversity. Marine organisms such as sponges, corals, and marine bacteria have evolved to produce unique secondary metabolites, many of which exhibit novel chemical structures and bioactivities not found in terrestrial organisms. For instance, the anticancer drug trabectedin (Yondelis) was originally derived from the marine tunicate *Ecteinascidia turbinata*. Similarly, marine-derived compounds such as ziconotide, a potent analgesic, have been approved for medical use, and many others are in clinical trials (Atanasov et al., 2021). Marine natural products often possess unusual structures and mechanisms of action, making them valuable for drug development. For example, bryostatin, isolated from marine bryozoans, has shown promise in treating cancer, Alzheimer's disease, and HIV by modulating protein kinase C (PKC) signaling pathways (Schaufelberger et al., 1991).

5.3 Techniques for Discovering New Secondary Metabolites

The discovery of new secondary metabolites for drug development has been significantly enhanced by advancements in several scientific disciplines, including genomics, metabolomics, and synthetic biology. These technologies allow researchers to identify, isolate, and characterize novel bioactive compounds more efficiently and accurately.

5.3.1 Genome Mining and Metagenomics

Genome mining involves analyzing the genomes of organisms to identify gene clusters responsible for the biosynthesis of secondary metabolites. Many of these gene clusters, especially in microorganisms, remain "silent" under standard laboratory conditions, meaning they do not actively produce secondary metabolites. By identifying these gene clusters through genome sequencing and then activating them using genetic or environmental manipulations, researchers can discover new compounds that would otherwise remain hidden. Metagenomics, the study of genetic material recovered directly from environmental samples, has also revolutionized the exploration of microbial secondary metabolites. By sequencing the genomes of unculturable microbes from diverse environments, scientists can identify new biosynthetic pathways and compounds that have potential therapeutic applications (Charlop-Powers et al., 2014). This approach has been particularly fruitful in the search for new antibiotics and anticancer agents.

5.3.2 Metabolomics and Analytical Chemistry

Metabolomics, the comprehensive study of metabolites within a biological system, allows researchers to profile the secondary metabolite content of organisms in detail. Using techniques such as mass spectrometry (MS) and nuclear magnetic resonance (NMR) spectroscopy, scientists can identify and quantify known and novel secondary metabolites. This high-throughput screening of natural products accelerates the discovery of bioactive compounds with therapeutic potential. Metabolomics is often coupled with bioassay-guided fractionation, where extracts from organisms are separated into fractions, each of which is tested for biological activity. The active fractions are then analyzed using MS or NMR to identify the compounds responsible for the observed activity. This process has led to the discovery of numerous drug leads, including new antibiotics and anticancer compounds (Han et al., 2015).

5.3.3 Synthetic Biology and Pathway Engineering

Synthetic biology has become an essential tool in the discovery and production of secondary metabolites. By reprogramming microbial systems, such as *Escherichia coli* or *Saccharomyces cerevisiae*, with biosynthetic pathways from other organisms, researchers can produce large quantities of bioactive compounds that are difficult or impossible to extract from their natural sources. This approach not only facilitates drug discovery but also makes the production of these compounds more sustainable. For example, synthetic biology has been used to reconstruct the biosynthetic pathway for the antimalarial compound artemisinin in yeast. This innovation has drastically reduced the cost and improved the scalability of artemisinin production, making it more accessible for malaria treatment (Paddon et al., 2013). Similar approaches are being applied to produce cannabinoids, opioids, and other medicinal compounds.

5.4 Challenges in Secondary Metabolite Discovery

Despite the tremendous potential of secondary metabolites for drug development, several challenges remain in their discovery and development.

5.4.1 Complexity and Structural Diversity

One of the main challenges in secondary metabolite discovery is the structural complexity and diversity of these compounds. While this complexity is often the source of their bioactivity, it also makes their isolation, identification, and synthesis challenging. Many secondary metabolites are produced in very low quantities, making it difficult to obtain enough material for testing and development. Additionally, their complex chemical structures often make synthetic production inefficient and costly (Wu et al., 2013).

5.4.2 Biosynthetic Pathway Complexity

The biosynthetic pathways responsible for secondary metabolite production are often highly complex, involving multiple enzymatic steps and regulatory mechanisms. In many cases, the genes encoding these pathways are not well understood, and the conditions required to activate them in laboratory settings are difficult to replicate. This has led to the underutilization of many promising secondary metabolites, particularly those produced by marine and soil microorganisms (Cragg & Newman, 2013).

5.4.3 Regulatory and Market Barriers

The development of new drugs from secondary metabolites is also hindered by regulatory and market challenges. The process of bringing a new drug to market is lengthy, expensive, and fraught with regulatory hurdles. Natural product-based drugs often face additional challenges, such as difficulties in ensuring consistent supply and quality, particularly when the compounds are derived from rare or endangered species. Furthermore, many pharmaceutical companies have shifted their focus away from natural product research due to the high costs and risks associated with drug discovery from natural sources (Zhao et al., 2024).

5.5 Future Prospects: Expanding the Frontiers of Secondary Metabolite Research

Despite these challenges, the future of secondary metabolite research for drug development is promising. Emerging technologies and interdisciplinary approaches are expanding the frontiers of natural product discovery, enabling researchers to access previously untapped sources of bioactive compounds.

5.5.1 Microbiome and Human Health

The human microbiome, composed of trillions of microorganisms living in and on the human body, is a new and exciting frontier in secondary metabolite research. These microorganisms produce a wide variety of metabolites that influence human health and disease. By exploring the secondary metabolites produced by the gut, skin, and oral microbiomes, researchers hope to discover new therapeutics for conditions ranging from inflammatory diseases to cancer (Donia & Fischbach, 2015).

5.5.2 Deep-Sea and Extreme Environment Exploration

Deep-sea environments, hydrothermal vents, and polar regions are extreme ecosystems that harbor unique microorganisms capable of producing novel secondary metabolites. Advances in underwater exploration and metagenomics are allowing scientists to access these extreme environments and discover new compounds with potential therapeutic applications. For example, the deep-sea bacterium *Thermomonospora* has been found to produce abyssomicins, compounds with potent antibacterial activity (Kroppenstedt & Goodfellow, 2006).

5.5.3 AI and Machine Learning in Drug Discovery

Artificial intelligence (AI) and machine learning are increasingly being applied to the discovery of secondary metabolites. These technologies can rapidly analyze vast datasets, including genomic and metabolomic data, to predict the bioactivity of unknown compounds and identify new drug candidates. AI-driven approaches are expected to streamline the drug discovery process, reduce costs, and increase the speed at which new secondary metabolites are brought to market (Stokes et al., 2020).

6. Integrating Plant-Based Medicines in Modern Healthcare Systems

6.1 Plant-Based Medicines and Their Role in Healthcare

Plant-based medicines, also known as phytomedicines or herbal medicines, have been used for centuries in traditional healing systems such as Ayurveda, Traditional Chinese Medicine (TCM), and Native American healing practices. These natural remedies have formed the backbone of healthcare in many cultures, providing treatments for a wide array of health conditions. Even in modern times, many pharmaceutical drugs have been derived from or inspired by plant secondary metabolites, demonstrating the enduring importance of botanical sources in medicine. As healthcare systems worldwide evolve and adapt to new challenges such as the rise of chronic diseases, increasing healthcare costs, and the growing threat of antibiotic resistance there is renewed interest in integrating plant-based medicines into conventional healthcare practices. The global trend towards personalized medicine, preventative care, and a more holistic approach to health has further fueled this interest.

However, the integration of plant-based medicines into modern healthcare systems presents several challenges, including standardization, regulatory oversight, scientific validation, and harmonization with conventional treatments.

6.2 The Role of Plant-Based Medicines in Modern Healthcare

Plant-based medicines are increasingly recognized for their role in addressing a range of health conditions, particularly chronic diseases, which are becoming the dominant health burden globally. The rise of non-communicable diseases such as diabetes, cardiovascular disease, and cancer has shifted the focus of healthcare towards long-term management and prevention, areas where plant-based medicines can play a critical role.

6.2.1 Treatment of Chronic Diseases

Many plant-derived compounds have demonstrated therapeutic potential in managing chronic diseases. For example, resveratrol, a polyphenolic compound found in grapes and red wine, has shown promise in improving cardiovascular health by reducing oxidative stress and inflammation. Curcumin, derived from turmeric (*Curcuma longa*), has potent anti-inflammatory properties and has been studied for its role in managing conditions such as arthritis, inflammatory bowel disease, and even certain cancers (Aggarwal & Harikumar, 2009). Similarly, ginseng (*Panax ginseng*) is widely used for its adaptogenic properties, helping the body resist stress and enhance overall well-being, particularly in aging populations (Reay et al., 2005). The use of these phytomedicines in conjunction with modern pharmacological therapies can provide a complementary approach to managing chronic diseases, offering holistic care that addresses both the symptoms and underlying causes of illness.

6.2.2 Preventive Healthcare and Wellness

One of the most significant advantages of plant-based medicines is their potential in preventive healthcare. As societies increasingly shift towards preventive strategies, phytomedicines offer an array of options for enhancing general health, boosting immunity, and preventing the onset of chronic conditions. For instance, echinacea is commonly used to support the immune system and prevent upper respiratory infections, while green tea polyphenols are associated with reduced risks of cardiovascular diseases and certain cancers due to their antioxidant properties (Singh et al., 2015). The preventive role of plant-based medicines aligns with the growing emphasis on wellness in modern healthcare. As more individuals take a proactive approach to their health, incorporating herbal supplements and plant-based remedies into daily routines has become common. Integrating these remedies into formal healthcare systems can promote wellness and reduce the long-term burden of chronic diseases.

6.3 Challenges in Integrating Plant-Based Medicines

Despite the growing recognition of the potential benefits of plant-based medicines, integrating them into modern healthcare systems presents several challenges. These challenges revolve around scientific validation, standardization, regulatory frameworks, and education of both healthcare professionals and patients.

6.3.1 Scientific Validation and Evidence-Based Medicine

One of the primary challenges of integrating plant-based medicines into mainstream healthcare is the need for robust scientific validation. While traditional knowledge provides valuable insights into the medicinal properties of plants, modern healthcare systems require evidence-based approaches to ensure the efficacy and safety of treatments. Many plant-based medicines have not undergone the rigorous clinical trials necessary for approval by regulatory agencies such as the U.S. Food and Drug Administration (FDA) or the European Medicines Agency (EMA). As a result, their use in conventional medical settings is often limited.

To overcome this barrier, there is a need for more clinical research on plant-based medicines. Randomized controlled trials (RCTs), meta-analyses, and systematic reviews are essential for demonstrating the safety and efficacy of these treatments in treating specific conditions. For example, clinical studies have provided strong evidence for the use of St. John's Wort (*Hypericum perforatum*) in treating mild to moderate depression, leading to its acceptance in some healthcare systems as an alternative to synthetic antidepressants (Linde et al., 2008).Collaborative efforts between researchers, pharmaceutical companies, and healthcare providers are necessary to accelerate the development of plant-based therapies that meet the rigorous standards of evidence-based medicine.

6.3.2 Standardization and Quality Control

Standardization is another significant challenge in the integration of plant-based medicines. Unlike synthetic drugs, which are chemically uniform, plant-based medicines can vary widely in their composition due to differences in growing conditions, harvesting methods, and preparation techniques. This variability can affect the efficacy and safety of the treatment, making it difficult for healthcare providers to recommend herbal medicines with confidence. To address this issue, the standardization of plant extracts is crucial. Standardization involves ensuring that each batch of a plant-based product contains consistent levels of active compounds. For example, standardized ginkgo biloba extracts used in the treatment of cognitive decline are required to contain specific concentrations of flavonoids and terpenoids to ensure their efficacy (Mahadevan & Park, 2008). Establishing quality control standards for plant-based medicines can enhance their reliability and safety, making them more acceptable within modern healthcare systems.

6.3.3 Regulatory Frameworks and Policy Integration

Regulatory oversight of plant-based medicines varies significantly across countries, creating challenges for their integration into healthcare systems. In some countries, such as Germany and Switzerland, herbal medicines are well-regulated and widely accepted as part of healthcare practices. The German Commission E, for example, has established monographs on the safety and efficacy of over 300 plant-based medicines, which are available as over-the-counter (OTC) treatments in pharmacies. In contrast, in many countries, including the United States, plant-based medicines are regulated as dietary supplements rather than drugs, which means they are not subject to the same rigorous testing and approval processes as pharmaceutical drugs. This lack of regulation can lead to concerns about product safety, quality, and therapeutic claims. To facilitate the integration of plant-based medicines into healthcare systems, harmonized regulatory frameworks are needed that ensure both safety and efficacy while allowing for innovation in the development of new therapies.

6.4 The Role of Healthcare Professionals in Integrating Plant-Based Medicines

Healthcare professionals, including physicians, pharmacists, and nurses, play a crucial role in integrating plant-based medicines into modern healthcare systems. Their acceptance and understanding of phytomedicine are key to ensuring its safe and effective use alongside conventional therapies.

6.4.1 Education and Training

One of the primary barriers to the integration of plant-based medicines is the lack of education and training among healthcare professionals. Many medical schools and healthcare training programs do not include sufficient coursework on herbal medicine or integrative medicine, leaving healthcare providers ill-equipped to advise patients on the use of plant-based therapies. As a result, patients often turn to alternative sources for information, which may be unreliable or biased. To bridge this knowledge gap, it is essential to incorporate education on phytomedicine into the curricula of medical and pharmacy schools. Training programs that focus on evidence-based herbal medicine, herb-drug interactions, and the role of plant-based medicines in preventive care can equip healthcare professionals with the knowledge needed to safely and effectively recommend these treatments. Continuing education for practicing healthcare providers is also critical in keeping them up to date with the latest research and clinical guidelines on plant-based medicines.

6.4.2 Communication with Patients

Effective communication between healthcare providers and patients is essential for the safe integration of plant-based medicines. Many patients already use herbal remedies alongside conventional medications but may not inform their healthcare providers due to concerns about disapproval or lack of interest. This can lead to potential herb-drug interactions, which may reduce the efficacy of treatments or cause adverse effects. Healthcare professionals should actively inquire about their patients' use of plant-based medicines and guide safe and appropriate use. For example, patients taking anticoagulant medications such as warfarin should be advised about the potential interactions with herbal supplements like garlic (*Allium sativum*) or ginkgo biloba, which may increase the risk of bleeding (Izzo & Ernst, 2009). By fostering an open dialogue, healthcare providers can ensure that plant-based medicines are used safely and effectively as part of an integrated treatment plan.

6.5 Public Perception and Acceptance of Plant-Based Medicines

The public perception of plant-based medicines has evolved over time, with growing interest in natural and holistic approaches to health. However, there are also misconceptions and concerns about the efficacy, safety, and regulation of herbal remedies. Public education campaigns that promote awareness of the benefits and risks of plant-based medicines are necessary to build trust and ensure informed decision-making among patients.

6.5.1 Addressing Misconceptions

One of the common misconceptions about plant-based medicines is the belief that "natural" equates to "safe." While many plant-derived compounds are indeed safe and effective, others can be toxic or cause adverse interactions with pharmaceutical drugs. For instance, the herb *Kava* (*Piper methysticum*), used for its anxiolytic effects, has been associated with liver toxicity in some individuals (Teschke,

2010). Educating the public about the potential risks and benefits of plant-based medicines is essential for promoting their safe use.

6.5.2 Promoting Evidence-Based Use

As more individuals turn to plant-based medicines, it is crucial to promote evidence-based use and discourage reliance on unproven treatments. Public health campaigns can highlight the importance of using standardized, well-researched products and consulting healthcare providers before starting any new herbal regimen. Encouraging the public to seek information from reputable sources, such as government health agencies, medical organizations, or peer-reviewed journals, can help reduce the spread of misinformation and ensure that plant-based medicines are used appropriately.

6.6 The Future of Plant-Based Medicines in Healthcare

As healthcare systems continue to evolve, the role of plant-based medicines is likely to expand, particularly in the areas of preventive care, chronic disease management, and integrative medicine. Advances in biotechnology, such as the use of genetic engineering to enhance the production of bioactive compounds, are also likely to play a role in the future development of plant-based therapies.

6.6.1 Integration into Mainstream Healthcare

The integration of plant-based medicines into mainstream healthcare systems will require ongoing collaboration between researchers, healthcare providers, policymakers, and patients. Evidence-based approaches, regulatory frameworks, and education programs are essential for ensuring that plant-based medicines are used safely and effectively in conjunction with conventional treatments. As research continues to uncover the mechanisms of action and therapeutic potential of plant-derived compounds, their role in modern healthcare will likely become more prominent.

6.6.2 Personalized Medicine and Phytomedicine

The future of medicine is increasingly focused on personalized approaches, where treatments are tailored to the individual based on their genetic makeup, lifestyle, and health status. Plant-based medicines are well-suited to this paradigm, as many herbs have been traditionally used to treat specific individuals based on their unique constitutions and health needs. Advances in genomics and metabolomics are likely to play a role in the future of phytomedicine, allowing for more targeted and effective use of plant-based therapies in personalized healthcare.

Conclusion

Secondary metabolites from medicinal plants play a crucial role in both traditional and modern medicine, offering a vast array of therapeutic benefits that address numerous health conditions. These bioactive compounds, including alkaloids, flavonoids, terpenoids, and phenolics, have long been utilized for their potent pharmacological properties, ranging from antimicrobial activity to anti-inflammatory and anticancer effects. In the context of modern healthcare, the integration of these natural compounds is not only desirable but essential as we face increasing challenges such as antimicrobial resistance, chronic diseases, and the need for more sustainable therapeutic solutions. Understanding the therapeutic roles of secondary metabolites, enhancing their production through biotechnology, and seamlessly integrating plant-based medicines into modern healthcare systems are crucial steps toward improving global health outcomes.

The therapeutic roles of secondary metabolites are vast and diverse, stemming from the unique bioactive properties that these compounds possess. Alkaloids, for instance, are known for their potent effects in treating conditions like cancer, pain, and infections. Vinblastine and vincristine, two alkaloids derived from *Catharanthus roseus*, are key chemotherapeutic agents used to treat cancers such as Hodgkin's lymphoma and leukemia. Terpenoids, another group of secondary metabolites, exhibit significant anti-inflammatory, antimicrobial, and anticancer properties. Compounds like paclitaxel (Taxol) have been groundbreaking in cancer treatment, particularly in managing ovarian and breast cancers. The impact of these compounds on human health cannot be overstated, and their continued study is crucial for advancing therapeutic approaches.

Flavonoids and phenolics are equally important in managing inflammation and preventing chronic diseases. Flavonoids, such as quercetin and catechins, possess strong antioxidant properties that neutralize free radicals and protect cells from oxidative stress, which is implicated in the pathogenesis of cardiovascular diseases, diabetes, and cancer. Phenolics, particularly resveratrol and curcumin, are well-studied for their anti-inflammatory and cardioprotective effects. Curcumin, derived from *Curcuma longa* (turmeric), is known for its ability to modulate key inflammatory pathways, making it a promising candidate for managing chronic inflammatory diseases like arthritis and inflammatory bowel disease. These compounds, with their ability to regulate inflammatory responses and protect against oxidative damage, hold tremendous potential for preventing and treating a range of chronic conditions.

In terms of antimicrobial properties, alkaloids, flavonoids, and terpenoids demonstrate significant activity against a broad spectrum of pathogens, including bacteria, fungi, and viruses. Alkaloids like berberine and sanguinarine have been shown to inhibit bacterial growth by interfering with microbial DNA replication and disrupting cell membranes. Flavonoids such as quercetin and catechins exert antimicrobial effects by inhibiting bacterial enzymes and disrupting cell wall integrity. Terpenoids, including thymol and carvacrol, are particularly effective against drug-resistant bacterial strains, making them valuable in the fight against antimicrobial resistance. These compounds provide a natural alternative to synthetic antibiotics and offer new avenues for developing antimicrobial therapies in an era of rising drug resistance.

Biotechnological advances have significantly enhanced the production of medicinal compounds, particularly those derived from plants. Traditional extraction methods often face limitations in scalability, yield, and environmental impact, but biotechnology offers innovative solutions through genetic engineering, tissue culture, and synthetic biology. Genetic engineering has allowed scientists to manipulate the biosynthetic pathways in plants and microbes to increase the yield of valuable secondary metabolites. For instance, transgenic plants have been developed to overproduce anticancer alkaloids such as vinblastine and vincristine. Similarly, metabolic engineering in microbes like *Saccharomyces cerevisiae* has enabled the production of artemisinic acid, a precursor to the antimalarial drug artemisinin, at industrial scales.

Tissue culture techniques, including callus cultures and hairy root cultures, provide another avenue for producing secondary metabolites in controlled environments. These techniques allow for the continuous production of bioactive compounds without the need for whole plants, making the process more sustainable and scalable. Hairy root cultures, in particular, have been used to produce high levels of withanolides from *Withania somnifera* (ashwagandha), which have immunomodulatory and anticancer properties. The application of synthetic biology has further pushed the boundaries of medicinal compound production by enabling the reconstruction of entire biosynthetic pathways in microbial systems. This has led to the production of complex compounds like opioids and cannabinoids in yeast and bacteria, offering a sustainable alternative to plant extraction.

The exploration of new secondary metabolites is critical for drug development, particularly as the pharmaceutical industry faces the challenge of developing novel therapeutics to address emerging health threats. Biodiversity remains a key source of these new compounds, with plants, fungi, and marine organisms offering untapped potential. Marine ecosystems, in particular, are rich in unique secondary metabolites with novel structures and mechanisms of action. Compounds like trabectedin, derived from marine tunicates, and bryostatins from bryozoans, have shown promise in treating cancer and neurodegenerative diseases. Genome mining and metagenomics have emerged as powerful tools for discovering new secondary metabolites, allowing researchers to identify biosynthetic pathways in microorganisms that were previously inaccessible. This approach has led to the discovery of new antibiotics and anticancer agents, offering hope in the fight against drug-resistant infections and cancers.

However, integrating plant-based medicines into modern healthcare systems presents several challenges, particularly in terms of scientific validation, standardization, and regulatory oversight. While traditional knowledge has long supported the use of plant-based medicines, modern healthcare systems require evidence-based approaches to ensure their safety and efficacy. Clinical trials and randomized controlled studies are essential for validating the therapeutic potential of these compounds. For example, the use of St. John's Wort in treating depression has been supported by clinical evidence, leading to its acceptance in some healthcare systems. Standardization is another critical issue, as plant-based medicines can vary in composition depending on growing conditions and extraction methods. Ensuring consistent levels of active compounds through standardization is essential for their safe and effective use in clinical settings.

Regulatory frameworks also need to evolve to accommodate the growing demand for plant-based medicines. In many countries, herbal medicines are regulated as dietary supplements rather than pharmaceuticals, which means they are not subject to the same rigorous testing and approval processes. Establishing harmonized regulatory frameworks that ensure both safety and efficacy will be essential for integrating plant-based medicines into mainstream healthcare. Healthcare professionals must also be educated on the safe and effective use of these treatments, as many patients already use herbal remedies alongside conventional medications. Educating healthcare providers on potential herb-drug interactions and promoting open communication between patients and providers will be key to ensuring the safe integration of plant-based medicines.

References

Achan, J., Talisuna, A. O., Erhart, A., Yeka, A., Tibenderana, J. K., Baliraine, F. N., & D'Alessandro, U. (2011). Quinine, an old anti-malarial drug in a modern world: Role in the treatment of malaria. *Malaria Journal, 10*(1), 144.

Aggarwal, B. B., & Harikumar, K. B. (2009). Potential therapeutic effects of curcumin, the anti-inflammatory agent, against cancer, cardiovascular diseases, and neurodegenerative diseases. *International Journal of Biochemistry and Cell Biology, 41*(1), 40-59.

Atanasov, A. G., Zotchev, S. B., Dirsch, V. M., & Supuran, C. T. (2021). Natural products in drug discovery: Advances and opportunities. *Nature Reviews Drug Discovery, 20*(3), 200-216.

Athar, M., Back, J. H., Tang, X., Kim, K. H., Kopelovich, L., Bickers, D. R., & Kim, A. L. (2007). Resveratrol: A review of preclinical studies for human cancer prevention. *Toxicology and Applied Pharmacology, 224*(3), 274-283.

Baur, J. A., & Sinclair, D. A. (2006). Therapeutic potential of resveratrol: The in vivo evidence. *Nature Reviews Drug Discovery, 5*(6), 493-506.

Charlop-Powers, Z., Milshteyn, A., & Brady, S. F. (2014). Metagenomic small molecule discovery methods. *Current Opinion in Microbiology, 19*, 70-75.

Cheng, S. C., Huang, W. C. S., Pang, J. H., Wu, Y. H, & Cheng, C. Y. (2019). Quercetin Inhibits the Production of IL-1β-Induced Inflammatory Cytokines and Chemokines in ARPE-19 Cells via the MAPK and NF-κB Signaling Pathways. *International Journal of Molecular Science, 20*(12), 2957.

Cragg, G. M., & Newman, D. J. (2013). Natural products: A continuing source of novel drug leads. *Biochimica et Biophysica Acta (BBA) - General Subjects, 1830*(6), 3670-3695.

Cushnie, T. P., & Lamb, A. J. (2011). Recent advances in understanding the antibacterial properties of flavonoids. *International Journal of Antimicrobial Agents, 38*(2), 99-107.

Del Rio, D., Stalmach, A., Calani, L., & Crozier, A. (2010). Bioavailability of coffee chlorogenic acids and green tea flavan-3-ols. *Nutrients*, 2(8), 820-833.

Delgado, A., Gonçalves, S., & Romano, A. (2023). Mediterranean diet: The role of phenolic compounds from aromatic plant foods. *Foods, 12*(4), 840.

Donia, M. S., & Fischbach, M. A. (2015). Small molecules from the human microbiota. *Science, 349*(6246), 1254766.

Dorsaz, S., Snäkä, T., Favre-Godal, Q., Maudens, P., Boulens, N., Furrer, P., Ebrahimi, S.N., Hamburger, M., Allémann, E., Gindro, K., Queiroz, E. F,. Riezman, H., Wolfender, J. L., & Sanglard, D. (2017). Identification and mode of action of a plant natural product targeting human fungal pathogens. *Antimicrobial Agents Chemotherapy, 61*(9), e00829-17.

Fiore, C., Eisenhut, M., Krausse, R., Ragazzi, E., Pellati, D., Armanini, D., & Bielenberg, J. (2008). Antiviral effects of *Glycyrrhiza* species. *Phytotherapy Research, 22*(2), 141-148.

Friedman, M. (2006). Overview of antibacterial, antitoxin, antiviral, and antifungal activities of tea flavonoids and teas. *Molecular Nutrition & Food Research, 51*(1), 116-134.

Glenn, W. S., Runguphan, W., & O'Connor, S. E. (2013). Recent progress in the metabolic engineering of alkaloids in plant systems. *Current Opinion in Biotechnology, 24*(2), 354-365.

González, R., Ballester, I., López-Posadas, R., Suárez, M. D., Zarzuelo, A., Martínez-Augustin, O., & Sánchez de Medina, F. (2011). Effects of flavonoids and other polyphenols on inflammation. *Critical Reviews in Food Science and Nutrition, 51*(4), 331-362.

Grassi, D., Desideri, G., & Ferri, C. (2010). Flavonoids: Antioxidants against atherosclerosis. *Nutrients, 2*(8), 889-902.

Gülck, T., & Møller, B. L. (2020). Phytocannabinoids: origins and biosynthesis. *Trends in Plant Science*, 25(10), 985-1004.

Han, J., Zhang, J., He, W., Huang, P., Oyeleye, A., Liu, X., & Zhang, L. (2015). Bioassay-guided identification of bioactive molecules from traditional Chinese medicines. *Methods in Molecular Biology, 1263*, 187-96.

Harrison, L. M., Kastin, A. J., & Zadina, J. E. (1999). Opiate tolerance and dependence: Receptors, G-proteins, and antiopiates. *Peptides, 19*(9), 1603-1630.

He, J., & Giusti, M. M. (2010). Anthocyanins: Natural colorants with health-promoting properties. *Annual Review of Food Science and Technology, 1*(1), 163-187.

Hodgson, J. M., & Croft, K. D. (2015). Tea flavonoids and cardiovascular health. *Molecular Aspects of Medicine, 31*(6), 495-502.

Izzo, A. A., & Ernst, E. (2001). Interactions between herbal medicines and prescribed drugs: A systematic review. *Drugs, 61*(15), 2163-2175.

Jang, M., Cai, L., & Udeani, G. O. (1997). Cancer chemopreventive activity of resveratrol, a natural product derived from grapes. *Science, 275*(5297), 218-220.

Jordan, M. A., & Wilson, L. (2004). Microtubules as a target for anticancer drugs. *Nature Reviews Cancer, 4*(4), 253-265.

Kolewe, M. E., Gaurav, V., & Roberts, S. C. (2008). Pharmaceutically active natural product synthesis and supply via plant cell culture technology. *Molecular Pharmaceutics, 5*(2), 243-256.

Kroppenstedt, R.M., & Goodfellow, M. (2006). The Family Thermomonosporaceae: Actinocorallia, Actinomadura, Spirillospora and Thermomonospora. In: Dworkin, M., Falkow, S., Rosenberg, E., Schleifer, KH., Stackebrandt, E. (eds) The Prokaryotes. Springer, New York, NY.

Lee, L., Hwang, S., Kim, W., Kim, J. H., Palsson, B. O., & Cho, B.-K. (2024). CRISPR-aided genome engineering for secondary metabolite biosynthesis in Streptomyces, *Journal of Industrial Microbiology and Biotechnology*, 51, kuae009.

Li, Y., Yao, J., Han, C., Yang, J., Chaudhry, M. T., Wang, S., & Yin, Y. (2016). Quercetin, inflammation, and immunity. *Nutrients, 8*(3), 167.

Linde, K., Berner, M., Egger, M., & Mulrow, C. (2005). St. John's Wort for depression: Meta-analysis of randomized controlled trials. *The British Journal of Psychiatry, 186*(2), 99-107.

Mahadevan, S., & Park, Y. (2008). Multifaceted therapeutic benefits of ginkgo biloba L.: Chemistry, efficacy, safety, and uses. *Journal of Food Science, 73*(1), R14-R19.

Man, S., Gao, W., Zhang, Y., Huang, L., & Liu, C. (2010). Chemical study and medical application of saponins as anti-cancer agents. *Fitoterapia, 81*(7), 703-714.

Marchese, A., Orhan, I. E., Daglia, M., Barbieri, R., Di Lorenzo, A., Nabavi, S. F., & Nabavi, S. M. (2016). Antibacterial and antifungal activities of thymol: A brief review of the literature. *Food Chemistry, 210*, 402-414.

Mipeshwaree Devi, A., Khedashwori Devi, K., Premi Devi, P., Lakshmipriyari Devi, M., & Das, S. (2023). Metabolic engineering of plant secondary metabolites: prospects and its technological challenges. *Frontiers in Plant Science, 14*, 1171154.

Paddon, C. J., Westfall, P. J., Pitera, D. J., Benjamin, K., Fisher, K., McPhee, D., & Newman, J. D. (2013). High-level semi-synthetic production of the potent antimalarial artemisinin. *Nature, 496*(7446), 528-532.

Pandey, K. B., & Rizvi, S. I. (2009). Plant polyphenols as dietary antioxidants in human health and disease. *Oxidative Medicine and Cellular Longevity, 2*(5), 270-278.

Pramod, K., Ansari, S. H., & Ali, J. (2010). Eugenol: A natural compound with versatile pharmacological actions. *Natural Product Communications, 5*(12), 1999-2006.

Reay, J. L., Kennedy, D. O., & Scholey, A. B. (2005). Effects of Panax ginseng on cognitive performance, mood and glucose regulation in healthy young volunteers: A double-blind, placebo-controlled, cross-over study. *Psychopharmacology, 19*(4), 357-365.

Renaud, S., & de Lorgeril, M. (1992). Wine, alcohol, platelets, and the French paradox for coronary heart disease. *The Lancet, 339*(8808), 1523-1526.

Ringman, J. M., Frautschy, S. A., Cole, G. M., Masterman, D. L., & Cummings, J. L. (2005). A potential role of curcumin in Alzheimer's disease. *Current Alzheimer Research, 2*(2), 131-136.

Ro, D. K., Paradise, E. M., Ouellet, M., Fisher, K. J., Newman, K. L., Ndungu, J. M., & Keasling, J. D. (2006). Production of the antimalarial drug precursor artemisinic acid in engineered yeast. *Nature, 440*(7086), 940-943.

Roberts, D. M., Gallapatthy, G., Dunuwille, A., & Chan, B. S. (2016). Pharmacological treatment of cardiac glycoside poisoning. *British Journal of Clinical Pharmacology, 81*(3), 488-495.

Saklani, A., & Kutty, S. K. (2008). Plant-derived compounds in clinical trials. *Drug Discovery Today, 13*(3-4), 161-171.

Saleem, S., Muhammad, G., Hussain, M. A., Altaf, M., & Bukhari, S.N.A. (2020). Withania somnifera L.: Insights into the phytochemical profile, therapeutic potential, clinical trials, and future prospective. *Iranian Journal of Basic Medical Science, 23*(12), 1501-1526.

Schaufelberger, D. E., Koleck, M. P., Beutler, J. A., Vatakis, A. M., Alvarado, A. B., Andrews, P., & Cardellina, J. H. (1991). The large-scale isolation of bryostatin 1 from *Bugula neritina* following current good manufacturing practices. *Journal of Natural Products, 54*(5), 1265-1270.

Seca, A. M. L., & Pinto, D. C. G. A. (2018). Plant secondary metabolites as anticancer agents: Successes in clinical trials and therapeutic application. *International Journal of Molecular Science, 19*(1), 263.

Semis, H. S., Gur, C., Ileriturk, M., Kaynar, O., & Kandemir, F. M. (2021). Investigation of the anti-inflammatory effects of caffeic acid phenethyl ester in a model of λ-Carrageenan-induced paw edema in rats. Human Experimental Toxicology, 40(12_suppl), S721-S738.

Senchina, D. S., Flinn, G. N., McCann, D. A., Kohut, M. L., & Shearn, C. T. (2009). Bloodroot (*Sanguinaria canadensis* L., *Papaveraceae*) enhances proliferation and cytokine production by human peripheral blood mononuclear cells in an in vitro model. *Journal of Herbs Spices and Medicinal Plants*, *15*(1), 45.

Shao, J., Yin, Z., Wang, Y., Yang, Y., Tang, Q., Zhang, M., Jiao, J., Liu, C., Yang, M., Zhen, L., Hassouna, A., White, W. L., & Lu, J. (2020). Effects of different doses of eucalyptus oil from *Eucalyptus globulus* labill on respiratory tract immunity and immune function in healthy rats. Frontiers in Pharmacology, 11, 1287.

Shishodia, S., Singh, T., & Aggarwal, B. B. (2005). Curcumin: Getting back to the roots. *Annals of the New York Academy of Sciences, 1056*(1), 206-217.

Siddiqui, M. Z. (2011). Boswellia serrata, a potential anti-inflammatory agent: An overview. *Indian Journal of Pharmaceutical Sciences, 73*(3), 255-261.

Singh, C. K., Ndiaye, M. A., & Ahmad, N. (2015). Resveratrol and cancer: Challenges for clinical translation. *Biochimica et Biophysica Acta (BBA)*, *1852*(6), 1178-85.

Singh, S., & Aggarwal, B. B. (1995). Activation of transcription factor NF-κB is suppressed by curcumin (diferuloylmethane). *Journal of Biological Chemistry, 270*(42), 24995-25000.

Smetanska, I. (2008). Production of secondary metabolites using plant cell cultures. *Advances in Biochemical Engineering/Biotechnology, 111*, 187-228.

Stokes, J. M., Yang, K., Swanson, K., Jin, W., Cubillos-Ruiz, A., Donghia, N. M., & Collins, J. J. (2020). A deep learning approach to antibiotic discovery. *Cell, 18*(2), 475-483.

Sun, Y., Xun, K., Wang, Y., & Chen, X. (2009). A systematic review of the anticancer properties of berberine, a natural product derived from *Berberis* species. *Anti-Cancer Drugs, 20*(9), 757-769.

Teschke, R. (2010). Kava hepatotoxicity. A clinical review, *Annals of Hepatology*, *9*(3), 251-265,

Thirumurugan, D., Cholarajan, A., Raja, S. S., & Vijayakumar, R. (2018). An introductory chapter: Secondary metabolites. In *Secondary Metabolites - Sources and Applications* (pp. 3-20). IntechOpen.

Tu, Y. (2016). Artemisinin A gift from traditional Chinese medicine to the world (Nobel Lecture). *Angewandte Chemie International Edition, 55*(35), 10210-10226.

Wani, M. C., Taylor, H. L., Wall, M. E., Coggon, P., & McPhail, A. T. (1971). Plant antitumor agents. VI. The isolation and structure of taxol, a novel antileukemic and antitumor agent from *Taxus brevifolia*. *Journal of the American Chemical Society, 93*(9), 2325-2327.

Wink, M. (2010). Introduction: Biochemistry, Physiology and Ecological Functions of Secondary Metabolites. In Annual Plant Reviews Volume 40: Biochemistry of Plant Secondary Metabolism, M. Wink (Ed.)

Wu, T., He, M., Zang, X., Zhou, Y., Qiu, T., Pan, S., & Xu, X. (2013). A structure–activity relationship study of flavonoids as inhibitors of E. coli by membrane interaction effect. *Biochimica et Biophysica Acta (BBA) -Biomembranes, 1828*(11), 2751-2756.

Zhao, J., Davis, L. C., & Verpoorte, R. (2005). Elicitor signal transduction leading to the production of plant secondary metabolites. *Biotechnology Advances, 23*(4), 283-333.

Zhao, L., Zhang, T., & Zhang, K. (2024). Pharmacological effects of ginseng and ginsenosides on intestinal inflammation and the immune system. *Frontiers in Immunology, 15*, 1353614.